ISLAND TINKERERS

History of Computing

William Aspray and Thomas J. Misa, editors

A complete list of the titles in this series appears in the back of this book.

ISLAND TINKERERS

INNOVATION AND TRANSFORMATION IN THE MAKING
OF TAIWAN'S COMPUTING INDUSTRY

HONGHONG TINN

The MIT Press
Cambridge, Massachusetts
London, England

The MIT Press would like to thank the anonymous peer reviewers who provided comments on drafts of this book. The generous work of academic experts is essential for establishing the authority and quality of our publications. We acknowledge with gratitude the contributions of these otherwise uncredited readers.

This book was set in Stone Serif and Stone Sans by Westchester Publishing Services. Printed and bound in the United States of America.

Library of Congress Cataloging-in-Publication Data

Names: Tinn, Honghong, author.
Title: Island tinkerers : innovation and transformation in the making of
    Taiwan's computing industry / Honghong Tinn.
Description: Cambridge, Massachusetts ; London, England : The MIT Press,
    [2025] | Series: History of computing | Includes bibliographical
    references and index.
Identifiers: LCCN 2024002681 (print) | LCCN 2024002682 (ebook) | ISBN
    9780262549387 (paperback) | ISBN 9780262380263 (epub) | ISBN
    9780262380270 (pdf)
Subjects: LCSH: Computer industry—Taiwan—History.
Classification: LCC HD9696.2.T284 T56 2024 (print) | LCC HD9696.2.T284
    (ebook) | DDC 338.4/7621390951249—dc23/eng/20240514
LC record available at https://lccn.loc.gov/2024002681
LC ebook record available at https://lccn.loc.gov/2024002682

10 9 8 7 6 5 4 3 2 1

To my parents, Jin-Rui Wu and Tung-Liang Cheng

CONTENTS

LIST OF FIGURES

INTRODUCTION

On September 21, 1999, an earthquake measuring 7.3 on the Richter scale hit Taiwan at 1:47 a.m., late morning on the US West Coast. The earthquake severely damaged power and water supplies throughout the island, and more than two thousand people lost their lives. Most semiconductor and computer manufacturing facilities in Taiwan quickly resumed operations. CNN journalists scheduled an interview on that day with Steve Jobs, then interim chief executive officer of Apple Computer, and asked him to comment on the impact of the earthquake on his company. Jobs noted, "The whole industry gets components from Taiwan." Apple sourced parts from Taiwan for its iBook, the company's popular line of laptop computers. A day after the earthquake, several US technology company stocks "swooned" due to the anticipated slowdown in the supply of computer and semiconductor products from Taiwan, and the Nasdaq composite index dropped 2 percent. On September 23, the *New York Times* predicted that the price of semiconductors would surge and the supply of motherboards, monitors, LCD displays, chargers, hard-disk drives, and modems would be affected, mostly due to power failures in Taiwan. The earthquake inadvertently demonstrated that, at the end of the twentieth century, Taiwanese computer manufacturers were crucial in supplying parts and finished computers to the world.[1]

A year before the earthquake, the Hollywood blockbuster *Armageddon* had revealed Taiwan's role as a key producer of high-end electronics. *Armageddon* portrays a NASA mission to stop an asteroid heading to Earth. As a nuclear bomb waits to be detonated on the asteroid, the crew fails to start the space shuttle engine to escape. In this scene, an American copilot dismisses a Russian cosmonaut attempting to fix the ignition. The copilot shouts, "You don't know the component!" The cosmonaut retorts, "Components, American components, Russian components, all made in Taiwan."

The ubiquity of Taiwanese electronics products at the turn of the twenty-first century offers a stark contrast to the earlier popular images of the island nation's technology. Taiwan was once a Japanese colony and the only territory where the defeated Chinese Nationalist leader Chiang Kai-shek, along with his anticommunist army and followers, could settle in 1949. Beginning in the mid-1960s, Taiwan hosted US and Japanese assembly factories on the island. In the late 1960s, the Radio Corporation of America (RCA) announced it was shutting down production at its television factory in Memphis, Tennessee. Workers believed that all production was going to Taiwan, as they helped ship test machinery to "RCA Taiwan Ltd." In reality, RCA planned to assemble only black-and-white TV sets there; color TV production in Memphis was relocated to Bloomington, Indiana. Nonetheless, Taiwan was gradually associated with the destination of offshore manufacturing. In the early 1980s, Taiwan—along with South Korea, Hong Kong, and Singapore—became known as one of the four "Asian tigers" for its miraculous economic growth, in part fueled by its electronics assembly sector. But being able to put together electronics did not immediately translate into innovative marketable technology. The *Wall Street Journal* scolded in 1982 that "Taiwan remained the center of Apple fakery," denouncing the unapproved manufacture of Apple computer knockoffs in Taiwan and their allegedly harmful presence in East and Southeast Asia.[2]

Now hosting Acer, Asus, Foxconn, Taiwan Semiconductor Manufacturing Company (TSMC), and a number of other prominent tech companies, Taiwan has transformed itself into a birthplace of digital gadgets, where desktop computers, laptops, servers, and semiconductor chips are developed or manufactured. This book chronicles the transformation over six decades by focusing on Taiwan's local technologists—a group of technology-savvy professionals, engineers, scientists, technocrats, technology users, and engineers-turned-entrepreneurs—leading Taiwan to become a country that calls itself the Silicon Island. *Island Tinkerers* demonstrates that these technologists made computers a workable, available, and manufacturable technology in Taiwan, despite a lack of financial and technical resources. Beginning in the late 1950s, technocrats, scientists, and engineers aspired to manufacture computers in Taiwan someday and embraced international technical assistance to enhance Taiwan's science and engineering education and industrial capacity. Since the mid-1960s, international corporations have sought out Taiwan and its local engineers to outsource transistor and integrated circuit assembly for military, aerospace, and consumer products. In the 1970s, the island supplied calculator and computer components, prompting the rise of engineers-turned-entrepreneurs, including Acer's founder, Stan Chen-Jung Shih (Zhenrong Shi). He and his competitors cultivated and supported generations of engineers for developing and manufacturing computers domestically since the early 1980s.

Taiwanese technologists popularized and manufactured computing technology in Taiwan by adopting two approaches. First, being immersed in Cold War politics and its enduring legacies, they sought technical assistance and resources from the United States and sourced computers from US companies. Second, characteristic of the bottom-up transformation, they strenuously grappled with newly available but "black-boxed" computers. Embracing "imported" technology, they developed strategies to adapt, modify, assemble, and work with computers in an innovative manner.[3] Their hands-on engagement with computers helped gain a better understanding of the technology, increased the technology's availability across the island, and explored the feasibility of related manufacturing endeavors.

This book reconstructs voices across the Pacific to better represent the perspectives of Taiwanese and international participants in shaping the development of computer manufacturing in Taiwan. In the United States, archival sources are drawn from the National Archives in College Park, Maryland; the United Nations Archives and Records Management Section in New York City; and the Charles Babbage Institute in Minneapolis, Minnesota. In Taiwan, archives of the Ministry of Foreign Affairs at the Institute of Modern History at Academia Sinica; Academia Historica; and university records at the National Chiao-Tung University (NCTU)[4] Museum and National Taiwan University have been critical to reconstructing ideas and plans regarding computers proposed by Taiwanese technocrats, scientists, and engineers.

To explore Taiwanese technologists' ideas and experience using, tinkering with, building, and manufacturing computing technology, I surveyed NCTU's alumni newsletter, the *Voice of Chiao Tung Alumni* (*Yousheng*), which contain news about alumni as well as essays, poems, and proposals written by technocrats, scientists, and engineers in Taiwan and overseas. I also scoured professional journals on computing and engineering, computer-hobbyist magazines, and relevant newspaper articles, all of which reflected discussions or debates among Taiwanese technologists, from both Taiwan and the United States. Annual reports published by US publicly traded companies that invested in Taiwan in the 1960s and 1970s were especially helpful in understanding the state of the electronics industry in Taiwan. Biographies of prominent engineers and technocrats in Taiwan also contain their thoughts about their careers, professions, colleagues, and viewpoints on the roles of the Taiwanese government and multinational electronic manufacturers.

Beyond archives and published sources, *Island Tinkerers* also draws on oral histories collected by the Charles Babbage Institute in Minneapolis and the Computer History Museum in Mountain View, California. I conducted oral history interviews with two US emeritus professors, whom the United Nations invited to teach at NCTU in Taiwan in 1962 and 1963; Taiwanese emeritus professors and retired engineers who were either students or young faculty members at NCTU in the early 1960s; and participants in the minicomputer-building project at NCTU circa 1970.

THE ISLAND AND THE COLD WAR

Owing to shared political and ideological interests, Taiwan and the United States maintained a close relationship throughout and after the Cold War. Taiwan is around the same size as the Netherlands, or slightly larger than the state Maryland. With the Philippines to the south, Okinawa to the north, and China to the west, Taiwan has been an interest of various empires since the seventeenth century. The Spanish established a fortress, and the Dutch operated a colony in Taiwan during the seventeenth century, before Qing controlled Taiwan. The island was then ceded to the Japanese empire in 1895. Generalissimo Chiang Kai-shek, ruling mainland China from 1927 to 1949, led China to fight Japan during World War II and reclaim mainland China from Japan in 1945. In 1943, US President Franklin D. Roosevelt, British Prime Minister Winston Churchill, and Chiang Kai-shek of China issued the Cairo Declaration, which granted Chiang sovereignty over Taiwan. Even though Roosevelt aided Chiang's China during World War II, the Truman administration was more wary of Chiang and did not actively intervene on his behalf in the civil war against Mao Zedong. Chiang lost the civil war and retreated with more than one million Chinese to Taiwan and several smaller islands close to the Chinese coastline in 1949. On his death in 1975, he had still not given up on retaking mainland China, and successive leaders of his Nationalist Party had the same ambition, no matter how unlikely they were to succeed. However, situated on the island of Taiwan, the government of the Republic of China (ROC) worked out unique and subtle political relations with another mainland—the United States—to ensure their national security.

Truman's reluctance to support Chiang's confrontation with Mao changed after the outbreak of the Korean War in June 1950, when the strategic military value of Taiwan became significant to the United States. Geographically, Taiwan is close to Korea, Japan, and the Philippines and was important to maintaining the Truman administration's military advantage in the region. Chiang's survival would help prevent the infiltration of Communists into other parts of Asia. Acknowledging the island's critical value, in 1951, Truman inaugurated the Mutual Security Program, which provided military assistance to Taiwan but restricted aid to defensive purposes. The US Military Assistance Advisory Group (MAAG) started operations in Taiwan in 1951, aiming to "reorganize, reequip, and retrain the ROC military." Between 1950 and 1966, Taiwan received the third-largest amount of military aid from the United States, behind France and Turkey but ahead of South Korea.[5] Beyond military aid, Chiang's Taiwan also received financial and technical aid from the United States. US economic aid to Taiwan from 1951 to the mid-1960s averaged 6.4 percent of Taiwan's gross national product per year. Behind Korea, Laos, and South Vietnam, Taiwan received the highest amount of overall US aid in Asia.[6]

The first and second Taiwan Strait crises especially demonstrate the close alliance between Taiwan and the United States. On September 3, 1954, the Chinese Communists opened fire on Quemoy Island (Jinmen or Kinmen). Two US military personnel, stationed there as MAAG consultant to the ROC, were killed on the same day. The series of military conflicts from September 1954 to February 1955 was known as the First Taiwan Strait Crisis. Chiang maintained military control over a couple of offshore islands after 1949 and deployed a significant number of forces there to defend them, despite their greater proximity to mainland China than to Taiwan. Among these ROC-governed offshore islands, Quemoy and Matsu islands are the largest and fewer than ten miles away from the coast of mainland China (figure 0.1). In late January 1955, President Eisenhower sent the Seventh Fleet to aid Chiang's evacuation of residents and soldiers from the Dachen Islands, several miles away from the coastline of Zhejiang Province. Commander-in-Chief of the US Navy and Commander of the US Pacific Fleet Felix B. Stump was authorized to bomb airfields in China if the Chinese Communists interfered with the evacuation. During the crisis, on January 29, the US Congress passed the Formosa Resolution, which authorized the president to "employ the armed forces of the United States to protect Formosa [Taiwan], the Pescadores [Penghu], and related positions and territories of that area." The related positions allude to Quemoy and Matsu, both islands Chiang Kai-shek exercised sovereignty over.[7]

The People's Liberation Army shelled Quemoy and Matsu on August 23, 1958, and created a blockade, marking the Second Taiwan Strait Crisis. Eisenhower again assisted Chiang. To break the blockade, the Seventh Fleet, with Eisenhower's authorization, escorted ROC supplies to the beach of Quemoy on September 29. At the same time, the Eisenhower administration deterred the People's Republic of China (PRC) with what was then the largest air-sea force ever assembled in US history in East Asian waters, and it prepared for the worst-case scenario. On October 6, the PRC dramatically decreased shelling after announcing a cease-fire. But when Secretary of State John Foster Dulles visited Taiwan on October 20, the PRC, in a symbolic show of force, escalated its artillery assault, if only for an hour. From late October 1958 to December 1979, PRC artillery fired on Quemoy on odd-numbered days (known as *danda shuangbuda* in Chinese), allowing Quemoy to receive supplies every other day yet reminding Taiwan and the

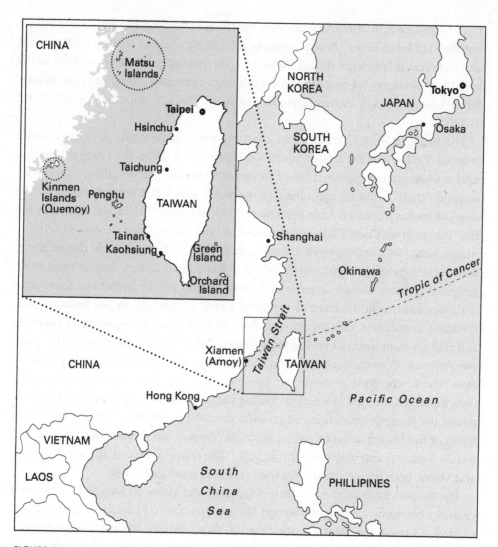

FIGURE 0.1
Map of Taiwan and neighboring countries. Map by Mary Reilly.

United States of its intention to take the island. In 1958 alone, half a million shells dropped on Quemoy, which made the shelling "the heaviest artillery bombardment in history."[8]

But the alliance was not entirely unwavering. While Taiwan's geographical location was strategically valuable to the military interest of the United States and many other countries, the PRC presented lucrative investment opportunities for countries looking for economic growth after fully developing their domestic markets. By 1982, the PRC's population was approximately a billion, fifty times larger than Taiwan's. With more and more national governments establishing formal diplomatic relations with PRC, Taiwan lost its seat to the PRC in the United Nations in 1971 and discontinued formal diplomatic relations with the United States in 1979. Despite the termination of official relations in 1979, the United States and Taiwan maintain close unofficial diplomatic relations primarily through the Taiwan Relations Act.[9]

Given the steadfast support for Taiwan from the United States in the 1950s and 1960s, when Taiwanese technologists strove to obtain knowledge and artifacts about electronic digital computing, they looked to the United States for technical, financial, and educational resources. At the macro level, the Cold War offered Taiwan access to ample international financial and technical aid. As "Free China," Taiwan was expected to exemplify the remarkable progress and wealth made possible by an anticommunist society. At the micro level, individual technocrats, scientists, and engineers aimed at improving the technological infrastructure of Taiwan, and they had to find ways to do so in the context of the Cold War. For sustaining the Republic of China on Taiwan, technologists eagerly fostered academic and technical exchanges with the United States from the 1950s onward. US companies invested in electronics factories on the island beginning in the 1960s, and Taiwanese technologists deepened technological relations with the United States. During and after the Cold War, Taiwanese technologists seized opportunities offered through collaborative relations with the United States and used institutional and technical improvisation to bring electronic digital computers and their manufacture to Taiwan.

American studies scholar Brian Russell Roberts has recently advanced the view that the United States should be seen as an archipelagic state, given its vast oceanic territory—in 1987, the United Nations recognized a 200-nautical-mile exclusive economic zone adjunct to territorial seas. Archipelagic American studies scholars challenge the tyranny of continentalism, in which islands are viewed as inferior. The school encourages scholars to view the world as archipelagic, abandoning the centrality of landmass, and advocates for more scholarly attention to borderwaters, instead of borderlands.[10] Taiwan, including its associated islands, demonstrates precisely the importance of an archipelagic approach to a nation's history. Considering Chiang Kai-shek had just lost vast lands to the PRC, Taiwanese technologists, starting in the turbulent 1950s, strove to make connections to the rest of world through borderwaters. Taiwan's archipelagic technologists introduced electronics and computers to the island of Taiwan and tinkered with them to reshape the industry there. Despite that Chiang and his followers often referred to Chiang's "lost territory" as the Chinese mainland and, worse, that some recent Taiwanese celebrities who need to guard their lucrative business in China tend to term the PRC "innerland," the Republic of China government on Taiwan after 1949 lacked sovereignty over the mainland.

Perhaps the 2019 book title of former Singapore diplomat Bilahari Kausikan speaks well to the archipelagic ethos: *Singapore Is Not an Island*. Becoming an independent nation-state in 1965, Singapore, a smaller and more recent nation-state than Taiwan, had to cope with international realities. Island states, archipelagic states, and small states, such as Taiwan and Singapore, often have to protect their sovereignties and secure resources unavailable to them. They often have to maximize their political, economic, cultural, or technological impact to transcend their relatively smaller territories. Kausikan describes how politicians and diplomats carved a path to transform Singapore from an extremely vulnerable to a powerful state.[11] He emphasizes the diplomatic strategies taken by Singapore to ensure its foreign policy interests. My book takes a different tack, highlighting strategies embraced by Taiwanese technologists and their plans to build sophisticated computing technologies on the island. The technology stories of islands, archipelagos, and small states, thus, are especially inspiring, as the recent publication of Victor Petrov's *Balkan Cyberia* has shown—Bulgaria, as the electronics manufacturing center of the Eastern Bloc, carefully navigated the shoals and shores of Cold War geopolitics.[12]

COMPUTERS AND GEO-TECHNOPOLITICS

Advances in computer technology are not only the accumulation of technological innovations but also a product of political discourses and societal changes.[13] The wartime demand for firing tables and ground-to-air defense set the stage for the later development of electronic digital computing in the United States.[14] During the Cold War, political discourse, strategic military concerns, and social changes coevolved with the science and technology of electronic digital computing in the United States. The proliferation of electronic digital computing accompanied the growing demand for defense needs as well as the containment ideologies associated with Cold War confrontations.[15] As a product of the Cold War, electronic digital computing conversely contributed to the emergence of new collaborative cultures, alternative uses of military technology, and novel ways of organizing technological systems.[16] The history of Cold War computing in the United States offers insights into how the social meanings and physical design of computing technology were co-constructed with the broader ideologies or discourses.[17]

Countries other than the United States fought the Cold War, asserting their own national interests, drawing on unique histories, and deploying specific strategies. The Cold War, in turn, also catalyzed the development of computers in those contexts. In socialist countries, computing was contingent on precarious political cultures. For example, in the Soviet Union, the social meanings of electronic digital computing were consistently redefined.[18] For Chile, historian Eden Medina has shown that a cybernetic system of computing technology was expected to carry out a socialist revolution in the early 1970s. Salvador Allende's government attempted to make use of electronic digital computers to build a computer network, known as Project Cybersyn, for the collection of real-time economic information collection and its subsequent monitoring and analysis. Although the project did not materialize, its history demonstrates the development of computing is contingent on political ideas. In places where the United States could exercise its political, cultural, or economic sway, the United States had its hand in the development of electronic digital computing one way or another. For example, the expansion of IBM in Europe was facilitated by the Marshall Plan in the 1950s. In India, local technical elites began the introduction of electronic digital computing from the United States in the early 1960s despite that India's nonalignment policy deliberately maintained independence from US influence.[19]

In the 1950s, coping with the Cold War, Taiwan's technologists quickly recognized electronic digital computing's potential to strengthen defense and industrialization. They sought to institutionalize the learning involved in cutting-edge electrical engineering fields, including servomechanism, electronic digital computing, and microwave signals. Through a UN technical-aid program, they established a training center on the Chiao-Tung University campus and taught electronic digital computing to state-owned enterprise engineers, military personnel, economists, and other Taiwanese. Taiwan's military was slightly behind academia in adopting electronic digital computing technology, and the island's history diverged from the US model of the military-industrial-academic complex, in which military-initiated research laid the foundation for industrial and academic participants.[20]

Economic competition between Taiwan and Communist China also emerged as another critical front of the Cold War. Taiwanese technologists aspired to own, explore, master, and manufacture electronic digital computers to broaden economic opportunities for domestic industry; it would be advantageous for the island to own the capability of manufacturing sophisticated technologies. It became an even more pressing issue after Taiwan lost its seat to the PRC in the United Nations in 1971 and the United States recognized the PRC in 1979. Taiwanese technologists' experimentation with building and mass-manufacturing of minicomputers and microcomputers reflected their deliberations over the future of the Taiwanese economy. By taking advantage of available and affordable integrated circuits and computer parts on the market in the 1980s, Taiwanese engineers-turned-entrepreneurs established profitable corporations and sold made-in-Taiwan products all over the world.

The alliance between the United States and Taiwan also sustained a set of personal networks that fostered technological relations between the two countries.[21] As early as the late 1950s, China-educated engineers and scientists who settled in Taiwan and their colleagues who settled in the United States formed social and professional networks across the Pacific. The networks helped bring visiting experts, donations, and the latest industrial news to Taiwan, all of which encouraged the expansion of the electronic digital computing system there. Starting in the late 1960s, Taiwanese technologists leveraged US investments in electronics assembly facilities in Taiwan, increasing trade between the two countries. The networks also facilitated connections with US-educated engineers and entrepreneurs, including those who were originally from China or Taiwan. Technological relations between the two countries nourished Taiwan's electronics, calculator, and computer manufacturing industries.[22] The cross-Pacific flows of knowledge, artifacts, and visions discussed in this book bring in new perspectives to long-standing scholarly interest in technological "diffusion."[23] I argue that the bottom-up force of local technologists popularized computing technology in Taiwan. The transfer of computing technology from the United States to Taiwan was less a process driven by foreign investment than it was the achievements of local technologists making new technology available, workable, and manufacturable in Taiwan.

DEVELOPMENTAL STATE AND BEYOND

The prevailing narratives of technological development articulated in postwar East Asian countries and in the academic community appear to follow the clear-cut stages of introduction, acceptance, and appropriation of foreign technology under the direction of a decisive and often authoritarian political leadership. The concept of "developmental state," proposed by Chalmers Johnson, Bruce Cumings, and Meredith Jung-En Woo-Cumings, has been the focus of an extensive scholarly community that emphasizes the role of East Asian governments in leading economic growth in the region.

Scholars have often attributed the developmental state with fostering economic miracles in places such as Japan, Taiwan, South Korea, Singapore, and, recently, mainland China.[24] The state has been a critical actor in making the Taiwanese economy grow. For the electronic assembly industry that powered the Taiwanese economy in the 1960s and 1970s, export processing zones, the brainchild of technocrats including K. T. Li (Kwoh-ting Li, Guoding Li), successfully attracted foreign capital investments and created job opportunities for the local workforce. For the computer and semiconductor sector, technocrats also laid out the blueprint of the Hsinchu Science Park and began, in the early 1980s, to offer start-up technology companies tax holidays, low-interest loans, and reduced rates for water and energy consumption.[25]

Historians of technology, however, have been studying engineers as active social agents in society, especially in its technological development and technopolitics.[26] Focusing on historical agents beyond the state, this book emphasizes the complementary contributions of mid-level professionals, including university-based scientists, industrial engineers, early computer users, overseas engineers and scholars, and engineers-turned-entrepreneurs to the formation of the domestic computer industry. This nexus cultivated collaborative opportunities with technocrats from the state. Mid-level professionals were eager to promote the installation of computers, study electrical engineering and computer science, and explore the potential of manufacturing advanced computing technology on the island. Their goals were consistent with technocrats' interest in fostering industrial development, and they enthusiastically welcomed the assistance of technocrats to achieve their goals.[27]

The developmental state school demonstrated a faithful interest in what East Asian countries possessed and could offer; scholars asked what Asians have done right to contribute to their economic prosperity. In contrast, an earlier generation of historians of science investigated insufficient or absent elements in Asian societies that slowed down scientific development there.[28] Recent decades have witnessed the prolific production of histories of technology in East Asia, indicating a complementary relationship between historians of technology and area studies scholars. Both groups of scholars have highlighted the prominent role of technology in understanding East Asian societies over time. For example, for the early modern era, Francesca Bray and Dagmar Schäfer have shown that technology was central to the gender and class politics of the time. To understand imperialism and colonialism in East Asia and their legacies in Southeast Asia, scholars simply cannot ignore the role of technology in this history. The use and deployment of technologies in early twentieth-century China has been a productive field too. The post–World War II era, characterized by the Cold War and its aftermath, illustrates East Asians' continuing commitment to technology.[29] This book contributes to the vibrant scholarship not only by adding one more sophisticated case study to the field but also by illustrating computing technology in Taiwan as a critical lens through which scholars can investigate technology and politics; in this case, the manufacture of computers by Taiwanese technologists was possible in part by their predecessors' careful leveraging of political, technical, and financial resources available to Taiwan during the Cold War. In light of the productive debate among Daiwei Fu, Fa-ti Fan, Warwick Anderson, and Michael M. J. Fischer about East Asian STS and "Asia as method," this book does not engage in the essentializing of Taiwanese characteristics. Instead, it commits to Fu and Fan's emphasis on power, politics, and understanding in light of the role of computers in "the messy politics and history of representation and practice," in Taiwan, as Anderson termed it. It also advocates, following Fischer, "more perspectives, located in different parts of the earth, on our bios and our polis."[30]

Historian of technology Brooke Hindle's emulation emerges as a critical concept in understanding the experience of Taiwanese technologists. Hindle argues that the concept of emulation best explains how nineteenth-century US inventors, mechanics, and artisans, such as Robert Fulton and Samuel Morse, used spatial thinking skills to comprehend and redesign European machines. American inventors were thus successfully emulating, not imitating, European machines from a relative latecomer's position.[31] This book offers a historical case that shares some similarities with that nineteenth-century US history—the production of technology from a peripheral location but beginning in the second half of the twentieth century. As latecomers, Taiwanese technologists did not participate directly in shaping early electronic digital computing innovations in the United States, Western Europe, or Japan. They were initially reliant on precariously available resources from US visiting experts, including Chinese Americans, and international aid agencies. But they also transformed this reliance into self-sufficiency through exploring, tinkering with, maintaining, and later prototyping computers and electronics at universities, electronics assembly factories, and start-ups.

In East Asian studies, scholars also challenge the perception of a "culture of copying." Anthropologist Rupert Cox advocated in 2007 for an investigation of acts of copying in their historical contexts, especially in relation to the definition of "innovation" and "tradition" in those contexts. The product of this advocacy is an edited volume, *The Culture of Copying in Japan*. Among the case studies in the volume, William H. Coaldrake's discussion of modern architectural models is especially illuminating. He points out that, traditionally, Japan did not have precisely scaled architectural models. After the 1873 World's Fair in Vienna, Japanese architects invested in adopting international standards to build and display Japanese architecture for the 1910 Japan-British Exhibition in London. These new models became representations of Japanese national identities and intended to challenge ideas of "Orientalism." The act of copying, in Coaldrake's research, is a Japanese response to the Western perception of Japan.[32] In the same vein, *Island Tinkerers* is interested in exploring how Taiwan's technology manufacturing was perceived and Taiwan's responses to that perception.

Taiwan is a successful case of a latecomer that managed to secure manufacturing advantages in its technology industries that other countries could not. This might give a misleading impression that the developmental state and technologists could easily achieve what they attempted. This book, however, is inspired by scholars of postcolonial studies of science and technology, standing with Cox and Coaldrake, who call for more analysis of the unequal exchanges, contradictory discourses, and violence that occurred during the circulation of science and technology in colonial or postcolonial societies.[33] Alongside success, Taiwanese technologists also encountered unpleasant misunderstandings, painful negotiations, and conflicts when interacting with their US and international counterparts. For example, chapter 2 describes the discrepancy between Taiwanese technologists and aid agency officials in the 1950s and 1960s, as the former strove to persuade the latter that science and technology education was essential to industrial development. In contrast, aid agencies expected to see aid result in an immediate impact on industrial development. Another discordance, illustrated in chapter 4, is that US military aid to Taiwan since the 1950s restricted the research and development capacity of the Taiwanese military. Chapter 6 reveals the different goals multinational electronics manufacturers and domestic engineers, managers, and factory workers held, despite all sides expecting electronics assembly factories to be profitable and prosperous. Chapters 8 and 9 focus on Taiwanese personal computer manufacturers as they began to sell their computers around the world in the 1980s. In response, world-leading computer manufacturers, especially Apple and IBM, contested Taiwan's strategies of developing compatible computers.

The concept of emulation clarifies that technological practice is never strictly "Taiwanese," "Chinese," or "American," even though historical actors make such nationalistic assertions. Postcolonial scholars have pointed out that advocating and highlighting the incommensurability between "Western knowledge" and "the Rest's knowledge" inevitably reinforces the separations between the two, because the separations were the essential part of Western exceptionalism and colonialism. For example, Itty Abraham has pointed out that self-reliance and autonomy were mere "watchwords" in the development of atomic bombs in India in the 1950s. These watchwords were imprecise, as India made use of technical expertise from Britain, Canada, the United States, and France.[34] Even as Taiwan became a powerhouse in developing and manufacturing computers in the late 1980s, this book demonstrates that scholarly attention to a uniquely

Taiwanese style of knowledge or practice should be directed to the acts of tinkering and improvisation in expanding the use, maintenance, and manufacture of computers.

TINKERERS

Historians of technology have advocated studying a diverse group of participants, including hobbyists, tinkerers, maintainers, amateurs, and mediators. Although these individuals were rarely inventors per se, they established unique, tangible relationships with technological artifacts. Studying beyond inventors enables a set of new ways to understand the myriad types of relationships between technology and society.[35] It also demonstrates a valuable characteristic of the history of technology: by studying historical actors' tangible explorations of the material features of technology, historians of technology can add a specific layer of observation to the scholarly community's understanding of technology, social change, and the agents of this change. Even in the neighboring community of the history of technology, historians of science have long acknowledged how an engagement with artifacts affects knowledge production, especially premodern artisanal knowledge and its assistance in the development of natural philosophy.[36] This book focuses on the tinkering activities of Taiwanese technologists who aspired to possess and explore technologies from abroad. It shows how Taiwan, as a follower and tinkerer, established successful enterprises and manufacturing sectors.

Over six decades, Taiwanese technologists tinkered; they engaged in a process of technology transfer in which acts of imitation, emulation, experimentation, and innovation overlapped with one another. Their improvised attempts and clever innovations were critical to their commitment to ensure the technological system was rooted in new soil. When the first two IBM mainframe computers arrived on the island in the early 1960s, NCTU engineers and scientists were confident that the maintenance of the computers could be conducted by an in-house technician instead of official IBM engineers. In the late 1960s and early 1970s, university-based graduate and undergraduate students ventured into building minicomputers and calculators from scratch. They probed, built, and manufactured them without direct support from the established computer industry or experienced engineers. In the 1970s, corporate engineers crafted calculator prototypes so that they could mass-produce for an export market. In the 1980s, computer manufacturers explored and created compatible Apple and IBM computer systems. The enthusiasm of Taiwanese technologists guided them to carry out research activities, embark on entrepreneurial journeys to make electronics components and products, and improve the quality of their manufacture through trial and error.

While *Island Tinkerers* emphasizes hands-on experience, the historical actors and their practices delineated in this book do not entirely constitute the garage-based entrepreneurship exhibited in the narratives associated with founders of contemporary Silicon Valley giants.[37] Tinkering with computers in Taiwan before the 1980s might not be comparable to the culture of hobbyists or hackers in the United States.[38] Though tinkering could be fun and playful, Taiwanese technologists also showed strong interest in transferring what they learned to locally mass-manufactured computer technology, a decidedly more purposeful goal. In doing so, they hopefully sought more opportunities to access less expensive, locally made technology.

The islandness of Taiwan especially prompted its technologists to build a transnational network for learning about computers, a culture of tinkering with computers, and a robust manufacturing sector for electronics. For other archipelagic states such as Japan and Malaysia, Lisa Onaga's and Clarissa Ai Ling Lee's research demonstrates that archipelagic scientists were determined in making science work within and beyond their archipelagos.[39] The American technological sublime—the concept proposed by historian David E. Nye—stresses the human feeling of awe inspired by technological achievements, from the Erie Canal, the first transcontinental railroad, and the Brooklyn Bridge to the Apollo mission. In Nye's analysis, the sublime can be attributed to the sense of evoking a visitor's emotional response when witnessing a large technological accomplishment. In contrast, the technological achievements in island, archipelagic, or small states are often magnificent at another spatial level—they are only possible because of a series of less-visible, transnational, borderwaters-crossing efforts.[40]

Taiwanese technologists' enthusiasm in an environment with many material, financial, and technical constraints was similar to that of Ronald K. Kline and Trevor J. Pinch's US farmers who converted their automobiles into tractors or used them to power their washing machines in the early twentieth century. They also resembled Simon Werrett's frugal early modern experimenters, who made use of the materials they possessed in their homes for experiments to learn about the natural world.[41] Enthusiasm for tinkering has been exhibited across the globe by users who had to cope with material constraints—for example, David Edgerton's technology users who appreciated "the shock of the old" and reinvented mature technology on their own terms.[42] Users' creative engagement with technology and their unauthorized modification of technology that manufacturers discouraged can be crucial to understanding these societies. A prominent example was tinkering with imported automobiles, which augmented transportation options for residents in various places including Tanzania and Thailand. In Taiwan, farmers also have put together cheap carriages and farming vehicles.[43] In China's republican era, entrepreneur Diexian Chen tinkered so his cosmetics products could compete with other foreign corporations, and Chinese radio users sourced parts to build and fix their radio sets despite that most radios were imported. Japanese tinkerers during the devastated post–World War II era looked for parts for their radios and televisions in the Akihabara area.[44] When it comes to the case of computers, Brazilian computer manufacturers built unauthorized Apple compatibles in the 1980s when the country permitted only computers manufactured by domestic firms. Since the mid-2000s, Ghana's shrewd internet café and small business owners have sourced computers and parts from machines recycled from Europe and the United States. Beyond computer hardware, free and open-source software programmers in Peru and iPhone hackers in Vietnam also developed their own strategies to establish technological relations with the global north.[45] In recent years, Chinese manufacturers also demonstrated a vibrant maker culture that stimulates business opportunities, despite that some consider the culture as pirating activities. The scholarly discussion of pirating activities in China, known as *shanzhai* culture, echoes Ravi Sundaram's concept of "pirate modernity," which reaffirms that the imitation and replication of an imported technology constitutes a significant part of the exploratory activities of technology users and engineers and their social and technological cultures. It might or might not further contribute to innovations protected by laws globally, especially those involving patents and copyrights. Imitation, however, guides historians of technology to reflect on what innovations are and what they mean to our academic enterprises.[46] In Taiwan, rather than engaging in wholesale copying of US sources, technologists tinkered with imported computing technology and experimented with manufacturing their own versions. Their creative engagement with computer hardware, software, and broader technology-supporting systems is seen as furthering laudable "innovations."

CHARTING COMPUTING IN TAIWAN

This book consists of ten chapters that unpack the histories of computing and computers in Taiwan from the 1950s to the twenty-first century. Chapter 1 centers on a group of Taiwan-based technocrats, engineers, scientists, and overseas Chinese professionals who directed their attention toward "electronics science" (i.e., electrical engineering). Many of these technologists were alumni of Chiao-Tung University, a prominent engineering and science university in China. They sought to establish Chiao-Tung University in Taiwan anew. Proactively mobilizing political connections on the island and support from overseas alumni at US universities and industries to train more engineers, in 1958, Chiao-Tung University opened its doors in Hsinchu. Seeking to underscore the university's importance, an alumnus penned a magazine article, "Atoms and

Electronics," that compared the fields of electrical engineering and atomic energy and advocated that the two fields should play critical roles in avoiding armed conflict with Communist China.

Chapter 2 delineates Chiao-Tung alumni's application for United Nations technical aid for the poorly equipped and understaffed university. Their efforts later contributed to the arrival in Taiwan of a UN-NCTU technical-aid program in 1962 and the island's first two electronic digital computers in 1962 and 1964. Taiwanese technologists and UN aid-agency officials had robust discussions over aid program goals. Taiwanese technologists lobbied for the acquisition of computers and computer experts that would improve basic science education in Taiwan and consequently trigger a "scientific takeoff." Technologists hoped that NCTU could soon move toward manufacturing electronic digital computers. In contrast, UN aid-agency officials preferred to see their assistance serve export-driven industry-based needs through the concerted practical training of Taiwanese engineers. Chapter 2 thus demonstrates that, in introducing computers and other technological artifacts to Taiwan, technologists explored and navigated the understructure of technical aid, similar to the way later technologists explored and navigated approaches to strengthening the infrastructure of the newly available IBM mainframe computers in Taiwan.

Chapter 3 shows how the first two mainframe computers in Taiwan, sponsored by the UN technical-aid program, became instrumental to local scientists and engineers and helped nurture the nation's first generation of programmers and computer users. With visiting experts and the precious computers, interested Taiwanese parties could now experience and explore cutting-edge computing technology. Given that the UN technical-aid program lasted for only a few years, technologists sought to strengthen their technical competence in the event that the aid program ended. To lower computer service costs, they substituted an in-house technician for IBM's official services support. William Wesley Peterson, a visiting professor from the United States, specifically urged NCTU to become independent from outside help. He helped make this goal a reality by encouraging and educating staff and students at NCTU. This chapter presents the ways technologists—in particular those at NCTU—adapted, modified, and worked creatively with the emerging technological system of IBM mainframe computers. The technologists' strenuous efforts illustrate their ability to strike a balance between making the best of US technical assistance while controlling and reducing long-term reliance on the United States.

Chapter 4 demonstrates how Taiwan's military forces began to use computers for data processing in the mid-1960s, a few years later than civilian universities and state enterprises on the island. The establishment of Chung-Shan Institute of Science and

Technology (CSIST), a military research institute dedicated to developing missile and, potentially, nuclear weapons, prompted Taiwan's military researchers to adopt Control Data Corporation (CDC) computers as research tools and to apply electronic digital computing techniques in weaponry research in the early 1970s. This chapter demonstrates that Taiwan's military coped with the constraints set by US military assistance and navigated solutions and opportunities to enhance its own capacity in utilizing electronics and computing to improve the development of aircraft, missiles, and even nuclear weapons.

Chapter 5 examines how several Taiwanese college and graduate students developed an avid interest in building minicomputers and calculators in the late 1960s. The graduate program in electronics at NCTU eschewed purchasing a minicomputer from the United States and instead built one from scratch between 1968 and 1971. NCTU students and faculty members studied the architecture of minicomputers and experimented with the construction, aiming to mass-manufacture them locally someday. Two other calculator- and computer-building projects emerged in the electrical engineering departments at National Cheng-Kung University in Tainan and National Taiwan University (NTU) in Taipei. All three groups believed that hands-on experience in building a minicomputer or calculator was critical to their journey in learning how to manufacture or mass-manufacture such technology. NTU participant Barry Pak-Lee Lam (Baili Lin) became a pioneer calculator manufacturer in the 1970s and did the same for laptop manufacturing in the 1980s. Since the 1990s, his company, Quanta Computer, has been one of the world's largest laptop contract manufacturers.

Chapter 6 focuses on the engineers-turned-entrepreneurs who, from the late 1960s to the 1970s, gained technical and management knowledge at prominent foreign-owned electronics factories and then began their own domestic electronics factories. In these plants, both expatriate and local engineers improvised to ensure overseas entities could function properly in Taiwan, where source materials were not always stably supplied and personnel who could support unique production technologies were not always available; technical support in almost every aspect could be precarious. Andrew Chew (Zaixing Qiu) was one of the era's engineers-turned-entrepreneurs. A Chiao-Tung alumnus, he began to work for Philco-Ford in 1966 and offered transistors for the Chiao-Tung minicomputer project described in chapter 5. He left the US corporation and founded his own company, Unitron, in 1969. He then hired Stan Shih, also a Chiao-Tung alumnus and Acer Computer's founder, to develop the first locally made calculators, released in April 1972.

In contrast to engineers, a large number of women factory workers tinkered with electronics to mass-manufacture and attain the standards set by international electronics manufacturers. Women assemblers produced high-quality work at factories, despite earning wages lower than their counterparts in the United States, Japan, and Hong Kong. Their handcrafted magnetic core memory units, transistors, and IC chips were distributed globally, and some even made their way into space. This chapter details women factory workers' skills, work conditions, and personal lives, as well as their lack of opportunities to become entrepreneurs, compared to their engineer colleagues, mostly men.

Chapter 7 focuses on a new group of engineers-turned-entrepreneurs, who produced calculators locally in the first half of the 1970s. After Chew's company Unitron rolled out its calculators, he envisioned his company making computers in a few years. While his vision did not materialize, Unitron and other calculator start-ups, especially Qualitron and Santron, prompted the rise of engineers and entrepreneurs, ready for the mushrooming Taiwanese computer and computer parts industry of the 1980s. Taiwanese calculator makers, with low-cost labor and highly educated engineering talent, captured a fifth of the global market for calculators in 1978. Unitron and Qualitron were Stan Shih's first and second jobs before he founded Multitech in 1976, known as Acer after 1987. Santron provided first jobs to the entrepreneurs of several prominent Taiwanese computer manufacturers. Barry Pak-Lee Lam, for example, the founder of Quanta Computer, began there.

Chapter 8 describes how Stan Shih, now a seasoned microprocessor engineer and fledgling entrepreneur, came to realize the dream shared by generations of technologists in Taiwan: manufacturing computers domestically. Shih and the US companies Franklin and Compaq, as well as other worldwide corporations, taking advantage of the availability of inexpensive and better-performing microprocessors, began to make Apple and IBM compatible computers in the early 1980s. This effort was challenged in the 1980s by Apple and IBM, which held patents and copyrights for their microcomputers. The chapter not only focuses on the technological achievements of Shih and his engineers but also critiques the images of invaders and counterfeiters portrayed by the US media and congressional hearings participants in late summer 1983. The misrepresentations indicate that the US computer industry, media, and analysts failed to make sense of Shih's success in the microcomputer market; they believed that the only way Taiwanese computer makers could succeed was from counterfeiting.

While chapter 8 discusses Apple Computer's responses to Shih's success, chapter 9 shifts to IBM's strategies to fight off the competition from Stan Shih's company, Acer. One of IBM's responses, among others, was to describe Taiwanese manufacturers of IBM compatibles as "a swarm of flies" or "scoundrels" that "misbehaved" by stealing IBM's technology.[47] Chapter 9 demonstrates that Shih and his fellow Taiwanese computer makers, such as MiTAC, became sought-after subcontractors for prestigious computer makers around the world in the second half of the 1980s. Acer became one of the top ten computer makers in the world in the mid-1990s and was one of the top five from 2004 to 2014. In 2006, Intel's CEO, Paul S. Otellini, noted that Stan Shih is "a big reason why your PC costs $1,000, not $10,000" in a *Time* magazine article titled "Asian Heroes: Stan Shih." Acer named its various computer models "Aspire" after 1995; the name speaks precisely to Shih's ambition of transforming from an engineer to the first computer manufacturer in Taiwan, a role that, since the 1950s, many of his fellow Taiwanese engineers also aspired to.[48]

Chapter 10 draws on the history of the fabrication of semiconductor chips in Taiwan, which is intertwined with the history of the island nation's computer industry. Chip manufacturing is an industry that requires experimenting with production conditions, an act of tinkering that engineers, researchers, and technicians executed. The chapter describes Morris C. M. Chang (Zhongmou Zhang)'s founding of TSMC and the multiple origins of the "dedicated foundry model." The model was independently created by Morris Chang and Robert H. C. Tsao (Hsing-cheng Tsao, Xingcheng Cao) at United Microelectronics Corporation (UMC). While Taiwanese technologists viewed computers as critical to aiding Taiwan's national defense and industrial development during the Cold War, various parties in the United States and Taiwan, at the time of this writing, deemed TSMC's high-end chips so valuable a resource that they have cultivated international support to help Taiwan in deterring a potential Chinese incursion.

For the Romanization system of Taiwanese and Chinese names, I have chosen to use the names of the historical actors as reflected in English-written archival documents or publications. These names are typically Romanized under the Wades-Giles system. I have also supplied the *hanyu pinyin* version of their names. In cases where authors did not leave an English translation of their names in sources, I have used *pinyin* to transliterate their names. As for the order of their last names and first names, I have followed English-language style: the surname follows the first name. I made exceptions for well-known historical figures such as Chiang Kai-shek, Chiang Ching-kuo, and Mao Zedong.

I choose to refer to many individuals discussed in this book as "Taiwanese," despite that some were born in China and emigrated to Taiwan after 1949. Some of them might not have ever considered themselves Taiwanese during their lifetimes, because of their faith that Chiang Kai-shek would reclaim China someday or their belief that Taiwanese are also Chinese. For example, many Chiao-Tung alumni mentioned in chapters 1 and 2 might have held such beliefs. But, since I focus on the historical encounters that occurred in Taiwan, I believe it is less confusing for readers if I use Taiwanese to refer to individuals participating in these encounters. For instance, at the time of this writing, it may be confusing to refer to a student who studied in a university in 1950s Taiwan as a Chinese student, even though this reference was not uncommon in Chinese or English texts of the 1950s. As the usage of the terms Taiwanese and Chinese have been highly political, I would like to note that I choose to use "Taiwanese" for its geographical connotation, and do not intend to surmise historical actors' national identities or assert my political opinions. Generally, most of the Taiwan-based technologists discussed in chapters 1 and 2 were born in China. Chapter 3 marks a transition, in that the newly opened NCTU admitted a few Taiwan-born students and hired a few Taiwan-born staff members. From Chapter 5 onward, readers will learn about an increasing number of Taiwan-born engineers, including Andrew Chew and Stan Shih, who joined a critical mass to build Taiwan's computer industry. However, exceptions may apply. A prominent one is Robert Tsao, who was born in China and emigrated to Taiwan with his parents when he was one year old.

Island Tinkerers shows that Taiwanese technologists and their computers defy the dichotomy of the West innovates and the East imitates. This false dichotomy has long concealed Taiwanese technologists' efforts over six decades to ensure that they could own, explore, tinker with, master, and manufacture computers. Building on Hindle's argument, the history presented in this book demonstrates that the ambitions of technologists to emulate US technological achievements and their own tinkering with computers led to Taiwanese computer manufacturers' incremental innovations, powering the supply of a large proportion of high-quality digital products to the world. Taiwanese technologists monitored the development of geopolitics to secure the resources for their technological projects and coped with misunderstandings by and negotiations with their US and international counterparts. As the following chapters show, Taiwan's present-day success in manufacturing computers and semiconductor chips draws from a larger and more complex trans-Pacific history of computing.

I EMBRACING ELECTRONICS, 1950s

1 NETWORK RESET: RESTORING A UNIVERSITY FOR ENGINEERING

On the evening of April 8, 1952, in the Taiwan Railway club in Taipei, around three hundred Chiao-Tung University alumni celebrated their alma mater's fifty-fifth anniversary. The 1952 event, during which guests were served tea and refreshments, was the Chiao-Tung alumni's third annual gathering in Taiwan. The reception desk was covered with a piece of rice paper on which one of the most complicated Chinese characters, *shou* (longevity), was calligraphically printed and where guests were invited to sign their names. An alumnus of the class of 1903, in his seventies, the oldest person attending the event, penned his name right next to the character to reflect what he saw as the great history of the university and the survival of its alumni. The event invitation was advertised in two major newspapers on the day; coincidentally, a vice president of one of the newspapers was a Chiao-Tung alumnus. Other alumni who attended included engineers from industries such as railways, telecommunications, petroleum refining, copper mining, fertilizers, steel, and electricity. Besides engineers, participants included officials in critical positions in the government of the ROC and a former mayor of Hangzhou, since the tenth century one of China's most prosperous cities.[1] Truly an illustrious gathering.

Founded in 1896 in Shanghai, Chiao-Tung University's predecessor, Nanyang Public School (Nanyang Gongxue), was one of the earliest Qing government (1648–1911) universities to reform and modernize the education system. Nanyang Public School also offered middle-school education in the early years. Focusing on fields including railway, postal services, and telegrams, it sought to cope with Western and Japanese imperialist challenges by introducing modern technical education to China. With the fall of the Qing dynasty in 1911, the Republic of China was established, and ten years later Nanyang Public School merged with several new colleges located in Tangshan, Beijing, and Shanghai to form Chiao-Tung University. Both *chiao* and *tung* refer to traffic, communications, and transportation—the Republic of China's Ministry of Transportation and Communications was known as the Ministry of Chiaotung. Taiwan Semiconductor Manufacturing Company's (TSMC) founder Morris C. M. Chang was an alumnus of the Nanyang Model High School, which was part of Nanyang Public School until 1927. With his background, some Chiao-Tung alumni considered Chang an "adopted" Chiao-Tung alumnus.[2]

Known as the "MIT of the Orient," Chiao-Tung University was one of the few college-level educational institutions to train professionals in critical engineering fields in pre–World War II China. According to the *China Press*, on July 1, 1930, Chiao-Tung University's Shanghai campus graduated forty-one students from its College of Railway Administration, twenty-two from Mechanical Engineering, and fifty-two from the College of Engineering. At the same time, sixty-one students from the university's Peiping College of Railway Administration and thirty-two from its Tangshan College of Civil Engineering received degrees. Graduates found viable employment at "well-known factories" and "important firms." At that time, the minister and vice minister of railways were the president and vice president of the university, respectively. Both facilitated the assignments of the majority of graduates to various communications and industrial institutions for postgraduate training or employment.[3]

About a thousand alumni of Chiao-Tung University left mainland China for Taiwan,[4] most fleeting to Taiwan with Chiang Kai-shek's Nationalist Party government in 1949 after it faced imminent defeat at the hands of its rival, the Mao Zedong–led Chinese Communist Party (CCP). The alumni were among the 1.2 million Chinese civilians and soldiers joining Chiang Kai-shek's exile in Taiwan. Those 1.2 million new residents represented a significant influx, numerically and culturally, as Taiwan had had a population of 6.2 million in 1945, and many residents spoke only Taiwanese and Japanese.

After 1949, Chiao-Tung's colleges came under the management of the People's Republic of China (PRC) government, turning it into a pivotal institution in Communist China's higher education system. Taiwan's Chiao-Tung alumni thus lost their connections with their alma mater in two ways. Spatially, they could no longer visit the campuses in Beijing, Shanghai, or Tangshan, or host events there. With the PRC at war with Chiang Kai-shek, Taiwan's Chiao-Tung alumni were not allowed to stay in contact with the faculty, administrators, or alumni who had stayed in mainland China.

Taiwan's Chiao-Tung alumni began to establish connections with one another as they settled on the island. They were predominantly government bureaucrats, technocrats, engineers, and scientists of all ranks. The executive committee of the alumni organization began to meet in Taiwan regularly after 1949. The first annual gathering open to all alumni probably took place in Taiwan as early as 1950. Regularly held reunions in Taiwan, the United States, Japan, and other places soon took place. In April 1952, the alumni founded the *Voice of Chiao-Tung Alumni* (*Yousheng*), a newsletter circulating among alumni in Taiwan and other overseas locations monthly or every two months. They shared the latest news of individual alumni and alumnae and renewed memories of their lives in China prior to the moment they left. Memories of pre–World War II college life, especially sports, and personal connections among alumni on the island were an important part of their lives when they migrated to the new society. As time went by, they also exchanged ideas on their scientific or engineering professions, expressed opinions on technological projects and engineering education, and made observations on the physical infrastructure and industries of other countries they had visited.

In 1953, the Alumni Association celebrated with even more grandeur, doubling the number of participants at the same venue and catering elaborate pork cutlets with rice. Alumni were invited to enjoy a series of performances and group activities in an indoor venue at the Taiwan Railway club while others wandered off to a night garden to listen to Chinese zither music. The garden was lit by Chinese lamps and decorated with beautiful seasonal flowers. An alumnus known for his dapper appearance performed a crosstalk (similar to stand-up comedy). After his performance, he walked about the venue selling boutonnières and corsages to raise funds to help an alumnus who had recently lost his eyesight. At eight o'clock, guests gathered to participate in lantern riddles, for which ten alumnae and women family members dressed up and carried lanterns to "decorate" the stage with a festive ambience; the term "decorate" was chosen by editors of the newsletters. This gendered arrangement coincided with another setting; earlier that evening, all alumnae were asked to register guests, serving as receptionists at the entrance of the club. For party favors and gifts for games scheduled for that evening, several alumni had been charged with soliciting sponsorships from wealthy enterprises. Delivered to the venue on a five-ton truck, the favors and gifts included cans of beef, fish, pineapples, soy sauce, loose tea leaves, soap, portable kerosene stoves, candles, butane lighter refills, DDT, BHC (an insecticide), magazines, aluminum stools, aluminum plates, men's undershirts, and rubber shoes, all precious consumer products for Taiwanese in the early 1950s.

These exuberant annual gatherings of diasporic Chiao-Tung alumni ironically also served as a bitter reminder that they were away from their alma mater and hometowns. As the alumni were planning for this splendid celebration, one alumnus optimistically envisioned that the 1954 annual gathering might be held on the mainland; Chiang Kai-shek proclaimed in 1953 that he would invade Communist China to "reclaim" the mainland.[5] To alumni's disappointment, the following year's gathering and those of the upcoming decades took place instead in Taiwan. Most alumni could not revisit China until 1987, when Taiwan residents were allowed to visit their divided families in China. One might compare these alumni to residents of East Germany, who as a rule could not travel freely to West Germany to visit their family members during the Cold War. For North and South Korea, travel between the two is still highly restricted today.

These gatherings also served as a critical venue for fostering Taiwan's Chiao-Tung alumni's personal and professional connections and their interest in contributing to the development of railway, telecommunications, mining, and other public utilities in Taiwan. The diasporic Chiao-Tung alumni eventually refounded in Taiwan what they saw as their glorious alma mater with an emphasis on telecommunications engineering. They facilitated the flourishing development of computing practices and the computer industry. The first two mainframe computers in Taiwan were installed at the refounded National Chiao-Tung University (NCTU) in the early 1960s under the auspices of a United Nations technical program. State-owned enterprise engineers, university-based scientists, military personnel, and other soon-to-be computer users across the island were then drawn to the NCTU campus to learn electronic digital computing. Students enrolled in its undergraduate and graduate programs became the first generation of computer engineers and scientists in Taiwan.

This chapter unpacks how Chiao-Tung alumni in Taiwan navigated the possibilities that emerged during the Cold War to promote their professions and industries. Chiao-Tung alumni in Taiwan mobilized political connections on the island as well as drew resources from overseas alumni in US universities and industries to open their alma mater in 1958 as a "new" place to train more telecommunications engineers for the Cold War. They developed a rhetoric of Chiao-Tung "usefulness" in two ways. First, the alumni acknowledged their eagerness to participate in the government's preparatory war efforts against Communist China. They argued that the new Chiao-Tung would be critical in training engineering experts for a potential "hot war" with the PRC. But after attempts to promote a bill to reopen Chiao-Tung in Taiwan, the alumni failed to extract sufficient resources from Chiang Kai-shek's government, which allocated financial resources overwhelmingly into direct military projects that would aid in his hope for reclamation of the mainland.[6] As I show, the alumni later leveraged enough support to set up a graduate program in electronics (i.e., electrical engineering) to initiate a successful refounding of their alma mater. They could do so because the alumni emphasized studying "electronics" to underscore the university's importance, explicitly compared electrical engineering with the field of atomic energy, and advocated the two fields' critical role in defending Taiwan and the "free world," the capitalist camp, during the Cold War. The alumni's annual gathering in 1956 was instrumental in creating a discourse to juxtapose the science of electronics and atoms.

Their motivations for reopening their alma mater were multiple. Many of the alumni indeed were patriotic and loyal to Chiang Kai-shek, as they had proven by following him to Taiwan, and they felt obligated to defend the island from enemy infiltration. Beyond patriotism, these engineers and technocrats contributed to Cold War–inspired military defense while also fostering and advancing the telecommunications professions. Their success in reopening a university founded in mainland China reflects their redefinition of Taiwan's geographical connotations; they connected alumni on the island with those overseas and built networks of professionals within and beyond the island.

CHALLENGES IN REPLICATING THE GLORIOUS PAST

From the perspective of Taiwan's Chiao-Tung alumni, their alma mater had "ceased" its operations in the "Republic of China on Taiwan" starting in 1949, and the government should "resume" running the university in Taiwan as soon as possible. In 1952, Guang Wang, a forty-seven-year-old alumnus and an accomplished official in the field of domestic water transportation, noted that Chiao-Tung University, if refounded in Taiwan, would contribute greatly to the reclamation of mainland China and the subsequent reconstruction of the country. The restored university would train a number of professionals for these national goals. According to Wang, the alumni should prepare to refound the university in Taiwan because their rival, the "Communist bandits" (a term commonly used in Cold War Taiwan) began to emphasize communications technologies in higher education. The Chinese Communists had recently resumed the operation of the Shanghai campus, expanded the Beijing and Tangshan campuses, and upgraded Wuchang's Marine School from a vocational school to an undergraduate college. Wang's proposal can be compared with the United States' support for the establishment of the Free University in West Berlin in the late 1940s. The Free University was meant to compete with the prestigious Frederick William University (Humboldt University after 1949), which fell under the control of the Communist government in East Berlin. The resurrection of Chiao-Tung University occurred much later, however—in June 1958—than the founding of the Free University in Germany.[7]

Wang's proposal was enthusiastically received by Chiao-Tung alumni, even though it would soon be seen as financially infeasible by the government. Inspired by Wang's proposal, in January 1953, the executive committee of the Chiao-Tung Alumni Association, whose members were elite bureaucrats, discussed strategies for refounding Chiao-Tung University in Taiwan. They argued that since the Ministry of Transportation and

Communications was now offering training courses to personnel of the Bureaus of Railways, Telecommunications, and Port Authorities, the ministry should consider expanding the current courses from training already employed engineers to providing college-level education. The ministry could then grant degrees in the name of Chiao-Tung University. The Ministries of Transportation and Communications and Education would be responsible for the operational funds for the first couple of years for the reestablished Chiao-Tung University. In the structure of the ROC government, the Ministry of Education oversaw the operation of all universities and dispersed to each annual funds approved by legislators. The executive committee of the Chiao-Tung Alumni Association submitted a bill to the National Assembly of the Chinese Nationalist Party that year and then the Executive Yuan (equivalent to the Cabinet or Council of Ministers).[8] But the Executive Yuan rejected the bill because it could ill afford such a long-term budgetary commitment. The lack of funding was due to military allocations for retaking mainland China.[9]

In 1952 and 1953, Chiao-Tung alumni were aware of war preparations and their indispensable role in a war against Chinese Communists. The alumni emphasized that engineers in the communications fields, including railways, roads, electricity, postal, water transportation, and so on, would be in high demand if Chiang Kai-shek initiated a large-scale military action to reclaim mainland China. These advances would be especially important for the operations of Chiao-Tung University in Taiwan for training more engineers. In August 1953, Gisson Chi-Chen Chien (Qichen Qian), then one of the vice ministers of transportation and communications, evaluated the possible shortage of communications engineers in the years to come. He stated that the demand for communications engineers would soon outstrip the 48,300 members presently in Taiwan. According to his data, each year three out of every thousand were expected to leave their positions because of retirement or death, and another three out of every thousand would resign or have their contracts terminated. In the event of a military action, some engineers would serve in the military. Though employers could enhance their engineers' qualitative performance on-site, increasing the quantity of communications engineers required formal school education.[10]

In May 1953, the executive committee of the Alumni Association met again. Alumni brainstormed on how to persuade the government to support Chiao-Tung's refounding. Gisson Chien argued that the Alumni Association should gain the necessary support from the Ministry of Education. The vice president of *Taiwan Shin Sheng Daily News* pledged to raise the public's awareness of the importance of training communications engineers and managerial personnel. Peilan Ye, an alumna who worked at the Bureau of Railways, advocated that the Alumni Association should plan to admit more women than had been enrolled in China. In September, the executive committee met again and suggested that three alumni holding top positions in the Chinese Petroleum Corporation (CPC), Taiwan Power Company (TPC), and the Ministry of Transportation and Communications should contact the premier and vice premier of the Executive Yuan to reiterate the significance of educating more communications engineers.[11]

From 1953 to 1958, three other universities "resumed" their operations in Taiwan, each for a different reason. In 1954, Soochow University, a Methodist private university founded in Suzhou (known as Soochow prior to 1949), convinced the Ministry of Education to permit its restoration in Taiwan because it had raised enough money to reopen the school. Soochow had secured its own funding by offering short-term training classes to the public, starting in 1949. It then began offering undergraduate degrees in law, political science, economics, accounting, and foreign languages.[12]

In June 1954, on the basis of a petition of the alumni of Chengchi University, the Minister of Education sought permission from the Executive Yuan to refound Chengchi University.[13] *Cheng* (zheng) and *chi* (zhi) refer to "governance" in Chinese. In 1927 the Chinese Nationalist Party had founded Chengchi's predecessor, a central party school (zhongyang zhengzhi xuexiao) based in Nanjing that trained political elites for the party-state. Chiang Kai-shek was the first president of the school, which began offering undergraduate degrees in 1932. In April 1949, five hundred alumni gathered for a reunion in Taipei. The large number of attendees reflected that many of its alumni followed Chiang to Taiwan. In July 1954, Chiang appointed a committee, which included his son Chiang Ching-kuo, to restart the university. In 1954, National Chengchi University set up graduate programs in public administration, civic education, journalism, and international relations. Chengchi University's direct political significance to the Chinese Nationalist Party's governance and legitimacy was crucial to the university's restoration.

Besides emphasizing the study of social sciences, Chiang Kai-shek was interested in the development of atomic energy, possibly for nuclear weapons development. He personally ordered the resumption of Tsing Hua University, a prestigious pre–World War II Chinese university, to advance the study of the field. Tsing Hua University's predecessor, Tsing Hua College, was set up in 1911 through the Boxer Indemnity Scholarship Program, which received money from Qing government war reparations to the United States for the Boxer Rebellion (1899 to 1901). The pre–World War II Tsing Hua University offered degrees in social sciences, humanities, natural sciences, and engineering. In June 1954, Taiwan's Legislative Yuan (equivalent to the US Congress) passed legislation to set up an atomic energy committee. In 1955, Chiang Kai-shek signed a bilateral agreement on peaceful use of atomic energy with the United States. He then met with the Ministry of Defense and decided to acquire and locate a nuclear reactor on the Tsing Hua University campus. Yi-Chi Mei (Yiqi Mei), a physicist and the former president of the university from 1931 to 1948, was recruited from the United States in November 1955 to prepare to purchase a nuclear reactor for teaching and researching and to develop a graduate program in atomic energy. In December 1955, the Executive Yuan set up a committee to begin formal preparation for refounding the university.[14] Among Chiao-Tung alumni participating in the restoration of Tsing Hua was S. M. Lee (Ximou Li), who, in 1918, received his master's degree in electrical engineering from MIT and was a dean of Chiao-Tung University when it relocated to Chongqing during the Second Sino-Japanese War (1937–1945). Lee joined the Atomic Energy Committee of the Executive Yuan in July 1955 and attended the first International Conference on the Peaceful Uses of Atomic Energy in Geneva in August.[15]

Taiwan already had universities offering electrical engineering education prior to the arrival of Chiang Kai-shek and the Chinese Nationalist Party. From 1895 to 1945, Taiwan was ruled by Japan, after the Qing government ceded the island to the Japanese at the end of the First Sino-Japanese War. Three Japanese-era institutions offered electrical engineering education during the colonial period: Taihoku Imperial University, Tainan Technical College, and Taipei Industrial School. In 1928 the Japanese colonial government established the Taihoku Imperial University, one of nine imperial universities supported by the Japanese empire. Taihoku Imperial University was renamed National Taiwan University in 1945 by the Chinese Nationalist Party and became the only university in Taiwan until 1954. Staffed with comprehensive faculties in the liberal arts, sciences, and medicine, National Taiwan University began to offer undergraduate degrees in engineering, including electrical engineering, only in 1943. According to historian Pi-ling Yeh, the late establishment of an engineering faculty in Taiwan was due to Japan's pre-1930s intention of using its colonies to supply agricultural resources to its metropoles. But after the 1930s, Japan began to take military or quasi-military actions to expand its influence in Manchuria, and the empire's demand for college-educated engineers and scientists increased. The new governor-general of Taiwan and navy admiral, Seizō Kobayashi, further implemented policies of Japanization (Kōminka) and industrialization in Taiwan after he took office in 1936. The establishment of the engineering faculty at Taihoku Imperial University was a legacy of the wartime emphasis on industrialization in Japan's colonies.[16]

Tainan Technical College (Tainan qongye zhuanmen xuexiao or Tainan kōgyō senmon gakkō) was renamed Taiwan Provincial Tainan College of Technology in 1945. Originally founded in 1931 as a vocational school offering electrical engineering education, it was elevated in 1956 from an engineering-centered college to a full-fledged university, National Cheng-Kung University. It is today known as one of the best universities in Taiwan. The third pre-Nationalist Taiwanese university, Prefectural Taipei Industrial Institute (Taibei zhouli Taibei gongye xuexiao or Taihoku shūritsu Taihoku kōgyō senmon gakkō) was set up by the Japanese colonial government in 1912. In 1923, it began to offer five-year or three-year programs in electrical engineering. After World War II, it was known as Provincial Taipei Institute of Technology (Shengli Taibei gongye zhuanke xuexiao), now known as National Taipei University of Technology.[17]

Even though the Chinese Nationalist government kept many of these Japanese-era schools, they replaced the Japanese faculty members, administrators, and staff with mainland Chinese. In 1947, Japanese professors dropped from 20 percent to 8 percent of the faculty members at National Taiwan University. The local-born and Taiwanese-educated were deemed unqualified to teach at the college level. Many did not speak Mandarin, and few had the opportunity to pursue higher education under Japanese rule. Generally, they did not enjoy as many promotion opportunities in higher education as the Japanese. Tsung-lo Lo (Zongluo Luo), the acting president of the National Taiwan University, sought to promote local Taiwanese scholars and personnel after 1945, but at the same time, he did not expect that the number of local Taiwanese scholars would be sufficient and began recruiting mainland Chinese scholars before he embarked on his trip from mainland China to Taiwan to take over the university from its Japanese management.[18]

Despite failing to reopen Chiao-Tung in 1953, the reestablishment of Soochow and Chengchi Universities in Taiwan in 1954 rekindled Chiao-Tung alumni enthusiasm for reopening their alma mater. More and more alumni proposed ideas on how to realize the re-founding plan. Vice Minister of Transportation and Communications Gisson Chien believed that the most feasible option for resuming university operations was setting up a graduate school of telecommunications. An alumnus also urged alumni with prestigious posts in ministries to bring up the proposal whenever possible with council members of the Executive Yuan. Nevertheless, several alumni thought of starting management courses, at graduate or undergraduate levels, because they would require less equipment, hence less money, to operate. In 1955, Qinbo Ye, a lecturer at Taipei Industrial School, wrote that Chiao-Tung alumni should consider offering correspondence education and form an engineering consulting firm. He regarded the two plans as ways to utilize Chiao-Tung alumni's irreplaceable talents and begin to function as a business organization immediately.[19]

Another alumnus suggested that the Alumni Association should consider raising funds on their own and making Chiao-Tung University a private university. With sufficient funding, it might be easier to persuade the Ministry of Education to permit the resumption of the university. The refounded university would still be able to train a good number of capable engineers to support Taiwan's railways, telecommunications, military, transportation, power plants, heavy industries, and chemical industry. Overseas Chinese who owned businesses in North America were also keen for capable engineers from Taiwan; the demand could not be met solely by the Taiwan Provincial Tainan College of Technology, as the alumnus indicated.[20]

FROM HOT WAR MEMORIES TO COLD WAR DISCOURSE

Chiao-Tung alumni's mobilization in the highest political echelons was effective, but I attribute another important strategy, developed in 1956, to their success in realizing their plan in 1958. What the alumni did was emphasize studying electrons and subsequently "electronics" to underscore the university's potential contribution to the Cold War. They linked electrical engineering with atomic energy and advocated the two fields' critical role in keeping the peace in Taiwan during the Cold War. They shifted focus from their past contribution during the hot war to their potential contributions during the Cold War. They also realized the plan because of intellectual and financial support from overseas alumni at US universities and industries.

The celebration of Chiao-Tung's sixtieth anniversary in 1956 fortuitously began a turning point that led to its reopening in Taiwan. On April 8, 1956, Chiao-Tung alumni hosted celebration events concurrently in cities across Taiwan. In Taipei, alumni attended an afternoon tea reception and evening dance party, gathering the night before to enjoy Beijing operas. The afternoon reception attracted two thousand alumni and family members in the Ambassador Hotel. Beyond these social events, some alumni sat in the morning for a formal ceremony in which the ministers of education, economic affairs, and transportation and communications lauded the university's accomplishments. A Chiao-Tung representative read Chiang Kai-shek's remarks to about a thousand attendees at the ceremony.[21]

Minister of Education Chi-Yun Chang (Qiyun Zhang), a historian and geographer, lauded the accomplishments of Chiao-Tung University in the Republic of China's history. As the ministry managed all universities, especially at the budgetary level, his favorable remarks excited the alumni. He noted that Chiao-Tung contributed to educating professionals in transportation and scientific management and promoting modern science and its applications. He first complimented Chiao-Tung's graduate program in industrial engineering. It was set up in 1926 as the first graduate program in China. Second, Chang brought up the role of Chiao-Tung alumni in building the railway between Zhejiang and Jiangxi Provinces in China around 1931. It was an unprecedented achievement because the railways received no foreign investment and were fully designed by local engineers. Third, he mentioned that the operation of the rails during World War II was only possible because of Chiao-Tung graduates, who made up almost all the railway staff. They were known for their wartime bravery and perseverance. They ensured that rails were functional even when under attack by the Japanese, and they remained at their posts until the very last minute after everyone else had retreated. The commendable spirit exhibited by Chiao-Tung alumni was remembered even after the government moved to Taiwan. He went on to discuss how the Ministry of Education began granting masters' degrees to students who completed courses offered by government agencies. The legacy of Chiao-Tung, he noted, could perhaps be preserved by the Ministry of Transportation and Communications' research department, which was offering courses to employed engineers; the ministry could explore ways to assist employed engineers in receiving master's degrees. Since vocational education was especially needed for the construction of Taiwan and the reclamation of mainland China, the ministry could consider expanding its training courses to nurture more engineers and fulfill these goals.[22]

Chang's remarks expressed his favorable attitude toward Chiao-Tung University's legacy in Taiwan's educational scene. Likewise, a prominent alumnus sensed the optimism of Chang's remarks and acted on the opportunity promptly. He was Hung-Hsun Ling (Hongxun Ling, see figure 3.2), president of Chinese Petroleum Corporation and former president of Chiao-Tung University. Seizing the moment, after the ministers offered their remarks, Ling walked to the stage to read a telegram from the New York Committee of the US chapter of the Chiao-Tung Alumni Association. The telegram said, "Best congratulations for the jubilee and proposed establishment of Chiaotung Electronics Institute for commemoration." Ling further elaborated that the bold proposal of a graduate program in electronics would contribute significantly to recent research on atoms and electronics and their military applications. Ling explicitly urged Chang to reconsider refounding Chiao-Tung University.[23]

Tsen-Cha Tsao (Cengjue Zhao), who sent the telegram to the celebration, was one of the leading advocates of restoring Chiao-Tung. Starting in 1953, he was the first chair of the US chapter. In the 1950s and 1960s, the US group kept close ties to Chiao-Tung's Taiwan alumni by subscribing to alumni newsletters, contributing newsletter articles, meeting regularly at alumni business and social events, and hosting Beijing opera performances cast with overseas alumni. A March 1956 survey found at least 133 alumni in the New York area.[24]

Tsao was a perfect example of Chiao-Tung alumni who had contributed to the operation of communications technologies in China during World War II, as Chang noted in his remarks. Born in Shanghai in 1901, he attended Nanyang Public School's elementary school and middle school and graduated in 1924 from Chiao-Tung's electrical engineering department in Shanghai. Supported by the Ministry of Transportation and Communications, he interned at radio, telephone, and telegram companies in England and Germany from 1924 to 1928. He pursued and received a master's degree from Harvard University in 1929, returned to China to teach at Zhejiang University from 1929 to 1932, and worked as the chief engineer for the Telephone Bureau of Zhejiang Province from 1932 to 1943. He held other government posts for wartime telecommunications from 1938 to 1945. After World War II, he worked until 1949 on reconstructing public utilities in Shanghai. He then moved to the United States that year, after the Chinese Communist Party took over China. In 1950 he began his US career by working as a systems engineer for the Consolidated Edison Company in New York. He is credited for reactivating the Chinese Institute of Engineers in the United States in 1953 and forming the Chiao-Tung alumni's US chapter the same year. Starting in July 1955, he also consulted for Taiwan Power Company in the field of atomic energy, where he had witnessed the growing attention to atomic energy from state-owned enterprises in Taiwan. When drafting the telegram in 1956, Tsao was a consultant at McGraw-Hill Book Company. In 1958, he began to work as a researcher at the Riverside Research Institute, and was affiliated with Columbia University. At the Riverside Research Institute, he participated in research projects contracted by the Army Rocket and Guided Missile Agency of the US Army Ordnance Missile Command, in Redstone Arsenal, Alabama.[25]

EMBRACING ELECTRONICS, FLYING WITH ATOMS

Despite his physical absence from the 1956 celebration ceremony, Tsao was a passionate supporter of the restoration project. His major contribution to the project was juxtaposing the study of electronics and atomic energy, and Ling broadcast Tsao's call in the 1956 ceremony. In the following two years, the narrative became an effective discourse that Chiao-Tung alumni utilized to highlight the importance of refounding their alma mater. In the ceremony, Ling read the telegram and asked for more public attention to electronics, to match its enthusiasm for atomic energy. Ling's enunciation was likely from his reading of a Tsao article, scheduled to be published after the April celebrations in *Essays for Celebrating Chiao-Tung's 60th Anniversary* (Jiaotong daxue liushi zhounian jinian zhengwenji), titled "Alma Mater in the Context of Engineering Science in This Pivotal Era" (Huashidai de muxiao yu huashidai de gongcheng kexue), in which Tsao reviewed recent progress in four fields of engineering science that exercised "a profound influence on human life": electronics science (i.e., electrical engineering), servomechanisms, nuclear engineering, and electronic digital computing. He described the trend of replacing vacuum tubes with transistors. Then he brought up the critical importance of the education and study of electronics. When elaborating on the application of electronic digital computers, he stated,

> Large-scale power plants in the United States could use digital computers to save labor and time in the process of printing bills for millions of customers every month. As for defense, because it is important to act swiftly in military actions, the application of digital computers in the military is noteworthy. Using analog simulators and digital computers, it only takes 0.0017 seconds to calculate the ballistic trajectory of a missile and to attack that missile.[26] (Original in Chinese)

Tsao concluded that Chiao-Tung University should resume its operation in Taiwan and set up a graduate program in electronics engineering, servomechanisms, and electronic digital computers.

This was the first time that Chiao-Tung alumni considered the future Chiao-Tung University as an institution that focused on electronic digital computers. Tsao's emphasis on computers and cybernetics could be credited to his understanding of cutting-edge research in electrical engineering in the United States. Recall that Tsao was a consultant for engineering fields at the McGraw-Hill Book Company. Moreover, Tsao likely attended a talk on a similar topic given by Yu Hsiu Ku, a professor of electrical engineering at the University of Pennsylvania who emigrated from China to Taiwan and then the United States in 1950. The talk was organized in New York City in 1956 by the Chinese Institute of Engineers, which Tsao helped reactivate.[27] Ku's specialty was in mathematics, electrical machinery, and modern control theory. In Ku's talk, he focused on the rapid development of automation, cybernetics, and computers, explicating the application of computers in missile guidance and in labor saving in offices.

Aware of the development of Tsing Hua University, Tsao's article strategically elevated the importance of electronics science to the level of atomic energy. Electronics science is a Chinese term that refers to the subfields related to communications, vacuum tubes, transistors, and other types of sciences of electronics in the broader field of electrical engineering. Tsao believed that if the government permitted the refounding of Tsing Hua University for atomic energy research, it should also consider designating another university to take care of electronics research. As Chiao-Tung was known for its pre–World War II prominence in electrical engineering education, it should resume its operation and play the important role in applying its research to military and economic application.

Tsao's point on the study of electronics for military applications as well as electronics for the use of atomic energy echoed news articles in the *United Daily News* (Lianhe bao) that had familiarized readers with the latest developments in military applications. Chiao-Tung alumni and the officials they would like to persuade were unlikely to miss news articles that introduced national defense and nuclear research in relation to computers, be they digital or analog. As early as 1953 and 1954, a news article a year in the *United Daily News* discussed how US atomic energy research institutes such as Oak Ridge National Laboratory were using computers. In 1954, a careful newspaper reader in Taiwan could learn that IBM had built a computer for the US Navy's Bureau of Ordnance for research on torpedoes and that Douglas Aircraft Company used computers to automate its factories.[28] In 1955, one could read a news article about the US Air Force's F-86 Sabrejets, stationed in Taiwan and carrying rockets with analog computing capability to intercept their targets. Two weeks before Chiao-Tung's 1956 anniversary ceremony, *United Daily News* reported that US Secretary of the Army Wilber M. Brucker had revealed a gigantic system of computers that could detect and shoot down incoming attacking aircraft with ten Nike Ajax missiles. The news article alluded to the Semi-Automatic Ground Environment (SAGE) system (see more discussion in chapter 4). Newspaper readers, including Chiao-Tung alumni and government officials, might sense the rapid development in defense technologies and wonder whether Taiwan's talent would be able to catch up with it.[29]

Furthermore, more than a year after the 1956 Chiao-Tung celebration, Katharine Hepburn and Spencer Tracy's new film *Desk Set* was released in Taiwan—we do not know when exactly it was first seen, but *United Daily News* published a film review about two months after the film was released in the United States. The film portrayed Hepburn leading the staff of a broadcast network library that believed their jobs were going to be replaced by a computer called EMERAC (Electromagnetic Memory and Research Arithmetical Calculator) that had been developed by Tracy. While not all readers of *United Daily News* would have watched the film, the review might have impressed readers with the room-size EMERAC blinking in the library. Those moviegoers and newspaper readers in Taiwan might have wondered how machines like EMERAC would change their lives.[30]

After the April 1956 ceremony, Chiao-Tung alumni were thrilled by Minister of Education Chang's remarks at the ceremony and the bold move by Ling and Tsao connecting electronic engineering with Chiang Kai-shek's favorite topic, atomic energy. The Chiao-Tung Alumni Association assigned several alumni to further press the issue with various ministers and influential figures. In September 1956, on his business trip to the United States, Gisson Chien made efforts to visit New York alumni Tsen-Cha Tsao and Yi-Chi Mei, the recently assigned president of the refounded Tsing Hua University, to obtain their continued support for reopening Chiao-Tung. Mei was on his trip to the United States to recruit scholars to teach at Tsing Hua and look for a suitable nuclear reactor to be installed on the Tsing Hua campus.[31] Tsao began to work on raising $10,000 US for purchasing equipment for the proposed graduate program in electronics. Enthusiastic about the refounding of Chiao-Tung, Mei promised to support the plan by drawing from the funds left over from the Boxer Indemnity Scholarship Program. Mei was in charge of the scholarship program, one of the key financial resources supporting Tsing Hua's operation at that time in Taiwan.[32]

Tsao's discourse further stabilized the rhetorical direction of the refounding movement and united the alumni about how to advocate for the refounding proposal in 1956 and 1957. For example, echoing Tsao's emphasis and tapping into interest in atomic science, the editor of the *Voice of Chiao-Tung Alumni* reiterated in its June 1956 issue that electronic devices were the basis for utilizing atomic energy and atomic bombs. The editor tapped into public interest in atomic energy to advocate for Chiao-Tung alumni's advantages in developing research on electrons and electronics—which were the central subjects of their telecommunications profession and now a symbol of that profession. The editor also mentioned that, in April 1956, Chang had pointed out that Chiao-Tung University had established the first graduate school in China three decades ago. Thus, now was an auspicious time to start a new graduate school in electronics in Taiwan.[33]

Gisson Chien also elaborated on the potential contribution of a refounded Chiao-Tung University to nuclear energy research and argued that electronics science researchers should profit from the widely recognized importance of existing nuclear research. In May 1957, the same month *Desk Set* was released in the United States, the *Voice of Chiao-Tung Alumni* published Chien's article "Atoms and Electronics." In the article, he recognized the importance of atomic energy research and noted the centrality of electronic instruments to atomic energy research and nuclear reactors. He also reviewed the recent application of computers and transistors in the United States to missile launching and missile detecting, aircraft navigation systems, and military-related data processing. He included examples of the SAGE project and equipment distribution planning in Fort Huachuca in Arizona. He then called for training more engineers in Taiwan through a graduate program in electronics. He pointed out that "our country's science" was "backwards." To promote science in this turbulent era might not sound realistic, but "science can help improve human beings' wealth," according to Chien. The entire country should try to "catch up with current advances and should thereby avoid being excluded from 'the scientific civilizations' [kexue wenming]." While Chien drew on nationalism to frame the importance of advanced electrical engineering research, he subtlety directed readers' attention to what a restored Chiao-Tung University could contribute to the defense of Taiwan.[34]

In the same month Chien published the article, the Chiao-Tung alumni lobby convinced the ministers of education, economic affairs, national defense, and transportation and communications to reopen Chiao-Tung University in Taiwan. By then, Minister of Education Chi-Yun Chang had expressed his favorable attitude for Chiao-Tung's refounding in various cabinet meetings. The ministers wrote to the Executive Yuan again to propose the refounding bill. Moreover, the minister of transportation and communications at that time was a Chiao-Tung alumnus, and the minister of economic affairs was originally trained as an engineer with a degree from Königlich Technische Hochschule zu Berlin, the predecessor of Technische Universität Berlin. Both ministers' educational background inclined them to support the refounding bill.[35]

In the proposal presented to the entire Executive Yuan, the ministers reiterated the commensurable importance between atomic energy and electronics research. They argued that the application of electronics research, such as radar, missile control, telecommunications, electronic digital computers, televisions, and servomechanisms, was crucial to industrial development and other progress in the military as well as transportation. In terms of the budget, the ministries of economic affairs, education, and transportation and communications would together contribute to the university's expected expenses for the first year. In addition, the US chapter of Chiao-Tung alumni, now pledged to raise another $100,000 US to acquire equipment for the university. The United Industrial Research Institute (lianhe gongye yanjiusuo) in the city of Hsinchu in Northwest Taiwan agreed to offer land and facilities to Chiao-Tung University alongside its original agreement to lease land to Tsing Hua University. As a result, Chiao-Tung University could collaborate with Tsing Hua University in the same city, Hsinchu.[36]

On November 1, 1957, Minister of Education Chang announced the approval of the founding of Chiao-Tung University's graduate program in electronics, the Institute of Electronics (Dianzi yanjiusuo). Former Chiao-Tung president Ling would serve as the director of the preparation committee for the program. Chiao-Tung alumni's efforts paid off after years of deliberation and mobilization. The ministry made the announcement at a luncheon on the day to honor Chen-Ning Franklin Yang (Zhenning Yang) and Tsung-Dao Lee (Zhengdao Li), who had become Nobel laureates in physics earlier that year, and it was the month after the Soviets launched the first Sputnik into space. Though Yang and Lee were not in Taiwan at that time, Yang's then mother-in-law and Lee's elder brother, who had migrated from China and lived in Taiwan after 1949, were invited to join the luncheon.[37] The president of Tsing Hua University, Mei, and Ling attended the luncheon, too. The two Nobel laureates studied with Mei while they were enrolled at the National Southwestern Associated University in Kunming, Yunnan, during World War II. Chang also announced that Mei was heading to the

United States again soon to recruit overseas Chinese scientists to teach in Taiwan. For example, Chang specifically suggested recruiting Chien-Shiung Wu (Jianxiong Wu), a women physicist teaching at Columbia University. He also looked forward to seeing more US-based Chiao-Tung alumni returning to Taiwan. Savvy in officialese, Ling took the opportunity to urge the government to consider distributing more funds for basic scientific research, though in fact it was more likely part of a strategy to call for more financial support from the state for the refounding of Chiao-Tung University.

Soon after news of the approval spread across Taiwan, alumni initiated a global campaign to raise funds to make this announcement a reality. They aimed to raise one million New Taiwan (NT) dollars, or $32,310 US (an equivalent of $354,000 US in 2024), to donate the funds during the next year's celebration of the university's founding. At the time, per capita income in Taiwan was $164 US, or $4,060 NT. One million NT dollars in 1957 would be equal to the annual income of roughly 200 average Taiwanese. To attain this goal, the Alumni Association arranged for a gathering of sixty-three alumni leaders right before Christmas Eve in 1957. Since there were around a thousand alumni in Taiwan, if each alumnus or alumna could donate $1,000 NT, they would be able to meet the targeted one million in funds. Like a military mobilization, all the registered alumni for the mid-December gathering were roughly divided into thirty-three units for this fundraising mobilization. The leader of each unit would then work with the unit members to maximize donations. Leaders of the overseas alumni associations in Japan, Thailand, Singapore, and Malaysia were urged to reach out to regional alumni, too. Hong Kong and Macao alumni also joined in this fundraising a month later. Chiao-Tung alumni's fundraising sought to alleviate the government's burden and allow the operation of the university to begin sooner than later. The alumni planned to raise enough money to construct new buildings on campus. The US alumni's $10,000 US donation from 1956 would be then used for purchasing instruments for teaching and laboratories.[38]

CHIAO-TUNG UNIVERSITY IN A BATTLE FOR THE FREE WORLD

In the months following the exhilarating news, the *Voice of Chiao-Tung Alumni* came to play an important role in uniting alumni worldwide for a successful fundraising campaign. Editors from December 1957 to February 1958 promoted the idea that electronics could guard the free world. Research in electronics and its applications contributed significantly to strengthening the capitalist camp's military defense against the communist camp. The operation of Chiao-Tung University's Institute of Electronics in Taiwan was critical to these efforts. To illustrate their point, the editors featured pictures of radar dishes, a stealth ultrasound bomber, and an aircraft integrated into a tactical air navigation system on the front cover of the newsletters. These images clearly constructed an idea that the future of Chiao-Tung University would be tied to the construction or installation of these defense technologies.

In December 1957, the editor of the *Voice of Chiao-Tung Alumni* furnished several articles to celebrate the rebirth of National Chiao-Tung University in Taiwan. An illustration of two 60-foot radar dishes graced the cover of the special issue (figure 1.1); the sentence below the illustration reads "Electronics have guarded the free world." According to the editor, the dishes were installed in a territory of the "free world," somewhere in Northern Europe, as a part of the telecommunications network to defend the frontier of the free world.[39]

The next issue, in January 1958, featured a front cover with a stealth ultrasound bomber evading two missiles. The bomber was equipped with a countermeasure system that was protected by an invisible electromagnetic shield that interfered with electronic-magnetic signals and subsequently blocked missile attacks. The front cover of the February 1958 issue featured an aircraft landing on an aircraft carrier as well as a sophisticated transmitter-receiver that consisted of seventy-eight vacuum tubes, both of which represented a part of a tactical air navigation system (figure 1.2). On the margins of this front cover was the following declaration:

> Only studies in the field of electronics promote atomic research; only the promotion of research in electronics fortifies national defense; only the use of electronics creates advanced industries; only the development of electronics creates advanced civilization.[40] (Original in Chinese)

The editor acknowledged that the practitioners of scientific research competed with one another, regardless of whether they belong to the capitalist camp or the communist camp. Chiao-Tung University would thus contribute to the Republic of China's "catching up" in the realm of scientific research. For example, in the *Voice of Chiao-Tung Alumni*'s December 1957 issue, the editor concluded, "We have to explain the uses of electronics science to the general public so that they can understand the importance of such research. They will be able to understand that our country has to catch up [with this research] as soon as possible [jiqi zhizhui]. It is time to set up the graduate program in electronics!" (original in Chinese).[41]

Tsen-Cha Tsao also reiterated that scientific competition was integral to the revived alma mater's new role. In February 1958, Tsao, on behalf of the US chapter, wrote to S. M. Lee, the new director of NCTU's graduate program in electronics, and Hung-Hsun Ling, the former president of Chiao-Tung University, to inform them of a couple of recommendations made by the US chapter in a recent meeting. The letter, also published in the *Voice of Chiao-Tung Alumni*, read,

FIGURE 1.1
An illustration of two 60-foot radar dishes graced the cover of the *Voice of Chiao-Tung Alumni* in December 1957. The sentence below the dishes reads, "Electronics have guarded the free world." Courtesy of National Yang Ming Chiao Tung Museum.

FIGURE 1.2
The front cover of the February 1958 issue of the *Voice of Chiao-Tung Alumni*, featuring a tactical air navigation system, which consisted of a fighter jet landing on an aircraft carrier, as well as a sophisticated transmitter-receiver that contained seventy-eight vacuum tubes. Courtesy of National Yang Ming Chiao Tung Museum.

The graduate program in electronics at National Chiao-Tung University and the graduate program in nuclear physics at National Tsing Hua University exemplify that Free China has resolved to emphasize scientific research and execute it. Since the Soviet Union launched a satellite [Sputnik 1], the United States and all free [non-communist] countries have understood that we need to train as many scientific researchers as we can, to fight for the freedom of human beings. We all have to collaborate to collectively achieve efficacy herein. If we can propose a concrete and reasonable plan [for running the graduate program at NCTU], we shall be able to elicit sympathy [tongqing] from the United States, sympathy that will perhaps enable us to obtain assistance from the United States.[42] (Original in Chinese)

Tsao argued that scientific research in noncommunist countries was a collective achievement. Chiao-Tung's graduate program had to "contribute to the research of the free world."[43] He not only acknowledged the ideological boundary between noncommunist and communist scientific research but also suggested that Chiao-Tung alumni rhetorically use that boundary to mobilize assistance from the United States, especially because the Soviet Union's achievements in 1957 significantly outpaced those of the United States.

In 1959, the Soviet Union and the United States held exhibitions in each other's country aiming at demonstrating their scientific accomplishments, with the famous kitchen debate between Nixon and Khrushchev taking place in July. Residing in New York City, Tsao visited the Soviet exhibition and witnessed the full-size replica of Sputnik 1, and in September, he discussed his views of the exhibition in the alumni newsletter. He noted that in addition to laundry machines and air conditioners, US scientific progress was represented by its display of an IBM RAMAC-305 computer in Moscow. On the global scientific competition, he concluded that the competitions among counties were now "decided by a country's advantages over others in science." He pointed out that a country had to produce "unmatchable achievements in science, applied science, and equipment [technology]" so that it could display its capability to destroy its enemy if attacked. The goal of deterrence was to achieve "peace."[44]

Chiao-Tung alumni attitudes toward the Cold War were complex. While some considered participating in a military conflict against the Chinese Communist Party in the early 1950s, the alumni did not directly comment in the *Voice of Chiao-Tung Alumni* on the Taiwan Strait crises in 1955 and 1958 (more in chapter 4), military conflicts between the ROC and PRC that occurred primarily on islands a few miles off the coast of mainland China that had been governed by the ROC since 1945. Their silence on these events may have been due to many reasons. For example, details of the conflicts were not revealed in the media, and alumni who played a role in the two incidents probably could not speak freely on telecommunications applications there. However, whether enthusiastic about defending "Free China" or war-weary, alumni might find telecommunications useful: the reopening of their alma mater would make fellow professionals aware of the need to strengthen defense; Taiwan's advances in scientific and technological realms would contribute to a deterrence strategy.

TRANSNATIONAL RECRUITMENT

In the winter of 1957, US alumni enthusiastically offered their suggestions for future curriculums and potential instructors, in addition to demonstrating their commitment to raising funds and sourcing laboratory equipment. Tsao was especially concerned about instructors. On December 12, 1957, he offered Hung-Hsun Ling and S. M. Lee a list of four alumni who worked in prominent companies or higher education in the United States and could be invited to teach in Taiwan for a semester or a summer. The first was Lan Jen Chu (Lancheng Zhu), a professor of electrical engineering at the Massachusetts Institute of Technology. He graduated from Chiao-Tung University in Shanghai in 1934 and obtained his PhD from MIT in 1938. During World War II, he headed the Chinese advisory specialist group to the United States Armed Forces in China. Chu specialized in ultra-high-frequency radio antenna design in his earlier career, but in the late 1950s Tsao recommended Chu as an expert in microwave communications.[45] The other three recommendations were from industry. Chao Chen Wang (Zhaozheng Wang, often known as C. C. Wang) was the chief scientist of the Vacuum Tube Engineering Section at Sperry Gyroscope. Wen-Yuan Pan was directing the television department at the Radio Corporation of America (RCA). Finally, Liangbi Ye was an alumnus chairing Westinghouse's microwave department. Among the four alumni, Chu, Wang, and Pan later visited Taiwan and taught at Chiao-Tung. Chu and Ernest Shiu-Jen Kuh (Shouren Ge, see below) formally served as consultants for NCTU around 1962; Wang and Pan participated in shaping the curriculum of Chiao-Tung in the 1960s, and they later headed the Industrial Technology Research Institute (ITRI), which initiated a technology transfer project with RCA on IC (integrated circuit) foundry production in the 1970s (see chapter 10).[46]

Two more alumni at US universities also came up with a list of prominent alumni for Ling and Lee to consider inviting to Taiwan—Ling-Yun Wei at the University of Illinois Urbana-Champaign, and Ernest Shiu-Jen Kuh at the University of California, Los Angeles. Ling-Yun Wei was born in China and worked as a telecommunications engineer in Taiwan from 1949 to 1956. He then went on to study with Nobel laureate in physics John Bardeen at the University of Illinois Urbana-Champaign, in 1956 and would later teach at University of Washington in Seattle in 1958. Wei contacted

Tsao to offer his thoughts on curriculum and scientific instruments to procure in September 1957. Tsao then passed Wei's suggestions to the *Voice of Chiao-Tung Alumni*. Wei proposed that Lan Jen Chu could offer suggestions on how and where to procure waveguide and transmission-related instruments, and Chao Chen Wang could do the same for microwave tubes and circuits. Wei then noted that although not a Chiao-Tung alumni, Hsu-Yun Fan—an MIT PhD, professor at Tsing Hua University from 1937 to 1949, and Purdue University professor—knew much more about the semiconductor field. Wei also pointed out that alumnus An Wang should be consulted about the field of magnetic core memory. Wang founded Wang Laboratories in Massachusetts in 1951 after receiving his PhD in physics from Harvard University. In the mid-1950s, IBM had bought Wang's magnetic core memory patent. Wang Laboratories was a successful computer and desk calculator firm, especially in the 1970s and 1980s. While editors of the *Voice of Chiao-Tung Alumni* republished the letters from Wei, they misspelled magnetic core as "Magnet, G Cove" because the field was then so unknown to them. Wang showed his support for Chiao-Tung by donating Wang Laboratories' calculator products in the second half of the 1960s. His investment in setting up a plant in Taiwan in 1967 would have long-term consequences on the growth of local electronics manufacturers (chapters 5 and 6).[47]

Wei further charted two directions for Chiao-Tung's graduate program. First, graduate-level courses should provide students sufficient knowledge to pursue PhDs overseas or intern in companies such as Bell Laboratories. Second, in doing so, the program would balance both academic performance and practical applications of electrical engineering knowledge. He urged a concentration on the fields of microwave communications and electronic circuits and components. He argued that the US and ROC Air Forces would be interested in applications of microwave communications. If possible, Chiao-Tung should consider officially sending graduates to intern in overseas factories or develop collaborative relations with overseas companies.

Ernest Shiu-Jen Kuh, a professor in the department of engineering at UCLA, wrote Tsao in January 1958 and suggested that the graduate program consider focusing on network theory and information theory, subjects that did not require a lot of laboratory equipment. He was willing to go to Taiwan to teach summer sessions, even though he had never been to Taiwan. Kuh was born in China and had interrupted his studies at the pre–World War II Chiao-Tung University to study at the University of Michigan in 1948 for his BS degree. Kuh warmly noted that Taiwan's Chiao-Tung University always recognized him as an alumnus even though he "did not even graduate from Chiao-Tung." He did keep his promise and paid visits to Taiwan and taught short-term courses at Chiao-Tung starting in the 1960s.[48]

Tsao also envisioned the new university as a conduit for exchanges between Taiwanese scholars and US experts. In addition to inviting professors from US universities to teach at Chiao-Tung, he believed that the reopened university should be proactive in sending talented junior researchers to study in the United States. He specified that an alumnus now working at the United Industrial Research Institute in Hsinchu could further his study in the United States and become a professor of instrumentation at Chiao-Tung. Sending scholars overseas and inviting scholars from overseas would ensure a supply of quality teaching faculty.[49]

Beyond fundraising, laboratory equipment sourcing, and visiting faculty recommendation, Tsao's New York network proved to be paramount in Taiwan's electronic digital computing. Tsao connected Taiwanese alumni with staff members of the United Nations, connections that became critical in the development of electronic digital computing in Taiwan (and the subject of chapters 2 and 3). After Tsao's letter sent to Ling and Lee on December 12, 1957, he immediately posted another two letters. He reported on his meeting with Fuyun Xu, a Chiao-Tung alumnus and ROC representative working at the Technical Aid Administration (TAA) at the United Nations in New York City. Tsao learned that the ROC's Ministry of Foreign Affairs was eligible to submit a proposal to TAA to request funds and recruit US experts with industrial and atomic energy expertise. Chiao-Tung could then invite the experts to spend time teaching there. The experts would be able to stay in Taiwan for at least a year, and the government could make a request to extend the visit for another year. After reading Tsao's letter, Ling assigned Gisson Chien the job of drafting and submitting a proposal. Chien, as the vice-minister of transportation and communications, would then coordinate with S. M. Lee, who soon became the director of the graduate program in electronics at Chiao-Tung, and the Ministry of Foreign Affairs, which sent out a proposal to the United Nations on behalf of the ROC government.[50]

CONCLUSION

In the fall of 1958, newly admitted students for the Institute of Electronics at Chiao-Tung started classes. Even though it was the only program at the university, it emerged as Taiwan's first master's program in a science and engineering field. The first cohort of graduate students held their classes at National Taiwan University, since the buildings of the new Chiao-Tung University was still under construction. In the first year, the 1958–1959 academic year, courses included electromagnetic wave, electron tube, electronics, electrical measurement lab, solid-state physics, and nonlinear theories. Courses for the 1959–1960 academic year included advanced circuit theory, microwave systems, solid-state physics, transistor electronics, electronic instruments, carrier systems, and television. These courses provided the critical foundation for the forthcoming UN support for courses on digital computers at NCTU.[51]

The plethora of courses at National Chiao-Tung University were underpinned by a story of contingencies in which cutting-edge electrical engineering education could flourish in Taiwan. The initial failure of the Taiwanese alumni in persuading Chiang Kai-shek and his Nationalist government did not dampen the enthusiasm for restoring the university. Using the annual alumni meetings and newsletters, the alumni in Taiwan leveraged their connections within the bureaucracy and with colleagues in the United States to raise funds, recommend personnel, and ruminate on political strategies. They observed how fellow Chinese universities utilized their histories, short-term courses, or atomic energy to facilitate the political and economic restoration of their institutions. Chiao-Tung alumni skillfully linked electronic engineering with atomic energy, associating their profession with the Cold War needs of Taiwan. Success begat success as Chiao-Tung alumni in the United States played a critical role in recommending experts, raising money, and connecting Taiwanese alumni with the United Nations in New York.

This process was not a plan fully executed by a top-down government agency. Chiao-Tung's history challenges the prevalent impression that the Chinese Nationalist government took the lead in the development of sciences in Taiwan. In terms of atomic energy, it was indeed the case. But for electrical engineering, the discipline home of computer science before the term computer science was coined, Chiao-Tung alumni's role was crucial. It was Chiau-Tung alumni who found themselves displaced in Taiwan. They engineered their careers, profession, and alma mater to fit into the wider geopolitics of the times. Chiao-Tung alumni knew their success would require the state's support: they needed at least several ministries to agree on reopening the university, either as a private or public institute, but they welcomed the government's financial support that would go to their alma mater should it resume as a public university. Many of the alumni were prestigious technocrats, who were able to reach into the highest echelons of the government. Their efforts to win the state's support for reopening the university laid the foundation for bringing and popularizing electronic digital computing in Taiwan.

The success of the alumni, in concert with other intellectuals, was salient in advocating for science during the same time period. Consider, for example, Hu Shih, a China-born but US-based intellectual who had studied under John Dewey and was a Cornell alumnus. Hu Shih was perhaps one of the Republic of China's most important intellectuals, many of whom made their way to Taiwan after 1949. Hu sought to persuade the

Chiang Kai-shek government to increase governmental funding in scientific research in the 1950s. He was worried that "Free China" was suffering from cultural, scientific, and economic "stagnation," and his worry could be traced back to his expectation—held since the beginning of the twentieth century—that the Republic of China would, or at least should, become a powerful state.[52] Compared to Hu Shih's lonely project, based on his social and cultural status, Chiao-Tung alumni's mobilization at the governmental level rested chiefly on those technocratic alumni who were serving in ministries or employed in state-owned enterprises, and those professionals who were working in US universities and industry. In contrast to the government-initiated resurrection of the pre–World War II Tsing Hua University in Taiwan, Chiao-Tung alumni used organization and mobilization to breathe life into their alma mater. Their selection of electrical engineering was slightly coincidental, but it reflected the view that electrical engineering was a much-needed discipline in the context of the Cold War. The island technologists' networking with overseas alumni provided the foundation for the internationally coordinated development of electronic digital computing in Taiwan through a UN technical-aid program, as chapter 2 shows.

2 NEGOTIATING TECHNICAL AID: "IMMEDIATE AND DIRECT" RESULTS OF SCIENCE AND ENGINEERING EDUCATION

Having successfully established the graduate program in electronics in 1958, Chiao-Tung alumni capitalized on the advantages of momentum by continuing their fund-raising among alumni. The graduate classes began after Mao ordered an attack on Quemoy Island in the Taiwan Strait on August 23, 1958; heavy shelling lasted for over a month. Vice Minister of Transportation and Communications Gisson Chien, assigned to work on the application for the United Nations Special Fund, devoted himself to bringing in more resources to the new university. Chien was born in China in 1900 and held a bachelor's degree in electrical engineering from Chiao-Tung in 1924. After graduation, he then worked for various telecommunications offices for the government of the Republic of China, in mainland China and in Taiwan. In 1958, Chien was Taiwan's representative at the International Telecommunication Union (ITU), an UN agency, and attended ITU meetings regularly. He discussed his interest in applying for the Special Fund with ITU administrators in person in spring 1959 and soon obtained ITU's support.[1]

With aid finally approved by the United Nations Special Fund office in 1960, Chiao-Tung–educated technocrats, engineers, and scientists established the Training and Research Centre for Telecommunications and Electronics on the NCTU campus. This UN-NCTU technical-aid program recruited US professors and distributed financial and material resources, including an IBM computer, among three overlapping institutions: NCTU's Institute of Electronics (a graduate program in electronics), the Training and Research Centre, and NCTU's computing center. The United Nations approved approximately $334,000 US, and the Taiwanese government had to contribute an additional $278,000 US to the project. The funding was substantial, given that the per capita income in Taiwan for the year 1961 was $153 US.[2] This chapter discusses how Chiao-Tung technologists worked with and negotiated with UN officials during the application and execution stages. The interactions between visiting professors and Chiao-Tung students are the subject of chapter 3.

The United Nations' technical-aid programs rested on the idea that experts could help a nation-state to attain the status of "take-off," evolving from an agricultural society to an industrialized society. Point four of Truman's Inaugural Address in January 1949 stressed the importance of setting up a collaborative program through the United Nations to make "the benefits of our scientific advances and industrial progress available for the improvement and growth of underdeveloped areas."[3] In the 1950s, the United Nations became a major facilitator of this effort, funding postcolonial nation-states' efforts to stimulate economic growth. The United Nations in 1950 started its Expanded Programme of Technical Assistance and Technical Assistance Administration (TAA), which was run by the Technical Assistance Board (TAB) of the Economic and Social Council.[4] Beginning in 1959, the United Nations launched the Special Fund to supplement the existing technical aid.[5] Countries such as the United States, Soviet Union, and Japan also provided economic and technical aid to countries worldwide.

In *The Priorities of Progress*, a 1961 pamphlet on the United Nations Special Fund, Managing Director Paul G. Hoffman stated that the Special Fund was geared toward a "full partnership between the advanced and the less developed countries in the mobilization of money, men and equipment for approved projects."[6] In particular, the Special Fund was supposed to assist less-developed countries' "accelerator projects" in order to spread "the 'seed' effects" of international funds. The pamphlet also listed three illustrative projects supported by the Special Fund: the "Volta River flood plain survey in Ghana," the "vocational instructor training center in Colombia," and the "central mechanical engineering research institute in India." In the 1963 pamphlet, *Target: An Expanding World Economy*, the first page was a picture of two mechanics inspecting the engine on an aircraft featuring the UN Special Fund's name and logo. Echoing this picture, the next page contained a section of texts titled "the United Nations Special Fund Helps to Prepare Countries for Economic 'Take-off.'" Beyond the United Nations, since the 1950s, Taiwanese officials had been taking advantage of US aid to obtain funds geared toward domestic-development projects, including constructing dams, improving railway infrastructure, and extending sewage systems. Other projects included nursing education, tideland reclamation, and petroleum geophysical surveys.[7]

The application for the UN-NCTU technical-aid program at the turn of the 1960s provides a window into how Taiwanese technocrats, engineers, scientists, and UN officials sought to define legitimate formats of technological exchanges for the island. The UN-NCTU technical-aid program was shaped by the negotiations between the two parties, who often held different ideas on the ends and means of building an electronics-manufacturing industry in Taiwan. The UN-affiliated officials and Chiao-Tung educated technocrats, engineers, and scientists adjusted their goals and plans from the 1958 initiation of the application to the UN approval in 1960. Despite of disagreement over some items, they eventually welcomed the arrival of the first digital electronic computers in Taiwan in 1962 and continued until the end of the program in 1965.

In their negotiation, Chiao-Tung–educated technologists exhibited an agency similar to that of the Indonesian technologists in Suzanne Moon's research.[8] Among all aid programs offered by superpowers in the late 1950s, Indonesian technocrats were shrewd to choose the one that promised their preferable technology. When identifying appropriate agricultural practices and machinery that would be sponsored through international technical-aid programs, they chose to pursue the machinery that could fulfill their goal of attaining a self-sufficient economy, rather than the takeoff-economy model promoted by US officials. They exercised a degree of agency in choosing what worked best for their country's agriculture. The concept of "mutual orientation" in Paul Edwards's work also helps illustrate how grant applicants and receivers, such as MIT scientist Jay Forrester, could persuade a funding agency to accept new ideas and discard traditional ones—in this case, the 1950s defense strategy of the US Air Force.[9] But the term "mutual" here probably better describes the consolidated US military-industrial-academic complex than it does the fragile relations between resourceful international aid agencies and less-resourceful countries during the Cold War.

In this chapter, we see how both Chiao-Tung alumni and UN-affiliated officials aimed at stimulating the growth of a robust industry in Taiwan. Chiao-Tung-educated technologists generally agreed with UN-affiliated officials' emphasis on creating connections between the university with the existing industry, but they paid more attention to the roles of electronics science and engineering education in the electronics industry. Initially, UN-affiliated officials were not particularly enthusiastic about Chiao-Tung's interest in basic science and engineering education, but later they were satisfied with the fact that the university-cased technical-aid program had trained engineers from various state-owned enterprises, which counted as "immediate and direct" contribution to industry.

This chapter begins with an analysis of the discussion between Taiwan-based technologists and UN-affiliated officials over basic science and engineering education, and its potential effect to industry, then reviews the two groups' negotiation of the relation between university and industry during the drafting of the aid program proposal, the revising of the Plan of Operation, and the evaluation of the execution of the program.

The difference between Taiwan-based technologists and UN-affiliated officials regarding their ideas about the ends and means of creating a robust electronics industry is not unusual. A similar tension can be found among Chiao-Tung overseas alumni. I then discuss the split among overseas alumni regarding whether NCTU should focus on educating students or aiding the industry. Finally, I delineate the arrival of the first computer on the island, one of the most welcome contributions to Taiwan from the UN-NCTU technical-aid program.

BOLSTERING SCIENCE AND ENGINEERING EDUCATION AT CHIAO-TUNG

After meeting with ITU administrators in spring 1959, Chien and his subordinates at the Bureau of Telecommunications began to draft a proposal to apply for support from the UN Special Fund. The bureau was a governmental agency in charge of telecommunications research to support the government's various levels of telecommunications projects. Chien and his colleagues then drafted the proposal with an eye toward improving the training of telecommunications personnel in Taiwan. In the June 1959 Chinese version of the draft proposal, Chien expected the technical-aid program to help train and educate Taiwanese scientists and engineers in fields such as, "long-distance direct dial-up telephone systems, electronics-assisted flight systems, cutting-edge radar systems, semiconductors, servo systems, electronic control systems, wireless frequency meters, solar-battery applications of telephone systems to rural areas, multiple-line microwave telecommunications, transistor circuits, and transistorized telegraph carrier systems" (original in Chinese).[10] The fields included in the 1959 draft proposal echoed recommendations made by Chiao-Tung's New York alumnus Tsen-Cha Tsao in 1956 to strengthen the development of electrical engineering, servomechanisms, nuclear engineering, and electronic digital computing.[11]

The Chinese-language proposal was translated by clerks into English in the same month, titled "The Expansion of Electronic Research Institute and Establishment of Telecommunication Advanced Training Centre." In this version of the proposal, Chien further specified a plan to set up a university research institute. The proposal shows Chien and his colleagues' belief that electronics research and electronics applications could help improve Taiwan's economy. But there were at least three cumulative steps leading to the realization of this idea: training a large number of domestic engineers and scientists, building an electronics-manufacturing industry, and achieving an industrial economy, as opposed to an agricultural economy. Chien believed that all three steps would necessarily rest on education on basic science and engineering involving university laboratories equipped with the "latest equipment" and facilities. Noting that Taiwan had vigorously pursued "economic and industrial developments" after 1949, Chien wanted to establish a university research institute for young scientists and engineers, and "the brilliant youth who [had] already received their bachelor's degrees," so that these to-be-trained "specialists in the electronics field" could "render service in all industries."[12] Chien further enumerated some key points:

1. The expansion of this research institute . . . will serve to lay a sound foundation for the basic scientific research and development in various branches of electronics in order to solve to-day' and to-morrow's urgent problems with which the various sponsoring agencies have been confronted.

2. In addition, advanced studies in the [sic] electronic and radio engineering are designed to enable the talented candidates to pursue their work leading to the advanced degree in science so that they would be proficient not only in modern theories as well as the up-to-date techniques in telecommunications, but also in the use of the latest equipment.[13]

The Chinese and the English version of the draft proposal, and of subsequent proposals, did not specify the means by which the trainees of this future research institute could contribute to or "render service" to industries. This connection between engineering education and the industrial was vague.[14] But Chien was certain that an invitation of foreign "distinguished physicists and three technical experts" to visit NCTU was one effective way to train the "brilliant youth." The English-version 1959 draft proposal stated,

After the invited research scientists come here to furnish the leadership for the electronic research and to advise us on matters of sound training system, it would not take [a] very long time for our country to develop a team of talented young specialists in this field. Moreover, it cannot be denied that recently a number of competent college graduates who could not afford to study abroad and did feel very much down-hearted will have an adequate place for doing research works.[15]

Chien and his contemporaries believed that it was important to offer college-educated youth an engineering education that could boast a faculty including distinguished foreign experts. Outfitted with impressive training, the young engineers would powerfully benefit Taiwan's various industries, by advancing the local manufacturing of devices or their components. The most specific example Chien was able to elaborate in the proposal was manufacturing components to improve telephone communications. Chien indicated that, once awarded the aid program, NCTU would work on improving Taiwan's current "small-scale production of the transistorized telephone carrier sets."[16]

Chien's interest in manufacturing electronics parts can be traced back to an earlier occasion, when Chien enthusiastically envisioned that the reopened Chiao-Tung university would be able to manufacture some electronic components domestically. In April 1958, Chien learned that Taiwan Sugar Corporation had rented "unit record equipment," or accounting machinery, from IBM. The machinery was not the type of mainframe computer that Chiao-Tung-educated technologists would like to acquire for the university, but Chien sensed the growing demand for computing devices and commented, "It would be great to have manufacturers of computing devices donate computers to the graduate program in electronics at National Chiao-Tung University. But, perhaps, in the near future, the graduate program may begin by purchasing parts to design and manufacture small-scale computing devices" (original in Chinese).[17]

Similarly, when the graduate program in electronics at NCTU celebrated its first anniversary in April 1959, S. M. Lee, now the director of the graduate program, also stated that "the graduate program will purchase parts to design and manufacture small-scale computing devices to meet researchers' demands."[18] Two years later, Lee envisioned an even tougher task for this graduate program. In 1961, after discovering that the United Nations had approved the technical-aid program, he told journalists that "three years later, Chinese scientists will be able to use computers we manufacture ourselves." This statement was his identification of one of the most positive effects of NCTU's rental of an IBM 650 computer through the UN-NCTU aid program. Once the university owned the computer, researchers and students there would aim to study the technology thoroughly and eventually manufacture the technology. Chien and Lee's understanding of the relationship between universities and industry was straightforward—the research and development at universities would be able to lead to the desired results in industrial development.[19] However, UN-affiliated officials had slightly different viewpoints and were interested in seeing Chien elaborate at great length on the connections between this proposed research institute and the electronics industry.

UN'S FOCUS ON IMMEDIATE AND DIRECT RESULTS FOR INDUSTRY

Working with the Ministry of Foreign Affairs in Taiwan, Chien sent the 1959 draft proposal to two entities for review—the International Telecommunication Union (ITU) and the United Nations. Established in 1865, ITU began helping the United Nations review and carry out technical-aid programs in the 1950s.

On January 12, 1960, after reviewing Chien's proposal requesting a technical-aid program to Taiwan, Jean Persin, a ITU senior counselor, wrote to the managing director of the UN Special Fund in New York City to express ITU's wholehearted support of the proposal.[20] Expressing his preliminary approval of the proposed UN-NCTU technical-aid program, Persin highlighted the implicit industrial applications in the proposal.

First, Persin noticed that Chien's proposal would train not only telecommunications engineers but also electronic physicists. Persin stated, "It is true that the aims of the institute go somewhat beyond telecommunication properly so called, since the institute would train electronic physicists apart from engineers, and would give them a general scientific education. But telecommunication is developing in such a way that ever closer cooperation is required between the scientific theorists and the practical engineers. In this sense the institute in question would meet the demands made of it." Second, Persin acknowledged that the proposed technical-aid program might be able to contribute to the "manufacturing of telecommunication equipment," for example, the components for the telephone carrier systems. Persin noted that the program then would be able to "find work for all graduates of the institute, so that their training would be put to good use."[21] While Persin did not disagree with setting up the research institute on a university campus, his presentation to the UN Special Fund was to highlight a university's contribution to bettering the telecommunications infrastructure.

The UN Special Fund formally approved the proposed technical-aid program in May 1960. Tsing-Chang Liu, director of the Treaty Department of the Ministry of Foreign Affairs in Taiwan, was informed of this decision by Sir Alexander MacFarquhar, regional representative for the UN Technical Assistance Board and Special Fund for the Far East.[22] Thereafter, ITU and government representatives in Taiwan worked together on finalizing the Plan of Operation for this aid program from July to November 1960.[23]

During the drafting of the Plan of Operation, ITU stressed that the proposed research institute at NCTU should "provide shorter training courses at a high level for scientists and technicians already engaged in the industry," in addition to offering engineers and physicists postgraduate courses. The phrase "already engaged in the industry" was then deleted by Taiwanese officials and S. M. Lee in August 1960. After V. R. Sundaram, chief of the Technical Assistance Department of ITU, visited Taiwan to finalize the Plan of Operation in November 1960, the promise to offer short-term courses for people from the industry was added back to the Plan of Operation. Sundaram probably persuaded the Taiwanese side to do so.[24]

The United Nation's emphasis on the connections between technical-aid programs and the industrial sector of the recipient country was related to the "efficacy" of aid programs. During one of his trips to Taiwan in 1961, T. E. Pigot, deputy regional representative for the UN Technical Assistance Board and Special Fund for the Far East, specifically pointed out that United Nations' technical aid was expected to bring about an "immediate and direct" effect. When clerks in the Ministry of Foreign Affairs received technical-aid proposals submitted by other ministries, they would review and comment on whether the proposal met the criteria of the Special Fund. They often brought up

Pigot's assertion: "Any project which does not lead to immediate and direct result for economic and social development will not be entertained by the U.N. Special Fund."[25]

This expected efficacy of proposed projects reflects the main purpose of the establishment of the United Nations Special Fund. The fund was designed to effectively supplement existing technical-aid and economic-aid programs. The *Yearbook of the United Nations 1958* stated, "The Special Fund is thus envisaged as a constructive advance in United Nations assistance to the less developed countries which should be of immediate significance in accelerating their economic development by, inter alia, facilitating new capital investments of all types by creating conditions which would make such investments either feasible or more effective."[26]

The processes of reviewing draft proposals and revising the Plan of Operation revealed that ITU, the United Nations, and the Taiwanese side were willing to acknowledge and incorporate each other's goals—the Special Fund's emphasis on practical industrial applications of the technical-aid program to industry and the Chiao-Tung alumni's avid interest in strengthening science and engineering education.

DEFINING UN-NCTU TECHNICAL-AID PROGRAM'S CONTRIBUTION TO INDUSTRY

After obtaining the funding from the Special Fund, NCTU established on its campus the Training and Research Centre for Telecommunications and Electronics. In this endeavor, NCTU invited three US professors to stay in Taiwan for one year, beginning in the summer of 1962. NCTU also hosted another US professor and a Bell Laboratories engineer to stay in Taiwan for a year beginning in the summer of 1963. From October 1962 to the end of 1963, NCTU offered six short-term training courses in fields such as microwave electronics, digital electronic computers, and modern communications systems, taught by these experts and by recent NCTU graduates.[27] (The courses and experts are discussed in chapter 3.) To recruit trainees for these short-term training courses, NCTU usually advertised through government agencies to personnel from the Ministries of Defense, Transportation and Communications, and Economic Affairs. In particular, the Ministry of Economic Affairs managed several state-owned monopolies, such as Chinese Petroleum Corporation, Taiwan Sugar Company, and Taiwan Power Company.[28] Because of NCTU's specific recruitment strategy, the trainees of these courses consisted mostly of graduate students, university scientists, and scientists and engineers from government agencies or from state-owned enterprises. Only a small portion of trainees hailed from private enterprises. The strategy was consistent with the way the Republic of China's technocrats designed their technical training during and immediately after World War II.[29]

In October 1963, J. N. Corry, regional representative for the UN Technical Assistance Board and director of Special Fund Programmes in the Far East, visited Taiwan and expressed his interest in knowing more on the employment of former trainees.[30] Rather than inform Corry of where trainees were working *after* their completion of courses in NCTU's Training and Research Centre, NCTU and Tsing-Chang Liu, in Taiwan's Ministry of Foreign Affairs, took an alternative, two-pronged approach: (1) they made a chart presenting a breakdown of trainees' jobs *before* the trainees had taken the courses, and (2) they explained that all post-course trainees were "back to their respective original posts."[31] The chart dealt with the 154 trainees who took the first six training courses between October 1962 and the end of 1963: regarding these trainees' pre-course vocations, thirty-six had been serving in the military, five had been employed by the National Security Bureau, twenty-two had been students at universities, nineteen had been working at Taiwan Power Company, fifteen at Chinese Petroleum Corporation, fifteen at the Bureau of Telecommunications, seventeen at other agencies under the Ministry of Transportation and Communications, eight at other government agencies, four at Taiwan Sugar Corporation, and two at Taiwan Fertilizer Corporation; a mere two trainees had been working at one of two private-sector enterprises. Nine trainees did not have a job or did not reveal their affiliations.[32] Engineers and researchers from state-owned enterprises made up the majority of the trainees.

In December 1963, Corry replied to Tsing-Chang Liu. In the letter, Corry declared that he and the managing director of the United Nations Special Fund were satisfied with the chart's information and that the United Nations should require such information upon the completion of an aid program. As Corry did not question why merely two trainees were from the private industry, perhaps he was content with the overwhelming percentage of trainees from state-owned enterprises. Moreover, Corry stated that he was interested, in the future, "of course in learning what happens to the post graduate students once they have secured their higher degree." He would like to know how many recipients of a master's degree would enter an industry and how many would become instructors or researchers. But this piece of information was not shared with Corry in the available archival documents. The master's degree students at NCTU after graduation were in general more likely to pursue advanced degrees in the United States or stay on campus to teach courses. The numbers of alumni who immediately became employed in industry domestically might be smaller than Corry had expected.[33]

Two examples illustrate that, after the UN-NCTU technical-aid program, NCTU administrators also agreed on the critical role of Chiao-Tung in advancing the industry. First, NCTU administrators considered the information Corry requested to be a demonstration of the university's accomplishment. They used a similar set of figures of trainees' employment to advertise the university's engineering education, in *The Previous Year's Engineering Accomplishments* published by the Chinese Institute of Engineers in 1970.[34] Second, S. M. Lee publicly acknowledged that the university's offering of training courses benefited the industry. In August 1964, on an occasion honoring two Chiao-Tung alumni, S. M. Lee pointed out that, before implementing the technical-aid program, he had not regarded electronic digital computers as capable of contributing to industry in the following ways. He stated,

> During the past three years, we used funding from the United Nations to obtain a considerable amount of equipment. We can proudly say that we have electronic computers in Taiwan now. NCTU was a forerunner in this effort. We did not envision that electronic computers could help industry in many different ways. . . . Now we train people who work in such fields as transportation and telecommunications and who work for such [sate-owned] enterprises as Taiwan Power Company and Chinese Petroleum Corporation. . . . In May, we held a workshop on the electronics industry, and currently, many companies, including the China Electrics Corporation and the Aluminum Corporation, are interested in working with us.[35] (Original in Chinese)

Lee's 1964 remarks on the relationship between university and industry are different from his earlier comments on the potentials of the graduate program to manufacture computers for scientists. After years of communicating and corresponding with the Special Fund officials, he now articulated the university's contribution in offering on-the-job training for the industry, broadening his original preoccupation with the manufacture of computers. Nonetheless, the manufacture of computers continued to be an aspiration owned by Lee and his fellow technologists at Chiao-Ting.

CHIAO-TUNG OVERSEAS ALUMNI'S INTEREST IN INDUSTRY

S. M. Lee and Chien often solicited advice from Chiao-Tung alumni who were residing in the United States. The objective of this collaboration was to format and design the training courses of the aid program and the graduate program; a couple of letters written from US alumni to Lee were published in the December 1962 issue of the *Voice of Chiao-Tung Alumni*. The discussions in the letters reveal that Chiao-Tung alumni were split regarding which roles the proposed NCTU research institute and the graduate program should play in Taiwan. Some Chiao-Tung alumni argued that NCTU should focus on conducting industrial research, and some expected that NCTU should lead the research in electronics science in Taiwan. The former held a similar idea to aid officials' focus on industrial application; the latter aligned with Lee and Chien's original emphasis of the bolstering of education on basic science and engineering.

Chao Chen Wang (C. C. Wang), a Chiao-Tung alumnus and then a Sperry Rand engineer specializing in vacuum tubes, belonged to the former camp. Recall in chapter 1 that, in 1957, Tsen-Cha Tsao recommend that Chiao-Ting invited Chao Chen Wang to teach at the university. Instead of accepting the invitation, Wang recommended that a Cornell professor in electrical engineering, G. Conrad Dalman, serve as a UN expert for the UN-NCTU technical-aid program for one year. Wang and Dalman were former colleagues at Sperry Gyroscope. Wang's recommendation was adopted, and Dalman embarked on his trip with his family in the summer of 1962 and stayed in Tiawan for a year (see figure 3.2). Months before Dalman's trip to Taiwan, Wang wrote to Hui Huang, a former general manager of Taiwan Power Company and then an International Monetary Fund (IMF) administrator, to discuss the possibility of starting vacuum-tube manufacturing in Taiwan. He pointed out that the future of the Taiwanese economy perhaps could rely on the manufacturing of electronics by taking advantage of the existing low-wage labor in Taiwan and potential technologies that NCTU might be able to develop. Wang urged Huang to discuss this issue with an influential economic bureaucrat, Chong-Rong Yin, as a way to raise capital. In particular, as Dalman used to work on vacuum-tube design for years, NCTU should take advantage of his one-year visit to Taiwan to start engaging in the small-scale manufacture of vacuum tubes on campus.[36]

When Wang brought up that NCTU should initiate research geared toward the manufacture of electronics, he discussed it with Tsen-Cha Tsao. Tsao then talked to Wang about the matter,

> It is difficult for Taiwanese college students with high-level engineering training to find jobs in Taiwan, because the industrial development in Taiwan is backwards. . . . I hope the graduate program in electronics may improve the level of industry in Taiwan. . . . If there would be an exodus of National Chiao-Tung University's graduate students to the United States, doesn't it run counter to outcomes that we would like to see? The Taiwanese economy would not benefit from these [US-based] graduate students. Thus, it is important to develop our electronics industry and our graduate program at the same time. The education and training offered by the graduate program should meet the local demand.[37] (Original in Chinese)

Wang and Tsao expected the graduate program at NCTU to experiment with the campus-based manufacture of vacuum tubes and to determine whether this production could be expanded to an industrial scale. There was a rationale for how the graduate program might benefit the Taiwanese economy: if the program helped develop its manufacture of vacuum tubes into an export-oriented industry, Taiwan might experience growth similar to that of Japan and Europe, which had had success in the export market. The students who graduated from NCTU would become contributors to the Taiwanese economy, instead of contributors to the exodus to the United States.

Earlier in 1962, Wang had also discussed his ideas with Ta-Chung Liu, a Chiao-Tung alumnus and an economist teaching at Cornell University.[38] Wang and Liu then coauthored a letter to the Hung-Hsun Ling and S. M. Lee recommending the university should pioneer research on manufacturing a chosen type of electronics—for example, vacuum tubes—and then help NCTU graduates receive employment opportunities in the chosen field.[39]

In contrast, David K. Cheng (Jun Zheng), an alumnus working as an electrical engineering professor at Syracuse University, considered the technical-aid program a chance to improve the "level of science" in Taiwan. NCTU originally considered inviting Cheng to serve as a UN expert for the technical-aid program, but Cheng's schedule did not permit such a visit. He wrote a letter to S. M. Lee, offering his suggestions on the technical-aid program in 1962, and indicated that he gave permission to publish the letter in the *Voice of Chiao-Tung Alumni* to circulate with all alumni. Generally, Cheng agreed that the university should pursue electronics manufacturing. But, he pointed out that "manufacturing without ressarch [sic] or at least development work does not offer enough challenge to people with postgraduate degrees." He advocated to "preserve, utilize, and develop our engineering talents [sic] in electronics" in Taiwan. The term to "preserve" referred to finding incentives that would encourage college graduates either to stay in Taiwan or to return to Taiwan after completing their master's or doctoral studies in the United States. He noted, "As conditions stand now, one is afraid to go back [to Taiwan] for fear that highly technical knowledge is not really needed there and that one [who remains in Taiwan] would soon fall behind [other countries' related experts] due to [Taiwan's] stagnation in to-day's world of rapid technological progress."[40]

In this letter, Cheng especially pointed out that many people might feel "hurt" by Purdue University's chemical engineering professor R. Norris Shreve's statement, made in November 1961 at a banquet hosted by Lee, that "there was no need for postgraduate engineering training in Taiwan." Cheng suggested that Shreve did not think the state of science, engineering, and industry in Taiwan was satisfactory. He also noted that those knew Shreve well "would agree that Shreve did not mean to slight Chinese." To consider Shreve's comment carefully, Cheng believed that it was important to improve "the scientific level" in Taiwan, and thus the island would be able to offer financially secure and intellectually challenging jobs to young engineers.[41]

Cheng's emphases on research and "stagnation" in science and technology, instead of on the industrial sector, were similar to another alumnus's attention to the opportunity of "scientific take-off." The notion of scientific take-off was embraced by many Chiao-Tung alumni. In April 1961, the *Voice of Chiao-Tung Alumni* featured an article by a Chiao-Tung alumnus, Qinbo Ye. Titled "One Step Further (*Gengshang cenglou*)," which means making progress in Chinese, the article proposed that his alma mater should establish a graduate program in applied mathematics with a division in electronic digital computing. He imagined that this division would be responsible for "the design of transmission lines for electric power companies and statistics for nuclear energy experiments." Ye commented that Taiwan is ready to pursue "scientific take-off." He then asserted that applied mathematics is an important step in Taiwan's attainment of scientific take-off, and that this step should follow the recent establishment of graduate programs in atomic physics at National Tsing Hua University and in electronics at National Chiao-Tung University.[42] The term "scientific take-off" also reflects trenchant similarities between this discourse adopted by Chiao-Tung alumni and the broader concurrent worldwide discourse regarding economic development, and its "take-off."

Cheng and Ye's emphasis on scientific development is close to Chien's and Lee's initial ideas that the UN-NCTU technical-aid program should nurture "a team of talented young specialists" through graduate-level education.[43] In line with the United Nations' and ITU's emphasis on practical applications of electronics, Wang, Tsao, and Liu believed that NCTU should conduct research necessary for manufacturing and exporting vacuum tubes on an industrial scale. In December 1962, Lee commented on the letters from Wang, Liu, and Cheng by praising these alumni's patriotism. Transitioning from his earlier emphasis on scientific research, he further acknowledged that NCTU had a responsibility to help develop Taiwan's electronics industry in particular and Taiwan's economy in general.[44]

FROM SUSPICIOUS CARGO TO THE HUMMING MACHINE

Obtaining access to an IBM computer for the UN-NCTU technical-aid program was first included in the draft of the Plan of Operation in August 1960. An electronic digital computer was standard equipment for a "modern electrical engineering school," as stated in the report produced and published by the United Nations in 1968 to chronicle what the UN-NCTU technical-aid program had accomplished. The report retrospectively noted that a computer would contribute to "the solution of research problems and since it was considered that training in computer applications, programming and operation would be useful to the country in general."[45]

The 1968 report described the computer as a "major addition" made "during the period of negotiation [between the United Nations and Taiwanese representatives in 1960]."[46] The "negotiation" might have referred to many aspects about how to execute the aid program, but, specifically, it alluded to a discussion about the permission granted the Taiwanese from the United Nations to rent a computer rather than purchase one. In November 1960, Sundaram visited Taiwan for finalizing the Plan of Operation. He told Taiwanese representatives that the Special Fund office would question why the Taiwanese side planned to rent rather than buy an IBM computer. He pointed out that the United Nations' aid programs aimed to help recipient countries obtain necessary equipment outright, but renting did not mean possessing. Nevertheless, at that time, IBM's standard approach to conducting business was to lease electronic digital computers or accounting machines to its customers. NCTU planned to follow the common practice, to rent a computer. During a meeting with Taiwanese representatives, Sundaram asked them whether they planned to continue to rent the IBM computer after the end of the technical-aid program. Lee answered that they would like to do so, as the computer was important to the future of NCTU. Sundaram thus promised to make an effort to persuade the Special Fund office of the importance of a computer to NCTU. But he also suggested that the Taiwanese side should prepare a backup wish list of research equipment, in case the Special Fund office would not agree to the rental of an IBM computer.[47] Three months later, in January 1961, Sundaram informed Taiwanese representatives that the United Nations had approved NCTU's rental of a computer.[48] With this agreement settled, an IBM 650 computer traveled across the Pacific Ocean and arrived at the NCTU campus in February 1962.

Chaowu Lin, an NCTU graduate student from 1964 to 1966, heard of the unconventional delivery of the computer to NCTU in 1962. When parts of the IBM 650 computer were being shipped to Hsinchu in 1962, the IBM 650 passed over a bridge guarded by several soldiers. At that time, the bridge was allowed only one-way traffic, so soldiers were sent to help control and manage the traffic. The driver of the truck loading the IBM 650 computer drove at an extremely slow speed to avoid breaking the gigantic and sophisticated machine. The soldiers by the bridge tried to figure out what the cargo was. "The soldiers were not very happy with the slow speed on a one-way bridge," according to an administrative staff member at NCTU at that time, Tang-Chin Kao.[49]

Even with the slow speed, the electronic digital computer reached Chiao-Tung campus. The memory units and the processor of the IBM 650 arrived in February 1962, and the peripherals, such as its card reader and punch unit, were scheduled to arrive weeks later.[50] In March, NCTU administrators and alumni began to prepare for the university's sixty-sixth anniversary celebration, scheduled in early April. These anniversaries drew a good number of alumni to visit their alma mater—in this case, the one they had reopened in Taiwan. The university planned to schedule the computer's debut during that year's anniversary celebration. In the March issue of the *Voice of Chiao-Tung Alumni*, editors advertised the annual event by focusing on the newly arrived electronic digital computer. On a page listing alumni's donations, the editor wrote in a text box, "Are you interested in seeing an electronic computer? Please join us to celebrate the birth-day of our alma mater next month."[51]

Partly because of Chiao-Tung alumni's enthusiasm for anything and everything hav-ing to do with their alma mater, and partly because of the general hoopla surround-ing the island's first electronic digital computer, the debut of the IBM 650 computer successfully attracted the attention of the anniversary celebration's attendees. On the sunny morning of April 8, 1962, eight hundred alumni and their families from Taipei reserved an express train of ten cars for the ninety-minute trip to Hsinchu. At the same time, another two hundred alumni traveled to Hsinchu from the central and southern parts of Taiwan. The celebration ceremony started at eleven o'clock, in the room where the computer was located. The room was crowded with alumni, students, professors, and invited guests, including the IBM manager in charge of IBM's Far East business operations, a Mr. Erwin. Gisson Chien and various well-known ministers and officials who were Chiao-Tung alumni showed up too. Around eleven o'clock, the minister of education "pressed a button to initiate the operation of the computer." Editors of the *Voice of Chiao-Tung Alumni* noted, "The computer started to hum and calculate prob-lems, which astonished the spectators."[52]

A ten-minute display of humming calculations kicked off the celebration, and then the spectators moved to a library and continued the celebration ceremony. Hung-Hsun Ling made a remark on the responsibility of NCTU to the wider Taiwanese society. He noted that, with resources from the UN-NCTU technical-aid program, NCTU could contribute to the development of electronics science in the free world. In addition to ministers and officials who praised their alma mater, Erwin gave a talk about the appli-cations of the IBM 650 computers. IBM opened a branch and started business opera-tions in Taiwan in 1957. At that time, Taiwan's IBM branch had at least two customers in Taiwan, the Taiwan Sugar Company and the Council for United States Aid (CUSA), renting IBM accounting machines and punched-card machines for their data pro-cessing tasks.[53] CUSA was in charge of US aid to Taiwan from 1948 to September 1963, liaising on behalf of the Taiwanese government with the US International Cooperation Administration, the predecessor of the United States Agency for International Develop-ment, known as USAID now.

The ceremony ended with remarks made by a Chiao-Tung alumnus, Wen-Yuan Pan, who then worked for RCA in the United States. In the 1970s, Pan became an important consultant to Taiwanese officials on a government-sponsored technology transfer proj-ect with RCA on IC foundry production in the 1970s (see chapter 10). After the cer-emony, IBM scheduled a film in the afternoon to introduce the IBM 650 computers to Chiao-Tung alumni. While some Chiao-Tung alumni remained on the NCTU campus to watch the film, others took a bus to the nearby National Tsing Hua University campus to visit Taiwan's first nuclear reactor, which the Taiwanese government had purchased from General Electric (GE) for educational and research-related purposes.[54]

CONCLUSION

The formation of the UN-NCTU technical-aid program on the island was a convergence of several different but intertwined ideas about engineering education and industrial development on the island. These ideas are different in terms of the ends and means of building an electronics-manufacturing industry. The island's technologists were interested in strengthening education for the development of the industry, as opposed to a group of overseas alumni and international aid officials' straightforward emphasis of the training of employed engineers.

Taiwan-based Chiao-Tung technologists aimed to use the technical-aid program to help institutionalize electronics-science research at NCTU, and they intended to invite US experts to educate Taiwanese students, engineers, and scientists. They believed that this type of engineering education would help create a large number of well-trained engineers and scientists, who would then build an industrial sector in electronics manufacturing. UN-affiliated officials, on the contrary, stressed the "immediate and direct" results of the aid program to the industrial development. Engineers from the existing industry should be the targeted trainees, from aid officials' point of view. They expected that the university would embark on significant collaboration with existing industries in Taiwan. Through negotiations during the process of proposal reviews and progress updates, Chiao-Tung–educated technocrats acknowledged the importance of creating connections with the existing industry, and UN-affiliated officials were satisfied with the fact that the UN-NCTU technical-aid program had been training engineers from the existing state-owned enterprises, instead of reaching out to other private industrial sectors.

Aid officials were keen to see visible outcomes of the aid program to the economic development of a recipient country. In a similar vein, technologists were eager to cope with an economic-development war against communist China in the context of the Cold War, by introducing electronics science, implementing engineering education, and developing an envisioned electronics-manufacturing industry. Moreover, as chapter 1 has illustrated, cutting-edge electrical-engineering research not only could strengthen Taiwan's defense and telecommunications infrastructure, but also could prompt a "scientific take-off." Taiwanese technologists recognized their responsibility to participate in the noncommunist camp's collective scientific research. Electronics science would help them succeed in the ideological, economic, and scientific races between the free world and the communist world.

Both aid officials and Taiwanese technologists enthusiastically acknowledged the potential of making some electronics products manufacturable in Taiwan, despite the variety of the products they brought up, ranging from vacuum tubes to computers. As early as 1958, Chien had noted that he would look forward to seeing NCTU graduate students design and explore the manufacturing of small-scale computing devices someday. Lee made a similar comment in 1959. In 1960, aid officials viewed Chien's plan to manufacture equipment for telecommunications systems as a concrete attempt to develop the electronics industry. Furthermore, in 1961, Lee optimistically anticipated the arrival of locally made computers within three years, if Chiao-Tung students could thoroughly study the forthcoming IBM computer. In 1962, Chao Chen Wang, as a vacuum-tube engineer and Chiao-Tung alumnus, urged NCTU to experiment on vacuum-tube manufacturing on campus. It took time to fulfill these aspirations. As the following chapters show, it would be a decade before Chiao-Tung students were able to build a calculator on campus. Nonetheless, NCTU members and Taiwanese technologists remained optimistic, and never stopped expressing their enthusiasm about tinkering with computers available to them and experimenting with the manufacturing of such technology.

II EMULATING HUMMING MACHINES, 1960s

3 TINKERING WITH A TECHNOLOGICAL SYSTEM: MAINFRAME COMPUTERS FROM AFAR

By mid-1961, the UN technical-aid program was ready to fund the Training and Research Centre for Telecommunications and Electronics on the NCTU campus. An IBM 650 computer, the electronic digital computer included in the budgeted equipment list for the UN-NCTU technical-aid program, arrived on the NCTU campus in February and debuted in a public display in April 1962. The computer was shared by three overlapping institutes on campus: the graduate program in electronics at NCTU, the Training and Research Centre for Telecommunications and Electronics, and NCTU's computing center. While chapter 2 focused on Chiao-Tung alumni's advocacy of electronics science and electronic digital computing to secure their access to the aid program, this chapter describes the process by which two visiting US professors, together with NCTU students, faculty members, operators, and technicians, worked together to establish and operate the computing center. Their effort was central to implementing the technical-aid program from 1962 to 1964. The center, with its IBM 650, and later an IBM 1620 computer, was able to provide a material and intellectual basis for academic learning in electronic digital computing and for expanding an infrastructure for computers in Taiwan in the following decades. Technologists actively explored and tinkered with the computers, establishing a technological system[1] of mainframe computers that maintained and supported the operation of the new technology.

Various groups of technologists contributed to setting up the NCTU computing center with the IBM 650 and 1620 computers and looked for resources to make them operable. The term "social groups," proposed by scholars of the social construction of technology (SCOT), is used to refer to these technologists, for emphasizing their contributions to the uses of a technology.[2] The social groups of this chapter included NCTU faculty, staff members, technicians, operators, and students, two US visiting professors, and IBM engineers. The installation of an IBM computer at NCTU neither naturally nor spontaneously engendered diverse uses of this new technology, such as engineering, accounting, and statistical analysis. Instead, networks of social groups helped begin and continue the momentum of electronic digital computing in Taiwan.

Different social groups worked together to keep the IBM 650 and 1620 computers functioning on an island far away from IBM's other overseas branches. The presence of the two computers helped in the completion of various research projects that could benefit from electronic digital computing. Enthusiastic Taiwanese computer users, operators, and technicians promoted the learning of programming and computer architecture, as well as the interest in manufacturing computers and computer parts.

The presence of the computers not only assisted in advancing the growth of computing activities, but also served as available objects for NCTU members to explore, probe, and tinker with. Recall that S. M. Lee told journalists that "three years later, Chinese scientists will be able to use computers we manufacture ourselves," after the aid program granting Chiao-Tung access to a computer was approved in 1961; he was probably confident that Chiao-Tung members would aim to study the computers thoroughly and eventually manufacture the technology.[3] Three examples illustrated their tinkering with the computers: First, while the IBM 650 was rented, NCTU purchased the IBM 1620, which created the opportunity for NCTU members to engage in maintaining the computer. Specifically, NCTU was able to substitute an in-house technician for IBM's official service support. Second, William Wesley Peterson, a visiting professor from the United States, specifically urged NCTU to become independent from outside help, alluding to support from IBM and technical aid. Third, NCTU members attempted to connect the IBM 650's peripherals to the IBM 1620 computer, despite no success.

OVERVIEW OF THE UN-NCTU TECHNICAL-AID PROGRAM

The first generation of early electronic digital computer users in Taiwan were instructors and trainees involved in the training courses that used the IBM 650 and 1620 computer, offered by the Training and Research Centre for Telecommunications and Electronics from 1962 to 1964. As the first electronic digital computer available in Taiwan, the IBM 650 was a relatively obsolete model in the US context. It was soon replaced by a newer model, the IBM 1620 computer, in 1964.

The operation of the IBM 650 in Taiwan overlapped with Massachusetts Institute of Technology professor Dean N. Arden's visit to Chiao-Tung from December 1962 to June 1963. Dean N. Arden (figure 3.1) was an electrical engineering professor who came to Taiwan with two Cornell University electrical engineering professors, G. Conrad Dalman

FIGURE 3.1
Visiting professor Dean N. Arden, shown in the *Yearbook of Training Course of Electronic Computers, February–June, 1963* (National Chiao-Tung University, 1963). The yearbook is from Chi-Chang Lee's personal collection.

(figure 3.2) and Henry McGaughan, in the summer of 1962. Arden's specialty was electronic digital computing. Dalman's specialty was electron devices, and McGaughan's was transmission systems. They taught graduate courses, advised students at NCTU, and helped offering training courses on behalf of the UN-sponsored Training and Research Centre for Telecommunications and Electronics. NCTU graduate students or recent graduates also participated in teaching these short-term training courses with the UN-sponsored experts. By using the IBM 650 computer, NCTU offered at least two training courses for fifty-two and thirty-six trainees, respectively, during Arden's yearlong visit in Taiwan.

FIGURE 3.2

Cornell electrical engineering professor G. Conrad Dalman (right); his wife Catherine Dalman (sixth from the left); Hung-Hsin Ling, a former president of Chiao-Tung University and the then president of Chinese Petroleum Corporation (fifth from the left); Dean N. Arden (second man from the right); and family members of Henry McGaughan, Arden, Dalman, and Ling. A photograph gifted by G. Conrad Dalman. The photograph was taken in 1963.

William Wesley Peterson (figure 3.3), an electrical engineering professor from the University of Florida in Gainesville, succeeded Arden in the summer of 1963. J. O. McNally from Bell Laboratories replaced Dalman and McGaughan to offer his expertise in electron devices. During Peterson's visit in Taiwan, NCTU offered four training courses; the first and second courses relied on the IBM 650 computer. Because NCTU acquired an IBM 1620 computer in February 1964, the third and fourth courses made use of the new machine (table 3.1).[4]

The majority of the new trainees at NCTU were from the military, universities, government agencies, and state-owned enterprises, partially because NCTU kept these state-related organizations well-informed and also gave them exclusive partial lodging and tuition waivers. Among the 126 people who took NCTU's programming courses from December 1962 to December 1963 utilizing the IBM 650, fourteen were from Taiwan Power Company (TPC), fourteen from CPC, and sixteen from

FIGURE 3.3
Visiting professor W. W. Peterson (right), operators Y. C. Chen (top) and Regina Lee working in front of the IBM 1620 at NCTU, 1964. *Source: Jiaoda yanjiusuo ba nian* (Eight years of the Institute of Electronics), a pamphlet published by National Chiao-Tung University in 1966. Courtesy of National Yang Ming Chiao Tung Museum.

telecommunications-related governmental agencies. Twenty-five trainees were from military units in fields such as logistics, intelligence, accounting, research, and the military's science and engineering colleges (table 3.1).

Before trainees attended the training courses, a few personnel from the military, TPC, and CPC were familiar with the concepts behind electronic digital computing that were applicable to their work but did not have access to physical electronic digital computers;[5] in this regard, NCTU's acquisition and NCTU's establishment of its first computing center in Taiwan were especially meaningful in providing opportunities for users to operate the new technology. Although it is not clear how the twenty-five

Table 3.1

An overview of trainees in the first seven training courses of electronic digital computing at NCTU

Course number	No. of trainees in each course	Computer model	UN expert	Date	Numbers of trainees by affiliations	
3rd	52	IBM 650	Arden	Dec. 1962 to Jan. 1963	CPC	9
					Military	11
					TPC	4
					Telecomm.	11
					Others	17
4th	36	650	Arden	Feb. 1963 to Jun. 1963	CPC	0
					Military	7
					TPC	0
					Telecomm.	0
					Others	29
5th	17	650	Peterson	Jul. 1963 to Aug. 1963	CPC	3
					Military	1
					TPC	5
					Telecomm.	3
					Others	5
6th	21	650	Peterson	Oct. 1963 to Dec. 1963	CPC	2
					Military	6
					TPC	5
					Telecomm.	2
					Others	6
8th	42	IBM 1620	Peterson	Mar. 1964 to May 1964	CPC	2
					Military	8
					TPC	5
					Telecomm.	4
					Others	23
9th	33	1620	Peterson	May 1964 to Jul. 1964	CPC	0
					Military	12
					TPC	4
					Telecomm.	5
					others	12
11th	38	1620	None	Nov. 1964 to Dec. 1964	CPC	6
					Military	1
					TPC	3
					Telecomm.	5
					others	23

Notes:

1. CPC refers to Chinese Petroleum Corporation; TPC refers to Taiwan Power Company; Telecomm. refers to government agencies of telecommunications and transportation.

2. I do not include the first, second, seventh, and tenth training courses, which were for microwave electronics and communications systems.

3. This table was organized by the author on the basis of the following three sources: *The Yearbook of Training Course of Electronic Computers, February–June, 1963* (National Chiao-Tung University, 1963); *The 11th Yearbook of Training Courses of Electronic Computers, November–December 1964* (National Chiao-Tung University, 1964), 19; and "Annex VI List of Training Courses," *Training and Research Centre for Telecommunications and Electronics, Republic of China: Report* (hereafter *1968 UN Report*), 51. There are discrepancies between the three sources, but I rely on mainly the information from *The 11th Yearbook*, as it provides more details about trainees' background.

military-affiliated trainees used the knowledge they learned at NCTU, the university became a source of knowledge for the military.

ARDEN AND LOCAL SCIENTISTS

During Arden's one-year visit in Taiwan, his offerings to NCTU trainees and graduate students included seminars on such matters as Boolean algebra, memory-circuit design, system-theory research, and system-optimization problems, and served as an adviser on three master's theses.[6] NCTU students gave him the nickname "A-Ding," a play on his last name, reflecting how Chinese speakers assimilated English pronunciation into a Chinese form by using available Chinese vowels and consonants.[7] Arden taught courses mainly on computer architecture, and was also familiar with computer programming. Before starting his professorship, he joined MIT in 1953 in the Division of Industrial Cooperation to help faculty members and students improve programs for an array of mathematical problems.[8] Arden offered similar assistance in programming to Taiwanese scientists and engineers. According to Arden, a group of scientists visited his office at NCTU to consult on the programming of a complicated numerical analysis of elasticity. After a discussion to identify the involved mathematical problems, Arden spent a couple of days writing up programs to solve them. When Arden invited the scientists to visit NCTU again and demonstrated how to make the IBM 650 calculate these problems, according to Arden, the scientists were amazed how fast the computer processed the problems.[9] I interviewed Arden about this whole sequence of events; however, as it had taken place more than four decades earlier, he could recall neither the names of the scientists nor the names of the scientists' affiliated institutes.

A glance at the report produced for this technical-aid program and submitted to the United Nations in 1968 can reveal the name of a scientist who might have received help from Arden. Among the research projects that partially relied on the two mainframe computers at NCTU, three projects about elasticity are named: "Stresses in a Perforated Circular Ring," "Stresses in a Perforated Quadrant Plate," and "Stresses in a Plate Having an Elliptic Notch under Tension."[10] The researcher for all three projects was a leading and prestigious scientist at that time: Chih-Bing Ling (Zhiping Lin), a native of China who had received a PhD in aeronautics from the University of London in 1937. He moved to Taiwan shortly after 1949 and worked as a researcher in Academia Sinica (the national academy of Taiwan). He gained a DSc in applied mathematics and elasticity from the University of London in 1959. In 1964, Ling left Taiwan and became a professor in the department of mathematics at Virginia Polytechnic Institute and State University.[11]

Ling also directly pointed out in one of his journal articles, published in July 1964, that he appreciated Dean N. Arden's "help in preparing Table 3 by using an IBM-650 electronic computer."[12] Ling had mainly relied on human computers prior to drafting this article. His coauthor of the 1964 article, Chen-Peng Tsai, had helped him in performing the numerical computations since 1952. Owing to the newly available IBM 650 computer, Ling transitioned from using human computers to using electronic digital computers for his research, and he did so with Arden's assistance.[13]

TAIWAN-BASED INSTRUCTORS AND STAFF

In a phone conversation in 2008, Arden summarized his trip to Taiwan by saying, "I showed them the power of the computer."[14] The statement is correct, as he was the first instructor in Taiwan to teach this subject. But several other groups also played critical roles in the introduction of electronic digital computing to Taiwan, upon the arrival of the IBM 650 computer. Particularly significant were two instructors, Henry Y. H. Chuang (Yinghuang Zhuang) and Hua-Ting Chieh (Huating Xie). Chuang was born in Yunlin, one of the several Taiwan-born students admitted to NCTU in 1958, the first class of the master's program in electronics after NCTU commenced operations in Taiwan. Chieh was born in China and moved to Taiwan in 1949 when he was a highschooler.[15] They both graduated in 1960, and taught training courses on electronic digital computing from 1962 to 1963 (figure 3.4). Before Chiao-Tung alumni could be certain about whether the United Nations would agree to fund NCTU's leasing of the IBM 650 computer, they raised a fund to support three recent NCTU graduates' electronic digital computer training in Japan. Chuang and Chieh were selected to visit IBM's Japan branch for six months, and then came back to Taiwan to teach electronic digital computing at the same time Arden was working in Taiwan. The two instructors' training in Japan indicates that Chiao-Tung alumni and NCTU were well prepared for the arrival of the IBM 650 computer.

Chuang remembered that his return to Taiwan from Japan was in October 1961, before the IBM 650 computer's arrival in 1962.[16] Chuang and Chieh taught applied mathematics, such as numerical analysis, and computer programming for at least two training courses at the Training and Research Centre, and their photos were placed right next to Arden's in the yearbook for the training course that lasted from February to June 1963. The photos featured each of them standing in front of blackboards with mathematical problems.[17] Chuang left Taiwan in 1963 and went to the United States to study electronic digital computing. He received a PhD in electrical engineering from North Carolina State University, taught at the Department of Applied Mathematics and Computer Science of

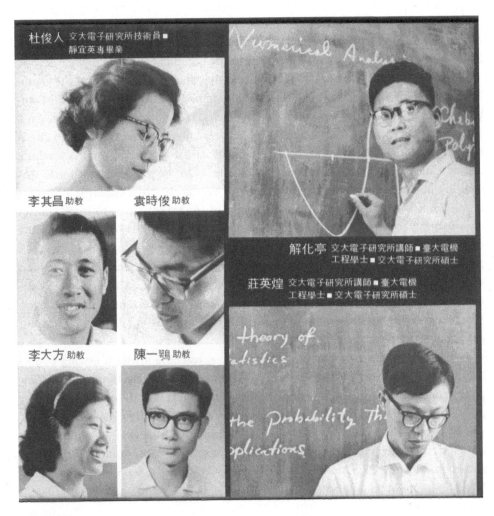

FIGURE 3.4

Top left: Junren Du, computer operator.
Middle left: Chi-Chang Lee, teaching assistant.
Bottom left: Dafang Li, teaching assistant.
Top right: Hua-Ting Chieh, instructor.
Bottom right: Henry Y. H. Chuang, instructor.
Source: The Yearbook of Training Course of Electronic Computers, February–June, 1963 (National Chiao-Tung University, 1963). The yearbook is from Chi-Chang Lee's personal collection.

the Washington University in St. Louis and then the Department of Computer Science of the University of Pittsburgh, and retired there. Chieh later earned a PhD in the United States and became a professor at Howard University beginning in 1967.[18]

A few new instructors who had just received their master's degrees from the graduate program in electronics at NCTU in 1962 also joined the teaching staff at the computing center, around the same time when Arden, Dalman, and McGaughan arrived in Taiwan. For example, Chi-Chang Lee (Qichang Li) completed his master's thesis on switching circuits in the summer of 1962 and started to work as Arden's teaching assistant thereafter (figure 3.4). Born in China in 1926, he joined the air force, specializing in weather research, before moving to Taiwan. He then obtained a master's degree from NCTU. After Arden left the center in 1963, Lee started to teach courses on switching circuits and computer operations; from 1965 to 1974, he directed the computing center. Like Arden's interactions with mathematician Chih-Bing Ling, Lee's assistance to variously situated scientists and engineers strengthened their efforts to solve programming-related problems. For example, around 1965, Chi-Chang Lee conducted a research project for the Taiwan Railways Administration, and in this endeavor, he used NCTU's IBM 1620 computer to calculate target speeds and distances for a train in various states of movement.[19]

In addition to Chuang, Chieh, Arden, and Lee, teaching assistant Dafang Li also helped facilitate the functioning of the two training courses in electronic digital computing (figure 3.4). She was one of two women staff members at the computing center prior to summer 1963. The other was a computer operator, Junren Du (figure 3.4). Li was likely pursuing a degree in electrical engineering. Her father was an administrator at NCTU. Du had graduated from Providence Junior College for Women (Jingyi yingzhuan) in Taichung City, founded by the Sisters of Providence. Soon the center would hire a few more women operators.[20]

SUPPORTING RESEARCH AT TSING HUA

NCTU alumnus Tseng-Yu Lee (Zengyu Li) was a frequent visitor at NCTU's computing center. Born in China in 1934, he received his master's degree from the Institute of Electronics at NCTU in 1962, the same year as Chi-Chang Lee. He worked—upon graduation—for chemical physicist Chi-Hsiang Wong at National Tsing Hua University. Lee went to the NCTU computing center to use the IBM 650 computers for performing least-squares calculations, Fourier synthesis, and difference synthesis to determine the arrangement of crystals' atoms. Apart from his own use of the IBM 650 computer, Tseng-Yu Lee also wrote and shared programming packages for calculating least squares and matrix summation with other computer users. With thousands of data points and hundreds of parameters, he used NCTU computers to approximate the locations of atoms. During an oral history interview, he commented that manual calculations would likely have been impossible.[21] National Tsing Hua University (NTHU) funded his use of computer time, and according to his memories of those years, he reserved NCTU computers for at least 150 hours while working on one of Wong's research projects, for which Lee published two papers, coauthored with Wong, in *Acta Crystallographica* in 1965 and 1969.[22] The latter article was also coauthored with a Taiwanese scholar, Yuan-Tseh Lee, who became a Nobel laureate in 1986 for his research on the dynamics of chemical elementary processes.[23]

Lee's use of computers was not unique at that time. As early as December 1949, the leading journal of crystallography, *Acta Crystallographica*, collected a series of papers to introduce the "the application of punched-card methods in crystalstructure analysis," papers that were "intended for the benefit of readers not familiar with the use of Hollerith and IBM machines."[24] From 1949 to 1965, hundreds of papers were published in the journal concerning crystal-structure analyses based on punched-card machineries or electronic digital computers, including IBM 650s and IBM 1620s.

Tseng-Yu Lee insisted that he had learned computer programming from reading a book, the title of which he did not recall. But he remembered that he had read the book in conjunction with an NCTU course on numerical analysis. When I asked how he had been able to learn programming on his own, he answered, "I could do debugging later."[25] Regardless of whether Tseng-Yu Lee learned electronic digital computing from reading a book, from studying with NCTU instructors, from practicing on electronic digital computers, or—most likely—from some combination of all three approaches, he became an important figure in electronic digital computing in Taiwan, following a career path similar to that of Chi-Chang Lee. Tseng-Yu Lee went to MIT as a visiting researcher in the Department of Mathematics in 1967, and Chi-Chang Lee also went to MIT as a visiting faculty member in the Department of Electrical Engineering during the same year.[26] They were housemates when they lived in Massachusetts. After returning to Taiwan in 1968, Tseng-Yu Lee became the director of the computing lab at Tsing Hua's Department of Physics. The computing lab was a newly established unit because of Tsing Hua's acquisition of an IBM 1130 mainframe computer, the first computer on the Tsing Hua campus.

COMPUTING FOR POWER ENGINEERING AND BEYOND

As briefly mentioned earlier, some TPC researchers were familiar with the potential of electronic digital computers because electricity-related engineering required complicated calculations. Prior to having access to NCTU's IBM 650 in 1962, engineers and researchers of the company used in-house analog computers, conducting manual calculations with slide rules and abacuses. They also discussed analog computing, and the possibility of using electronic digital computing in the company's technical journal—*Monthly Journal of Taipower's Engineering*.[27] For example, TPC engineer Yuan-Kuang Chen (Yuanguang Chen) wrote an article in the 1962 issue concerning the application of analog computers to the coordination of hydraulic and thermal power plants' electrical supplies; Jia-Hui Zhou's article in the 1963 issue discussed the same type of coordination problems in relation to a Japanese company's use of an IBM 650.[28] These two researchers examined the coordination problem in creating models for forecasting supply and demand of electricity. The coordination problem was important because both the number and the intensity of Taiwan-bound typhoons are unpredictable and significantly affect the amount of rainfall.

Despite their awareness of the potential of electronic digital computers, TPC engineers had only indirect access to this technology until the company rented an IBM 360 computer in 1966. Researchers, engineers, and even accountants from the company had enrolled in the NCTU training courses to study the uses of the IBM 650 computer since 1962. In particular, most of these TPC trainees belonged to one of two major research-and-development departments in the company: the Department of Planning (qihua) and the Department of High-voltage Research (gaoya yanjiusuo).[29] After completing their training courses at NCTU, many of these personnel continued to visit NCTU's computing center to work on their research.

Among these TPC trainees was Yuan-Kuang Chen, who exemplifies the exchanges of electronic digital computing expertise that took place between NCTU and TPC. Prior to NCTU's acquisition of the IBM 650, Chen published two papers, in 1960 and 1962, on analog computing in *Monthly Journal of Taipower's Engineering*.[30] He then enrolled in an NCTU training course in the summer of 1963. That December, he wrote another article in the journal, on how to program an IBM 650, helping to introduce electronic digital computing to his colleagues.[31] When TPC decided, in 1965, to rent an IBM 360 computer, the latest electronic digital computer from IBM at that time, several researchers and engineers who had attended NCTU training courses wrote articles for *Monthly Journal of Taipower's Engineering* to explain programming principles and to "help our company prepare for the arrival of the IBM 360."[32]

TPC personnel's experience with NCTU's computing center contributed to the establishment of its own computing center. Other institutions followed a similar trajectory when they rented IBM's electronic digital computers after 1964. For example, National Taiwan University (NTU) sent recent graduates to learn programming at NCTU, and in the summer of 1964, NTU formed its own computing center after renting an IBM 1620 mainframe computer.[33] In contrast, National Tsing Hua University acquired an electronic digital computer for its computing center only much later, in 1968. Perhaps because the Tsing Hua campus neighbored the Chiao-Tung campus, the former university could temporarily rely on the latter for its computing needs. As the hub of the expertise of electronic digital computing in Taiwan, NCTU benefited subsequent Taiwanese computing centers in the 1960s, as the above cases demonstrate.

Chinese Petroleum Corporation (CPC) expressed enthusiasm for the IBM 650 soon after it arrived in Taiwan in February 1962. Yiqian Zhan, the head of the Department of Exploration and Drilling, led twenty CPC personnel on a visit of NCTU right away.[34] The following year, fourteen trainees from CPC took NCTU's programming courses. After the 1620 computer arrived in 1964, William Wesley Peterson specifically offered two three-day seminars, titled "Computer Applications," to seventy CPC engineers and staff in July 1964.[35] With access to the two IBM computers at NCTU, CPC staff conducted several research projects: "Analysis of Reformed Oil," "Computation for Fuel under Different Temperature," "A Geophysical Problem," and "Stratigraphy [sic] and Sedimentation."[36] Pei Jan Lee (Peiran Li), a CPC geologist, enrolled in NCTU's training course in electronic digital computing in December 1962, used the IBM 650 computer for his research on sediments and submitted his article to CPC's research journal, *Petroleum Geology of Taiwan*. In Lee's paper, he pointed out that he had not obtained satisfactory research results until a digital computer had become available in Taiwan. He stated, "A few years ago, the author noticed that polynominal approximation can be used to filter out some local components of a sedimentary rock, but to compute the polynominal of best fit by the least-squares method was very tedious. The writer stopped studying this method to avoid spending time and money until the high speed digital computer became available in Taiwan for research work."[37] NCTU's IBM 650 computer was instrumental to early 1960s researchers in state-owned enterprises, including TPC and CPC.

THE IBM 650 COMPUTER IN THE UNITED STATES

The 650 computer in Taiwan was IBM's Magnetic Drum Data-Processing Machine; by that time, this computer was a relatively obsolete, if beloved, model in the United States. The acquisition was announced in July 1953, and its first delivery was in December 1954. By January 1961, there were 1,250 IBM 650 computers manufactured and delivered. IBM stopped manufacturing the model in 1962.[38] The popularity of the IBM 650 computer on US campuses might explain Chiao-Tung alumni's choice of this model. For example, Cornell University obtained its first IBM 650 computer in 1956 for its computing center, where faculty and students used the computer until 1959 for instructional or research purposes related to numerical analysis, statistical analysis, and so on.[39] Another example is Duke University, which obtained an IBM 650 in 1958 to establish a computing center.[40] The Statistical Research Laboratory at the University of Michigan rented an IBM 650 in 1956; there, researchers grew so accustomed to and were so pleased with the IBM 650 that university researchers found the IBM 704 computer—the computer at the university's newly established computing center—to be inefficient and difficult to program.[41] On the Cornell campus, in 1958, the Animal Science Department obtained its own IBM 650 computer for their research on dairy-records processing.[42]

In the United States, university computing centers were also an important place to expand the technological system of electronic digital computers. According to historians William Aspray, Bernard O. Williams, and Atsushi Akera, in the early 1950s, electronic digital computers at US universities were prioritized for work related to defense contracts and industrial problems, as they were sponsored by the military for research purposes. At MIT, for example, IBM and physicist Philip Morse worked together to install IBM's then-latest computer, the IBM 704, thus establishing MIT's Computational Center in 1955. IBM and MIT agreed to have the Computational Center work as a regional facility that would provide computer training and computer time for New England–area colleges. IBM knew well that a university was a perfect place to expand computer use and was keen to help universities establish computing centers. From the mid-1950s to 1959, the company donated its IBM 650 to more than fifty universities. In exchange for these computers, universities had to offer courses in data processing and numerical analysis.[43]

In addition to IBM, the National Science Foundation (NSF) understood the potential of computing centers. Beginning in the 1950s, NSF provided grants to support the operations of university computing centers; its support for computing centers peaked in the early 1960s. In the United States, the number of university computing centers increased from forty in 1957 to four hundred in 1964.[44]

The costs of operating computing centers, including acquiring or renting computer facilities, were an important issue for US university computing centers. In 1963, all federal agencies, including the NSF, supported "about half of the cost of computing on campus."[45] Just as US universities had to rely on outside funding for computing facilities in the 1950s and 1960s, NCTU was fortunate to have secured financial support for Taiwan's first computing center by acquiring the technical-aid program from the United Nations in 1962. Nevertheless, no evidence shows that Chiao-Tung–educated technocrats studied US universities' attempts to gain access to computers before applying for the technical-aid program.

WELCOMING A NEWER COMPUTER

In 1963, a class of trainees at the NCTU Training and Research Centre put together a yearbook in which is recorded a remark from Arden describing his teaching in Taiwan (figure 3.1). Cognizant that Taiwanese trainees still had much to learn, Arden summarized his experiences at NCTU:

> It has been a pleasure to have the privilege of helping to introduce computers and computer application to Taiwan. The students in the training course at Chiao-Tung University have impressed me by their desire to investigate the new and difficult technology of computer programming. Although the technique of programming has been only partially acquired and insight into the many interesting areas and methods of application is still incomplete, there is no doubt that continued practical use of the partial knowledge will perfect the skill and increase the capacity for the successful use of computers in Taiwan.[46]

While Arden looked forward to the completeness of introducing programming to Taiwan, William Wesley Peterson, the second UN expert in electronic digital computing, believed that the independence of the NCTU computing center was going to be the best achievement of the UN-NCTU technical-aid program. Peterson materialized his idea of independence through educating staff and encouraging NCTU to opt out of IBM's maintenance contract. NCTU replaced IBM's maintenance contract with an in-house technician. Through purchasing rather than leasing a computer, NCTU was able to fully utilize the four-year funding from the United Nations. In hiring an in-house technician to maintain the IBM 1620 computer, NCTU aimed to work independently from IBM, "to operate the centre without outside help,"[47] and save on maintenance expenses.

Peterson was thirty-nine years old when he visited Taiwan. He had worked for IBM before joining the faculty of the University of Florida.[48] He saw the independence of the computing center as the foremost goal of his trip to Taiwan. Peterson's view was that the staff of the computing center should become independent from outside help. The *1968 UN Report* further stated,

> By the end of Dr. Peterson's mission the programming and teaching staff, the operators and maintenance personnel of the Institute's computing centre had developed such a proficiency that he was able to recommend no further outside assistance be provided in the meantime. His view was that they should best be left to operate the centre without outside help for several years and thus gain experience and confidence. Subsequent results have proved the correctness of his recommendation.[49]

Peterson also encouraged NCTU members to better know the industry. He had participated in a workshop held by the Institute of Electronics at NCTU, and discussed possible future cooperative projects between the graduate program and the industry. J. O. McNally, a specialist in vacuum-tube manufacturing and another UN expert visiting NCTU at the same time, was invited to this workshop. McNally was then sixty-one years old and worked for Bell Laboratories. He was Dalman and Chao Chen Wang's colleague at Sperry Gyroscope before Dalman joined Cornell.[50] In a workshop McNally and Peterson attended in 1964, McNally urged workshop participants to pursue possible cooperative projects between the university and industry, because he found that this was exactly the direction that higher education in the United States was moving following World War II, at Stanford and MIT, for example.[51] Peterson echoed McNally. The *Voice of Chiao-Tung Alumni* rephrased what Peterson stated, "If NCTU is a university that teaches merely mathematics, there is no need for it to have some connection with the outside world. But, as it is a university for applied science, it has to pay attention to applications, and it has to stay in touch with the industry. Otherwise, professors will not be able to carry out their teaching, because they don't know what the society wants and what to study."[52] Peterson and McNally viewed NCTU as a promising place to follow the trajectory of universities in the United States.

THE MAINTENANCE OF THE IBM 1620 COMPUTER

Arden's time teaching at NCTU relied on an IBM 650 computer, whereas Peterson used both the IBM 650 and the newly acquired IBM 1620 computers. From the time of its installation, the IBM 650 computer was useful. In fact, the computer was under such use as to experience many breakdowns, for which NCTU had to call IBM engineers for maintenance. As the IBM 650 computer relied on vacuum tubes, it generated a considerable amount of heat during its operation, which made it extremely unstable when it had been run too long or in a space lacking air-conditioning. The *1968 UN Report* attributed the necessity of air-conditioning to "the tropical conditions existing in Taiwan," but, in fact, air-conditioning was a must for US users of 650 computers too.[53] The short-term solution to this problem was to replace overheated and broken vacuum tubes with spare ones. In the Taiwanese case, and for others all over the world, the replacement of vacuum tubes was usually performed by IBM engineers. For NCTU in 1963, the long-term solution was to replace the entire computer.

Before Arden left Taiwan, he submitted a proposal and persuaded the United Nations to allow NCTU to replace the IBM 650 with a transistor computer—an IBM 1620. The request was granted by the summer of 1963, but the computer did not arrive until

February 1964.[54] The *1968 UN Report* stated, "Arden's most important contribution to the development of the computer centre was his recommendation for the replacement of the equipment. He provided a complete list of the equipment to be purchased and assisted in the negotiations with the IBM Company which led to the placing of the necessary order."[55] At that time, NCTU and Arden decided that "an IBM-1620" should be "purchased outright."[56] It appeared to be an unusual decision, compared to IBM computer users in other countries. IBM's standard way of doing business at this time was to lease its customers computers and charge them monthly leasing and service fees covering the provision of programs and maintenance.

The funding period of the UN-NCTU technical-aid program might have been the main reason for NCTU to have made this unusual decision. The UN-NCTU technical-aid program was going to end in 1965, making it wiser for NCTU to decide to purchase the computer outright in 1963. Recall, in chapter 2, United Nations–affiliated individuals did not consider rental a common option for a technical-aid program. Sundaram then promised to advocate for Taiwanese technologists on their request to rent the IBM 650 in 1960, instead of purchasing it.

This decision to hire an in-house technician was made in the second half of 1963, though the precise time is unknown. It is difficult to estimate how often IBM's customers opted out of IBM's maintenance contract at that time. However, as electronic digital computing was a far more common activity in the United States than in other parts of the world at that time, a US customer had a better chance of managing without IBM's maintenance-service plan, as this customer could draw on other available resources. In Taiwan, on the other hand, there was only one computer available, and NCTU members and trainees had begun to learn electronic digital computing only in 1962. To opt out of IBM's maintenance contract in 1963 was an audacious decision for NCTU.

There were several additional reasons for them to do so. First, this decision was consistent with Peterson's idea of independence. Peterson wanted to assure that the computing center would be able to function independently after the end of the technical-aid program. According to the *1968 UN Report*,

> As the time approached for the delivery of the IBM-1620, Professor Peterson undertook a study of the equipment on the basis of the available documentation and even before it had arrived, developed a capability for its immediate use. By the time it [the IBM 1620 computer] arrived, not only had the teaching staff of the Institute [the graduate program in electronics at NCTU] developed a competence in programming the equipment, but also two girls had been trained as operators. Moreover, so much had been learned about computer maintenance by working on the IBM-650 that Professor Peterson was able confidently to recommend that the Institute itself should take over the maintenance of the 1620 type equipment immediately after the end of the 90-day guarantee period. This has since been done with complete success.[57]

Second, it might have been a unanimous decision by NCTU administrators and instructors at this time, because IBM's maintenance-service plan was an extremely costly expenditure. Chi-Chang Lee recalled that NCTU had negotiated hard with IBM to purchase the IBM 1620 computer, instead of renting it monthly. Chi-Chang Lee and Tseng-Yu Lee, when both emeritus professors in their eighties, explained to me in vehement tones how expensive IBM's maintenance services had been and how arrogant IBM salespeople had been at that time. They noted that when IBM's Taiwan branch sent an engineer to check or fix computers at NCTU, the service fee would be charged from the time the engineer left the IBM office, instead of from the time the engineer arrived at NCTU. They also pointed out that customers could not bargain with IBM because it monopolized the Taiwanese market in the 1960s. They were also unsatisfied with the sales tax charged by IBM's Taiwan branch.[58]

Third, IBM had Taiwanese employees to take care of maintenance and installment, but they were not systems engineers. Chin-Chi Kao (Jinji Gao), who was hired by NCTU to maintain the IBM 1620 in 1964, pointed out that Peterson spent two-thirds of his time on the IBM 1620. When I asked whether the university could have requested and paid IBM engineers to help, Kao answered, "No. IBM's Taiwan branch didn't have any systems engineer or software engineer available. It had merely maintenance engineers. Perhaps there were one or two engineers that knew systems, but they might have not known [compliers] well. If Peterson would not have been there, the machine [the IBM 1620] would not work at all."[59] Peterson might have sensed the limited technical support offered by IBM and decided that he should train an in-house maintenance engineer.

Fourth, Peterson might have been confident that he had the capability to make the computer work and to train NCTU members in general. He likely believed an in-house engineer would maintain the computer well. According to Kao, for the IBM 1620 computer, Peterson came up with methods to separately compile sections of a larger amount of program codes. These methods were particularly useful when Yun-Tzong Chen, an instructor and an NCTU recent graduate, had to write a larger amount of codes for an economic-planning project.[60] Peterson had previously worked on a project to create a "programming system" for the IBM 650 in the Statistical Laboratory at the University of Florida, circa 1959,[61] and was well known for his textbook, *Error-Correcting Codes*, published in 1961.[62] His confidence and competence in programming and codes and his employment at IBM allowed him to believe that he and NCTU members could work on maintaining the computer together.

Fifth, the reliability of vacuum-tube computers is intrinsically poorer than transistorized computers.[63] The poor reliability of vacuum tubes could create difficulties for users or engineers seeking to identify a given problem. NCTU members, Peterson, and

Arden might acknowledge that it should have been easier for Taiwanese engineers to identify the problematic transistors of the IBM 1620 computer than the problematic tubes of the IBM 650.

Eventually, for the reasons discussed above—including financial concerns such as the technical-aid program's funding structure, the assessment of Peterson and NCTU members, their perception of IBM's technical service, and the characteristics of transistorized computers as opposed to vacuum-tube computers—NCTU members and visiting US professors decided to purchase the IBM 1620 computer and hire an in-house engineer to maintain the computer.

The in-house engineer Chin-Chi Kao is a native of Taiwan. Kao graduated from the Department of Electrical Engineering at Provincial Taipei Institute of Technology.[64] Before joining the computing center at NCTU in January 1964, Kao had worked at manufacturing palm-held transistor radios for China Electronic Products Corporation, a Taiwanese manufacturer of loudspeakers and radios, beginning in September 1963.[65] Kao recalled that it was the first Taiwanese company to export electronics products.[66]

Kao was outstanding. To obtain the technician position at NCTU, he first passed a number of tests, including a written test on binary circuits, a test on troubleshooting radio circuits, and a written English test. He was interviewed with Chu-Yi Chang, a professor and academic dean at NCTU. Kao recalled that Chang handed him a magazine in English and asked him to translate some paragraphs. He remembered that he spotted the term "reliability" in the magazine. Though it was a relatively new term for him and his contemporaries, he had discussed the concept of reliability with his supervisor when working on radio circuit designs.[67]

Kao began to work at NCTU following the New Year holidays of 1964. When the IBM 1620 computer arrived on the NCTU campus in February, Kao started to get to know the IBM 1620 by working with IBM engineer De-Yang Cheng. Cheng was sent to install the computer and took care of its maintenance during the first three months following the installation. Kao recalled that he picked up Cheng's expertise in maintaining the computer through observing Cheng installing and, later, on the job. Cheng, a fellow alumnus of the Provincial Taipei Institute of Technology, presented, explained, and tested the functions of different circuit cards and their corresponding slots in the wiring panels. The most important part of the training was learning how to read the block diagrams on the maintenance manuals that came with the computer, such as the *Customer Engineering Reference Manual: 1620 Data Processing Systems*, a copy of which is still kept at NCTU's library.[68] In addition to these face-to-face interactions, Kao occasionally phoned Cheng to ask questions.

While Kao could not remember all the details of a job he had held almost five decades before our interview, he did recall some common malfunctions of the IBM 1620 computers. Each of the abovementioned circuit cards consisted of, at most, six transistors. During the first few months following the arrival of the IBM 1620 computer, Kao identified some malfunctioning circuit cards and had them replaced with spare cards. In the beginning, the most frequent problem was malfunctioning of the electromagnetic relays in the computer's input-output unit that supported the standard-equipped typewriter. For the IBM 1620 at NCTU, operators entered commands via this typewriter, which also served as an output device. In addition, the computer had a paper-tape punch unit as an output device, and a paper-tape reader an input device, which was connected to the computer via a long cable. Aligning the photocells in paper-tape reader's reader head was a major part of Kao's monthly maintenance.[69] The computing center functioned well without direct maintenance service from IBM. Kao's maintenance did not involve a direct modification of the computer's design, but it can be understood as an act of emulation, by Brooke Hindle's definition. He carefully studied the operations of the IBM 1620 computer and joined a collective effort to run the computing center "without outside help."

Kao left NCTU in 1966. Beginning in September 1966, he worked for General Instrument's Taiwan factory, which manufactured television tuners, capacitors, and transistors. It was the first US investment in Taiwan's electronics industry (chapter 6). From then until the end of 1966, Kao returned to NCTU every weekend to assist and train a new engineer, Ji-Kwan Zho, who succeeded Kao, since NCTU had decided to continue to hire in-house engineers to take care of the maintenance of their mainframe computers.[70]

One particular episode reveals unsuccessful attempts by Taiwanese users to tinker with the two NCTU electronic digital computers. When a US-based computer engineering expert, Ju Ching Tu (Rujin Du), visited NCTU for about one year beginning in late summer in 1965, Academic Dean Chu-Yi Chang discussed with Tu the feasibility of connecting the IBM 650's card reader and punch unit to the IBM 1620. At that time, National Taiwan University had already obtained an IBM 1620 computer with the card-read-and-punch unit.[71] NCTU's paper tape punch unit now looked less convenient and slightly obsolete. Perhaps because the IBM 650 was not functioning well at that time, Chang intended to recycle its old input and output devices to improve those of the IBM 1620. According to Kao, Chang proposed this project as a research topic for graduate students attending one of Tu's courses. However, Chang's plan resulted in no real action when Tu left Taiwan in early summer of 1966. The plan probably failed, since it is not mentioned by any of my informants and no historical record discussed it. Kao believed that it was an infeasible plan.[72]

Peterson was not able to use the new IBM 1620 computer until the sixth month of his one-year stay in Taiwan, but his teaching was not hindered by the absence of the new computer. Like Arden, he offered training courses for different types of engineers and scientists (see table 3.1), provided graduate-level courses, and advised graduate students. Specifically, he offered a master's level course, Error Correction Coding, in October 1963, and served as adviser for a master's thesis titled "The Simulation of Control Systems [*sic*] on Digital Computer."[73]

After leaving Taiwan in the summer of 1964 and resettling at the University of Hawai'i, Peterson maintained contact with instructors and professors in Taiwan. In particular, Chi-Chang Lee helped him publish a textbook in Taiwan, based on class notes from one of his eight-week training courses.[74] This book was published only in Taiwan, and he tailored it to Taiwanese readers. When he wrote Lee a letter to discuss logistics of publishing, he proposed that he could "have the notes typed in good form so that it can be published by photo-offset, by the same process used for standard pirating of books in Taiwan." Clearly, he was well informed about the pirating of English books in Taiwan, but remained willing to have his work published there.[75] His teaching activities in Taiwan were not limited to the textbook and courses that could be identified by individual names and start dates. Beyond courses and textbooks, he emphasized the training of a computer-maintenance technician, teaching staff, as well as women computer operators.

WOMEN COMPUTER OPERATORS

With his ideas for helping the computing center to operate independently, Peterson prioritized staff training above other obligations. Peterson arranged two women operators to be trained before the computer arrived. As the *1968 UN Report* stated, "two girls had been trained as operators." The term "girls" chosen by the authors of the *1968 UN Report* precisely demonstrated gendered practices in the NCTU computing center at this time. "Computer operators" refers to people who used and operated paper tapes, punched cards, printers, and similar devices to input data into computers, or to collect output data from computers for college students, engineers, or scientists. Mainframe computer operation had been a predominantly women's occupation.[76] Historian David A. Grier has pointed out that, in 1944, the term "girls" replaced "computers" when scientists were talking about computing: "girl-years" and "kilogirl" were new terms referring to new types of computing labor in the United States—the simplification and generalization of computer operators through emphasizing only their sex.[77] At NCTU in the 1960s, it was common to see "girls" used to refer to "operators," as well. When the director of the computing center, Chi-Chang Lee, wrote letters to Peterson, he often referred to "girls" when he meant "operators." Among the nine letters that Lee collected privately, the following two excerpts show how he described, in English, the reorganization of operators in 1965 and 1966, respectively:

> The computer is still very busy now. Everything is going well. The only trouble is that we don't know whether Ms. Chen can go to Hawaii or not, so that we can't employ another girl in advance.[78] . . .
>
> We have two girls, Miss Han and Miss Ma, working in our computing center. The former is working very well and the latter cannot work independently so far.[79]

From reading the excerpts, one can see that Lee talked as if operators were a part of the physical facilities connected to the computer. When identifying operators as a mechanical part of a computing center, Lee might reveal two types of ideas behind this language usage. First, operators' jobs were mechanical and, thus, replaceable. It reflects a linguistic process of de-skilling. Second, as if it were a "spare part," the operator as an occupation was as critical as the computers themselves in computing centers, as computers were expensive and, to some extent, rare, in Taiwan and many societies in the early 1960s. The explicit description of women computer operators as girls contradicted the viewpoint that they were important to the operation of computers. The contradiction has been illustrated by historian Jennifer Light in her analysis of women computer operators working on ballistic firing tables at the University of Pennsylvania during World War II.

The two operators who worked with Peterson at that time were Y. C. Chen (Yazhi Chen) and Regina Lee (Jingxing Li) (see figure 3.3).[80] As one may imagine, key-punching was one of the most important tasks for computer operators. In 1964, when economic bureaucrats came to NCTU to utilize the IBM 1620 computer for an economic-planning project, operators had to handle a significant amount of statistical data. In addition to Chen and Lee, two more women workers, according to Kao, were temporally transferred from a governmental agency to NCTU to help with data-entry in the evenings.[81]

Beyond key-punching, operators sometimes obtained programming knowledge, since they reproduced programming codes from pieces of paper, written by people requesting to use computers, to punched cards or paper tapes. Chiong-Yuan Han (Qiongyuan Han), an operator who succeeded Chen in 1965, noted that Tseng-Yu Lee taught her how to recognize errors in programming codes. With this knowledge, if Han identified incorrect codes in college students' assignments, she could help them fix these mistakes. Chen also confirmed that identifying incorrect codes was precisely a part of her routine job.[82] Women operators' knowledge contributed to the computing center, and their capabilities should be seen as significant contribution that enabled the center to operate well.

NCTU emeritus professors recalled that Peterson was especially satisfied with Chen's work performance; eventually, she went to study at the University of Hawai'i and obtained an undergraduate degree there, after Peterson returned to the United States and moved from the University of Florida to the University of Hawai'i. Two NCTU graduate students of the class of 1964 of the Institute of Electronics followed Peterson to pursue their PhDs there, as well.

Another critical group of women participants in the computing center and at the university were UN experts' women family members. Peterson's then wife, Marion Peterson, conducted weekly English classes without receiving a salary from NCTU.[83] The class of 1964 yearbook listed Marion Peterson's name, picture, and her bachelor's degree in psychology from the University of Michigan—the class of 1964 clearly recognized her educational contributions.[84] "She also introduced us to American culture," said Chin-Long Chen, who followed Peterson to study at the University of Hawai'i in 1964, obtaining the first PhD from the Department of Electrical Engineering there.[85] Marion Peterson's English classes or her introduction to US culture might not have been a necessary element in keeping up the everyday operation of the NCTU computing center, but she was an important member of the center nevertheless.

CONCLUSION

Technologists at NCTU adapted, modified, and worked creatively with the emerging technological system of IBM mainframe computers on the island Taiwan. The technologists' strenuous efforts illustrate their ability to strike a balance between making the best of US technical assistance while controlling and reducing long-term reliance on the United States. Two decisions made by NCTU members and UN experts—purchasing the IBM 1620 computer and hiring an in-house engineer to maintain the computer—demonstrate their determination to better utilize the funds from the United Nations. In so doing, NCTU aimed at reducing maintenance expenses and working more independently of IBM. To opt out of IBM's maintenance plan was an innovative move, as IBM was known especially for its technical support at that time. Bucking conventional practices, the university forwent IBM's services and educated its in-house technician to maintain the computer, allowing it to save on costs and focus more on its mission of training technologists for the university, industry, and the military.

As an UN expert, Peterson considered that the best scenario for a technical-aid program was that the recipient country or organization becomes independent and no longer need to receive aid. For the NCTU computing center in the early 1960s, Peterson's ideas of independence could be related to Taiwan's and NCTU's independence from three different sources of assistance: the United Nations' funds, UN experts, and IBM. Nevertheless, it was either difficult or unreasonable to block the three different sources of assistance in Taiwan. First, at the same time in the United States, universities sought funds from the US National Science Foundation and others to support their computer uses and computing centers; it was reasonable for the NCTU computing center to continue to receive financial support from the United Nations. Second, Arden and Peterson left NCTU after their one-year visits, but electronic digital computing knowledge continued to flow through other channels from the United States to Taiwan. Some students who went to study in the United States visited Taiwan or returned to Taiwan after the end of the UN-NCTU technical-aid program (as following chapters show). Taiwanese universities continued to utilize electronic digital computers and textbooks from the United States. Third, it could have been difficult for NCTU to be entirely independent from IBM, the dominant mainframe computer company in Taiwan and the most powerful mainframe computer manufacturer in the world at that time. The frequent exchanges of computing knowledge between the United States and Taiwan, however, did not often occur in the military aid programs between the two countries, as we learn in chapter 4.

4 GRAPPLING WITH MACHINES: LATE ADOPTION OF COMPUTERS IN TAIWAN'S MILITARY ALLIANCE WITH THE UNITED STATES

Taiwan's electronic digital computing practices began on the NCTU university campus. In the United States, in contrast, the military initiated and facilitated the collaboration of research and development activities on electronic digital computers in the growing military-industrial-academic complex.[1] Through an analysis of Taiwan's defense strategies and its military research activities, I find that even though US military assistance, operating through the Military Assistance Advisory Group (MAAG), helped Taiwan gain access to state-of-the-art military weapons, beginning in 1951, it offered both promises and limitations for the Taiwanese military.[2] After the first two Taiwan Strait crises, in 1954 and 1958, Taiwanese military personnel had access to US-aided advanced weaponry systems that adopted analog computing devices. But Taiwan's military did not begin systematically training R&D personnel in electronic digital computing until the 1960s, because analog computers were bundled components of larger weaponry systems, the operation of which did not require mastering up-to-date computing techniques. The role of computing power in military applications was thus overshadowed by the weapons systems in which the computing mechanisms were embedded. Furthermore, US military advisers adopted the philosophy that the Taiwanese military should strengthen its ability to service and maintain newly acquired technological artifacts rather than pursue an ability to manufacture them independently. The peculiar nature of the US military alliance with Taiwan, especially in the 1950s, ironically slowed the Taiwanese military's exploration of electronic digital computing for its R&D agenda. Instead, NCTU's early 1960s UN technical-aid program played a more influential role than the military in expanding Taiwan's "technological system" of computers, as Thomas P. Hughes termed it.[3] For these reasons, despite the turbulent 1950s, the

ROC military did not own electronic digital computers for military-purposed research and development until 1969.

More specifically, the Taiwanese military began pursuing digital computing only in 1962, when military units sent personnel to NCTU to learn computing on the IBM 650 computer (chapter 3). In 1963, the US Central Intelligence Agency (CIA) arranged the installation of an IBM computer at an office of Chiang Kai-shek's son, Chiang Ching-kuo, to aid his collaboration with the CIA, possibly sharing reconnaissance data about the PRC, collected by ROC military personnel. Officially, Taiwan's first military computer, used for data processing, was an IBM 360 series, leased in 1967 by the ROC Army Logistics Command (*gongying siling bu*). The second computer, for the research and development of weapons, was a Control Data Corporation (CDC) computer—the CDC 3300, installed in 1969 at the Ministry of Defense's new research center—the National Chung-Shan Institute of Science and Technology (CSIST, *zhongshan kexue yanjiu yuan*).

In this chapter, I describe US military assistance to Taiwan and its limitations, as the assistance did not prompt the ROC military to research and develop military technologies, despite the existential threat from the PRC. To contextualize the frequent contacts and substantial collaboration between US and Taiwanese military personnel, I discuss the structure of MAAG. I then review the weaponry systems the United States sent to Taiwan right before and after the Second Taiwan Strait Crisis in 1958 to illustrate the bundled military technology transfers and the integration of analog computing techniques in missiles and aircraft at the time. I also discuss the Taiwanese military's attempts to expand its R&D units and how MAAG discouraged such plans. An example that illustrates the situation was MAAG- and CIA-imposed restrictions for Taiwanese personnel to access advanced avionics while maintaining the reconnaissance aircraft supplied by the CIA and flown by ROC Air Force pilots.

I then discuss, chronologically, the development of the military and semi-military uses of computers in Taiwan from 1963 to the 1970s: Chiang Ching-kuo's IBM computer installed in 1963 for reconnaissance missions, the Logistics Command of the ROC Army's IBM computer obtained in 1967 for data processing, and the use of CDC computers at CSIST for the development of nuclear weapons and missiles after 1969. I also include the growth in IBM and CDC computers in Taiwan at that time, as various corporations and government offices, alongside the ROC military, began to adopt mainframe computers. Based on published oral histories, biographies, and archival documents in Taiwan and the United States, this chapter stresses the agency of a wide range of Taiwanese actors including ministers, military officers, engineers, and personnel in maximizing the access to US military technologies and the military usage of electronic digital computing from the 1950s to the 1970s.

MAAG: THE STRUCTURE OF US MILITARY AID TO TAIWAN

ROC weapons acquisitions through US military aid during the Cold War can be divided into two rough categories. The first consisted of relatively cutting-edge aircraft, missiles, and other weapons, transfers of which mostly occurred because of specific military crises in Taiwan. Chiang Kai-shek and ROC ministers of defense played a critical role in negotiating the procurements, but it was up to the US government to make the decisions about when and what types of military technologies would be offered to Taiwan. For example, prior to the Second Taiwan Strait Crisis, Minister of Defense David Ta-wei Yule (Dawei Yu) sensed that large-scale military conflicts between the ROC and PRC were imminent in the next few years. Born in 1897 in Zhejiang, China, he served as minister from 1950 to 1951 and a second time from 1954 to 1965. He held graduate degrees from Harvard University and Friedrich Wilhelm University (currently known as Humboldt University of Berlin). Between 1955 and 1957, Yule requested the United States supply Taiwan with several state-of-the-art military technologies to deter the PRC. In 1955, when he discovered that PRC-fired shells were from Russian 152 mm (6-inch) howitzers, he broached supplying eight-inch howitzers with MAAG advisers. In January 1956, he bought up the request again with Secretary of the Navy Charles Sparks Thomas and Chief of Naval Operations Arleigh Burke. But the transfer of eight-inch howitzers did not materialize until late 1957, when intelligence showed that the PRC was shipping armaments to Fujian to prepare an attack on Quemoy and Matsu.

Another example was the Nike Hercules. On October 26, 1957, Yule proposed in a meeting with the chief of the US Taiwan Defense Command, Austin K. Doyle, that the United States supply Taiwan with the Nike Hercules, a surface-to-air missile. Doyle discussed the request with Burke, who noted that he would consider the proposal. But the Nike Hercules arrived in Taiwan only after the outbreak of the Second Taiwan Strait Crisis.[4]

To complement the cutting-edge technologies developed by the US military, a second type of military assistance consisted of mundane military equipment transfers. The US military assistance's major objectives in Taiwan at that time were "to assist in organizing, training, and equipping GRC [Government of the Republic of China] military forces in order to maintain and increase the effectiveness of such forces in the defense of Taiwan, Penghu, and the GRC-held off-shore islands." The US government sought to encourage ROC self-defense capabilities rather than giving it the power to attack the mainland unilaterally. In this regard, the Military Assistance Program would provide, for example, spare parts, communications electronics, soon-to-retire aircraft, ships, combat vehicles, and munitions that required regular resupply.[5]

To better utilize the cutting-edge technologies and standard equipment, MAAG repeatedly urged Taiwan's military to focus on maintaining existing technologies, instead of on the research and development or the acquisition of new technologies. MAAG and ROC military officers often disagreed on the priority of these approaches. From the perspectives of MAAG advisers, they intended to ensure the transferred technologies were in good hands, but from the perspectives of ROC military officers, they were keen to own advanced weaponry.

Military aircraft, for example, require extensive preflight and postflight preventive maintenance and periodic inspections. If the ROC Air Force received a batch of new aircraft, crews had to acquire new knowledge and techniques to operate the planes. When the first batch of F-84G Thunderjet fighter-bombers arrived in Taiwan in 1953, the military officer Yongzhao Li (known as Y. C. Lee in English documents) paraphrased the head of MAAG, Major General William C. Chase: "ROC Air Force pilots are well-trained and will not encounter any significant difficulties to fly these jets, but I envisioned the mechanics will face significant challenge. It is the same in the United States."[6]

From ROC Minister of Defense David Yule's perspective, MAAG's harping on the ROC military's insufficient maintenance did little to resolve the military problems facing Taiwan. When Admiral Felix B. Stump, commander in chief of the United States Navy and commander of the United States Pacific Fleet, visited Taiwan in September 1956, he met with Yule, who took the opportunity to express his dissatisfaction with MAAG advisers. The advisers believed the ROC Air Force was incapable of properly conducting aircraft maintenance and hence decided to postpone some aircraft transfers until winter 1958. Yule had anticipated imminent military conflicts in the years to come but found that MAAG advisers failed to consider the possibility seriously and thus gave little heed to transferring military weapons and technologies to Taiwan. Yule pleaded with Stump for more strategists—he implied that MAAG advisers were not strategists because the drumbeat on maintenance impaired the defense of Taiwan. He paid frequent visits to ROC Air Force bases to ensure the quality of service and maintenance tasks and hoped by doing so he could earn the trust of MAAG advisers and find an opportunity to access more timely US military technologies.

MAAG made the same evaluation of the ROC Navy, too. In December 1960, after viewing an ROC military exercise, MAAG commander Chester A. Dahlen, who had arrived in Taiwan only in November, commented to Yule on the inadequate quality of ship maintenance. Dahlen considered the ROC Navy's operations inefficient, which he believed was due to neither the equipment on the ships nor the training of personnel, but to inappropriate service of the ships. While Yule acknowledged the problem, he also thought that the overemphasis on maintenance was an excuse for delaying or denying his requests for procuring additional ships at that time. He noted that he had a hard time persuading the United States to transfer or sell its decommissioned ships, especially destroyers.[7]

MAAG's focus on maintenance also affected the Taiwanese military's fairly limited investment in research and development. The size of its military research and development offices shrank after Chiang's retreat in 1949. The Aviation Research Institute, Hangkong yanjiu yuan, set up in 1941 in the Republican Era employed around one hundred domestic researchers. British scientist and sinologist Joseph Needham and Chinese mathematician Qian Xuesen were appointed as two of twelve overseas-affiliated experts. But after 1949, the ROC government reorganized the research institute into an office under the management of the Aviation Industry Bureau (gongye ju), a sign that the ROC Air Force had limited personnel and resources to work on research and development after 1949.[8] Furthermore, around 1954, to strengthen Taiwan's capacity for aircraft maintenance and service, MAAG suggested the reorganization of the Aviation Industry Bureau as the Aviation Technical Bureau (jishu ju). The bureau's research and development projects of jet fighters were halted in 1955. Aircraft manufacturing factories in Taichung, Qingshui, and Gangshan were subsequently repurposed as maintenance factories (xiuhu chang). Colonel Yongzhao Li was directing the maintenance office (xiuhu chu) for the Combined Service Forces in 1955, and he testified that fewer resources would be available for research and development after the reorganization. In a proposal in 1967, Li specifically noted that, "US military aid was solely for defense purposes. It did not aid in our reconstruction [of manufacture facilities for aircraft in Taiwan]."[9]

Despite constraints imposed by MAAG, the ROC Air Force found alternatives to increase R&D capacity. For example, around 1955, MAAG had opposed an ROC Air Force proposal to expand an existing office to a larger factory to enhance understanding of avionics and air fleet maintenance. Yongzhao Li believed the opposition was unreasonable. In 1955, when briefing commander-in-chief of the ROC Air Force, General Shuming Wang, Li appealed to Wang and proposed establishing a factory to catch up with fast-evolving developments in jet electronics. Wang immediately gave the green light to Li and promised to liaise with MAAG in case of any questioning. With Wang's full support, the factory was operational in three months.[10]

Li was a key figure in reviving the ROC military's production and assembly of aircraft in the late 1960s. Born in 1915 in Jiangsu, China, he began his military career in 1936 in an aircraft assembly factory in China and, in the 1950s and 1960s, developed Taiwan's military aircraft maintenance capacity. In March 1969, the air force promoted Li to direct its Aero Industry Development Center (Hangkong kongye fazhan zhongxin), and he led the center from 1969 to 1982. The center was the predecessor of the Aerospace Industrial Development Corporation (Hanxiang hangkong gongye gufen youzian gongsi), which continues to conduct research and development as well as manufacturing military technologies in Taiwan to this day. One of his most notable contributions was manufacturing aircraft in Taiwan by drawing on assistance from US companies and US military aid. In 1967, as a major general, Li proposed a coproduction project to manufacture Bell UH-1H helicopters. MAAG, the US State Department, and consultants evaluated the proposal, offering little support initially, but over time, the head of MAAG, Major General R. G. Ciccolella, became enthusiastic about the project and lent his support. The project was approved in 1969. One hundred and eighteen UH-1H helicopters were eventually manufactured in Taiwan, the last batch of which were decommissioned in 2019.[11]

THE ANALOG MISSILE SUBLIME

The Second Taiwan Strait Crisis in 1958 and the tension preceding the crisis created opportunities for Taiwan to receive a series of advanced defensive weapons developed in the United States. In May 1957, the United States deployed Matador missiles in Taiwan, with the first test-firing there in May 1958. As Matador missiles was nuclear-capable, they heightened Mao's concerns about the US military presence in East Asia. Matador missiles, which could carry conventional or nuclear warheads and had a range of six hundred miles, threatened mainland industrial cities such as Shanghai and Wuhan. According to *New York Times* journalist John W. Finney, they were the first surface-to-surface guided missiles installed in East Asia. South Korea also gained access to the Matador missile in 1958. The Matador, controlled by sophisticated electronics, was the first all-weather and night-flight missile available to Taiwan. According to Finney, its accuracy was based on "electronic instructions sent from the ground." An "advance guidance station," known as the AN/MSQ-1, was necessary for "directing the missile on its course."[12]

The Second Taiwan Strait Crisis began on August 23, 1958, when the People's Liberation Army fiercely shelled Quemoy and Matzu and set up a blockade. At least three state-of-the-art weapons were shipped to Taiwan. First, on September 17, the US Navy shipped three eight-inch howitzers to Quemoy, publicizing the transfer so that the PRC knew that the howitzers were capable of firing both nuclear and nonnuclear shells.[13] Second, Operation Black Magic brought Sidewinder missiles to Taiwan, installing them on one hundred ROC Air Force F-86 Sabre jets. On September 22, the air-to-air infrared-homing missiles shot down four to ten PRC MiGs, the first use of Sidewinders in combat. Third, the US installed the Nike Hercules missile system in Taiwan in October 1958. It was a surface-to-air system and the first antiaircraft missile system acquired by Taiwan. According to Bob Mackintosh, a member of the Second Missile Battalion 71st Artillery who was sent to Taiwan to operate the Nike Hercules, his assignment marked "the first expeditionary missile battalion that went overseas on a combat mission." While the Nike Hercules was nuclear-capable, according to Mackintosh, no nuclear warheads were stored in Taiwan at that time. He noted, when the crisis subsided, "the Taiwanese took over their missile sites and continued to use them for decades to come."[14]

The crisis created the "missile sublime" in Taiwan, a term I borrow from historian of technology David E. Nye's concept of American technological sublime. But the missile sublime rested on analog computing devices rather than digital devices.[15] Historian James Small and Paul Edwards have noted that the speed of electronic analog computers could easily meet the demand for real-time applications and handle arduous computation and control mechanisms for military application in the 1950s. The emergence and adaptation of electronic digital computing in military research and applications, however, did not immediately replace electronic analog computing. The first digitalized (transistorized) guided missile system in the United States was only implemented in the Minuteman Intercontinental Ballistic Missile Program brought online in 1962.[16] In Taiwan, 1950s missile guidance systems were based primarily on analog systems. Sidewinder missiles incorporated infrared homing. Matador missiles were controlled by ground guidance stations and used radio signals and analog computing at that time; the Nike Hercules missile system worked similarly.[17] The cutting-edge missile systems that ran on Taiwan's analog guidance system did not prompt the military sector to pursue research and development of digital military technologies. The Taiwanese military did not conduct a systematic and effective study of analog computing mechanisms in the 1950s, either.

One of the prominent military technologies that began utilizing electronic digital computers in the United States was the Semi-Automatic Ground Environment (SAGE) system, the first part of which became operational in June 1957 at McGuire Air Force Base in New Jersey. SAGE required operators to convert radar signals from analog formats to digital formats in order to monitor Soviet nuclear bombers and send aircraft or missiles for interception or an offensive attack in real time, as illustrated in the films *Dr. Strangelove or: How I Learned to Stop Worrying and Love the Bomb* and a similar film, *Fail Safe*.[18] Had ROC military leaders noticed the ongoing construction of the SAGE system prior to 1957, they would have realized that the United States encountered different types of military threats than Taiwan did. The ROC military had to be wary of potential attacks on offshore islands just a few miles from the PRC coast, while the United States faced the threat of Soviet bombers or missiles traveling over thousands of miles. Proximity to the mainland forced Taiwan to counter a bombardment almost instantaneously. By the time a radar system in Quemoy had detected an artillery attack from a military base in Fujian, those shells would have already fallen on Quemoy. With the PRC yet to complete the first locally manufactured Dongfeng series of ballistic missiles, digitalization of radar signals would have been of little help in the 1950s. The first of the series, Dongfeng 1 (DF-1, a licensed reproduction of Soviet missiles), was completed and tested only in 1960 and had a range of 500 kilometers, which slightly undershot the distance to US military bases in Japan. Prior to the DF-1, the ROC military had no immediate need for a digital computing technology to detect and intercept artillery shells aimed at its offshore territories.[19]

MILITARY TECHNOLOGIES AS INTEGRATED SYSTEMS

The operational nature of a large and complex weaponry system makes less visible individual analog computing devices. To operate a missile in the 1950s, one had to learn the radar system, the fire control system, and the maintenance program to regularly inspect the hardware. Over time, the ROC military became much more reliant on holistic technological systems for deploying and maintaining missiles, aircraft, and navy ships. Analog computing components in a missile system were essential but liable to be overshadowed by the large missile system itself.

Take the Nike Hercules missile system, for example. It consisted of several separate mobile components: the missiles themselves, launchers, missile radar tracking, computers, target ranging radar, and low- and high-power acquisition radar systems. Some were larger than the size of a cargo van. In a *United Daily News* story covering a public demonstration of the Nike Hercules system in 1961, a journalist revealed to Taiwanese readers that its computer system crunched numbers to locate the targeted position. ROC priority was immediate access to the entire Nike Hercules missile system. A plan to train military researchers to master analog or digital computing devices would not have been immediately useful. The missile system with its deterrence function could seem mightier than sophisticated computers alone.[20]

The MK 1 fire control computer, an analog computing system installed on US Navy ships that had been procured by the ROC Navy in the 1950s, illustrates the same point, as well. A 1956 document held at the State Department's Bureau of Far Eastern Affairs listed US Military Assistance Program items scheduled to be transferred to ROC forces in the years to come. An "MK 1 computer" and its spare parts were included under the section of an unspecified navy ship to be transferred in 1957. While the ROC Navy was unlikely to ignore the importance of analog computing to the naval fire control system in the 1950s, the MK I fire control computer had to be transferred across the Pacific Ocean with the navy ship that installed it. During the 1950s, it might have been more practical for the ROC to procure ships than to mobilize military personnel to research and develop the workings of fire control systems. It only began to develop fire control computer systems in the 1970s.[21]

RESTRICTED ACCESS TO AVIATION ELECTRONICS

Even though Taiwanese and the US military personnel frequently interacted in Taiwan, restrictions could be imposed on the Taiwanese military personnel's access to certain technologies. These restrictions were to ensure US military advantages. An example is the maintenance of US aircraft in Taiwan, which was subcontracted to Air Asia, a subsidiary of Civil Air Transport Incorporated (CAT). CAT was one of only a few companies that provided commercial flights in Taiwan in the mid-1950s. Among CAT's initial funders were Claire Lee Chennault and the CIA. The CIA's investment was for aerial reconnaissance. During the Vietnam War, Air Asia hired more than four thousand personnel to meet the high demand for maintenance and service in East and Southeast Asia. When US military aircraft landed in Taiwan for maintenance, Taiwanese personnel were sometimes prohibited access to the aircraft's electronics systems. Only US personnel could work on classified or higher-level technical maintenance.[22]

Avionics was one of the areas in which US personnel wanted to sustain their technical advantages. For example, after MAAG transferred new military aircraft to Taiwan, a group of US personnel would be assigned to visit and share their knowledge of and techniques for the maintenance and service of the new aircraft with the Mobile Training Detachment (MTD), their Taiwanese counterparts. Historian Yu-Ping Lin noted the importance of the MTD to the ROC Air Force, but she also pointed out that MAAG encouraged MTD personnel to master only the use and maintenance of the transferred aircraft; it discouraged MTD personnel from obtaining or developing knowledge about aircraft that were not found in Taiwan. For example, MAAG believed that MTD personnel had no need to know anything about the maintenance of the latest avionics installed on new jets.[23]

Restrictions were also imposed on the maintenance of CIA reconnaissance aircraft. In the 1950s, due to the limitation of PRC's radar detection, it was still likely that ROC Air Force pilots could fly into the PRC without being detected or shot down. Pilots had to fly at either extremely high or low altitudes to evade radar and surface-to-air attacks. Taking advantage of this, from the mid-1950s to 1974, the CIA worked with the ROC Air Force to conduct reconnaissance missions on the status of PRC nuclear weaponry development. These missions were not revealed to the public until the 1990s.

The 34th and 35th Squadrons of ROC Air Force, known as the Black Bat and Black Cat Squadrons, were specifically designated to execute reconnaissance missions assigned by the CIA. The Black Cat Squadron was responsible for high-altitude reconnaissance, and the Black Bat executed low-altitude reconnaissance. The CIA provided aircraft in most cases and sent US personnel along with their Taiwanese counterparts to maintain these aircraft. Taiwan and the United States then shared the intelligence—for example, photos of PRC military bases, ballistic missile test sites, and nuclear-weapon sites. They also continuously gathered signal and electronic intelligence to identify PRC radar stations and missile locations. The CIA interpreted the radar data before 1961, and US and Taiwanese personnel cooperated on the interpretation after 1961. The collection of signal intelligence could be used for evasive measures during reconnaissance missions. The US Air Force and CIA offered Taiwan U-2s and various types of aircraft for reconnaissance starting in the late 1950s, and sometimes customized them with special electronics to avoid detection and gather signal data. For example, the ATI repeater-jammers used for the Black Bats' P2V-7 aircraft were a brand-new technology developed by the US Navy and first used in reconnaissance planes on Black Bat missions.[24]

Such collaboration became tricky in maintaining state-of-the-art reconnaissance aircraft. Taiwanese maintenance personnel could be restricted from accessing the aircraft. An example can be found in the biography of Xianghe Xie, who was assigned to the 4th Squadron of the sixth group of the ROC Air Force and was responsible for reconnaissance missions in the 1950s. Xie shared that the ROC Air Force built a "Blue Room" in 1957 for the squadron that was supposed to fly two RB-57A planes to execute reconnaissance tasks. Accompanying US support personnel arrived in Taiwan in 1957 with the two planes. No other personnel than the US support group was allowed access to the Blue Room.[25] A Black Cat Squadron member, Chungli Johnny Shen (Zongli Shen), revealed that only US technicians (likely Lockheed ground crew and technicians) conducted U-2 maintenance.[26] However, restrictions were loosened over time. According to sociologist Chin-Shou Wang, who collected oral histories of Black Bats, the CIA granted more involvement to Taiwanese in electronics maintenance as time went by.[27]

In sum, analog computing devices played an important role in the advanced military technologies transferred from the United States to Taiwan in the 1950s. The maginificent military technologies did not necessarily encourage the ROC military to develop a program to master electronic digital computing. But the goals of US military assistance to Taiwan were not entirely about giving Taiwan the best military technologies to defend itself. Instead, MAAG emphasized maintenance programs over advanced weaponry. MAAG's preference in maintenance also excluded the ROC military from allocating resources into research and development. Furthermore, the US military and

CIA placed restrictions on the ability of Taiwanese military personnel to gain a full understanding of the most advanced avionics. Hence, military assistance from the United States discouraged ROC forces from developing R&D programs, especially those about studying or utilizing analog computing devices and electronic digital computers.

CHIANG CHING-KUO'S "PERSONAL" COMPUTER

Collaboration between the CIA and Taiwan on reconnaissance tasks also prompted an installation of an IBM computer at Chiang Ching-kuo's office in 1963. The computer was to be secret, known and accessible to only a small number of people. Chiang Ching-kuo succeeded Chiang Kai-shek in 1972, as president of the ROC and leader of the Chinese Nationalist Party. In 1962, Chiang Ching-kuo was the chair of the Veterans Affairs Council and the secretary-general of the ruling Chinese Nationalist Party. He was also in charge of collaborating with the CIA on reconnaissance missions. In that year, Chiang Ching-kuo persuaded the CIA to supply two C-123 long-range transport planes for his father's plans to airdrop secret paramilitary teams into the PRC to incite local resistance. The PRC was experiencing a severe famine in 1960 caused by the excesses of the Great Leap Forward. The CIA's head officer in Taipei, Ray S. Cline, helped coordinate the aircraft supply. Cline and his Washington colleagues were also interested in making these planes available for future missions in Vietnam; Chiang Ching-kuo concurred with Cline. The Kennedy administration also agreed to offer "joint training and planning for 200-man airdrops," according to historian Jay Taylor. In 1962, Cline was promoted as head of the CIA's Directorate of Intelligence and left Taiwan for Washington. His successor, William E. Nelson, met with Chiang to continue discussions on the use of two C-123 planes. In a June 1963 meeting, Nelson offered Chiang an IBM computer, the model of which was not specified in the surviving documents. Nelson told Chiang in July that IBM planned to send three US IBM personnel to install the computer. A few Taiwanese would be selected to operate the computer and would receive training in the United States. An additional IBM staff member would be transferred from Tokyo or Okinawa to Taiwan to assist in operations and service. All the involved personnel would receive a background check but would not be allowed to contact staff at IBM's Taipei office. Nelson planned that this computer would remain unknown to the public. Surviving documents do not state exactly how the computer would assist in the data processing of the intelligence collaboration between Chiang Ching-kuo and the CIA, but the computer may have assisted in reconnaissance missions. CIA reconnaissance of the military establishments and relevant transportation lines in China created detailed records about locations and distance between nuclear-weapons related sites. As seen in a January 1964 report, the CIA surveyed every railway close to the PRC's Lanzhou's Gaseous Diffusion Plant, which had produced weapon grade U-235 the year before. Therefore, the CIA, assisted by Chiang's office, considered the IBM computer a good tool to receive and transfer the intelligence in digital and tabulatable formats.[28]

COMPUTERIZING ARMY LOGISTICS

In 1967, the ROC Army's Logistics Command (gongying siling bu) became the first military office to install a computer. It had already adopted IBM electronic accounting machines in 1964 and recruited seven army personnel to operate the machines. The navy's Logistic Command and the Ministry of Defense also had IBM electronic accounting machines installed by 1965. With the growing demand for data processing, the army's Logistics Command began using an IBM system/360 model 20 on August 2, 1967, to replace its former IBM electronic accounting machines.[29] The ROC's Joint Logistics Command (Lianhe houqin siling bu) showed interest in installing an IBM system/360 model 20 in 1965. The Joint Logistics Command was one of the units that dispatched several students to take courses offered under the UN-NCTU program, but for unknown reasons the installation plan never came to fruition in the mid-1960s.[30]

The army's Logistics Command's IBM 360 computer was used primarily to manage military supplies. In the few news stories covering the army's adoption of the computer, journalists often presented it as the ROC military's move to an era of scientific management. Shifting to computerized data, the army even aimed at "sending these data overseas via telecommunications technology" and "working with MAAG in a more collaborative manner." Taiwan intended to digitalize data to comply with the US military's data formats, following the same logic that the CIA employed when it requested that Chiang Ching-kuo install an IBM computer for intelligence collaboration. In August 1967, a few weeks after the installation at the army's Logistics Command, eight visitors from South Korea, including military officers, MAAG advisers, and an IBM representative arrived in Taipei to learn more about how the Logistics Command used computers for managing its tasks.[31]

IBM'S AND CDC'S GROWTH IN TAIWAN

The army's Logistics Command installation of the IBM 360 was also slightly later than a few state-owned enterprises that had begun to use the same model of computer. As of 1965, there were three IBM computers in Taiwan: an IBM 1620 installed at NCTU and one at NTU and an IBM 1440 in service for accounting in the state-owed enterprise,

Taiwan Sugar Company, which was also one of IBM's first users of accounting machines in Taiwan in the late 1950s. After 1965, the cutting-edge IBM 360 series became the mainstream for Taiwanese IBM customers. Between 1965 and 1967, Taiwan Power Company (TPC), Chunghua Information Processing Center, and Formosa Plastics Corporation adopted the IBM 360 series. In 1968, Taiwan Sugar Company, the Business Administration Department at NTU, the Chunghua Postal Service, and an unnamed military office also planned to install the IBM 360 series computers in their offices. In 1969, the Ministry of Economic Affairs scheduled an IBM 360 installation in its southern Taiwan office, which at the same time would provide service for four state-owned enterprises including Taiwan Aluminum and Chunghua Engineering. Another series of IBM computers that was popular in Taiwan at that time was the IBM 1130, which was less expensive but suitable for engineering professionals and university researchers. IBM 1130s were rented by the Kaohsiung Oil Refinery in 1967, an engineering research center at Cheng Kung University in 1967, the Center for Population Studies at NTU in 1968, and the National Tsing Hua University in 1968. More private enterprises became IBM customers at the turn of the 1970s. In 1969, Tianli, a large private trading company that imported parts for machineries, televisions, and appliances from Japan, the United States, Germany, and Sweden leased an IBM computer, model number unknown, to manage its sales, inventory, and accounting. In 1970, TPC and Taiwan Sugar Company, the two state-owned enterprises that led the trend of computerization, installed new IBM computers again to replace their old ones. The Executive Yuan (the equivalent of the Cabinet) was scheduled to lease an IBM 360/40 system in 1970, primarily to assist the Directorate General of Budget, Accounting, and Statistics. But its computer time was expected to be shared with other Executive Yuan offices. The IBM computer was recommended by Dr. Ruth M. Davis, who pioneered in creating software for US public service offices, and visited Taiwan briefly as a consultant for the Executive Yuan.[32]

Control Data Corporation began doing business in Taipei in 1967. Similar to IBM, Control Data took advantage of Taiwan's growing computer market for government offices, thriving industries, universities, and state-owned enterprises. Four CDC computers in Taiwan in 1970 were quite representative in these areas. The first CDC 3300 computer in Taiwan was installed in January 1969 in the Ministry of Finance to computerize eight million tax records. The second CDC 3300 computer, at the Chung-Shan Institute of Science and Technology (CSIST), aided the state-owned defense industry. After the CSIST installation, National Taiwan University's Electrical Engineering Department installed a CDC 3150 computer that same year. A prosperous export company at the time, Far Eastern Textile, leased a CDC 3150 computer in 1970.[33]

A survey found that, as of 1970, computer installations in Taiwan included twenty IBM computers, four CDC computers, one NCR computer, one Sperry Rand UNIVAC, three NEAC computers, and at least one FACOM computer. The NEAC computer was developed by the Nippon Electric Company (NEC), and Fujitsu developed the FACOM computer; both were Japanese companies. The IBM and CDC installations drew more attention from journalists than the others did. Thus, it is not clear at which companies or organizations the remaining computers were located. For example, there was no public information on where the 1970 UNIVAC was installed. But the *Economic Daily News* (*Jingji ribao*) noted that two additional UNIVAC 9400 computers were installed for Taiwan's China Airlines to handle its real-time reservation and ticketing system in April 1971. Taiwan's thirty computers in 1970 did not represent an extremely enthusiastic adoption of computers, compared with other countries of similar economic performance. For example, Chile had about fifty computers in 1968, a number lower than that of Brazil, Argentina, Colombia, and Venezuela.[34]

Historian James Cortada noted that by 1974 there were thirty-five IBM installations in Taiwan; Control Data installations in Taiwan grew from four sites in 1970 to thirty-five in 1978. In January 1971, Taiwan Railroad ordered a CDC 3150 computer. In November of the same year, one of the Taiwan Railroad computer center's staff attended a linear programming conference organized by Control Data in its new office, which had a large classroom designed for such events. Twenty-nine state-owned and private businesses sent staff members to attend the conference, which taught attendees how to use CDC computers and mathematical models to optimize the allocation of human, financial, and material resources in a company. Control Data offered four such conferences in the winter of 1971. Its market share reached 31.5 percent in 1971. Among the thirty-five CDC installations as of 1978, Control Data's customers included several universities, Chia Hsin Cement Corporation, the National Security Bureau, China Shipbuilding Corporation, the Bureau of Telecommunications, and Jixian, a private technology company, though the specific models of some of these CDC computers are unknown. Based on an Executive Yuan survey in 1979, Taiwan was home to twelve computers worth over $1 million US each, sixty-four computers worth $160,000 US to $1 million US each, and 206 computers worth $40,000 US to $160,000 US each.[35]

CHUNG-SHAN INSTITUTE OF SCIENCE AND TECHNOLOGY

Officially established by the ROC government in July 1969 as a military research institute, the CSIST was one of Control Data's most prominent customers in Taiwan in the 1970s. Its preparatory office began operations in April 1965. CSIST consisted of three departments: nuclear energy, material science of missiles and rockets, and electronics for missile guidance, known, respectively, as the first, second, and third departments (*diyisuo*, *diersuo*, and *disansuo*). In 1969 CSIST installed the CDC 3300 computer, which was intended to support nuclear weapons development there. The electronics (third) department tended to use their TR-48 analog computer in the early 1970s, instead of the CDC computer.[36]

The establishment of CSIST was intimately related to Chiang Kai-shek's desire to develop nuclear weapons. His interest in nuclear weaponry was especially reinforced after the PRC's detonation of its first nuclear device in October 1964. In 1963, Chiang sent Administrative Deputy Minister of Defense (guofangbu changwu cizhang) Chun-Po Tang (Junbo Tang) to meet in person with and invite Israeli nuclear scientist Ernst David Bergmann to visit Chiang at his Sun Moon Lake residence in Taiwan. Bergmann's visit later that year likely led Chiang to believe that it could be feasible to initiate his nuclear weapons development project. He needed a research and development institute, however, to carry out the project secretly and avoid scrutiny from the United States or any international actors involved in monitoring nuclear weapons proliferation. Chiang then authorized setting up a preparatory office for the CSIST in 1965 and had the CSIST officially inaugurated in 1969.[37]

The president of the newly formed CSIST was Chen-hsing Yen, then president of the National Taiwan University; the civil appointment was meant to underplay CSIST's military ambitions. The vice president of CSIST, however, was Chun-Po Tang, who directed military weapons research and development in missile and nuclear fields. Tang became the president of CSIST from 1975 to 1982. It is worth noting that the US embassy in Taipei reported that Tang was hesitant at first because he was concerned that the research and development of a nuclear weapon would be so costly that it would go "beyond GRC's resources."[38]

Tang attended one of Chiao-Tung University's college-preparatory schools in China before he joined the ROC Military Academy (huangpu junxiao). Hence, he was considered an alumnus by the Chiao-Tung alumni community in Taiwan. He studied artillery and electrical engineering at the University of Cambridge in England in the 1930s. The US Embassy in Taiwan referred to Tang as a Cambridge-trained mathematician. He left CSIST in 1982 in part because tests of nuclear weapons at CSIST conflicted with the agenda of the Ministry of Foreign Affairs, which was excluded from the secrecy of the nuclear weapons project and had to deal with growing pressure and questioning from the United States. On January 8, 1988, when then ROC President Chiang Ching-kuo was hospitalized and on his deathbed, CSIST's director of nuclear research, Hsien-yi Chang, left Taiwan for the United States with the CIA's help—some referred to the incident as a "forced defection." The International Atomic Energy Agency and US representatives subsequently arrived in Taiwan to seal, demolish, and remove any remaining nuclear research facilities or materials. Nuclear weapons development in Taiwan has halted since.[39]

Even though CSIST sought to make military weapons, its choice of the CDC 3300 computer to begin research work in 1969 was a conservative decision. Control Data could have offered the more powerful 6000 series (supercomputers) or the high-end 3000 series, such as the CDC 3400, 3600, 3800 computers. Released in November 1965, the CDC 3300 is a "smaller" mainframe computer, according to Control Data's Robert M. Price, who held key positions at Control Data from 1961. Price considered CDC's strength to be supplying computers for scientific calculations, though the 3000 series also targeted commercial or industrial users. In a news article introducing the CDC 3300 computer in Taiwan, Control Data's Taiwanese manager specified that the large memory storage and the multitasking operation system implemented on CDC 3300 computers were attractive features. Moreover, according to the manager, Control Data offered its Taiwanese customers unlimited hours of usage, whereas other computer companies typically allowed customers a total of only 182 hours per month and charged additional fees for overtime hours. The special treatment Control Data offered its Taiwanese customers was part of its effective international expansion in the 1960s; the corporation set up offices in places including India, Australia, Mexico, South Korea, and Japan. It also established manufacturing facilities in Hong Kong and South Korea.[40]

In Taipei, Control Data maintained a cordial relationship with the ROC military. When Control Data's Taipei office decided to move to a new building in June 1971, the minister of defense was invited to cut the ribbon at the more than five-hundred-guest celebration. The new Taipei office received another order from the CSIST in 1972, a CDC Cyber 72/14 computer.[41]

On May 16, 1972, the *Minneapolis Star* reported CSIST's order for a new CDC computer for "scientific research applications." The news article, on page 44, consisted of only twenty-four words, but the $1.2 million computer deal was significant enough to make it to the largest newspaper in Control Data's home state. The deal created an opportunity for ROC Vice President Chia-kan Yen (Jiagan Yan) to receive Paul G. Miller, senior vice president and group executive of marketing of Control Data, and Marvin R. Swenson, vice president of Pan American Pacific marketing operations of Control Data. Swenson was probably in Taipei to sign the contract with the CSIST, but the Taiwanese media coverage at that time noted only that he was there to research challenges to information management in Taiwan, without mentioning the military nature of the computer purchase.[42]

The CDC Cyber 72/14 computer was one of the Cyber 70 Model 72 series, which became available in March 1971. It was a successor of the 6000 series, early supercomputers. It was likely that CSIST chose the cyber series because the CDC 6600 and 7600 computers, among other computers, were selected for installation at the Los Alamos National Laboratory. One of the most ambitious projects that used CDC Cyber 70 computers around the same time was the US Air Force Logistics Command's attempt to build a "massive unified database and real-time computer network for highly complex logistics," though the project failed and was reshaped as a batch-processing-only system. The Cyber 70 series was also adopted for business and academic users in many countries, such as Austria, Israel, and Poland. A CDC Cyber 72/14 computer was installed in Ben-Gurion University of the Negev in Israel after a thorough evaluation conducted by its staff and faculty from 1972 to 1973. In Poland, a CDC Cyber 72/14 computer was a central component of "a multiprogramming system called Cifronet that [would] serve remote terminal equipment at eleven academic, research, and planning establishments."[43]

In the case of Taiwan, the CDC Cyber 72/14 at CSIST was connected through a "dedicated line," supposedly a leased telephone line, with the computer center of the National Tsing Hua University in 1973. Control Data's "Cybernet" techniques, developed in 1968, made this connection possible. The Cybernet was a type of timesharing system. Students and researchers would access the CSIST computer through terminals.[44]

The 1974 CIA National Intelligence Survey of Taiwan noted that CDC Cyber 72/14 computer was mainly reserved to "handle the HAMMER nuclear computer code for simulated nuclear reactor operations." The previous CDC 3300 computer was then reassigned to be "used mainly for data processing for the armed forces." In this survey, the CIA also stated that "the Republic of China has an extremely limited military research and development capability." The CIA had reservations about the progress made in the fields of "air, ground, and naval weapons," even though the CIA was clear that CSIST intended on developing nuclear weapons and missiles.[45]

Despite the CIA's critical evaluation of CSIST, the CDC Cyber 72/14 computer supported scientific computing on the CSIST campus. In his autobiography, nuclear energy researcher David Li-Wei Ho mentioned that he frequented the computer center at CSIST circa 1973 or 1974, when he was pursuing a master's degree in physics at the Chung Cheng Institute of Technology (Zhongzheng ligong xueyuan), a military academy. During his time as a master's student and an air force captain, Ho interned at CSIST, where he developed his thesis in CSIST's first department. He worked on experiments on and computer simulations of neutrons with CSIST's heavy water nuclear reactor, which had been purchased from Canada in 1969. Ho wrote FORTRAN programs with punched cards for his simulations. When commuting between the computer center, the nuclear reactor, and the campus of Chung Cheng Institute of Technology, Ho had to guard his punched cards from rain damage. If the cards were damp and subsequently slowed down the reading, the operators at the computer center would grumble. Ho later received a PhD in nuclear engineering from Iowa State University in 1981, where a CIA agent approached him right after he began his studies. The agent targeted international students in the same PhD program to monitor nuclear weapons development in their home countries. Ho chose not to work with the CIA and returned to CSIST for nuclear weapons development in Taiwan after his graduation.[46]

Another alumnus of Chung Cheng Institute of Technology reported that he worked for the computer center at CSIST for a few years until 1975. Jenn-Nan Chen, an applied mathematics major, was recruited by Sanqian Wang, an adjunct instructor at Chung Cheng Institute of Technology. Similar to Chin-Chi Kao at Chiao-Tung in 1964, Chen's job was to maintain and operate the CDC computers. Sponsored by CSIST, Chen attended Stanford University to pursue a master's degree in operations research from 1975 to 1976 and a PhD degree in computer science at Northwestern University from 1983 to 1986. He then returned to work for CSIST until 1996. Later, he left to work at corporations and teach at a private university.[47]

CSIST purchased its third computer from Fujitsu in the early 1980s, even though Control Data appeared to be doing well in Taiwan. In March 1978, Vernon Sieling, one of Control Data's vice presidents visited Taiwan to promote the computer network connected with a Cyber 74 computer, which cost $9 million and had been set up just months before. Control Data intended to attract various customers that would connect their terminals at their own offices to the Cybernet. Sieling noted that he saw the Taiwanese computing market expanding with the demands for computing in state-led, large-scale construction, such as airports, highways, shipbuilding, petroleum exploration, and nuclear energy. In 1980, Control Data won over two more customers: National Tsing Hua University leased a CDC Cyber 172 computer, and the Kaohsiung Harbor acquired a Cyber 170–720 computer.[48]

Control Data's supercomputer market share, however, was gradually challenged by Japanese companies in the US market in the 1980s with Fujitsu and NEC the two major Japanese companies then entering that market. In 1981, an anonymous Taiwan- or China-born nuclear engineering professor at UC Berkeley, whose name David Ho did not want to reveal but referred to as "Professor C," flew to Japan to meet with Fujitsu's CEO to acquire a Fujitsu supercomputer for CSIST. Eventually, Professor C persuaded the CEO to make the deal, despite the CEO's initial hesitation. Ho used the Fujitsu computer for seven years, ending in 1988 when the nuclear weapons development at the

CSIST halted. Moreover, in 1983, Ho paid a visit to Professor C's laboratory at UC Berkeley and accessed the supercomputer Cray II at Boeing to model nuclear reactions for his research. He simulated the process of nuclear chain reactions and the impacted area of an explosion. The available commercial software for such a simulation was designed for nuclear power plants, and nuclear weapons development use was prohibited. But Ho and his colleagues at CSIST would sometimes tweak the available input columns to work out desired computations. They sometimes had to send the data to consulting companies that then ran the computations on software that was unavailable to CSIST. Ho noted that the nuclear weapons development in Taiwan was a patchwork or bricolage (*pinzhuang*) as researchers had to find unconventional solutions to work on their research. Hsien-yi Chang was Ho's supervisor at CSIST and oversaw the computer simulations there until he left for the United States with the CIA's assistance in 1988. Chang also noted in an oral history interview that computer simulations were critical for understanding and analyzing the small-scale nuclear explosion tests in Taiwan that began in the 1970s. In sum, the CDC computers and the Fujitsu computer were indispensable tools for CSIST's nuclear weapons development program.[49]

MISSILE COMPUTING AT CSIST

In the late 1950s, Matadors, Sidewinders, and Nike Hercules missiles proved useful in deterring PRC attacks. While a few sources point to a small number of preliminary rocket experiments conducted starting the late 1950s at the Aviation Research Institute, CSIST was in a much more advantageous position to research and develop missiles. But, by 1966, the gap between China and Taiwan in terms of missile development was probably too wide for Taiwan to catch up after almost two decades of procuring weapons from the United States. In a report of the US embassy in Taipei in April 1966 on CSIST's interest in nuclear weapons research, well-known historian Cho-yun Hsu told the reporting officer that CSIST, in its planning stage, had run "into difficulties in developing missile capacity." According to Hsu, the CSIST was interested in "developing its own missile tracking radar," as "the tracking radar supplied by the US is adequate tracking only aircraft." It was likely CSIST sought to develop weapons to counter the PRC's newly developed DF missiles. The severe lag in missile research and development meant that CSIST relied heavily on overseas experts to make progress. For example, in 1968, Israeli nuclear scientist Ernst David Bergmann returned to Taiwan and had discussions with Chun-po Tang and Kuang-wei Han (Guangwei Han) on the types of missiles CSIST might begin with.[50]

Kuang-wei Han began to work as the chief of the third division of the third department of the CSIST in its preparatory stage in July 1966. He was responsible for researching

and developing missile guidance systems. Born in Shandong, China, in 1930, he graduated from the Naval Engineering Academy in Taiwan in 1955. From 1956 to 1957, he
served as an engineer, maintaining the communications system on the *Tai Kang*, which
was the decommissioned USS *Wyffels*, which had been transferred to the ROC Navy after
World War II. He then pursued a PhD at the US Naval Postgraduate School, Monterey,
California, from 1958 to 1962. For his doctoral research, he conducted experiments analyzing the phase space of a feedback control system, which contained a relay-switching
computer functioning as an adder. After returning to Taiwan in 1963, he taught at the
Naval College and Chiao-Tung and witnessed the IBM computers installed there. He
visited the Electrical Engineering Department at the University of California, Berkeley,
for a year in 1965.[51]

In the summer of 1966, Han led his research team to survey information on missiles
all over the world. He stayed in ROC's missile battalion for two weeks to study manuals of the two US-aided guided missiles, the Nike Hercules and MIM-23 Hawk. The
latter was a surface-to-air missile system installed in Taiwan in April 1964. Han soon
realized that these missiles were state-of-the-art and technologically beyond what the
nascent CSIST could model. To cultivate the team's technical capacity, he had to choose
another missile for his researchers to study and build. In 1968, Han met with Bergmann
and presented him with a proposal to build a missile to model after the design of the US
Army's short-range Lacrosse missiles, which Bergmann affirmed as a workable project.
Bergmann bounced between Taiwan and Israel to allow the ROC military to connect
with Israel weapon developers. In 1972, the ROC military also went ahead and ordered
several Gabriel II missiles, short-range surface-to-surface missiles made in Israel.[52]

From 1971 on, Han devoted himself to the development of a missile system for
navy battle ships—the Hsiung Feng I missiles. During the early stages of development,
a TR-48 transistorized analog computer was his team's major tool for simulating the
performance of the early prototypes of the missiles. The TR-48 computer was made by
Electronic Associates Inc. in New Jersey and became available on the market in 1961.
But Han soon chose to opt for digitalizing Hsiung Feng I's H-930 fire control system
(wujin ixing). The choice indicates the team's adoption of electronic digital computing.
Han specified that it was a milestone "to digitalize the ROC military's weapons system."
He also recalled that some engineers preferred analog techniques and challenged his
decision. The chief engineer sent by the ROC Navy to help this project had a dispute
with him on whether to use digital or analog techniques for a specific device in the
missile system. Han opted for digitalization, and the navy engineer opted for analog
techniques. Han did not specify the engineer's name in his memoir.[53]

From 1974 to 1981, Han led the development of Hsiung Feng I's fire control system, including hardware and software for detecting and aiming at targets and launching missiles from navy ships. Missile fire control system development was coordinated by CSIST, Honeywell, RCA, and the ROC Navy. RCA built the radar system. Honeywell was in charge of fire control software and hardware on navy ships, except for the simulator. Han insisted that the CSIST alone should work on the simulator, as it is the essence of a fire control system. He believed that it was the only way for his team to learn how the entire fire control system worked. It required considerable work from Han and his team to coordinate with the contractors, but it was invaluable for CSIST researchers to learn the latest digital techniques for fire control. It also justified ROC researchers' acquisition of as much know-how as possible from Honeywell and RCA. Honeywell considered it an unusual request but concurred. Computing was a critical part of this project, as Han mentioned that, for a test in October 1975, his team relied on both assistance from the second (material science of missiles and rockets) and third (electronics for missile guidance) departments to compute and estimate the needed numerical data. Overseas Chinese lent their help too. For example, Yijin Liu, a China-born, MIT-trained visiting scholar at National Tsing Hua University, passed on his advice and made critical numerical estimations for Han's team while he was in Taiwan from July 1971 to June 1974. Liu had worked as a researcher for developing missiles at the Lockheed Corporation in the United States. Another China-born, MIT-trained Lockheed engineer, Hua Lin, visited CSIST several times between 1968 and 1975 to offer suggestions.[54]

Engineer Mingjie Zheng designed the hardware of the computer inside the Hsiung Feng missile. For the missile's homing system, CSIST adopted a hybrid computer, partly analog and partly digital. Zheng indicated that precision was a major challenge when he was designing the device. Zheng obtained a bachelor's and a master's degree in electrical engineering from the National Cheng Kung University and held a PhD in electrical engineering from the University of Florida. He began to work at CSIST in 1973.[55]

The Hsiung Feng I's fire control system was based on a semi-distributed computing network connecting the radar and the fire control system on each ship. Han noted that a trade-off of a distributed system was insufficient computing power, which might occur when the number of navy ships increased. Eventually, the fire control system was completed. According to Han, Honeywell's Frank R. Wiemann, who oversaw the H-930 project and was promoted after its completion, thought highly of CSIST's work; Wiemann even proposed to Han that the ROC military and Honeywell should market the H-930 fire control system to other countries. But higher-level officials in the military rejected the idea.[56]

Han's insistence on having CSIST researchers and engineers gain know-how of weaponry design and manufacturing also reflected on his strong interest in having the Ministry of Defense purchase Gabriel II missiles from Israel in the early 1970s. Han believed that, similar to weapons procurement from the United States, Taiwan's deal with Israel would grant Taiwanese personnel opportunities to visit Israeli military manufacturers and sites and learn from them. Nevertheless, many critics considered Hsiung Feng I missiles a reverse-engineered product of Gabriel II missiles. Han disputed this in his memoir, explaining that both missiles looked similar because the ROC Navy decided to make both Hsiung Feng and Gabriel fit the same launch devices on navy ships. Some part specifications for the Hsiung Feng missiles thus had to be the same as that of Gabriel missiles. In 1978 Han presented a disassembled Hsiung Feng missile to the Israeli ambassador to Taiwan to prove that Hsiung Feng missiles were not an imitation of Gabriel missiles. He further added that the Gabriel II missiles ordered in 1972 arrived only after Han's team began testing Hsiung Feng I prototypes.[57]

CONCLUSION

Counterintuitively, despite Taiwan's frequent military conflicts with the PRC and its alliance with the United States, its military forces did not see access to electronic digital computers in the 1950s as a priority. This was partly because ROC military research and development activities decreased during the turbulent years of the 1950s, particularly after the ROC's massive defeat and retreat from mainland China to Taiwan in 1949. The decline of such activities was also exacerbated by US MAAG advisers. Instead of encouraging the manufacture of self-sufficient military aircraft and weapons, they emphasized Taiwan's maintenance of available military technologies. The establishment of CSIST, a military research institute dedicated to making missile and nuclear weapons, allowed the ROC nuclear weapons researchers to adopt CDC computers in the 1970s as their research tools. Missile engineers began to show a strong interest in digitalization when the development of the Hsiung Feng I missile began in the 1970s. Taiwan's case clearly differed from the US case of a growing military-industrial-academic complex that was built upon cutting-edge computer research and development bolstered by military funding.

ROC forces were able to begin their computerization of data processing in the mid-1960s. The UN-NCTU program offered opportunities for military personnel to learn electronic digital computing in the first half of the 1960s. In 1965, the Joint Logistics Command had expressed interest in installing an IBM 360 computer. In 1967, the ROC Army's Logistics Command soon acquired an IBM 360 computer for handling material supplies and financial accounting. The move to computerization also reflected the ROC Army's interest in being consistent with their US counterparts, who offered a good number of military supplies to Taiwan's army. In a similar vein, Chiang Ching-kuo's IBM computer, installed in 1963 at the CIA's suggestion, reflected the agency's interest in pushing Chiang to digitalize intelligence data for collaboration.

Only after 1969 did ROC military personnel at CSIST gain access to electronic digital computers—Control Data computers—to further the goals of developing missiles and nuclear weapons. Even though the missile research at CSIST originally relied on an analog computer, efforts to develop the Hsiung Feng missile guidance system were digitalized, as a collaborative project between CSIST, Honeywell, RCA, and the ROC Navy. Participants in the research and development of nuclear weapons and Hsiung Feng missiles noted the "patchwork" nature of their endeavor—they improvised and adjusted their goals with the available and accessible technical resources.

The history of military uses of electronic digital computers in Taiwan, despite differing from that of the United States, has been contingent on both the constraints placed and opportunities offered by US military aid in the 1950s and 1960s. The constraints and opportunities continue to be characteristic of Taiwan's weapons procurement from the United States. For example, after Taiwan purchases a weapon system from the United States, Taiwanese military personnel receive training from US military personnel, as they did in the 1950s. Weapons procurement remains dependent on negotiations between the two countries, but the United States, as the exporter, can decide what and what not to sell to Taiwan. International conflicts, such as the war in Ukraine, also factor into the willingness of the United States to aid in Taiwan's defense. Taiwan's CSIST, the Aerospace Industrial Development Corporation, and the Ministry of Defense continue to improvise and adjust their goals with available technical resources to research and develop weapons to defend Taiwan against threats from China. Even though Taiwanese military researchers can benefit greatly from Taiwan's increased capacity for manufacturing computer products, they have to advance in various fields, such as material science and systems engineering, to build military technologies they intend to supply self-sufficiently. The "patchwork" nature of manufacturing military technology seems to remain salient in Taiwan from the Cold War to today. In chapter 5, I focus on another type of "patchwork" in the late 1960s and early 1970s: students at Chiao-Tung University, National Taiwan University, and National Cheng-Kung University experimented with putting together available parts to build minicomputers and calculators.

5 MANUFACTURING HOPE: EXPLORATIONS IN MAKING MINICOMPUTERS AND CALCULATORS FROM SCRATCH

Because of this project, participants have obtained practical knowledge and experiences that cannot be gained by merely reading books. The project has displayed its multiple effects. When UCLA electrical engineering professor Ernest Shiu-Jen Kuh visited last year, he commented that this project demonstrated the most practical and effective training method.

—Jong-Chuang Tsay, April 1972

This chapter examines how three groups of Taiwanese college and graduate students developed an avid interest in mass-manufacturing computers in the late 1960s. I contextualize and contrast the student-initiated hardware tinkering projects to two sets of contemporaneous proposals about the future of the electronics industry in Taiwan, including proposals that were articulated by Chiao-Tung–educated technocrats Gisson Chi-Chen Chien and S. M. Lee, as well as UN-affiliated officials and experts. Taiwan's domestic electronics manufacturing sector took off in the 1970s, but only after three promising campus-based projects and their sponsors pursued the unlikely goal of manufacturing minicomputers. The projects were based at NCTU, National Taiwan University, and National Cheng-Kung University. These three sets of social groups deliberated on the future of the electronics industry on the island.

One of the three campus projects was at the Institute of Electronics, the graduate program in electronics at NCTU, which eschewed the decision to purchase a minicomputer from the United States and instead built a minicomputer from scratch between 1968 and 1971 (figure 5.1). NCTU students and faculty members studied the architecture of minicomputers and experimented with manufacturing them domestically. They aimed at building a minicomputer, but the final product they came out with was closer to a programmable calculator with a memory unit of 1,248 bits, with their

FIGURE 5.1
The NCTU minicomputer. Courtesy of National Yang Ming Chiao Tung Museum.

own design of logic units and magnetic core memory units. They also attempted to construct these units and spent a significant amount of time and energy to wire each component of the calculator.

Preceding chapters have described the aspiration of Taiwanese engineers, scientists, and technocrats to own and use electronic digital computers, including an IBM 650 and an IBM 1620. As noted in chapter 2, when S. M. Lee, director of the graduate program in electronics at NCTU, learned about the United Nations' approval of Taiwan's application in 1961 for a technical-aid program to bring an IBM computer to Taiwan, he commented that, with help from the UN, NCTU could be manufacturing electronic digital computers in the near future.[1] Lee's prediction came to fruition, if much later. In this chapter, we explore how a group of early Taiwanese computer users, mostly students, attempted to build the artifact themselves after shifting their focus from the ability to own and use computers to the ability to manufacture them. The construction reflected the ethos of NCTU, where students, staff, and faculty members continued to tinker with computing technology after gaining access to mainframe computers (discussed in chapter 3).

Two other calculator- and computer-building projects emerged in the electrical engineering departments at the National Cheng-Kung University (NCKU) in Tainan and National Taiwan University (NTU) in Taipei. The former completed a calculator, and the latter a minicomputer. NCTU, NTU, and Cheng-Kung's projects included countless trial-and-error attempts to build a technological artifact with components that were not manufactured by or assembled in a conventional way. Their production of the calculator or minicomputer was different from the methods of existing successful manufacturers and was an informal production of a technological artifact. Building a technological artifact from scratch is a distinctive activity at the borders of replication, emulation, and creativity, involving large inputs of diverse knowledge. The three groups' attempt to build a minicomputer was a practice of adapting, modifying, assembling in an innovative manner, and working creatively with technologies.

While not directly related, the NCTU project took place only two years after the Amateur Computer Society (ACS) was set up in the United States. The editor of *Electronics* magazine, Stephen Gray, circulated *ACS Newsletters* in 1966 as a space where hobbyists could share their experience making home-built computers, as well as the necessary knowledge and technical documents for such projects. But such a project was only accessible at the time by a resource-rich social group in the United States. The least expensive computers available then still cost $10,000 US.[2]

To build a computer from scratch is particularly relevant in the East Asian context. Tracing how a computer was built and what types of epistemological input contributed to the process of building has become a "genre" within the history of East Asian computing. In the 1980s, Hidetosi Takahasi and Sigeru Takahashi each published a description of their participation in the design and building of several computers in 1950s Japan.[3] They emphasized hardware design and the components they chose, with relatively little interest in social contexts, such as the backgrounds of participants, funding sources, or the uses of these computers.[4] Recently, historian Chigusa Kita has written about how the seat reservation system for the Japanese National Railway was designed and began operation in 1959.[5] She has proposed the concept of "technological mimesis" to discuss a group of Japanese designers, trained at MIT, that decided to use Bendix G-15 computers manufactured in the United States, and further adopted the technology to the company's technical needs.

Members of the three campus projects in Taiwan held the belief that hands-on experience in building a minicomputer or similar technology was the path that would lead them to the know-how needed to mass-manufacture such technology. It was also considered the best path given the fact that no domestic companies were making computers at that time. The projects also benefited from the wider emerging electronics industry in Taiwan. In particular, the NCTU minicomputer project was especially aided by an alumnus who worked at Philco-Ford's plant in Taiwan and a friend who managed Wang Laboratories' Taiwan factory. Philco, originally a radio manufacturer, became an early manufacturer of transistors and was acquired by Ford Motor Company. Wang Laboratories, a pioneer in word processing and office technology, was based in the Boston area and founded by An Wang, a Chinese immigrant. Several NCTU master's graduates were recruited to work as top-level engineers at Philco-Ford's and Wang Laboratories' plants that assembled TV and radio sets, transistors, integrated circuits, and various semiconductor products at the export processing zone in Cianjhen (or Qianzhen), Kaohsiung.

Starting in 1966, Taiwanese government officials began to run this special economic zone to entice foreign investments with greatly reduced tax rates, preferential land pricing, and an abundant supply of inexpensive labor. Home to many electronics assembly factories, the export processing zone became a symbol of the future of Taiwan's electronics industry and, thus, the site that two experts from the United Nations visited to assess the industry in Taiwan in the late 1960s. United Nations experts witnessed an emerging market for local electronics companies to tap into the expanding demand for materials and service from plants owned by foreign investors. Frank Roy Gustavson, an American UN expert, acknowledged the critical role of Chiao-Tung in this growing market, and S. Skoumal, a British UN expert, argued that domestic companies and the government should form a trust to take the lead in bolstering the industry. While the UN experts' visits were the last before the Republic of China (Taiwan) lost its seat in the United Nations to the government of the People's Republic of China, they offered firsthand observation of the beginning of the making of world powerhouse of computers and computer parts.

CHIAO-TUNG STUDENTS' AMBITION

In May 1964, when the UN-NCTU technical-aid program was set to end, S. M. Lee, the director of the NCTU graduate program, and Gisson Chi-Chen Chien, director of the Directorate General of Telecommunications, submitted an application for a new technical-aid program to the United Nations Special Fund section, seeking technical and financial assistance to set up pilot factories to experiment with manufacturing "electron tubes, transistors, and other components" at NCTU.[6] S. M. Lee had reenvisioned the NCTU campus, from a place to start up computer manufacturing to a laboratory for transistor manufacturing. Chien and Lee aimed to set up an "Applied Research Center in Electronics Technology" at NCTU. Following a pattern similar to the successful application for the UN-NCTU technical-aid program, Chien and Lee sought more educational resources for the NCTU campus. They specified in the 1964 proposal that this newly proposed program would invite experts to teach at NCTU in electron tubes, passive network components, circuit design, computer logic and design, and transistors and other semiconductor devices.

The UN Special Fund section commented, however, that their proposal was "a little pre-mature."[7] S. M. Lee presented the proposal to John N. Corry, resident representative and director of the Technical Assistance Board and director of Special Fund Programs in the Far East in July 1964. Lee emphasized that the new project would train technicians and engineers "who will be absorbed by local electronics industries." But Corry doubted Taiwan would need so many engineers if the electronics industry "were not highly developed in Free China [Taiwan]."[8] Corry's colleague's reply to the application was more explicit. Resident Representative and Director of Special Fund Program in China Knut H. Winter stated directly, "Although there is every possibility that within three or four years there will be a large electronics industry, it is doubtful that the current status of the industry is such that we could make a very firm prediction as to the possible requirements for people trained in the electronic fields."[9]

Although these UN officials questioned the potential for an electronics industry in Taiwan, a computer-building project at NCTU was emerging, with an aim far higher than the UN official's evaluation. In the summer of 1968, NCTU graduate students and instructors began to discuss how to manufacture an electronic digital computer. The final product they came out with was a programmable calculator in 1971. The product was often referred to as a "small-scale" computer or the NCTU "minicomputer" by participants, although the NCTU machine did not achieve the computational power demonstrated in the minicomputer market at the time. This project lasted until June 1971. The main funding source for the project was the National Science Council, Taiwan's equivalent of the US National Science Foundation. Using floating-point decimal numbers, the programmable calculator performed addition, subtraction, multiplication, division, computation of square roots, Fibonacci numbers, and Bessel functions. Participants included a teaching faculty Chao-Chih Yang and PhD students Ching-Chun Hsieh (Qingjun Xie) and Jong-Chuang Tsay (Zhongchuan Cai). Oher participants included fourteen graduate students and three technicians. Tsay and Hsieh contributed to a large portion of the computer-building project. Their extraordinary attempt to build a computer at Chiao-Tung defied UN-affiliated officials' cautious forecasting of the growth of the electronics industry in Taiwan.[10]

Hsieh and Tsay initiated the minicomputer building project as graduate students and gained support from NCTU faculty and administrators. Hsieh was born in China in 1941 and moved to Taiwan with his family in 1949. He was admitted to the PhD program in electronics at NCTU in 1966 after receiving a master's degree there. He was clearly a star student at NCTU. His 1971 doctoral degree was only the second one awarded by NCTU. Tsay is a native of Taiwan. He obtained a BS in electrical engineering at the National Cheng-Kung University in 1966, and a master's degree from the Institute of Electronics at NCTU in 1968. He was then admitted to the PhD program and graduated in 1975. In their oral history interviews in 2007 and 2009, respectively, Hsieh and Tsay stated that the project provided NCTU graduate students with an opportunity to study the architecture of electronic digital computers.[11]

As PhD students, Hsieh and Tsay offered courses for undergraduate students and graduate students before they received their terminal degrees. Hsieh completed his master's thesis, "Implementation of Error Correcting Codes by Linear Sequential Circuits," with a visiting Fulbright scholar, Ju Ching Tu, who held a PhD from the University of Michigan at Ann Arbor and had worked for Lockheed Corporation and the Canadian Northwest Telephone Company. Hsieh described Tu as so patriotic that he offered as many courses as possible to students at NCTU. Tu specialized in computer hardware, a valuable contribution to the program at that time, and taught courses in computer architecture, switching circuits, coding theory, and information theory. As Hsieh was Tu's outstanding pupil, Hsieh continued his courses on computer architecture and switching circuits after Tu's Fulbright fellowship ended.[12]

As a young and ambitious educator, Hsieh set up a laboratory of digital circuits for his teaching while pursuing his PhD. This was probably not a common practice in the United States and other parts of Taiwan, but it shows that Hsieh's expertise and leadership was so recognized by NCTU faculty and administrators that the university allowed him, as a graduate student, to set up a laboratory. Hsieh considered using vacuum tubes to teach digital circuitry. Tubes would guarantee successful experiments, but by the mid-1960s they were simply not state of the art. For conducting experiments in his course, Hsieh fortuitously secured integrated circuits from a NCTU alumnus and close friend, Andrew Zaixing Chew, who is a native of Taiwan and began to work in 1966 at a Philco-Ford-financed assembly plant in Kaohsiung, Taiwan. The plant assembled small-scale and medium-scale integrated circuits. After meeting with and learning from Chew that the plant was assembling integrated circuits in Taiwan, Hsieh told Chew that integrated circuits would greatly assist his teaching, but he had no budget to procure them. A few days later, Chew brought Hsieh a suitcase full of integrated circuits. They became critical assets for conducting experiments for teaching. The ambitious prodigy soon sought to construct an artifact based on the integrated circuits instead of simply conducting experiments on them. Computers came to mind, and he began leading a project to build a computer with his fellow graduate students.[13]

Hsieh's ambition included an exploration of possible industrial applications, which aligned with Chien and Lee's interest in connecting the university with the electronics industry. In April 1967, Hsieh published an article in *Science Bulletin National Chiao-Tung University*, elaborating on his design of a preliminary general-purpose computer. The publication preceded Hsieh's acquiring of integrated circuits from Chew and was a precursor of the minicomputer project he worked on with Jong-Chuang Tsay. Hsieh and four other NCTU students used P-N-P transistors to set up the switching circuits for this preliminary computer. Although these transistors were not suitable for the computer, they were the only available components. The lack of resources was so severe that the computer that he had planned on in his 1967 paper did not have a memory unit. Hsieh noted that he would look for financial resources to procure one in the future. Despite such constraints, his purpose was not only to design an experiment for students to learn the operation of computers but also to conduct an experiment to explore the potential for manufacturing general-purpose computers in Taiwan. Hsieh pointed out that the experiment was necessary for developing an electronics industry in Taiwan.[14]

Hsieh's article in the *Voice of Chiao-Tung Alumni* about a conference trip to Canada also demonstrates his confidence regarding what he and his colleagues at NCTU could achieve in Taiwan. In late September 1967, Hsieh took his first overseas journey, a ten-day trip to Toronto, to present a paper on logic circuits at the Canadian IEEE's International Electronics Conference. At the conference, Hsieh met Dr. P. M. Thompson at the University of Ottawa, who was impressed by Hsieh's intelligence and actively recruited Hsieh to study in his program. Thompson, anticipating that most electrical engineering students would be unable to find satisfactory jobs in Taiwan, was amazed by Hsieh's description of recent progress in Taiwan's electronics industry. Hsieh told Thompson that five alumni of Chiao-Tung's master's program in electronics had begun their promising careers in Taiwan. Chew and two recent graduates were working for Philco-Ford as chief engineer and product engineers. Additionally, two graduates worked for the Philips plant in Taiwan as a product manager and a quality control manager. Hsieh also caught up with an old friend, Jianying Chen, a PhD student at the University of Toronto. Hsieh and Chen were undergraduate classmates at National Taiwan University. They stayed up late chatting and boldly concluded that Chiao-Tung University's laboratories of microwaves, servomechanism, transistors, laser, and electronics seemed to be much more updated than those of the University of Toronto at that time, thanks to United Nations financial and technical aid to Chiao-Tung.[15]

Like many engineers who traveled abroad, Hsieh was dazzled by the modernization of the city of Toronto. He shared with readers of Chiao-Tung's alumni newsletters how he got around the city with no effort. Polite drivers yielded to pedestrians, and one of them even rolled down the window to say thank you when Hsieh let the driver go first. Pedestrians crossed the streets only within crosswalks. Buses ran frequently, bus routes were well-designed, and the city was built using a grid plan, all of which seemed to be the opposite of how he got around in Taipei. His admiration might remind readers of narratives written by travelers from the colonies to metropolises in the early twentieth century.[16] At the same time, Hsieh was absolutely confident that his home country would soon be able to advance to the level of Toronto in terms of electronics manufacturing. At the conference, he toured the booths set up by electronics companies to showcase and promote their latest products. The small but bustling booths reminded Hsieh of the prosperous Chung-Hua Shopping Arcade in Taipei, where Taipei city dwellers of the 1960s were able to find various commodities, including used and new electronic products. Hsieh told his readers that, now, Taiwanese companies were unable to manufacture the cutting-edge products displayed at the booths at that conference, but he was confident that students at Chiao-Tung knew how these products worked; hence, it would not be overly challenging for them to craft a prototype in a laboratory on campus.[17]

Hsieh and Tsay also represented a unique group of young students of science and engineering at Taiwanese universities who aimed to contribute to local industry, as opposed to their fellow classmates who chose to pursue advanced degrees in the United States and find employment at big tech companies such as IBM, Bell labs, Raytheon, and Lockheed. To advance their aspiration, Hsieh and seven other students at NCTU's program in electronics organized the Abnormal Club, the acronym of which was ABC, making a deliberate choice to use the same acronym as for American-born Chinese. Abnormal Club members aimed at studying and finding employment in Taiwan instead of heading to the United States for further studies and employment. They played sports and chess as they discussed their career plans. Chew was a key member of this club. The aforementioned two recent graduates employed by Philips belonged to the club, too. Xiaofan, one of Hsieh's friends in college who later pursued a graduate degree in the United States, wrote that after witnessing the accomplishments the eight members of the club achieved in 1972, he regretted that many students chose to leave Taiwan and simply worked as "technical staff" at big companies in the United States. He used the Chinese term for "literati" to refer to the technical staff. Powerless like literati under the absolute rule of emperors, technical staff were cogs in the wheel in big companies in the United States. These overseas students from Taiwan were "sewing wedding dresses for others," a Chinese phrase indicating doing things in vain or contributing without being recognized. Xiaofan, likely a pen name, decided to return to Taiwan right before he published the essay and worked closely with Abnormal Club members, holding workshops to assist industrial engineers.[18]

Tsay shared Hsieh's ambition to explore the possibilities for turning a university lab into the cradle of an electronics industry, despite that Tsay was not in the ABC club, whose members were more senior than Tsay's class. An article published by Tsay in 1971 in the science-education magazine, *Kexue yuekan* (Science Monthly), reveals several reasons why Tsay and other participants decided to build a computer from scratch. He recalled:

> The invention and application of the electronic digital computer after WWII has significantly changed industrial production and contributed to other scientific fields. Scientific research in our country is behind [*luohou*]; our citizens [*guoren*] merely obtain knowledge of computers through books and journals, which emphasize theories and seldom mention design and architecture. National Chiao-Tung University imported an IBM 650 in 1962 and, to some extent, introduced relevant theories and practices to our country. Currently, there are about 30 electronic digital computers in Taiwan, but this is a number far behind those of advanced countries such as the United States and Japan. To implement good engineering education and scientific research, we cannot be satisfied with merely the ownership of technology and the knowledge of using technology. We have to further pursue the manufacture and the design of technology. To help the national economy, to nurture technologists for the country, and to lay the foundation for the industry, we have to research and manufacture our own and avoid relying on foreigners.[19]

Tsay argued that it was critical for Taiwan to move from a stage of owning a technology, a stage of using it knowledgeably, to a stage of manufacturing that technology. This shift involved the logic that a *country* had to own this set of capacities—owning and knowing how to use and manufacture the technology—which would eventually improve the country's technologies as well as its economy and industry.[20] Based on this set of capacities, countries could be categorized as *advanced* or *behind*. Tsay's phrase "relying on foreigners" implied two types of dependence. First, if some Taiwanese companies or individuals manufacture computers domestically, they would not only strengthen Taiwan's industrial sector but also improve its economy by exporting domestically manufactured computers, or at least by decreasing the overall expense of importing computers. Second, domestically manufactured computers would showcase the country's technological achievements.

MINICOMPUTERS IN TAIWAN

Hsieh and Tsay knew the most feasible general-purpose computer they could make was a minicomputer instead of a mainframe computer. Retrospectively, their goal of building and mass-manufacturing minis was realistic. Digital Equipment Corporation (DEC) was the first company to make minicomputers, in November 1960. DEC was also one of the most prominent companies in this industry, known as the IBM of the mini-computer industry. Its first commercially successful minicomputers were PDP-8 computers, released in 1965. Minicomputers were popular among scientists and researchers for two reasons. First, they were much cheaper than mainframe computers. Users could buy these machines instead of renting them. Their lower price allowed for installation at a greater number of offices and factories. Second, minicomputers responded to their operators more quickly than using terminals connected to a time-sharing mainframe computer. As early as 1968, *Economic Daily News*, a Taiwanese business newspaper, published two news articles discussing the increase of minicomputer manufacturers and consumers in the United States. One article pointed out that the cost of a minicomputer ranged from $10,000 US to $20,000 US, but a mainframe could cost up to $5 million US. The early 1970s witnessed a minicomputer boom. In 1972, 22,000 minicomputers were shipped worldwide. The number of minicomputer installations increased 75 percent in that year.[21] Hsieh and Tsay likely heard of minicomputers, given the PDP-8's success in the mid-1960s, or read the *Economic Daily News* article and decided the minicomputer was a much more manageable goal. While they eventually built a calculator instead of a minicomputer, their ambitions were not far behind those of US manufacturers. For example, Hewlett-Packard began making minicomputers in 1967; Wang Laboratories announced in 1969 its plan to release its first minicomputer, the Wang 3300, which debuted in March 1971.[22]

Hsieh recalled that there was no way to buy a minicomputer in Taiwan in 1968. Although Hsieh could not have predicted such a trend, by 1970 the sale and marketing of minicomputers began booming in Taiwan. DEC set up a sales and service office in Taiwan in 1973, after it had invested in building a 73,000-square-foot core-memory stringing plant a year earlier in Tachi, Taiwan, employing hundreds of workers.[23] Hewlett-Packard began promoting its minicomputers and desk calculators in Taiwan, setting up a sales office in October 1970. It donated two minicomputers to Chiao-Tung university in 1973. Students and faculty members used these two minicomputers to develop and experiment on a Chinese character input system. Wang Laboratories began to sell minicomputers in Taiwan in 1971. Sampo, a Taiwanese company, joined the market in 1972. Sanyo, a Japanese company, also made its minicomputers available in Taiwan in 1972, the price of which was $150,000 NT. National Cash Register (NCR) did so in 1974.[24] Had Hsieh and Tsay's team planned to conduct reverse engineering of a minicomputer, they might have encountered difficulties and delays in ordering a minicomputer from overseas in 1968. But even if minicomputer manufacturers had

had sales offices in Taiwan in 1968, Hsieh did not have the resources to purchase one. Hsieh could not have easily obtained or procured the necessary components in Taiwan for the minicomputer-building project. He stated that it would take a long time and a good amount of money to import and buy units such as memory cards or memory boards from US companies.

TIMELY SUPPORT FROM THE EXPORT PROCESSING ZONE

For the minicomputer building project, Hsieh and Tsay secured access to key components, integrated circuits and a magnetic core memory unit, from two factories at the newly established export processing zone in Taiwan in 1966. The first was Philco-Ford, where Chew worked, and the second was Wang Laboratories. Hsieh and Tsay obtained the components through high-level managers in Taiwan; the companies' US headquarters might not have known about it.

Philco-Ford's integrated circuit assembly plant was in the export processing zone in Kaohsiung. Taiwan, followed by several other developing countries, created export processing zones to increase foreign investment and boost employment rates in the 1960s.[25] The zone offered international companies better tax rates, discounted land leases, and access to inexpensive labor. When the Taiwanese government set up the first export processing zone in 1966 in Kaohsiung, the largest city in southern Taiwan, Philco, founded in 1892 and acquired by Ford in 1961, was one of the first thirteen companies to establish factories there. At that time, Philco-Ford made TV and radio sets for the consumer market and electronics for the US government's defense and space program. It produced the US Army's Shillelagh missile and the supply of microcircuits for the computer on the Apollo command spacecraft.[26]

In Taiwan, Philco-Ford set up two factories. The first was a plant to manufacture television sets in Tamsui. The second was Kaohsiung Electronics, located in the export processing zone. In 1966, Philco-Ford acquired General Microelectronics (GME) in the United States, and the former GME offices and personnel were probably involved with the new assembly lines set up in Kaohsiung. GME was an especially valuable addition to Philco-Ford's business, as GME was an important "training grounds for engineers and center for the diffusion of MOS technology," as historian Ross Knox Bassett has pointed out. The metal-oxide-silicon (MOS) technology was critical for making semiconductor memory and the microprocessor possible. Frank Wanlass, a key figure in pioneering the development of MOS products, had left GME to join General Instrument (GI) in December 1964. GI also set up a plant in Taiwan in 1964, though it kept its MOS technology production in the United States until 1971.[27] Kaohsiung Electronics had manufactured and assembled transistors, integrated circuits, and semiconductor products, but did not implement the MOS technology in Taiwan until the end of the 1960s.[28]

Because of Chew's knowledge of the latest products at Kaohsiung Electronics, he could informally provide integrated circuits, diode-transistor logic units, for Hsieh's teaching at NCTU and the NCTU programmable calculator. When speaking of Chew's generous offer of a suitcase of integrated circuits, Hsieh laughed and clarified that it was technically illegal for Chew to do so. Chew's employer, Philco-Ford, was permitted only to export these integrated circuits manufactured in Taiwan. If Philco-Ford wanted to sell integrated circuits to Taiwanese consumers, the company had to apply for the government's permission and be subjected to hefty import taxes.[29] Chew would not have informed his supervisors in the United States about his supply of transistors to an instructor of his alma mater.

Wang Laboratories' Taiwan factory provided Tsay and Hsieh with a customized core-memory unit to build the computer. Wang Laboratories had been making desktop electronic calculators for the US market since 1964.[30] It set up a factory to manufacture printed circuit boards for calculators in Taiwan's export processing zone in October 1967 and planned to begin assembling calculators in 1968. The factory put together four thousand printed circuit boards per month in 1968.[31] The factory soon began making magnetic core memory units in Taiwan. Magnetic core memory production was highly labor intensive, and the major reason Wang Laboratories set up plants in Taiwan was to access inexpensive labor. The Wang 700 calculator, known in the marketplace as a top-of-the-line calculator and announced in 1968, was the last successful magnetic core memory product of Wang Laboratories. The memory unit that Hsieh and Tsay obtained probably shared the same source components used in the Wang 700. Although Wang Laboratories manufactured parts in Taiwan and displayed its products at various electronics shows, it did not set up a subsidiary there to officially promote sales of finished products until August 1971.[32]

An Wang's background was a prime example of Chiao-Tung alumni's success in the United States. He obtained his BS in electrical engineering in 1940 from the pre–World War II Chiao-Tung University, located in China. He earned his PhD in physics from Harvard University in 1948. He worked at Howard Aiken's Computation Laboratory at Harvard University after graduation, tackling a major problem associated with the computer memory field. Although it was possible to store information by magnetizing a toroid or core of magnetic materials, he and other researchers were not sure how to read the information without demagnetizing the magnetic core. Wang then developed and patented "Pulse Transfer Controlling Devices," which rewrote information back into the magnetic toroid after reading the information from a magnetic core delay line. The concept was implemented in the Harvard Mark IV computer, was well-known by his contemporary researchers, and contributed to the development of magnetic core systems, whose pioneers included Jay W. Forrester at MIT. IBM later purchased the patent from Wang and manufactured magnetic core memory units for its computers from ferrite coils, instead of toroid, including the IBM 360 system, announced in 1964. Combined with his successful business in calculators after 1964, Wang's reputation and fortune soared in the later 1960s. Scientists, engineers, government officials, and journalists in Taiwan admired his achievements in manufacturing and mastery of a scientific field that was state of the art. Among Taiwan's Chiao-Tung community, he was known as an outstanding alumnus. Taiwan particularly welcomed Wang Laboratories' investment, which was culturally deemed a "Chinese" investment instead of an investment of United States or Japanese capital.[33]

Wang assigned a Wang Laboratories vice president, Yuanquan Feng, who was fluent in Mandarin Chinese, to manage the plant in Taiwan. Feng was born in Shanghai in 1930, and moved to Taiwan in 1950, but eventually graduated from Purdue University with an undergraduate degree in the 1950s, and worked for RCA, Sylvania, and Admiral in the United States. He then set up a factory in Hong Kong for Fabri-Tek to make magnetic core memory, before joining Wang Laboratories. Feng was thirty-eight years old in 1968 when Hsieh and Tsay began the minicomputer-building project. Shortly after Tsay and Hsieh's group began to meet, Feng was invited to the group meetings at Chiao-Tung campus. He was probably instrumental in persuading An Wang to donate a Wang Laboratories model 360 desk calculator to the university, worth $3,300 US. Feng offered his expertise, participating in the periodic meetings, and eventually provided the project with a customized core memory unit. Hsieh and Feng became acquainted as time went by. Hsieh noted that when he flew to the United States to begin his stint as a visiting graduate student at MIT in 1969, Feng picked him up from the airport in Boston. Feng was likely visiting Wang Laboratories' headquarters close to Route 128 in Massachusetts. Hsieh recalled that they were stopped by a police car as Feng inattentively made a right turn on red. Hsieh told the story to illustrate his friendship with Feng, but it also speaks to Feng's willingness to offer help to Chiao-Tung graduate students.[34]

In contrast to his enthusiastic support for Hsieh and Tsay's computer building project, Feng wrote an article for *Economic Daily News* on New Year's Day in 1970, stating that it was unnecessary to mass-manufacture computers in Taiwan at that time. He noted that "the degree of industrialization in Taiwan" was not advanced enough to support the technical know-how necessary for manufacturing computers. It would be more economical to import computers than manufacture them in Taiwan. Given that the use of computers in Taiwan prevailed among large companies, universities, and several government offices, Feng offered four recommendations regarding computer use in Taiwan: First, the government should consider training more personnel on how to operate computers. While there were at least two dozen mainframe computers installed in Taiwan, they seemed underutilized because of a lack of sufficiently knowledgeable staff members to incorporate computers into their work. Second, Taiwanese organizations and government offices should purchase smaller computers instead of large mainframe computers, which were expensive and underutilized across Taiwan. Third, the government should lower the import tax for computers (around 30 percent at that time) to encourage more purchases and wider use. Fourth, the government should welcome more "foreign" computer manufacturers to set up plants in Taiwan. These plants would then employ local engineers, who would then learn how to manufacture computers. Feng's role as a manager of an international computer manufacturer allowed him to develop an analysis that emphasized cost and profit, and thus advocated for established US computer makers. He did not hesitate to promote the import of minicomputers from overseas. While Feng believed Taiwanese could obtain know-how from plants set up by international computer manufacturers, graduate students Hsieh and Tsay, in contrast, were keen to gain expertise for computer manufacturing in a university setting in Taiwan.[35] Although Feng did not think that mass-manufacturing computers in Taiwan was an urgent task, he never hesitated to support Hsieh and Tsay's project.

FROM OWNING TO BUILDING COMPUTERS

To build a minicomputer, Hsieh organized periodic meetings with Tsay, Yuanquan Feng, visiting professor Chao-Chih Yang, and fourteen graduate students at NCTU to study computer infrastructure and brainstorm on its design and execution. Three technicians eventually also participated in the construction of the computer.[36] In contrast to Feng's firsthand experience in hardware manufacturing, Yang's specialty was memory allocation in computers. Yang was a Chiao-Tung alumnus, earning his master's degree from NCTU's program in electronics in 1962. Yang's undergraduate degree was from the Naval College of Engineering in Taiwan. After returning to Taiwan from Northwestern University, where he received his doctorate, in 1971 he wrote a book discussing the status of computerization in Taiwan and translated a book the same year about artificial intelligence authored by Donald G. Fink, who was the first executive director of the Institute of Electrical and Electronics Engineers (IEEE) in 1963. Yang showed his enthusiasm for educating the public about novel computer-related knowledge and technology in Chinese.[37]

According to Hsieh, the group read journal articles published by the Association for Computing Machinery (ACM) and the IEEE. It also consulted internal technical documents from companies such as IBM, Bell Laboratories, and AMD, acquired through informal personal connections. For example, Tsay cited an article published in 1965 in *Fairchild Application Bulletin*, for his article on the design of the memory unit published in 1969. Another corporate publication that Tsay cited in his publications was *Computer-Circuitry Considerations*, from Philips' Electronic Tube Division. Chiao-Tung alumni hired by electronics manufacturers provided either internal technical documents or latest research articles to minicomputer-building project members. Philips' documents were likely from Chiao-Tung alumni who worked for Philips' plant in Taiwan, established in the export processing zone in 1966. Feng, at Wang Laboratories, shared technical documents with the minicomputer-building group, too.[38]

Hsieh undertook the logic design of the minicomputer from the beginning. He noted that, while he had briefly considered vacuum tubes as the basis of the switching circuits, he made use of transistors to develop a preliminary plan. Fortunately, the Chiao-Tung group could use the integrated circuits brought by Chew at Kaohsiung Electronics to make the final product. Hsieh noted that these integrated circuits saved the group time in completing the logic design of the calculator.[39]

Hsieh left NCTU from 1969 to 1971 for MIT to work on his dissertation, creating the opportunity for Tsay to take over the leadership of the computer-building project. Hsieh's visit to MIT was arranged by electrical engineering professor Lan Jen Chu, an alumnus of the pre–World War II Chiao-Tung University in China who frequently visited the National Chiao-Tung University after it resumed operation in Taiwan in 1958 (chapter 1). Hsieh studied at MIT with Chung Laung (Dave) Liu and Francis Fan Lee, both of whom held PhDs from MIT, became faculty members at the Department of Electrical Engineering there, and participated in Project MAC, which allowed computer users from various locations to connect to a time-sharing system. Lee taught at MIT from 1965 to 1987. Liu taught at MIT from 1962 to 1972 and then joined the University of Illinois at Urbana-Champaign as a professor of computer science. After he retired, in 1998, from the associate provost position at the University of Illinois, he returned to Taiwan to become president of the National Tsing Hua University.[40]

During Hsieh's time at MIT, Tsay played a leading role in the design and production of memory units in Taiwan. Having obtained magnetic cores from Feng, Tsay planned to construct a three-dimensional, $12 \times 13 \times 4$, 50-mil ferrite-core memory. In the beginning, Tsay wanted to manufacture the core memory on the NCTU campus, but he soon encountered challenges. Both Tsay and Hsieh noted that no one on the team was able to wire lines, the diameter of which was one thousandth (10^{-3}) of an inch, through magnetic cores. Each of the 624 magnetic cores was supposed to be threaded by four lines (figure 5.2), and the scale of each magnetic core and line was too small to manipulate without factory facilities, such as those at Wang Laboratories. Feng then generously agreed to manufacture a magnetic core memory consisting of 1,248 bits for the project (figure 5.3).[41]

Wang Laboratories' Taiwanese factory workers, who were likely mostly women laborers, did the actual customized manufacturing for Tsay. The assembly of magnetic core memory was extremely labor-intensive and required good eyesight. An Wang noted in his autobiography that "the work required people with a good deal of experience at fine, detail work." He added that "Taiwan has an abundance of people well suited to this type of job." The majority of the workers in such assembly lines were women, in the United States and in Taiwan. According to scholar Daniela K. Rosner and her colleagues, women were hired in Waltham, Massachusetts, for threading and weaving the magnetic core

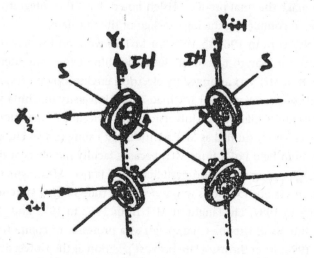

Fig. 2 Driving direction of Cores (the direction indicated is as Fig. 1.)

FIGURE 5.2
Threaded magnetic cores. *Source:* Jong-Chuang Tsay, "Design and Experimentation of a 4X4X2 Memory," a 1969 research report submitted to the National Science Council in Taiwan.

FIGURE 5.3
Magnetic core memory of the NCTU minicomputer. *Source:* Jong-Chuang Tsay, "Woguo diyibu
xiaoxing dianzi jisuanji jianjie jiaotong daxue shouzhi chenggong" [National Chiao-Tung University's success in building the first domestic electronic digital computer], *Kexue yuekan* 21 (September 1971): 49. Reprinted with permission from *Kexue yuekan*.

memory for the Apollo Guidance Computer (AGC). The design of AGC began in 1961.
The Raytheon Corporation, as a contractor, subsequently manufactured memory core
units for the computer. The director of the AGC project, Richard Battin, recalled that
engineers at that time referred to the women workers as "Little Old Ladies," and hence
the manufacturing method as "the LOL method." Two of Rosner's coauthored publications feature a portrait of a African American woman, employed by Raytheon, focused
on wiring ferrite cores while sitting by a desk with bright light to enhance her eyesight.

According to a United Nations' report surveying Taiwan's electronics industry in the
fall of 1967, plants in Taiwan employed 452 Taiwanese workers to work on magnetic
core assembly lines, though the survey was taken before Wang Laboratories' factory
began to operate in Taiwan. Taiwan provided a high-quality labor force, according to
Frank Roy Gustavson, a UN expert, who stated, "Example, a 17 to 20 year old girl has
optimum eye sight to thread 0.07 mm wire in 20 mil diameter (OD) memory core."[42]

As Wang Laboratories' manager in Taiwan, Feng faced a dilemma similar to Chew's. He first offered individual magnetic cores because, considering Wang Laboratories' best interest, it was inappropriate to give away a finished core memory, and there were regulations in the export processing zones. Knowing the difficulties Tsay encountered in threading the cores, Feng probably decided to request that an engineer and factory workers, likely women, customize the memory unit for Tsay. He may have developed a strong attachment to the project and Chiao-Tung students and, like Chew, simply decided to ignore any legal regulation. Hsieh noted that, had he had more resources or envisioned the challenges early on, he might have obtained semiconductor memory units instead.[43]

After installing the memory unit, Tsay further redesigned the logic unit with two other graduate students. What daunted them in the project's final year was the wiring. Because of a lack of wiring experience, their efforts at connecting wires among the parts were not organized enough for testing (debugging) and practical operation. He worked alone on the wiring starting in 1970, after most participants became busy working on their own theses for graduation. There were more than a hundred output wires in the control unit alone. These wires were not marked with numbers or codes, and some were soldered to other parts of the calculator, making it difficult to identify problems when the machine malfunctioned. After months of testing and identifying problems, he rewired the entire calculator, which took another three months.[44] In an article for *Kexue yuekan*, Tsay wrote about the computer-building project, providing four pictures to emphasize how complicated the computer wiring was (figure 5.4).

In the same article, Tsay twice brought up that this project provided him and other group members opportunities to learn things that they could not have gotten from books. In detailing the process of organizing wires, he specified that the noisy signal and poor connection were the most difficult problems to identify, and that such problems could not be solved by looking up information in books. He also recognized the valuable assistance from the three technicians who participated in this project. He pointed out that the project trained participants to a great extent: "Because of this project, participants have obtained practical knowledge and experiences that cannot be gained by merely reading books. The project has displayed its multiple effects. When UCLA electrical engineering professor Ernest Shiu-Jen Kuh visited last year, he commented that this project demonstrated the most practical and effective training method."[45]

Tsay emphasized that the knowledge and practices that were not included in books constituted an important part of the minicomputer-building project. This is what science and technology scholars have referred to as tacit knowledge.[46] Taiwnese

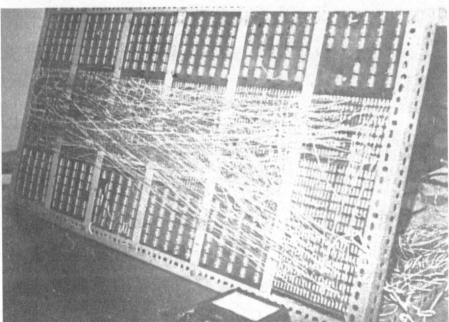

FIGURE 5.4
Wires of the NCTU minicomputer. *Source*: Tsay, "Woguo diyibu xiaoxing dianzi jisuanji," 53.
Reprinted with permission from *Kexue yuekan*.

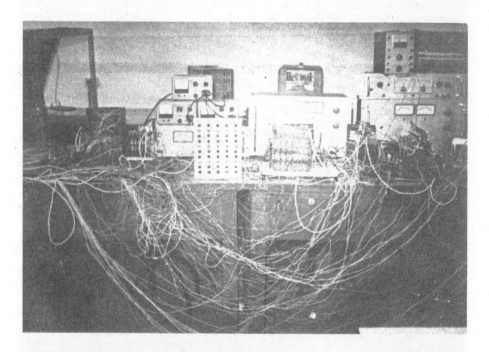

FIGURE 5.4 (*continued*)

technologists knew that tacit knowledge was critical in cultivating future engineers. Chien and Lee sought to enhance educational resources to train "talented youth" who would work with visiting experts to gain hands-on experiences with imported artifacts and might eventually become industrial engineers. In contrast, Feng advocated for training industrial engineers in manufacturing plants. Tsay's approach was a middle ground, emphasizing nurturing more native technologists while carrying out research plans, as his project had already done:

> The logic design and circuit design of this computer was done by our citizens [*guoren*]. The process of building this computer has helped members at NCTU be confident about the manufacturing of computer hardware on our own. From now on, we should continue to follow the spirit we had when our country started the research and development of expertise on electronic computers. We should embark on larger research plans to train more technologists for our country.[47]

In 1975, Tsay received his PhD degree and was hired as an associate professor in the Department of Computer Engineering at NCTU after a two-year stint at the Center for Informatics Research at the University of Florida. In Florida, he worked with Julius T. Tou on a computer program to typeset Chinese-language newspapers, which he was committed to from 1973 until the mid-1980s. Tou had graduated from the pre–World War II Chiao-Tung University in Shanghai in 1947 and obtained a PhD degree in electrical engineering from Yale University in 1952. He was nominated to be an academician to the Academia Sinica in Taiwan in the 1970s and led the establishment of the Institute of Information Science there. In the 1970s, Tsay began a new project to build a programmable robot.[48]

Hsieh became an associate professor and acting director of the Department of Computer and Control Engineering at NCTU after returning from MIT to Taiwan in 1971, and he received his doctoral degree in 1972. Part of his research at MIT under Chung Laung (Dave) Liu's supervision was published in 1972, focusing on periodic sequences, which could be applied to fields such as "coding and decoding implementation, data processing, numerical analysis, data compression, etc." Like Tsay, Hsieh began shifting his research in the 1970s to investigating how computers systems could process Chinese characters, which was a pressing issue for scientists and engineers in Taiwan at the time. Information processing in the Chinese language was a long-term commitment in Hsieh's career. Hsieh left NCTU to join the Institute of Information Science at Academia Sinica in 1983. While he was the director of the Computing Center and built up the computer network for the campus of Academia Sinica from 1984 to 1990, he also pioneered, as early as 1985, digitizing a series of twenty-five published Chinese historical books by the courts from the Han dynasty to the Qing dynasty, an early foray into the digital humanities.[49]

Among the fourteen graduate students who participated in the minicomputer-building project, several of them worked for Wang Laboratories because of either Feng's recruitment or the students' growing interest in the company as the project evolved. Some ended up working in Wang Laboratories' US headquarters. Hsieh amusingly noted that, in return for the customized magnetic core memory, he and Tsay paid back Wang Laboratories by sending a group of excellent engineers there.[50]

Given that the minicomputer-building project was not entirely fulfilled, and the UN financial and technical support proposed in 1964 remained unaccomplished, Chiao-Tung University continued expanding its computing power by forming physical computer networks with other computers in Taiwan. In April 1971, Chao-Chih Yang, the director of the newly formed Department of Computer and Control Engineering, reported that the Chiao-Tung would install IBM's 2770 data communication system to connect two other computers located in Taipei. Chiao-Tung users in Hsinchu would be able to remotely use the two Taipei computers through dedicated telephone lines. The first connection was to an IBM 1130 computer at the Research Institute of the Directorate General of Telecommunications. This computer supported remote batch processing. The second connection was to an IBM 360/40 at the Executive Yuan (equivalent to a cabinet or council of ministers), supporting remote job entries. Furthermore, Yang declared that Chiao-Tung would purchase a Hewlett-Packard computer for the university's telephone switch system.[51] Hewlett-Packard also donated two minicomputers to Chiao-Tung in 1973 for a research project on developing a Chinese character input system. The computing power at Chiao-Tung was impressive and comparable to most universities in other countries at that time.

CELEBRATING THE COMPLETION

Soon after its completion, NCTU publicized the programmable calculator and the university's intention to manufacture computers on an industrial scale. On July 17, 1971, *Economic Daily News* published a story on this first "domestically manufactured" "computer." It would have been much more precise if Chiao-Tung members and journalists had referred to it as electronic digital calculator. But in Taiwan at the time, people referred to a computer as *diannao*, for electronic brains, or *dianzi jisuanji*, for electronic computer and electronic calculator—both terms broadly referred to technologies ranging from IBM's general-purpose electronic computers to desk calculators. *Economic Daily News* proclaimed that NCTU's "first domestically manufactured computer" was a "third-generation computer [that] uses integrated circuits, as the IBM 1130 computer does."[52] It was an overstatement, as its computing power was lower than that of the IBM computer.

Despite the overstatement, Chu-Yi Chang, dean of studies at NCTU, was optimistic about upscaling NCTU's minicomputer manufacturing to an industrial level. Several government agencies, including the National Science Council and the Ministries of Economic Affairs and Education, considered future collaboration with NCTU to test the possibility of developing mass-production capability for computer-manufacturing. They would explore ways "to mass-manufacture different types of computers with reasonable prices for both domestic and international markets." Chang was especially excited about the fact that the NCTU product cost $500,000 NT ($16,000 US), lower than the price of purchasing one from the foreign computer companies—$800,000 NT ($26,000 US). The news article also claimed that Tsay expected the university to begin to work on the manufacturing of commercial minicomputers in the following years.[53]

But the aspiration of Chang, Tsay, and the *Economic Daily News* journalist was not realized. In an 2007 interview, Tsay shared that he had not emphasized to interested investors his "confidence" in the realization of the mass-manufacturing of the NCTU machine because he knew that "advanced" computers such as the IBM 360 were already available on the market in 1971.[54] Moreover, if government officials had seriously evaluated the mass-manufacturing of minicomputers, they would have taken into consideration that the integrated circuits and magnetic core memory of the NCTU computer were gifted from Philco-Ford and Wang Laboratories, respectively. Knowledge of the origins of these parts would have made the project sound less impressive to these officials. Officials would have had to consider whether NCTU or an allied company would generate profits manufacturing minicomputers, if the company had to buy key parts from other suppliers. Chang was nonetheless confident in 1971 that setting up a pilot plant exploring minicomputer manufacturing would lead to a thriving Taiwanese economy.[55] First, Chang believed that the linear trajectory of building an industry included research in universities, setting up pilot plants, and establishing an assembly line. Chang's idea of a pilot plant was connected to his colleagues S. M. Lee and Gisson Chien's 1964 proposal, in which Chiao-Tung would set up a pilot plant on campus to make electron tubes and transistors. Second, Chang firmly believed that the demand for computers would increase rapidly over the years. A domestic computer manufacturing industry would be useful for the thriving economy of Taiwan.

Though the mass-manufacturing of minicomputers did not occur in Taiwan, the minicomputer-building project demonstrates developments following NCTU's acquisition of electronic digital computers and the ending of the technical-aid program established at NCTU in the mid-1960s. To Tsay, the minicomputer-building project was an extension of previous efforts in which NCTU members learned to understand and use their first two computers. Believing that a country had to own its set of capacities—owning and knowing how to use and manufacture the technology—his minicomputer-building project fulfilled the goal.[56] Furthermore, Tsay emphasized that "the logic design and circuit design of this computer was done by our citizens [*guoren*]."[57] In an article published in the *Voice of Chiao-Tung Alumni*, in 1972, Hsieh also explained how the computer was designed and manufactured by the team, instead of having purchased components from abroad.[58] But Wang Laboratories and Kaohsiung Electronics, companies that were owned primarily by non-Taiwanese investors, had contributed critical components to build the NCTU minicomputer. This case demonstrates the development of a discourse of self-reliance and autonomy.

The celebrated programmable calculator was disassembled at a later date, because another student or scholar at Chiao-Tung needed the embedded integrated circuits for other experiments, according to Tsay.[59] It affirms that the components of the NCTU machine were a scarce resource. But the spirit of building computers flourished. Hsieh and Tsay's interest in building computing devices domestically was also shared by two other student groups a year after Hsieh and Tsay began their project. Hsieh complained in 1972 that recent media coverage on the computers built at NTU and National Cheng-Kung University had misled the public, leading to the belief that both were built earlier than the one completed at Chiao-Tung. Hsieh emphasized that that was not the fact, and that all parts of the NCTU "minicomputer" were domestically made.[60]

CONCURRENT CAMPUS-BASED PROJECTS TO BUILD COMPUTING DEVICES

On January 25, 1972, a few weeks before the lunar new year and six months after Tsay completed the programmable calculator at Chiao-Tung, *Economic Daily News* revealed that electrical engineering professor Benyuan Huang had led undergraduate students, including a senior named Wang Tang, to build a calculator at the National Cheng-Kung University. The calculator was named "Cheng-Kung I," which sounded auspicious, as "Cheng-Kung" literally means "success" in Chinese. Huang's project had begun in 1969, only a year after Hsieh and Tsay's project. Huang's research was funded by the National Science Council, as Hsieh and Tsay's project was. The calculator was able to process addition, subtraction, multiplication, and division for figures up to sixteen digits. Huang emphasized that the computing speed of the calculator was impressive because it used large-scale integration (LSI) circuits. The calculator would take only eighty microseconds to execute an addition. Perhaps because Huang had read news articles about Tsay's "third-generation" computer, he emphasized that his calculator was one of the "fourth-generation" computing devices. Huang noted that third-generation computers relied on solid-state integrated circuits, but the fourth-generation computers adopted LSI circuits, as the latter was an improvement over the former and exhibited stronger computing power.[61]

The news article gave the impression that the professor was the major contributor to the project. But Tang, then an undergraduate senior, has recently written that he actively participated in that project. Tang explained in a 2020 article in the newsletter of an overseas Taiwanese association in San Diego, California, how he joined the calculator project.[62] As a senior at Cheng-Kung in September 1971, Tang had completed all required credits and had no more courses to take. He obtained four integrated circuits specifically for calculators and decided to ask Professor Huang to be his adviser in a project building a calculator in Huang's laboratory. He noted only that he fortuitously collected the integrated circuits without specifying where he had gotten them. Tang's interest originated in his hobby making audio amplifiers. His friend, Muyuan Chen, who later founded the large electronics manufacturer Yageo Corporation, taught him how to make printed circuit boards at home.[63] Around the Cheng-Kung campus, Tang persuaded other students to join him in making homemade audio amplifiers with homemade printed circuit boards. Many of Tang's friends picked up the hobby. Shengqiang Dai, one of these hobbyists, founded a company called CTT, Inc., in San Jose, California, and has been making amplifiers since 1981.[64]

Tang also secured components such as a power supply and a device to display numbers for the calculator. But he had a hard time finding reed relays. Fortuitously, Tang found ten reed relays when the campus computer center gave away items they found no longer useful. Tang then completed the building of the calculator by the end of 1971, and the university made it public on January 24, 1972. The next day, when the news story about the Cheng-Kung I calculator was published in *United Daily News*, the minister of education summoned Tang and arranged for him to fly from Kaohsiung to meet him in Taipei that same day. Upon reading the news article, an entrepreneur even contacted *United Daily News* excitedly to express his interest in offering funds to mass-manufacture the calculator, though that business plan was not realized. For his final semester at Cheng-Kung, the first half of 1972, Tang interned at the military research institute, CSIST, where he had an office desk at the computer center, in which a CDC 3300 computer was installed (see chapter 4). For his internship, he wrote a thesis about the calculator he built. After fulfilling his mandatory military service, he went to study in the United States in May 1974 and obtained his master's degree and PhD from the University of Texas, El Paso, and Iowa State University, respectively. His first job after graduation was working at Analytic Sciences Corporation, a Boston-area research contractor for the defense industry.[65]

Neither Tang nor Huang specified where Tang had procured the integrated circuits for the calculator in 1972, but Huang did secure sponsorship for his research from

Kaohsiung Electronics the next year. After Tang left Cheng-Kung University, Huang continued to work on designing a more powerful calculator as well as developing a computer system to control and optimize the operation of a power plant in 1973. For these new projects, Huang secured circuits of medium-scale integration (MSI), large-scale integration (LSI), and extra-large-scale integration (ELSI) from Kaohsiung Electronics. Unlike Chew's informal support for Hsieh based on their alumni connection, the sponsorship from Kaohsiung Electronics was likely a formally arranged collaboration. As early as March 1970, Kaohsiung Electronics was already committed to working with the National Science Council to explore the manufacturing of general-purpose computers to aid in industrial development in Taiwan.[66]

Three months after the story of the Cheng-Kung I calculator was covered in newspapers, NTU also announced that several members of its electrical engineering department were going to complete a "small-scale" computer in a few months. The participants included professor Sing Chang (Xing Chang), PhD student Albert Lau (Qiyi Liu), and master's students Barry Pak-Lee Lam (Baili Lin), Sayling Wen (Shiren Wen), and Jicheng (James) Sun. Chang also began the project in 1969, as Huang started his. But the actual design and construction of the minicomputer seemed to begin only in spring 1971, after Chang invited a group of students to work together in his solid-state electronic laboratory. Lam was the chief designer of the machine. He was born in Shanghai in 1949 but moved to Hong Kong at the age of one. He went to Taiwan for his undergraduate and graduate education. In his master's thesis he noted that he found "the job hard but interesting." He especially noted that Wen and Sun participated in helping him "build the machine." The computer was named "The National Taiwan University Experimental Computer—NTUEC-1000." However, when Lam submitted his thesis in June 1972, the computer was not finished and was scheduled to be completed only that summer. Unlike the NCTU programmable calculator's decimal system, the completed NTU's minicomputer used the binary system.[67]

When he was a student at NTU, Barry Pak-Lee Lam had heard about minicomputer manufacturers, such as the Digital Equipment Corporation (DEC). With Professor Chang's encouragement, he decided to try building an experimental minicomputer prototype. He then secured integrated circuits and other components from used electronics markets. He used transistor–transistor logic (TTL) integrated circuits for the logic circuits, a marked improvement over the diode–transistor logic units used in Chiao-Tung's programmable calculator. Lam had access to technical documents as Tsay did—he cited *Designing with TTL Integrated Circuits* from McGraw-Hill's Texas Instruments electronics series, published in 1971. In contrast to the used items, the magnetic core memory unit of Lam's computer was brand-new and extremely expensive. Zhao

Xu, head of the electrical engineering department at NTU, was enthusiastic about the computer-building project. Xu reached out to his contacts to look for a memory unit in Hong Kong or the United States for this project. Eventually he spent $3,000 US to procure an RCA-manufactured memory unit of 4K storage. At the time, that level of spending for the memory was a huge investment for an electrical engineering department.[68]

The completed minicomputer became a pride of the university and a springboard for other creations. Lam recalled that the minicomputer, locked in a display cabinet, was shown to special guests. Since it was stored in a cabinet, the computer was never used to execute computations; in the 1972 newspaper coverage of the computer, Lam noted that its purpose was only for research and teaching. It lacked "characteristics" that could be mass-manufactured on assembly lines, but Lam was keen to find ways to scale up for mass-manufacturing. He and his best friend, Sayling Wen, did achieve that goal and became billionaires for doing so, though it was more than a decade later. After learning about Wen's participation in the computer-building project, an entrepreneur invited Wen, in 1972, to manufacture calculators at the entrepreneur's start-up, Santron. Lam also joined Santron immediately, though Lam and Wen left the start-up and, in 1973, founded Kinpo Electronics to design and manufacture handheld calculators. Lam founded Quanta Computer to manufacture notebook computers in 1988. Wen later became a vice president of Inventec, which has been manufacturing notebook computers since the 1990s.[69]

Lam was asked in 2011 whether he began his design of the computer from scratch or copied an existing computer design from a company such as DEC; Lam answered, "We copied it, but our structure was still simpler." It's unclear which computer, exactly, Lam meant to imitate. DEC built a plant in Taipei only in 1972 and set up a sales office in Taipei only in 1973. Lam likely had access to some products from Hewlett-Packard, which had set up a sales office in Taiwan in October 1970. Wang Laboratories, well-known in Taiwan, began its sales business in Taiwan in August 1971. Or Lam might have referred to a corporation's technical documents through personal connections. Lam cited a document from Hewlett-Packard, "2116 Computer Maintenance Course," in his master's thesis. The 2116 model was the first minicomputer designed by Hewlett-Packard in 1966. It also marked the company's first use of integrated circuits. The document was likely acquired by Lam or an NTU member through participating in courses offered by Hewlett-Packard in Taipei, perhaps a three-week Hewlett-Packard workshop in Taipei in August 1971. Covering both the software and hardware of the company's 2114B computer, the workshop was taught by Ernie Poblacion, who flew to Taiwan for this workshop. The Hewlett-Packard 2114B computer, a "cost-reduced version" of the 2116A model, was released in 1968.[70]

The three projects of computer- or calculator-building reflected the strong interest in crafting a computer shared by students and scholars across university campuses in Taiwan. They witnessed on campus and heard of in classrooms the powerful computers manufactured by IBM, Control Data, and DEC, Wang Laboratories, and Hewlett-Packard and devoted themselves to making similar technologies on their own. Their curiosity motivated these projects, tinkering with and opening the black box of a sophisticated computer. Students of all three projects also showed ingenuity when it came to finding materials, skills, and information to work with. They also successfully built a supportive environment in their respective universities to gain support from alumni, professors, and administrators.

Students of all three projects also further attracted financial investment or stimulated continued research in transforming the student projects to industrial projects. The dean of studies at Chiao-Tung University was proud and hoped to explore opportunities to work with industry. Government officials approached Tsay to see if they could bring Tsay's project to an industrial level of production. The head of the Electrical Engineering Department at National Taiwan University generously allocated his available funds to purchase a rather expensive memory unit for building a computer. Professor Huang at Cheng-Kung University further worked on developing computer systems and calculators for industry after his brilliant student Tang graduated. An entrepreneur expressed interest in investing in mass-manufacturing the Cheng-Kung I calculator, and another entrepreneur invited Wen and Lam to work on a start-up making calculators. These students and their sponsors' view of the future of the domestic electronics industry was promising and optimistic. Although the technologies they manufactured on campus did not show them an immediate path to mass-manufacture general-purpose computers in Taiwan at that time, participants played important roles in the later development of computer science education and computer manufacture in Taiwan.

VARIED PATHS TO A DESIRED ELECTRONICS INDUSTRY

University laboratories and international and local companies all held differing expectations of the electronics industry in Taiwan and made different technological products to realize their goals. University laboratories envisioned promising plans and experimented with limited resources to build computers from scratch. Plants in the export processing zone and beyond supplied low-labor-cost computer memory units and semiconductor components for calculators, computers, and various products globally. UN experts in the late 1960s visited these plants and offered their evaluation about how to nurture the local electronics industry, given the growing presence of multinational manufacturers' plants.

Recall that Lee and Chien had contacted the United Nations Special Fund office to explore setting up an "Applied Research Center in Electronics Technology" at Chiao-Tung University in 1964. But that plan was never realized, as administrators of the United Nations' Special Fund program were more focused on the status of the electronics industry in Taiwan. Resident representative and director of the Special Fund Program Knut Winter then suggested that Taiwanese applicants consider sending the proposal to the Special Industry Services Program under the United Nations' Industrial Development Organization (UNIDO), which would review requests for short-term assistance that did not fit into the current available programs.[71] Lee and Chien did, and UNIDO sent two experts in 1967 and 1970 to review the electronics industry in Taiwan. As it turned out, the two expert visits *were* the technical assistance offered from the Special Industry Services Program to Taiwan.

The first UN expert, Frank Roy Gustavson from the United States, stayed in Taiwan from August to October 1967 "to assess the need for an electronics industry for engineering, management, and probably also for training." While Gustavson expressed strong support for Lee and Chien's proposal, no archival document suggests the United Nations helped establish a new center on Chiao-Tung campus to aid the electronics industry. Gustavson was a successful engineer-turned-entrepreneur and had worked as a UN expert for the International Atomic Energy Agency (IAEA) in Vienna. He was a UN expert in Seoul from November 1965 until he visited Taiwan in August 1967.[72] The second UN expert, S. Skoumal from the United Kingdom, paid a two-month visit to Taiwan as an adviser on the development of the electronics industry in 1970. His first name was not recorded in surviving documents. Skoumal was a manager of Marconi Company in Chelmsford, Essex, right before he headed to Taiwan.[73]

Gustavson visited thirty-five corporations, interviewed several government officials, including K. T. Li, then minister of economic affairs, and met managers of foreign and domestic companies. In his report, he focused on the thirty electronics companies with more than a hundred employees. Among these large companies, thirteen were locally owned, three were joint ventures, and fourteen were foreign owned; the annual sales in these three categories were $16 million, $3.6 million, and $152 million US, respectively. Philco-Ford and GI fell into the third category. Many of these foreign-owned plants were in the export processing zone in Kaohsiung. Gustavson noted that the major products of all electronics companies were components or finished products for TV and radio sets. But Gustavson also specified that Tatung, a prominent local company, was making magnetic core memory. He did not note that Tatung was making them for IBM. Overall, 452 of the 9,500 factory workers of the surveyed companies were assembling magnetic core memory units. Gustavson did not include Wang Laboratories in his survey because the company had just set up its plant to make printed circuit boards sometime in October, and Gustavson completed his report on October 23.[74]

Of the thirteen local companies and three joint ventures, Gustavson surveyed, eleven were assembly plants for radios and TV sets for the domestic market. Eight domestic companies were making capacitors, resistors, and speakers, which he referred to as "cheap and dirty" components, as opposed to components of "consumer quality" and of "industrial and military quality." He believed that with additional technical knowledge, management, and financial resources, domestic companies making consumer products could significantly improve their quality. While he noted that no company supplied industrial and military quality components, he believed that the market could be cultivated by setting up an industrial research park in Hsinchu. The industrial research park was a proposal of K. T. Li's, whose idea was to encourage foreign companies to set up offices there and hire local engineers to conduct research and development.[75]

Gustavson's report called for more UN support to assist Taiwan in rapidly developing its electronics industry. He believed that domestic companies, which supplied only 2 percent of electronics to foreign-owned companies in Taiwan, should consider tapping into this growing market of electronics by supplying more components to foreign-owned plants if the latter could source them locally instead of internationally. Advocating for Lee and Chien, Gustavson believed that their proposal of the Applied Research Center in Electronics Technology would help domestic companies enhance the "quantity, quality, [and] service" of their products sold to foreign companies. He recommended the United Nations offer its immediate support to Taiwan.[76]

During Skoumal's visit to Taiwan in 1970, he also met with government officials and managers of both domestic and "expatriate" companies. Skoumal noticed that some Taiwanese he talked to seemed to consider expatriate companies a part of the domestic electronics industry. He believed that this was an inaccurate categorization, even though he agreed that the expatriate companies offered extremely valuable "on-the-job training in up-to-date techniques and methods" to local engineers (further discussed in chapter 6). Recall that Wang Laboratories' Taiwan manager Yuanquan Feng also emphasized that Taiwanese engineers could have learned much at factories owned by foreign computer manufacturers.[77]

Like Gustavson, Skoumal believed that the home industries in Taiwan should increase their supply of components and other services to the expatriate industries. Arriving three years after Gustavson, Skoumal witnessed the even more dramatic growth of expatriate electronics companies in Taiwan: these companies consumed $60 to $70 million US worth of components and materials in 1969, reflecting growth of 70 percent from the preceding year. But still about 10 percent of components were imported, because Taiwanese companies could not meet the quality desired or did not attempt to offer the supply. He suggested that the Taiwanese companies should tap into that market. Among the statistical data about the 118 domestic companies he had access to, half of the companies employed fewer than fifty people, and 5 percent employed more than five hundred.

Both Gustavson and Skoumal saw a shortage of engineers in the electronics industry in Taiwan. According to Gustavson's 1967 report, 304 electrical engineers worked for the thirty-five companies he surveyed, only six of whom held master's degrees. By September 1967, in their plants in Taiwan, GI had hired a hundred engineers with bachelor's degrees in EE, among its 2,800 employees; Philco's television plant hired fifty among 1,050; and Philco's Kaohsiung Electronics hired sixteen among five hundred. Additionally, about fifty foreign engineers were working in the thirty-five companies, mainly on short-term assignments. Besides engineers, 9,500 workers were employed in these companies. However, in 1967, Taiwanese universities graduated about 288 electrical engineers with undergraduate degrees, and about half left the country to pursue advanced degrees in the United States. Among the remaining 150 engineers, fifty chose to work in "power stations, telephone communications, instrumentation in textile mills, etc."; one hundred chose the electronics industry. Gustavson estimated the growth of the overall electronics industry would be 30 percent per year, and that in five years, the industry would require 1,300 engineers. Should electronics become more sophisticated, the number of engineers needed might have to increase to two thousand, which would require four hundred engineers to join the labor market every year, a number greater than the universities in Taiwan could supply at that moment.[78]

Skoumal also perceived that in 1970 Taiwan lacked a sufficient number of engineers for the local electronics industry. He believed that local companies should consider the employment of another seven hundred production engineers to improve product quality and two hundred R&D engineers to improve design capability. He found that most local companies focused on assembly and did not hire design engineers. The profit margins earned by local companies, about 20 percent of the sale price of their products, did not get redistributed for R&D investment or sales and administration costs. Skoumal believed that the local companies should increase their investment in the two fields to meet future challenges. For example, he noticed that, among the many joint-venture companies, those jointly owned by Japanese tended to depend on Japanese designs. The dependence would make these companies uncompetitive and unsustainable.[79]

Gustavson advocated for allocating more resources to Chiao-Tung's computing center. He reviewed the computer time log and found that, from 1966 to 1967, the IBM 1620 computer was up for ten hours every day to support campus research activities, government and industrial research projects, and training courses; the three types of activities accounted for 70, 23, and 7 percent, respectively, of overall computer time. Gustavson recommended that the United Nations consider bringing a new and more powerful computer to Chiao-Tung. His recommendation was even more ambitious than that of Chiao-Tung's dean of studies and IBM's Taipei office; the former only looked for funding to rent more peripherals and the latter suggested leasing a card reader to increase the utility value eight to ten times. The recommended assistance from the United Nations never arrived, but, as mentioned, Chiao-Tung's computer center was going to connect to two additional IBM computers installed in government offices in 1971, which would greatly expand Chiao-Tung's computing power.[80]

Gustavson further specified fourteen projects that the proposed Applied Research Center for Electronic Technology at the NCTU could begin to work on in a year. He argued that the proposed center should explore the manufacturing of printed circuit boards, silicon planar transistors, composition resistors, ferrite production for IF coils for radio, and television receivers. International manufacturers' Taiwan plants needed these products. For example, Taiwan supplied 60 percent of IF coils to the world market. Three hundred tons of ferrite was imported to Taiwan to make IF coils every year. Two local companies, Pacific Wire and Cable and Tatung, told Gustavson that they would welcome any aid on ferrite production, including research assistance from universities.[81] Skoumal, however, did not emphasize NCTU's possible role in strengthening the electronics industry. Instead, he believed that the government and local companies should set up a trust to coordinate among the local companies. The trust should undertake marketing for local companies, coordinate R&D activities, improve the quality of products to meet the expatriate companies' standards, and provide resources to improve business management.[82]

By the time Skoumal made the trip to Taiwan, he saw local Taiwanese companies assembling desk calculators. While he considered assembling desk calculators a growing and lucrative field, as opposed to TV and radio sets, he warned that the emergence of MSI and LSI products would possibly reduce the demand for assembly. He then suggested that the Taiwanese companies think about automating the assembly operations; Philips' Taiwanese manager also recognized and worked on it (see chapter 6). Skoumal also urged Taiwanese companies to consider the development of testing equipment, as demand was growing. Testing equipment did pose a problem to Chew at Philco-Ford (see chapter 6).[83]

In October 1971, the government of the ROC (Taiwan) lost its seat in the United Nations to the government of the PRC. Skoumal's visit was one of the last UN-supported projects in Taiwan. While his suggestions about setting up a trust would not be adopted, government officials in Taiwan took over the initiative and actively looked for profitable fields for the domestic electronics industry in the second half of the 1970s: two predominant, large-scale projects included executing the RCA CMOS technology transfer project (more in chapter 10) and setting up the Hsinchu Science Park. The local electronics industry, as Gustavson and Skoumal expected, did expand. A group of engineers-turned-entrepreneurs who had previously worked at factories owned by international electronics manufacturers established their own companies to assemble electronics and calculators. Many of them further evolved into manufacturers of computers and computer peripherals in the 1980s (the emergence of domestic electronics entrepreneurs is discussed in the chapters that follow).

CONCLUSION

From 1968 to 1971, a Chiao-Tung team experimented with the construction of a minicomputer from scratch, despite the availability of comparable minicomputers from various US suppliers. The members of this laboratory were motivated by the idea that they might acquire, on their own terms, knowledge of manufacturing electronic digital computers. They aimed to enhance Taiwan's domestic mass-manufacturing capacity. Conversely, their project benefited from foreign electronics manufacturers in Taiwan's export processing zone—they obtained integrated circuits from Philco-Ford's Kaohsiung Electronics, and Wang Laboratories' Taiwan factory, which generously custom-made a magnetic core memory consisting of 1,248 bits for the project. Eventually, Chiao-Tung's laboratory did not build a minicomputer, but it put together, in 1971, a programmable calculator, though it was never mass-manufactured.

From 1969 to 1972, two other similar projects were carried out at National Cheng-Kung University and NTU. Members of the two projects shared the ambitious goal of exploring the knowledge of mass-manufacturing minicomputers in Taiwan. On Cheng-Kung campus, Tang as a student and Huang as a faculty member built a calculator together, and Huang soon embarked on researching how to automate and optimize the performance of a power plant with a computer he was going to construct. NTU participants Lam and Wen built a minicomputer as they had planned, though it was displayed in a cabinet and was not fully utilized. Lam and Wen joined the calculator manufacturing industry after graduation.

The three computer- and calculator-building projects, I argue, show a continuity with the earlier efforts of NCTU members and Taiwanese technologists to learn about, understand, and tinker with their computers. After acquiring the ownership of the computer technology, they sought to master manufacturing the technology. Students of electrical engineering programs learned computer architecture and attempted to construct a workable computer on campus. They intended to explore the tacit knowledge involved in building that technology in a laboratory and move on to mass-manufacturing it in a plant. Given the lack of ready-to-use and over-the-counter computer components on the island, these young tinkerers sourced the components from various channels such as used component markets, foreign electronics manufacturing plants, and overseas suppliers. They also had to draw on available networks to obtain technical documents or gain the know-how that did not exist in textbooks or available publications. Using components from different companies and even from markets of used items further increased the level of difficulty of these projects. Tsay noted the laborious process of threading the magnetic cores and his never-ending quest to rewire the programmable calculator. All these young tinkerers were participating in projects similar to what an R&D department would do at a computer or electronics manufacturer, but the university groups were not supported by a group of R&D colleagues in a full-fledged company. They had no access to a company library, and they entered the field with no prior experience. But the young tinkerers ambitiously offered themselves as an R&D division for Taiwan's domestic electronics industry.

Melvin Conway, a manager at Sperry Rand's Univac Division in 1968, wrote "How Do Committees Invent?". His argument was later dubbed Conway's law, in a widely circulated book, *The Mythical Man-Month: Essays on Software Engineering*.[84] Conway recapped his major thesis on his homepage: "Any organization that designs a system (defined broadly) will produce a design whose structure is a copy of the organization's communication structure."[85]

Conway's law may look familiar to historians of technology who are aware of the social construction of technology (SCOT).[86] Designers' and users' cultural beliefs, economic concerns, and political ideas can be transcribed into technological artifacts. The components of the artifacts or systems built by the three university groups also further reveal the networks formed by the builders. With components from varied sources, including new and used items, custom-made and imported parts, island-made and overseas products, the builders connected the heterogeneous components and made them compatible and work with each other. The young tinkerers were ambitious in acquiring resources from their potential contacts, hired by the international electronics manufacturers they aimed at replicating in Taiwan in the near future. They incorporated a spirit of experimenting in campus settings with a transnational circuit of electronics manufacturers.

Taiwanese proposed several distinct ways to engage with manufacturing electronic digital technologies in the 1960s. Throughout that decade, Lee and Chien sought financial and technical support from the United Nations to enhance Chiao-Tung University, which then would stimulate the growth of the electronics industry. At the same time, foreign-owned, joint-venture, and locally owned factories located in the export processing zone and beyond were assembling transistorized radios, TV sets, transistors, integrated circuits, semiconductors, and other products on the island. Beginning in 1968, the participants of the three campus-based computer-building projects set out to build artifacts and to explore ways of fabricating computers on university campus, aiming to cultivate a computer manufacturing industry in Taiwan. Gustavson and Skoumal took short trips, in 1967 and 1970, respectively, to evaluate the international and domestic electronics manufacturers in Taiwan and recommended strategies to allow domestic companies to tap into the rapidly growing production of electronics brought to the island by the foreign investors. UN's technical assistance to Taiwan came to an end in 1971, but the influence of US electronics manufacturers on Taiwan's technical capacity continued through their financial investment and training of Taiwanese engineers in Taiwan.

III TECHNOLOGY INSCRIBED, 1970s–THE PRESENT

6 ASSEMBLING ELECTRONICS: WOMEN'S MEMORIES, MEN'S FACTORIES

In December 1966, Taiwan inaugurated the Kaohsiung export processing zone to attract and host international manufacturers seeking to set up factories in Taiwan. Prominent international electronics makers such as Philco-Ford, Philips, Hitachi, and General Instrument (GI), enthusiastically built factories in the processing zone and took advantage of inexpensive labor and tax breaks in Taiwan since the mid-1960s. The export processing zone was the brainchild of two Taiwanese politicians, Vice President Chia-kan Yen and Minister of Economic Affairs K. T. Li. At the inauguration ceremony, they unveiled a memorial that stood at the entrance of the zone. *United Daily News* described the statue with "a Hercules-like man pushing the giant wheel forward [dalishi tuidong julun]." This Herculean man in the form of a naked ancient Greek sculpture appears to be steering a rudder clockwise with all his might (figure 6.1). The statue attributed the man's muscularity to steering the rudder, which, against all odds, gave Taiwan its economic growth. In reality, it was predominantly the thousands of Taiwanese women factory workers who would join the zone and work there in the years to come that propelled the prosperity of Taiwan. They, together with the hundreds of college-educated engineers who supervised the assembly of electronics, made Taiwan a world market leader.[1]

This chapter explores how the factory-based engineers, mostly men, and the women assembly workers of international electronics manufacturers came to produce components for computers and calculators in Taiwan in the second half of the 1960s. Even though they started working for international electronics manufacturers, many soon participated in building an emerging domestic electronics industry in Taiwan. When experimental projects making computers thrived on college campuses, as described in chapter 5, many international electronics manufacturers began taking advantage

FIGURE 6.1

Vice President and Premier Chia-kan Yen unveiling a statue at the inauguration of the Kaohsiung export processing zone. Under the statue, in Yen's calligraphy, the marble pillar is engraved "Productivity for the country." *Source:* "Fu zongtong jian xingzheng yuanzhang Yan Jiagan zhuchi Gaoxiong jiagong chukou qu jiemu" [Vice President and Premier Chia-kan Yen unveiled a statue at the inauguration of the Kaohsiung export processing zone], December 3, 1966, p. 1, doc. No. 006-030203-00001-115, "Minguo wushiwu nian Yan Jiagan fuzongtong huodong ji (yi)" [Vice President Chia-kan Yen events in 1966, Part 1], Yan Jiagan wenwu 1966–1972 [President Chia-kan Yen files, 1966–1972], AH. Reprinted with permission from Academia Historica, Taipei, Taiwan.

of the island's strong political alliance with the United States, low-cost labor, and tax breaks to make electronics products, including TV sets and parts, computer memory, transistors, and integrated circuits. They hired a predominantly women workforce that, lacking better-paid job opportunities, accepted lower wages but exhibited excellent assembly skills to craft high-quality electronics. Taiwanese engineers also utilized and improvised their engineering knowledge to attain the quantity and quality goals set by international electronics manufacturers. Inexpensive Taiwanese labor, including women factory workers and college-educated engineers, mostly men, granted international electronics manufacturers a niche in globalizing their market.

This chapter discusses the contingencies and agency exercised by expatriates, local Taiwanese engineers, and women factory workers in shaping the development of an electronics assembly industry since the mid-1960s. Many engineers and workers subsequently worked for computer and computer parts manufacturers in the 1980s and 1990s. Beyond material and tangible tech products, US and other international electronics manufacturers left intangible legacies in the manufacturing capacity of the domestic electronics makers that had emerged since the 1970s. In Taiwan's electronics factories in the 1960s and 1970s, expatriate and local engineers improvised and innovated to ensure factories could function in Taiwan, where source materials were not always readily available and maintenance technicians of various machineries were not always stationed. Technical support in almost every aspect could be unstable or precarious. In their careers at international electronics manufacturers, many engineers-turned-entrepreneurs had gained both technical and business knowhow. They then leveraged what they knew to set up their own companies and enter the booming electronics market. In the 1970s, many launched companies to work on IC packaging (the final stage of IC production, which assembles components on an IC) or the manufacture of electronics parts. Subsequently, in the 1980s and 1990s, they graduated to become world-leading computer peripheral manufacturers. Table 6.1 lists international manufacturers that established factories in Taiwan, starting in the mid-1960s, and the corresponding manufacturers set up by their former Taiwanese engineers.

The skills and knowledge of women factory workers ensured the smooth operations of international electronics manufacturers' offshoring factories. On one hand, they were asked to assemble and replicate the products that manufacturers requested, while, on the other hand, they frequently engaged in tinkering—figuring out the most successful way to put together or build an electronics product and standardize bodily and technical practices to meet production demands. Overseas employers knew that Taiwanese women assemblers produced high-quality work but did not offer wages equivalent to their counterparts in the United States, Japan, and Hong Kong. Their

Table 6.1
Taiwanese electronics manufacturers with a founder or a CEO who once worked for international manufacturers' Taiwan factories in the 1960s–1970s

International electronics manufacturer	Year factory set up in Taiwan	Taiwanese electronics manufacturing company with a founder or a CEO who previously worked for the international electronics manufacturer on the left
General Instrument	1964	Dee Van
Philco-Ford	1966	Unitron, Pantronix, Lite-On, and Universal Microelectronics
TRW	1966	Delta Electronics
RCA	1967	Simplo Technology (a CEO)
Wang Labs	1967	Dee Van
Ampex	1970	Simplo Technology (a CEO)
Texas Instruments	1970	Lite-On
Taiwan Showa	unknown	Chyao Shiunn Electronic Industrial Ltd.
Electronic Memories and Magnetics Corporation	1971	Powertech Technology (a CEO)

The left column lists international manufacturers that set up factories in Taiwan since the mid-1960s and trained many Taiwanese engineers who later became entrepreneurs, whose companies are included in the right column. Many of these new and local companies worked on IC packaging or the manufacture of electronics parts in the 1970s, and they subsequently became world-leading computer peripheral manufacturers in the 1980s.

handcrafted magnetic core memory units, transistors, and IC chips were distributed globally, and some even made their way into space. Well-trained women factory workers subsequently became the domestic electronics manufacturers' valuable labor force. This chapter delineates women factory workers' skills, work conditions, and personal lives alongside their lack of entrepreneurial opportunities. As Mar Hicks has pointed out, the British government had been reluctant to enlist trained women to the workforce during the 1960s and 1970s. As a result, the country "lost its edge in computing," as Hicks' book title suggests. In Taiwan, women factory workers were welcomed to the electronics manufacturing workforce, but their wages were low and the working conditions unsatisfactory. Although women labor force in the 1960s and 1970s was highly utilized and not "discarded" by Taiwanese governments and entrepreneurs, the women were also "programmed" to prevent their having opportunities to start their own electronics manufacturing businesses.[2]

A survey of annual reports from US manufacturers with factories in Taiwan in the late 1960s and the early 1970s forms the basis for my discussion. I also construct a history of electronics manufacturing in Taiwan from the late 1960s to the early 1970s through biographies of entrepreneurs and professional managers and oral history interviews with and literary works on women factory workers. In this chapter, I begin with an introduction of the variety of electronics products that international manufacturers chose to assemble in Taiwan. Survival documents show that the production of these products gave an opportunity for Taiwan's women factory workers to compare their execution of assembly jobs to computers. I then delve into women factory workers' working conditions, experiences and their contribution to the industry. The third section is about Taiwanese engineers, mostly men, who were hired to manage factories with limited resources. In contrast to women workers, some of them became entrepreneurs by employing the knowledge and skills they obtained when managing others' factories.

THE ELECTRONICS HIGHWAY

By the 1970s, Taiwan was well-known worldwide as a place to which international manufacturers could outsource TV set assembly. At the end of 1964, the first assembly lines in Taiwan began making or assembling components for TV sets and transistor radio sets, including capacitors, UHF television tuners, television deflection yokes, and the like. Television manufacturers including GI, Philco-Ford, RCA, and Philips opened their Taiwanese facilities for TV set assembly in 1964, 1966, 1969, and 1970, respectively. By 1971, five among the largest US television set makers—Admiral, Motorola, Philco, RCA, and Zenith—were assembling TV sets or making TV components in Taiwan. The island's low-cost labor ensured these companies could compete with Japanese companies in the US and Southeast Asian markets. In 1971, the stretch of Provincial Highway No. 1 between Taoyuan and Hsinchu that housed US TV and electronics manufacturers RCA, Zenith, Arvin, Ampex, Bendix, and Corning was dubbed the "Electronics Highway" (Dianzi gonglu).[3]

Beyond TV and radio sets, in 1967, Taiwanese factory workers were already making computer memory products and transistors. By the end of the 1960s, international manufacturers had assigned their Taiwan factories to assemble semiconductor electronics and ICs. For example, GI's Taiwan subsidiary began in November 1964 to assembled automobile radio sets and black-and-white and color TV sets. By 1967, it had installed a full-scale semiconductor facility in Taiwan. Philco-Ford set up two factories in Taiwan in 1966. The first, located in Tamsui and completed in the summer of 1966, assembled TV sets. The other factory, Kaohsiung Electronics, began with transistor assembly in December 1966, gradually expanding to semiconductor electronics and eventually IC assembly in 1968.[4]

The products made in Taiwan were not necessarily "old-fashioned kind." It was not the age or simplicity of production that determined whether a commodity would be manufactured overseas. Instead, the quantity of focused labor proved a more influential factor in offshoring electronics production. For example, in 1967, RCA and Philips set up operations in Taiwan primarily for magnetic core memory assembly, even before they added TV set assembly to their operations. Some technology, such as metal oxide semiconductors (MOS), was cutting-edge and could easily be transferred to Taiwan with the proper machinery imported from overseas.[5] Table 6.2 shows the variety of electronics made or assembled in Taiwan in the late 1960s.

OFFSHORING MAGNETIC CORE MEMORY PRODUCTION TO TAIWAN

One of the earliest computer parts manufactured in Taiwan in the late 1960s was magnetic core memory. At least 452 Taiwanese factory workers worked for multinational corporations assembling magnetic core memory units in Taiwan by October 1967,[6] and that number would grow in the years that followed. Before semiconductor memory proliferated in the market, magnetic core memory units were the mainstream random-access memory for computers and calculators. These offshored jobs created employment opportunities for women factory workers and inspired local entrepreneurs to begin the same type of manufacturing businesses. Magnetic core memory production prompted women to consider their assembly work in relation to the workings of a computer. Taiwanese viewed magnetic core memory production as a symbol of Taiwan's capability to fabricate an advanced technology.

But, in reality, memory production was both tedious and skilled because it required a worker to use a microscope to thread thin wires through a tiny magnetic core. The tedium was a primary reason that international manufacturers moved the production overseas. Other types of electronics assembly also required a microscope for assembly, when ordinary eyesight would not suffice. But wiring magnetic cores resembled sewing, thus creating a gendered image of the job. To elaborate the feminization of the labor force in electronics manufacturing, digital studies scholar Lisa Nakamura has written about Fairchild's factory, which was inaugurated in 1965 in Shiprock, New Mexico, where the Navajo Nation resides. She focuses on Fairchild's brochures, in which the "inherent flexibility and dexterity" of Navajo women was featured. The brochures claimed their weaving skills enhanced their ability to assemble miniaturized semiconductor products. East Asian women factory workers were similarly portrayed as perfect workers of "nimble fingers" for magnetic core memory units in the late 1960.[7] Recall that UN expert Gustavson used Taiwanese women worker's quality of work in

Table 6.2
Electronics products assembled or made in Taiwan, 1964–1971

International electronics manufacturers	Year of set up in Taiwan	Products
General Instrument	1964	Rectifiers, diodes, transistors, MOS circuits, MOSFET, and TV sets
Philco-Ford	1966	Transistors, ICs, TV sets, and diode-transistor logic integrated circuits
RCA	1967	Solid-state devices, memory planes, TV and radio sets, and integrated circuits
Philips	1967	Magnetic core memory, rectifiers, capacitors, integrated circuits
IBM* (subcontracting)	1967	Magnetic core memory
Hitachi	1967	Magnetic core memory
Wang Labs	1967	Magnetic core memory, calculators
Ampex	1970	Magnetic core memory planes, printed circuit boards
Texas Instruments	1970	Diode-transistor logic transistors, transistor-transistor logic ICs and MOS chips, calculators
Motorola	1970	Black-and-white TV sets (ICs and other semiconductor products made in Hong Kong and South Korea)
Electronic Memories and Magnetics Corporation	1971	Magnetic core memory

This table lists a variety of electronics made or assembled in Taiwan, including radio and TV sets, rectifiers and capacitors, diodes, transistors, magnetic core memory, and integrated circuits. The companies on the left set up factories in Taiwan. This list may not include all US electronics manufacturing facilities in Taiwan at that time.
*IBM subcontracted the manufacturing of magnetic core memory units to Tatung, a local Taiwanese company. I include IBM here to offer an overview of magnetic core memory production in Taiwan at that time.

manufacturing magnetic core memory as an example in illustrating Taiwan's valuable labor force.

RCA, Hitachi, Philips, and IBM were the first four companies to hire Taiwanese workers, predominantly women, to make magnetic core memory. They represented investor interest from the United States, Japan, and the Netherlands. Their magnetic core memory production began around 1967. RCA set up new manufacturing facilities in Taiwan and in Mexico City in 1967. RCA's Taiwan's factory was expected to ease production pressure, as its US manufacturing capacity was "strained." By 1969, RCA had a 433,000-square-foot facility in Taiwan to produce consumer products, solid-state devices, memory planes, and integrated circuits.[8] The Japanese company Hitachi also began making magnetic core memory in Taiwan for its computer products in 1967.[9] By October 1967, Tatung, a Taiwanese company, reported that it had manufactured magnetic core memories for IBM for about a year. Tatung's total labor costs for this department were $10 million NT (an equivalent of $323,100 US in 1967, $3.06 million US in 2024). Tatung made 126,000 magnetic core planes a year.

To offshore part of its magnetics core memory production, IBM conducted a survey in Taiwan to find the company with satisfactory "management, experience and a workforce" at the lowest cost. While all the products were for export, several of Tatung's magnetic core memories were displayed in 1967 in the annual electronics exhibition in Taiwan, serving as "the badge of honor" Tatung earned for reaching a valuable deal with IBM, which planned to manufacture more memory components in Taiwan the next year. K. T. Li announced in November 1968 that IBM was going to make "electronic memory circuits [dianzi jiyiwang xianlu]" in Taiwan. The deal would worth $2.7 million US in Taiwan.[10]

Of the 452 Taiwanese factory workers who assembled magnetic core memory units in Taiwan in 1967, 180 worked for Philips.[11] Philips' then chairman of the board, Frits Philips (a grandson of cofounder Frederik J. Philips and son of cofounder Anton Philips), was invited by the Taiwanese government to evaluate whether his company should invest in manufacturing there. Subsequently, Philips decided to set up a subsidiary, Jianyuan Electronics, in Kaohsiung, though many of the company's board members did not welcome the decision. The two characters "Jian" and "yuan" were carefully chosen, referring to "to build," and "components."[12] A business management scholar, Kie-han Yap, seemed to facilitate Frits Philips' pre-1966 trip to Taiwan.[13]

K. van Driel (also known as Deli Fan, his Chinese name) was assigned to manage Jianyuan Electronics in 1966. He arrived in Taiwan in October to interview forty applicants for engineering positions at the new company. Two Chiao-Tung graduates, Huang San Samuel Wang and Chaowu Lin, were hired and sent to the Netherlands to receive months of training, possibly on various types of production. Both were Abnormal Club members, close friends of Hsieh and Chew (see chapter 5). Lin worked as a product manager and became vice president for Philips Taiwan in 1996. In February 1967, van Driel recruited technicians and trained them to begin to make magnetic core memory in two months. With the plant's steady operation, Philips then decided to make rectifiers (devices that convert alternating to direct current), capacitors for TV and radio sets, and integrated circuits in Taiwan. It also began its business in IC packaging in 1967.[14]

In 1968, a journalist touring Philips' factory in Kaohsiung with van Driel shared with his readers a feeling of wonder about the sophistication of assembling memory planes, a type of magnetic core memory, and he noted that the work was carefully and perfectly done by women. He stated,

As a lay person, I was amazed by every corner of the company. For example, I saw a plane of 16 square inches, which consisted of 16,000 pieces of enameled wires threaded by women workers. At first glance, it looks like a piece of steel plate. But, under the microscope, it is clearly a delicate web of filament. . . . Testing equipment that costs four million New Taiwan dollars serves as the most important part of the manufacturing. If a memory plane has any defects, the testing equipment's display will blink. A memory plane would meet the quality standards if the display has no activity at all.[15]

The second wave of new computer memory factories, which were formed after 1968, included at least four prominent US companies—Wang Laboratories, DEC, Electronic Memories and Magnetics Corporation (EMM), and Ampex. Wang Laboratories began to shift its magnetic core memory production from the United States to Taiwan in 1969.[16] In 1972, the minicomputer manufacturer DEC set up a factory to string its core memory units in Taiwan.[17]

EMM and Ampex, along with other companies, received licenses from MIT to use magnetic core memory technology after Jay Forrester developed this new type of memory.[18] As a defense contractor, EMM, which decided to move to Taiwan in 1971, was one of the leading memory suppliers. At its opening ceremony in February 1972, guests toured the factory, where a good number of women in light blue uniforms were making memory stacks and ferrite core arrays using microscopes. Similar to the journalist's tour at Philips' Taiwan factory, a group of Taiwanese women workers and their microscopes were required to demonstrate the expanding capacity of electronics manufacturing of international corporations. EMM intended to hire seven hundred employees by the end of year, and 1,500 to its maximum capacity.[19] Ampex began to prepare its operation in Taiwan in 1968 and completed its Taiwan factory around 1970. By 1973, the company made magnetic core memory planes and printed circuit boards in Taiwan.[20]

EMM and Ampex had made magnetic core computer memory in Hong Kong but expanded its assembly facilities to Taiwan after widespread strikes, protests, and riots instigated by a 1967 labor dispute at a plastic flower factory in Hong Kong.[21] The 1967 incidents, combined with low labor costs in Taiwan—the monthly salary of a factory worker in Taiwan was $20 US, almost half that of a worker in Hong Kong— prompted many companies to consider additional investment in Taiwan. *Business Week* in July 1968 described the move: "Another factor is Communist fomented turmoil that has hit Hong Kong and South Korea, the island's chief Asian rivals for new investment.

The fact Taiwan is effectively buffered from Red China by the U.S. Seventh Fleet makes it attractive to skittish investors."[22]

Along with EMM and Ampex, other "skittish investors" included Motorola, Texas Instruments, and Arvin, all of which set up factories in Taiwan after 1967. In contrast, less anxious manufacturers—for example, Fairchild, Lockheed, and Control Data—continued to manufacture a portion of their magnetic core memory in Hong Kong. Control Data staff even named Hong Kong their "core house."[23] Fairchild, the pioneer in transistor manufacturing, was one of the earliest US semiconductor firms to set up an assembly factory in East Asia. It established a transistor assembly plant in Hong Kong in 1963 and South Korea in 1965. The idea of offshoring came from Robert Noyce, who had invested in a radio firm in Hong Kong and realized that the low wages there would significantly help Fairchild compete in the US market. His belief in Hong Kong workers' good work ethic also contributed to the decision to set up plants there. By 1966, the number of Fairchild employees in Hong Kong had reached four thousand. They not only assembled transistors but also tested transistor dies, transistors, and wafers.[24]

LOCAL MAGNETIC CORE MEMORY PRODUCTION—UNITRON (HUANYU) AND DEE VAN

At least two domestic start-ups in Taiwan made threading magnetic core memory their major business. The magnetic core memory's labor-intensive process, low requirement for expensive machinery, the availability of a women labor force, and founders' working experience at multinational manufacturers' Taiwan factories allowed the two Taiwanese companies to enter the field. Manufacturing magnetic core memory served as a way for the companies to begin their business and look for the next viable opportunity. Unitron (Huanyu), one of the first domestic electronics manufacturers (discussed in detail in chapter 7), was set up in 1969. Zaixing Andrew Chew (see chapter 5) left Philco-Ford to become a Unitron founder. His strategy for running the new five-hundred-employee company was to diversify Unitron's products. Magnetic core memory was one of the early products. The founder of Acer, Stan Shih, joined Unitron in 1971, and the first job Chew assigned to Shih was managing magnetic core memory production. Shih noted in an oral history interview that he was aware that the eyesight of workers was key to production. He knew little about the assembly line, however, and for several months worked hard to learn from the foreman or forewoman and the section chief every day.[25]

In contrast to Unitron's mass-manufacturing scale, in 1972, twenty-eight-year-old Xingdao Zheng founded Dee Van and hired five women workers to make magnetic core memory. Sometimes Zheng would have to wire the cores, too. He faithfully believed he could make a profit from the market share of much-in-demand magnetic core memory units. In a year he was short of needed funds to import raw materials for making memory cores, which was likely due to the oil crisis. He then shifted to making rectifiers and AC adapters for local start-up calculator manufacturers and made a fortune from it. His knowledge of magnetic core memory units was probably from his work at Wang Laboratories, and at GI he learned about rectifiers. Around 1967, Zheng worked at GI's electronics factory as a transistor technician, where he was allegedly the most skilled among the four technicians there. Working at GI for two years, his salary rose to $2,200 NT. But he was so determined to join Wang Laboratories' Taiwan subsidiary in 1969 that he took a lower monthly salary of $1,800 NT. He believed that he could learn more about computers at Wang Laboratories, and the knowledge was worth the lower salary. He was right, as he started his own company by using what he learned about how to make magnetic core memory there.[26]

Based on newspaper records that contained in-depth interviews with women factory workers, I argue that the magnetic core memory industry offered women workers an opportunity to consider their relationship with computers, despite that their work did not require computers per se. In contrast, IC assembly workers did not seem to immediately associate what they were working on with the "electronic brains," the way people have referred to computers in Taiwan since the 1950s. Magnetic core memory technology gave women workers a cybernetic conceptualization of the relationship between computers and their jobs. Working at Unitron for two years, Yingjiao Wei was just over twenty years old in 1972. A journalist at *Economic Daily News* interviewed her after Unitron's first calculator became well-known in Taiwan. With a junior high school education, Wei first worked for RCA's factory in Taiwan. She had worked on electronics assembly and packaging and eventually on magnetic core memory planes there. She then left for Unitron and continued to wire memory planes at the new company. She decided to move to Unitron because the company was building a factory in an area neighboring her hometown. At Unitron, she was promoted to the quality control department for testing memory planes and calculators. Asked about how to perform her job, she pondered and answered, "It is important to adjust our mentality to think of our work as the operation of a computer with a memory unit." The journalist was puzzled, "A computer?" Wei elaborated,

Yes, like a computer. After many years of employment, someone may still make the same mistake that he or she made ten or twenty years ago in workplace. That person might have been hard working. But what is absent is that he or she should recall carefully the mistakes he or she made, think thoroughly what caused the mistakes, and remember the causes and results of the mistakes by heart.[27]

The journalist concluded, "The use of memorization was her key to success." Wei obviously shared her reasoning based on her job on the quality control in the assembly line. She believed that human beings could try to be as meticulous as a computer to avoid mistakes. With years of immersive experience at both computer memory manufacturing and factory-style quality control, she utilized the metaphor of computer memory technology to explain the nature of her job.[28]

A similar analysis was offered by a textile factory worker around the same time. Yuli Liu, a forewoman working in a textile factory for twenty-one years, responded to a journalist's query, "How do you work with younger workers?," by explaining that she aspired to be a model worker for everyone to learn from. She added, "New employees should act like the beads of an abacus that follow instructions. They should observe [how jobs are performed] and practice. After gaining some experience, they should act like a computer that automatically memorizes and stores working experiences. They ought to be autonomous. They should then eventually be considerate and become excellent models for new employees."[29] Computer memory, in the forewoman's narratives, worked as a metaphor to depict workers' memorization and thorough performance of tasks in a factory. The abacus worked as a metaphor to describe workers following assigned steps. The forewoman, though not in the electronics industry, borrowed the computer memory metaphor to understand the nature of factory jobs. While it is unclear where Liu learned how a computer works, the existence of a strong computer memory manufacturing industry encouraged women factory workers, as a potential labor force of that industry, to become informed of the workings of computer memory and computers.

MANUFACTURING TRANSISTORS AND ICS FOR COMPUTERS AND CALCULATORS: PHILIPS, PHILCO-FORD, GI, TEXAS INSTRUMENTS, AND MOTOROLA

Beyond magnetic core memory, several international electronics manufacturers made or assembled other computer and calculator components in Taiwan. In 1967, Philips assigned Ludo van Bergen (also known as Peihan Fang, his Chinese name) from the Netherlands to start a new IC plant in Taiwan, in his early forties.[30] Van Bergen oversaw the IC departments for two years, and then the magnetic core memory department for three years before he was assigned to other overseas Philips factories. Recall that the name of Philips' Taiwan subsidiary was Jianyuan. The work of Jianyuan was IC packaging, which is the last step of IC fabrication. Philips shipped silicon wafers to Jianyuan. Then Jianyuan employees cut the wafer into chips. Factory workers, mostly women, next assembled the chips into a case and wired the chips into the case with gold wires, a process known as wire bonding. After that, the IC package had to be molded—covered with plastic materials for protection. Philips' Taiwan plants were well-organized, neat, and attracted many visitors, especially when officials wanted to demonstrate modernized, high-tech, model electronics factories. A photo of Philips' women factory workers working in the factory space was included in a motion picture that accompanied the ROC's national anthem, which, in the 1970s, was usually shown in theaters prior to the movie. Playing the anthem was mandatory for theaters until the late 1990s. Overall, by September 1970, Jianyuan already had 850 employees in Taiwan, which was before its groundbreaking ceremony for two new factories to produce TV cathode ray tubes and glass.[31]

Since the 1960s, women factory workers, predominantly, assembled ICs. For example, long before Fairchild began its operations in East Asia in 1963, it had hired women assembly workers. Similar to the case with magnetic core memory, electronics manufacturers considered hiring women for reducing the cost of labor intensive assembly jobs. The trade publication *Solid State Journal* featured Fairchild in 1960 and published two photos in which assembly and testing of transistors was carried out by women workers in Fairchild's transistor plant in Mountain View, California. In one of the photos, women workers mounted transistors to headers, functioning on circular platforms. The article then detailed how assembly workers then attached gold wires from the transistor die to posts on the platform and welded the wires. It also emphasized that the entire process had to be done under microscopes because a transistor die is smaller than a pinhead. To lower the labor cost, Fairchild then set up a new plant in Portland, Maine, in 1962, as "wages for unskilled workers in Maine were only half of those in the Bay Area," according to historian Christophe Lécuyer. Beyond Fairchild, women were an important presence in transistor production at companies such as Texas Instruments. Taiwan Semiconductor Manufacturing Company (TSMC) founder Morris Chang noted in his memoir that his first job at Texas Instruments in Dallas, in 1958, was to manage the fabrication of a germanium NPN transistor for IBM. The production line had three shifts, seventy operators in total. Chang noted that the hourly rate for operators was $1 US (equivalent to $11 in 2024). He did not specify whether the operators were mostly women, but his description noted that one of the operators was a woman. She sat at the end of the production line and her job was to test final products. Chang noted that the group paid close attention to her announcement on the result of yield every day, as the entire production line was working together to increase the yield from zero to over 30 percent.[32]

Philco-Ford's subsidiary Kaohsiung Electronics began making transistors in December 1966. It then gradually expanded to the assembly of semiconductor electronics and eventually IC in 1968. In 1969, it had 1,200 employees; the number of employees reached 1,500 the next year, and the plant produced three million pieces of electronics per month.[33] A recent collection of twelve oral history interviews with women workers at Kaohsiung Electronics indicates that the factory covered all steps of production, from slicing silicon crystals to manufacturing a finished chip. The informants' work varied: die attachment, wire bonding, molding, electroplating, trimming, printing marks on the top of the chip, and inspection. While the facilities of Kaohsiung Electronics were sold to GI in 1971 and later other companies, some employees remained working there for different companies. (Philco-Ford sold the facilities when it decided to discontinue its microelectronics operations globally, but it retained its ownership of its Tamsui TV assembly plant in Northern Taiwan.) The informants had experience on a wide range of products, from assembling transistors in the late 1960s to working on ICs into the 1990s. Interviewees were skewed toward a group of long-serving employees, but precisely because of their long employment, several were extremely knowledgeable about every step of the process. One informant took such good notes that even her supervisor borrowed them.[34]

GI's subsidiary in Taiwan was named Taiwan Electronics. A full-scale semiconductor facility for making rectifiers, diodes, and transistors was set up in its plant in 1967, the second year of GI's operations in Taiwan. In addition to consumer electronics, it produced semiconductors for the F-111 aircraft and the Mark 46 torpedo.[35] Among foreign-owned electronics manufacturers, Texas Instruments and GI were two that tended to assign their Taiwan plants to manufacture or assemble the latest product lines. After GI's production of metal–oxide–silicon field-effect transistors (MOSFET) became profitable in the United States, GI immediately assembled and tested the products in Taiwan, as it anticipated a huge potential in the semiconductor calculator and computer memory market. By 1972, GI's MOSFETs had supplied "industrial and military markets," as well as the "TV tuner market," a highly profitable field.[36]

With growth of employees almost doubling every year, workers hired at the GI plant in Taiwan increased from 1,300 in 1966 to 10,000 in 1970, reaching 17,500 in 1974. Even after the oil crisis, it continued to grow. But GI sometimes laid off and rehired workers and recruited new workers to deal with the fluctuating market after the oil crisis. By 2007, GI had employed 200,000 workers in its more than forty years of operation in Taiwan, close to 1 percent of the population of Taiwan. Between 1965 and 1970, the size of its plants grew ten times, from 40,000 to 445,000 square feet. In 1972, GI claimed that its Taiwan operations was the world's largest producer of rectifiers.[37]

Both Philco-Ford and GI were prominent companies in the United States at that time. When, in 1968, computer manufacturer Sperry Rand sought to secure its advantage in the transition from plated wire memory (a variation of magnetic core memory) to semiconductor memory, it considered acquiring a semiconductor firm to gain critical knowledge in the field. The company rated Texas Instruments, Motorola, and Fairchild as first-tier semiconductor companies. Philco-Ford and GI, in addition to National Semiconductor and American Microsystems, were rated second-tier firms. Among Sperry Rand's seven candidate semiconductor firms, four companies—Philco-Ford, GI, Texas Instruments, and Motorola—set up factories in Taiwan.[38]

Texas Instruments and Motorola slowly moved some of their production to Taiwan after the 1967 incidents in Hong Kong. In 1969, Texas Instruments announced it had bought land for a facility to produce aerospace components and computer products. In 1970, Texas Instruments was making diode–transistor logic (DTL) transistors and announced that it would soon fabricate transistor–transistor logic (TTL) ICs and MOS chips in Taiwan. In May 1972, it was planning to make calculators in its Taiwan factory and export them. In contrast, Motorola focused primarily on black-and-white TV production in its Taiwan factory, which began to operate in 1970. Motorola had made Hong Kong the center of its semiconductor production, and manufactured ICs in Korea as early as 1968.[39] While Fairchild did not set up a factory in Taiwan, it left a legacy in Taiwan in a subtle manner. In the mid-1960s, Frank Yih, a chief engineer hired for managing Fairchild's Hong Kong plant, moved to Taiwan to take a new job offered by Philco-Ford—to manage its subsidiary, Kaohsiung Electronics.[40]

GI and Philco-Ford, two of the earliest US electronics manufacturers to invest in Taiwan, left various legacies on the island. Their success encouraged more US companies to set up factories in Taiwan. In a 1968 *Business Week* article, "Taiwan Creates a Place for Foreign Money," the magazine reported that the number of GI employees in Taiwan was the largest among foreign-owned electronics manufacturers in Taiwan.[41] As for Philco-Ford, it contributed to the minicomputer building project at Chiao-Tung (described in chapter 5). Several of Philco-Ford's early employees, including Chew, left the company and formed or joined local electronics manufacturing start-ups, many of which later morphed into computer peripheral manufacturers in the 1980s. But at Philco-Ford, occupational health problems caused by chemicals used in the factory surfaced in 1972 and shocked Taiwanese society, prompting state responses to improve safety and health issues in factories as well as civic advocacy for women factory workers rights, though with limited effects.

LIFE OF WOMEN IC ASSEMBLERS

Women laborers were not an entirely new category in Taiwan in the mid-1960s. As early as 1960 and 1961, two popular Taiwanese songs sung by an eleven-year-old, Fenlan

Chen, depicted young women working at blue-collar and factory jobs. Born in Taiwan in 1948, Chen became a well-known singer at the age of eight. The first song, "dapin de gongren" (Assiduous Workers), described seventeen-year-old women working to support their families in assembly jobs, menial work, and finally an unspecified blue-collar job that soils clothing at the end of the day. The other song, "gunü de yuanwang" (An Orphan's Wish), tells the story of an orphan leaving the farming town where she grew up to look for a job in Taipei. She intended to find a job at a factory, the front door of which was pasted with a recruitment placard. "An Orphan's Wish" was the theme song for a popular movie of the same name. Menial and factory work was most likely to be found in canned food factories or garment factories in the early 1960s, instead of electronics manufacturing plants, which emerged in Taiwan only after 1966. But the two songs were reissued on Chen's albums in 1967, 1969, and 1970, in response to her new listeners—the rapidly growing population of women factory workers. While "Assiduous Workers" is little known in present-day Taiwan, "An Orphan's Wish" has been covered many times by several singers and is well-known among present-day Taiwanese.[42] Listeners of Chen's albums in the 1960s, however, were subject to more challenges at work, including unsatisfactory dormitories, the culture of treating corporate managers as respected family members, exposure to harmful chemicals used in electronics manufacturing, mandatory overtime, and lack of opportunities for promotion or becoming an entrepreneur.

To employ a large number of factory workers—mostly young women workers from farming towns, who were new to factories in urbanized areas—GI, Philips, and Philco-Ford had to offer places for workers to live. But those dorms were insufficient and in unsatisfactory condition. Catholic priest Edward J. Wojniak, of the Society of the Divine Word, wrote in December 1965 that about 1,300 young women workers were hired by a dozen factories in his neighborhood in Xindian, a town close to Taipei city. But the rental market could not meet the surging demand in the area. Many workers had no choice but to live with nine to eleven workers in a small traditional Japanese-style room with tatami mats. He was deeply concerned with their quality of life and potential health risk, especially the high likelihood of contracting diseases. In addition to no places being available to rent, workers could not afford a decent place to live. The average salary for a women factory worker was $12 to $15 US per month (this 1965 figure was lower than *Business Week*'s 1968 estimate), Father Wojniak then embarked on a years-long campaign to raise money to construct a modern building complex that could accommodate over a thousand women workers. The first building was completed in 1968. Many of the residents of the building complex turned out to be workers at GI.[43] GI donated $1 million NT (an equivalent of $25,000 US at that time) to Father Wojniak's project in 1966, showing that GI did not offer its employees sufficient accommodations. In Kaohsiung, the export processing zone set aside land to build a multistoried dormitory to accommodate up to 3,600 women workers in three hundred rooms. Minister of Economic Affairs K. T. Li attended the groundbreaking ceremony in November 1968, but when the building was completed is unclear. A newsreel revealed that, by March 1971, more than 1,800 women workers lived in the dormitory. Children showing up in the newsreel were probably visiting their mothers during the weekends.[44]

New types of "benefits" offered by these large electronics factories displayed the companies' healthy finances and their intention to construct an image of companies as families. Many electronics manufacturers would organize short vacation trips for young workers. The *Economic Daily News* covered Philco-Ford's annual trips, indicating they were a new and trendy phenomenon. In 1973, Philips began to offer employees a bottle of milk every day, and GI's IC plant in Kaohsiung, along with many other plants, quickly followed. The chief manager of GI's Kaohsiung plant in 1973 was an American, whose Chinese name was Weilian Ke, indicating his English first name must have been William. William was probably beyond his middle age. Interviewed in 1973, the journalist emphasized that William's hair and eyebrows were turning gray. William decorated the reception room of the plant with an ink wash painting of a hen tending several chicks. The painting's artists were a dozen women students at the vocational school from which GI recruited interns and employees. The students even glued their headshot photos at the bottom of the painting, which was meant to be a present to thank William for an internship opportunity. William told a journalist, "They said that I am like a hen." He continued, "I have seven children. I do like to think of them [factory workers] as my children." William explained to the journalist that he found women in southern Taiwan were more economical and less open to new ideas. Instead of issuing extra financial compensation for workers to buy nutritious food, GI decided to give out free milk, vitamins, and fish oil capsules, to strengthen their workers' health and "eyesight." That way, workers, especially those from farming families who usually did not keep cows or drink cow's milk, would be more willing to take these extra nutritional supplies, according to William. He even phoned an official to find out which brand of milk was the most suitable for Taiwanese. He must have considered the higher occurrence of lactose intolerance in East Asia.[45]

Free milk and supplements were probably an intended though ineffective remedy. More seriously loomed the deaths of three women Philco-Ford workers and two workers at other plants in 1972. Seven of the three thousand workers of Philco-Ford's Tamsui factory, which made and assembled TV components, became sick in 1972.

Their symptoms included serious skin problems, fever, and liver damage. *Economic Daily News* publicized the news about Philco-Ford's sick workers on October 21; on October 31, Qiying Chen, a nineteen-year-old women worker whose job was packaging coils, died from liver and gallbladder damage after a month-long hospitalization. She had just begun to work in the Japanese-owned Mitsumi Electric plant in Kaohsiung in September. Her sixteen-year-old sister also worked there. The plant made parts for TV and radio sets, hiring 2,600 women workers in 1972. Around the same time, another women factory worker at a plant right next to Mitsumi also died from liver damage. By then at least three hundred women factory workers at Philco-Ford decided to quit to secure their health. A later investigation showed that the deaths and illness at Philco-Ford and Mitsumi were related to trichloroethylene, an industrial solvent used to degrease metal equipment. While the deaths prompted governmental officials to inspect all the manufacturers and suggest replacing trichloroethylene with trichloroethane, the regulation failed to eradicate the subsequent emergence of illnesses in Taiwan's electronics industry.[46]

One of the most well-known subsequent cases was a widespread illness caused by organic solvents among workers in RCA's plants in Taoyuan. RCA's TV assembly began in Taiwan in 1969, but RCA was acquired by GE in 1986 and by Thomson in 1988. The RCA plants were closed in 1991. Between 1975 and 1991, its plants in Taiwan violated occupational solvent regulations eight times. The plants that failed to meet toxic air exhaust standards were later found to have dumped untreated wastewater, and never remedied the pollution. In 1972, the number of RCA's employees in Taiwan was 6,400. But, in the 1990s, more than two hundred former RCA workers died from, and more than a thousand employees were diagnosed with, cancer. The number continued to grow every year. Former workers filed a class-action lawsuit in 2004 against RCA, GE, and Thomson. In 2018, Taiwan's Supreme Court ruled in favor of 262 of the more than five hundred plaintiffs and ordered RCA to pay more than $500 million NT in compensation. But the Supreme Court assigned the High Court to retry the case for 246 other plaintiffs. The most recent verdict delivered by the High Court ruled that only twenty-four of the 246 plaintiffs had illnesses that resulted from exposure to toxic materials when employed by RCA.[47]

WOMEN FACTORY WORKERS IN MOVIES AND LITERARY WORKS

In the mid-1960s, readers began to see Taiwanese factories and their women factory workers appearing in short stories and essays in newspapers and literary magazines.

Scattered literary works were published between 1966 and 1976, and their numbers increased, sometimes in anthologies, after the late 1970s. Ching-chu Yang (Qingchu Yang) is the pen name of the acclaimed writer Ho-hsiung Yang. Born in Taiwan in 1940, Yang was a blue-collar warehouse worker at the Chinese Petroleum Corporation's Kaohsiung Oil Refinery. He was Taiwan's most famous short story chronicler of factory workers' lives. He began working in the oil refinery in 1961 after his father, a refinery firefighter, died putting out a fire on a petroleum tanker. Yang published his first short story in 1970, but he became more prolific after 1976. His short stories were sometimes first published in newspapers, to which many working-class readers had access. Many of his stories revealed unsatisfactory working conditions: forced overtime, the piece-rate system, and the low salaries that women factory workers relied on, as well as that colleagues and managers, mostly men, earned much better salaries and enjoyed swift promotions. Yang also discussed the tensions women workers felt between their ties to their natal, farming, and often patriarchal families and their new lives of economic independence away from home. Some stories focused on women factory workers' expectations of and the subsequent disappointment with romance and marriage, as their jobs put them on an unequal footing with the men they were dating.[48]

The settings of Yang's stories, however, were mostly in textile factories or in generic factories, with no elaboration of the products the protagonists made. Electronics manufacturing was not often the setting for these works, but Yang and another writer, Xiang Ye, who was once a factory worker, situated a small part of their work in electronics manufacturing. Yang's "Qiuxia's Sick Leave" (Qiuxia de bingjia) portrays Qiuxia, a factory worker at a Japanese-owned electronics plant, who was briefly hospitalized after she fainted and was subsequently injured in the bathroom. The fainting may have been from anemia, stomach discomfort due to irregular mealtimes, or fatigue from long working hours. Her brother, a factory worker at a textile plant and a former council member of his union, navigated the bureaucracy of his sister's factory to request paid leave, to which she was entitled by law. "Qiuxia's Sick Leave" was originally published in 1976; the next year Yang wrote another short story about electronics manufacturers. "Our Own Manager" (Ziji de jingli) describes the husband of a factory worker who was injured in the electronics etching process as he struggled to request that his wife's company issue the necessary paperwork for her health insurance claims. With the paperwork completed, the health care she received in the hospital would be partially covered by the occupational health insurance run by the state. But the factory's chief manager Ling denied her claim to health insurance and insisted the company had no responsibility for her injury. The factory was a subsidiary owned by a Western company, and the top person at the subsidiary was a Westerner. The expatriate later realized that the chief manager had covered up the incident. The expatriate then gave his word to the occupational safety and health officials that the company would shoulder the responsibility for the incident and improve the ventilation system to exhaust fumes in the company's three factories. Yang used the expatriate's remarks to conclude the short story, "I don't understand why a Chinese manager would treat his own people in such a manner? I don't understand!" Yang chose to connect the electronics manufacturing plant to the work injury, which was not a major theme in his other short stories. The stories were also possibly inspired by the series of deaths and sickness from trichloroethylene in 1972. These stories show Yang's concerns about the ineffectiveness of the occupational health insurance associated with a mid-level-management-imposed bureaucracy in large electronics manufacturers.[49]

In "Qiuxia's Sick Leave" and "Our Own Manager," it was the brother or the husband who fought for the woman's rights as a worker. But, in two other stories in Yang's 1978 anthology, he portrayed two women workers in textile and food processing factories bravely challenging their supervisors' unreasonable treatment and humiliation. And their actions made improvements in their factories. In Yang's books, he included a questionnaire for his working-class readers to answer and mail back to him. Through this channel he could know more about factory workers' lives. He noted that he had heard from readers in their letters sharing similar stories about how workers challenged authorities and improved working conditions in their factories. While many researchers have concluded that Taiwanese factory workers were more docile than those of other countries,[50] individual-initiated resistance did exist in the 1960s and 1970s.[51] In Yang's writings from 1983 to 1985, his protagonists tended to be women workers who switched jobs between electronics manufacturing, textile plants, department stores, restaurants, and construction sites. His protagonists frequently complained that electronics manufacturing jobs were the most monotonous jobs they had taken.[52]

Poet Xiang Ye worked briefly in quality control in an electronics manufacturing plant in Kaohsiung in the early 1970s. She later pursued a career in newspaper and magazine publishing and now teaches creative writing. Starting in the late 1970s she published six poems about women factory workers. The themes varied from remaining delightful despite working the night shift to being powerless and exhausted by the traffic jams in the export processing zones during morning rush hours.[53] One of Ye's poems, "On the Production Line" (Shengchanxian shang), was published in 1978 as a foreword to Yang's *Womanland of Factory Women*. It is a celebration of Yang's anthology and a salute to women factory workers:

A piece of blue work suit
A myriad of machineries
Ready to go

Several thousand watts of light bulbs
The brightest is the boss's keen eyesight
Never miss a thing
Meet the production target. Go, go, go.

This tweezer is dull. Reject.
A gold wire on this transistor is broken. Re-do it.
After magnifying it 30 times under the microscope, do the 80 times magnification
No merest blemish
No tiny dent

The air conditioner placed right next to a furnace of 500 Celsius degree
Sends the breeze to us, caught in-between
Some bring goods in, some deliver stuff out
Some sit and examine the quality of products
Some sneakily snack a dried sour plum[54]
What is common among us are our arms stretching out
Flying and moving like an awakening snake from hibernation

Do we have enough for the next shipment?
Not yet.
Stay later to work more. Go, go, go.[55]

In the poem, Ye describes the tight working schedule and the overtime required in electronics manufacturing jobs. The only obvious resistance was to snack without being caught. In a 1978 romance film, *Yige nügong de gushi* (with an official English title, *Fly Up with Love*, figure 6.2), one of the women factory workers, whose role was to create comedic relief in the film, also sneaks snacks while working.[56] The poem vividly captured factory workers' fast work pace, though it does not clearly reveal whether workers approved of the pace. "Go, go, go," the last lines of the second stanza, and the last stanza could be read as a supervisor's requests, but they could also be interpreted as words uttered by the author or a factory worker.

Ye's poem portrays the detail-oriented nature of electronics assembly and its quality control aspects. It also confirms the widely perceived importance of eyesight to this job, as pointed out by GI's chief manager William, or Weilian Ke, in 1973 and UN expert Gustavson in 1967. Furthermore, through describing the nature of the work, Ye, at the same time, exhibits a woman's pride and confidence in her job. "No merest blemish" and "no tiny dent," could be interpreted as factory workers' commitment to the quality of their products. The feeling of pride, however, was probably best described

FIGURE 6.2

Film *Yige nügong de gushi* (Fly Up with Love). The literal translation of the film's name is "a story of a factory woman," directed by Shu-Sheng Chang (Shusheng Zhang), produced by the Central Motion Picture Corporation in 1978. Reprinted with permission from the Central Motion Picture Corporation.

as "tinged with growing bitterness." Factory women's strong work ethic, and their docile personality, was not compensated with better wages.[57]

Ye's poem was published in 1978, but as early as 1973 the pride of factory work was voiced by a worker. One of Unitron's women workers told a journalist that she thought the title "women workers" (*nügong*) should be abandoned. Liqi Lin, chief of Unitron's quality control department, explained the importance of her job to manufacturing desks or handheld calculators. She refused to describe anyone in the factory as a member of "women workers." Instead, she insisted that she would refer to someone as miss or mister so and so. She often called her colleagues sister so and so too, believing that "assemblers" might be a better term than "women workers." She further noted that everyone in the company was entitled to government occupational health insurance, which indicated that they were all workers, even though their specialties varied. Andrew Chew, the co-founder and vice-manager of the company, also supported Lin's reasoning. Both Lin and Chew believed that it would eventually benefit the management of the company if everyone showed each other respect by using proper titles.[58]

WOMEN WORKERS' LACK OF ENTREPRENEURIAL OPPORTUNITIES

Despite economic independence, pride, and confidence in their jobs, women factory workers did not enjoy the same opportunities to become entrepreneurs as did men in the same factory. In Yang's short story "Guipabi shuibengshan" (Turtles Climbing Up the Cliff, Water Washing Away Soil on the Slope), he described a textile factory woman, Qinglan, overhearing a visitor and a college-educated supervisor's conversation. The two men discussed the disparate prices and quality between German and Japanese embroidery machineries, the daily volume of fabric production for each machine, and the sale prices of various fabrics. They concluded that each machine could make a profit of about $10,000 NT every day for the factory owner, while the monthly salary of women employed in the factory was about $2,000 NT per person. That evening, Qinglan identified over thirty machines in the factory and did some math. She soon wrote a letter to her farming parents, saying, "If we have one such embroidery machinery, each month we will be able to make a fortune, or to win a jackpot." At the end of the short story, Qinglan, of course, did not have the funds to start such a business. The college-educated supervisor turned out to be a sympathetic person and encouraged Qinglan to leave for other factories for better working condition. The more workers that would leave for other factories, the more likely the boss would maintain a reasonable level of wages to keep his workers. Qinglan then jokingly urged the supervisor to set up a new plant where he would offer better salary to his workers, and it would be a factory where workers could take ownership of. When conversing with Qinglan, the supervisor used the contrast of turtles climbing up the cliff to water washing away soil on the slope to indicate that factory workers earned their wages as slowly as turtles, whereas the boss made money like the sudden movement of a mudslide. At the end of the story, Qinglan was hospitalized after falling to the ground from an elevated level during her night shift. She then returned home to recover from her concussion.[59]

In Lydia Kung's ethnography of Taiwan in 1974, she detailed women's work and family by conducting her fieldwork in three plants owned by an anonymous US electronics manufacturer. The plants assembled TV sets and solid-state components. Among four thousand workers employed by the plants, most assemblers were women, but only three troubleshooters and one forewoman were women. Troubleshooters had positions higher than assemblers, testers, and quality-control workers. A foreman or forewoman was responsible for eighty to a hundred workers. These jobs were the highest positions women could hold in the plants. Kung found that many factory women workers had no ideas about the future, nor would they talk about it. When they talked, they commented that "fate and destiny" "will take their own course." Kung's ethnographic observations explained why women entrepreneurs were rarely seen at that time. Women factory workers did not get promoted. Many did not know what to do next, because of youth, lack of role models, or expectations to marry soon.[60]

But, in the exceptional plot of the 1978 romance film, *Fly Up with Love*, the film's protagonist, an IC factory woman named Chin-mei, started a small business in her family's living room. At the beginning of the movie, Chin-mei and her women colleagues worked on chips with microscopes, brushes, and soldering guns in an IC plant, though the film never specified what their jobs exactly were. The movie was shot in EMM's Taiwan factory, though the setting presented the company as owned by a local businessman. In the movie, Chin-men was dating Ko-ti Lin, the adopted son of the company owner—assigned to work on testing chips as a prelude to managing the company. After quarrelling with Lin's fiancée and quitting the plant, the protagonist, who has no family assets to inherit, took advantage of a new government policy to encourage entrepreneurship and applied for a bank loan to start a small business. She procured machines to make plastic flowers alongside her parents in their living room. She also hired friends from the factory, presumably on a part-time basis. While the woman aspired to be and eventually became an entrepreneur, in this film the man aimed at something higher and nobler. Despite wishing otherwise, Lin majored in electrical engineering in college at his adopted father's urging: nature, not electrical engineering, is his calling. In an excursion to a lake arranged by the IC plant, Lin even told Chin-mei that it was especially beneficial for people staying in the factory all day to spend time outdoors. Lin believed that getting closer to nature was always what he wanted to do. He then chose to leave the electronics factory owned by his adopted family to pursue his dream, raising cattle on a ranch owned by his friend. The film valorized pursuing dreams, but the man lucidly understood the technical knowledge and factory system and expressed his criticisms, while women factory workers focused on succeeding in managing their basic economic needs.

Fly Up with Love acknowledged Taiwan's women factory workers' key contribution to the island's electronics industry and its ability to supply the global market since the second half of the 1960s. The 1978 acknowledgment was late and was made only in a movie, instead of in an official setting.[61] Still, the movie had the owner of the electronics manufacturer chide his daughter when she looked down on women factory workers as her fiancé fell in love with one. The owner of the manufacturer told her daughter that he could not run the company without these workers, and the rapid economic growth of the country should be credited to these women factory workers. The movie provided the public with a lesson in how women factory workers were key to economic growth—an argument more acceptable in 1978 than in the 1960s.

ENGINEER MEN, IMPROVISING AND INVENTING AT OFFSHORING FACTORIES

The supervisor in Yang's short story about Qinglan was expected to leave the factory and form his own company to change the game. In reality, at electronics factories such as GI and Philco-Ford, factory women's engineering colleagues and supervisors, mostly men, often with university or technical college degrees, seized opportunities to own new manufacturing facilities. For example, Xingdao Zheng, the aforementioned founder of AC adapter manufacturer Dee Van, was once a competent transistor technician at GI, and a technician and then a salesman at Wang Laboratories.

At Philco-Ford's Kaohsiung Electronics, many prominent Taiwanese electronics and computer manufacturing entrepreneurs learned how to scale up the manufacture of electronics components and apply what they learned to their start-ups. In March 1966, Philco-Ford appointed Frank Yih (Shouzhang Ye), a thirty-one-year-old University of London and MIT-trained Shanghainese manager of Kaohsiung Electronics. Yih soon arrived in Taiwan to interview master's students at Chiao-Tung on its campus and decided to hire Andrew Chew as the chief engineer and Gibson Wang (Jisong Wang) as a product engineer. Both planned to obtain their master's degrees in electronics in the spring of 1966. Another Chiao-Tung student in the same program was recruited the next year. Chew and Wang flew to the United States for a six-month training in May 1966. Chew was impressed with the well-arranged workplace and rigorous work culture at Philco-Ford. The trip changed his previous impression of Americans. In contrast to the demeanor of the off-duty US servicemen whom he often saw in his hometown port city of Keelung, he was surprised by US engineers' willingness to share with him every detail of the production of electronics, even though he was simply a stranger from afar. Returning to Taiwan, Chew oversaw electronics production of a factory, earning a monthly salary of $12,000 NT, which was then thirty times the salary of a factory woman, or fifteen times that of a teacher.[62]

While international companies poured financial resource into plants and facilities in Taiwan, most top-level engineers in the plants were locals, due to both the high cost of stationing a US engineer overseas and the high capability of local engineers. A door to door trip from the United States to Taiwan may take about twenty-four hours in 2024; it would have taken longer in the 1960s. Also, most US engineers would not be able to speak Chinese or Taiwanese. In addition, the labor-intensive assembly of transistors and other electronics could be arranged in a way that did not require much intervention from engineers or managers at US headquarters. From the perspective of international companies, it only made sense to outsource those production processes that would require minimal supervision from headquarters. In this context, local engineers played an important role in making the factory work.

Chew's autobiography emphasized his extraordinary skills solving unexpected technical factory problems. Chew noted that even though engineers and technicians were well-trained, material conditions in Taiwan sometimes failed to meeting production demands in his factory. Improving the quality of various source materials and the condition of equipment was the toughest part of his job as chief engineer at Philco-Ford. For example, for a time, he could not find a domestic supplier to provide heating devices with precise temperature control for soldering wires. He then sought technical suggestions from a military arsenal and eventually hired talent from the arsenal to manufacture the heating device in-house. Another challenge was securing supplies of nitrogen. Growing up with a humble background by the port of Keelung, he thought of sourcing nitrogen from workers there, who used oxygen-fuel torches for cutting metal materials and could collect nitrogen as a by-product of the work. Chew noted that other electronics manufacturing factories such as Philips had similar problems, but he boldly looked for alternative sources. He explained that he could do so because the assembly work in Kaohsiung Electronics did not required nitrogen of specific quality, whereas Philips' management culture somehow asked for more rigorous procedures to procure nitrogen.

The physical distance between the United States and Taiwan also forced Chew to improvise. On an occasion of unexpected damage to an IC testing device during shipment from the United States, Chew found that it would take six months to send it back, fix it, and receive the testing device again in Taiwan. The device was supposed to test twenty-four functions of a new IC that Chew was going to deliver to customers. Chew then worked with his engineers and found a solution: they made twelve different testing devices, each of which was able to test two functions of the IC. Though it required more steps and labor to go through the testing, these locally improvised testing devices solved the unanticipated problem and ensured on-time delivery. The resourceful engineer Chew eventually left Kaohsiung Electronics, even though Philco-Ford offered him opportunities to transfer to the United States, which would have been an incredible opportunity for many Taiwanese at that time. As a member of the Abnormal Club, he chose instead to participate in founding Unitron in 1969. The company, one of the first electronics manufacturing start-ups in Taiwan, included IC packaging, testing, transistor fabrication, and magnetic core memory assembly services.[63] He soon led Unitron to develop the first calculator made in Taiwan and later worked on semiconductor computer assembly for the GDR government in East Germany (see chapter 7).

Bruce C. H. Cheng (Chonghua Zheng) was another engineer-turned-entrepreneur who had extensive experience at a US-owned factory and founded his own company to supply the local computer industry in the 1980s. Cheng moved with an uncle from China to Taiwan in 1948 when he was twelve years old. He graduated from the department of electrical engineering at National Cheng-Kung University in 1959. His first job was a five-year stint at Air Asia. He enjoyed his work on aircraft maintenance, especially avionics. But over time he developed concerns about servicing newer aircraft with only limited experience and technical support. Looking for a new job, he was immediately hired by the US electronics manufacturer TRW when the company decided to set up a factory in Taiwan in 1966 to make capacitors and convergence yokes for TV sets.[64] He then went to Illinois—Watseka and later Chicago—for training. Cheng noted in his biography that the workers in Watseca knew their factory might close if the factory in Taiwan could soon operate successfully. Watseka workers, with a sense of gallows humor, laughed that if Cheng and others learned fast, the time for their unemployment would arrive sooner. But the workers remained enthusiastic about teaching Cheng and his fellow Taiwanese colleagues every part of the production, and they generously took turns inviting Cheng and others to each of their homes.[65]

One aspect of his time at TRW illustrates the legacy foreign-owned electronic manufacturers had on domestic companies. To save expenses in shipping source materials overseas to Taiwan, TRW authorized Cheng and his colleagues to look for and work with Taiwanese companies to procure source materials including reels and plastics. It usually took a long time, however, for local companies to come up with products with precise measurements and the proper quality. Cheng believed that contracts with local companies prompted Taiwanese suppliers to become capable and then versatile and flexible with customer requests, a key element in supporting the growth of Taiwan's computer industry in the 1980s.

But Cheng noticed that his quality-control work no longer required him to keep abreast of developments in his technical fields. He also found himself powerless when TRW headquarters laid off the best-paid and most experienced workers when orders were dropping. After the factory's orders grew again, the company would soon hire workers with less experience and lower pay to ensure a sufficient labor force. Cheng began to consider leaving TRW.[66]

One day in 1971, instead of being driven to work, Cheng rode his bike and unexpectedly spotted factory space for rent. He then inaugurated Delta Electronics. He hired fifteen employees to make coils and IFTs (intermediate frequency transformers, TV components), products similar to TRW's, but he promised his former TRW supervisor that he would not approach TRW's customers for three years. With the company steadily growing with high-quality products, TRW even purchased parts from Delta Electronics when needed. In the 1970s, Cheng's major customers were RCA and Zenith. Delta also worked with North American Philips to make laser disc players and supplied parts to Taiwan's Philips, meaning he had bested Philips' Japanese suppliers. The rise of the PC market around 1980 prodded Delta Electronics to grow even larger. Delta developed EMI (electromagnetic inference) filters and invested in making technical advances in power-switching supplies; both are essential parts for a personal computer.

Starting in 1978, Cheng worked with Acer (then known as Multitech) to supply its first product—the microcomputer "Micro-Professor I"—and became a key supplier of power components to other Taiwanese computer manufacturers thereafter. Delta also became a contractor for IBM in the 1980s and for Apple and Tesla recently. In 2020, company sales reached $10 billion US, a result of a COVID-19 surge in consumer demand.[67]

Similar to Chew and Cheng, Y. C. Lo (Yiqiang Lou) had experiences at Philips solving technical problems at an offshoring factory. While Lo did not start his own company, his improvisation and wisdom running the Philips factory was replicated by other local engineers. Not all such wisdom and experience would be recorded in biographies such as Chew's and Cheng's. But Y. C. Lo, not surprisingly, appeared in print, too. In 1996 he became the first Asian board member in the company's 150-year history, one of six members of the management board.

Lo, a physics graduate from National Cheng-Kung University, worked for Chiao-Tung as a physics instructor right before he was hired by van Bergen as an entry-level engineer in 1969. In no time van Bergen was satisfied with Lo's management of IC production. When van Bergen took a new position to reorganize the neighboring magnetic core memory factory in 1972, he told the Dutch headquarters that there was no need to send a Dutch manager to Taiwan to oversee the IC plant; Y. C. Lo was more than capable.[68]

For the IC packaging process, the factory needed purified water and nitrogen. During the first three months of Y. C. Lo's employment at Jianyuan, he and van Bergen had to examine malfunctioning water purification machinery designed by a US company and then nitrogen density monitoring machinery designed by a German company. Van Bergen insisted that Lo and he should locate problems, as it would be impossible to fly engineers from overseas locations every time a machine had a glitch. It took them weeks to solve each machine malfunction. Soon after the two episodes, a new problem emerged: Lo found that the yield of chips suddenly went down; 40 percent of chips were defective. Van Bergen and Lo carefully looked into every part of the assembly line and found no explanation. Lo then removed the plastic that covered a faulty IC chip and found the gold wires were distorted. He then reexamined the molding facility and realized a loose screw unexpectedly disrupted the flow of plastic. These examples show how local engineers were critical to the performance of the overseas plants of international electronics manufacturers. Jianyuan was Philips' first IC packaging plant outside Europe, but Jianyuan's yield eventually reached 92 percent, much exceeding the 72 percent rate requested by headquarters. Van Bergen was proud that, when Jianyuan and a Philips' British plant each sent an IC sample to their client IBM to review, IBM found that Jianyuan's sample was qualified while the other sample was not.[69]

One of Y. C. Lo's remarkable achievements was initiating a project to automate the process of wire bonding. In 1972, Zhexiong Chen, one of Acer-founder Stan Shih's classmates at Chiao-Tung, joined the engineering staff at Jianyuan, where he found many other Chiao-Tung alumni. Lo encouraged Chen to lead a team to explore ways they could automate some sections of the production process. Chen designed a computing system that used Texas Instruments' arithmetic logic unit chip 74S181 to compute the coordinates with which a machine would be able to execute automatic wire bonding. The machine increased productivity significantly. Without the automated system, each operator with a wire bonding desk could solder wires for a hundred chips per hour, on average. But the computerized automation system would allow operators, mostly women, to supervise two automated wire bonding desks, each of which bonded 150 chips per hour. Overall, the productivity was tripled. The operators' wire bonding skills thus became less relevant—and so did their eyesight, according to Chen. The first automated wire bonding system was named the "QQ line." QQ referred to *quaiquai*, a word that means "disciplined" and "well-behaved" in Chinese. Ironically, according to Lo, even though the team achieved significant increases in productivity though automation, Philips' Dutch headquarters had no interest in supporting more installations of these automated systems. Headquarters did not think the automated system could be tailored to the international enterprise's various IC products. A few years later, however, when the IC market's expansion surged, headquarters was finally willing to invest funds to automate all of the 136 wire bonding desks in Jianyuan. In 1981, Philips began to automate bond wiring for all its plants across the globe. Lo's biographer Manpeng Diao noted that this example showed how "local technologies" flow in a reverse direction to contribute to an international enterprise.[70]

Y. C. Lo also authorized Chen to work on automating electroplating after the automatic wire bonding system was online. Since the beginning of the operation, the variety of ICs that Jianyuan worked on required no electroplating. At that time, automatic electroplating systems were used in other types of ICs, and in some of Philips' European factories, and it seemed to be gradually evolving into a mainstream method. Headquarters, however, believed that because labor costs were low in Taiwan, it was unnecessary to invest in machinery to automate IC packaging. Despite headquarters' lack of interest in automating projects at Jianyuan, Lo was ambitious about the development of Jianyuan and strove to expand the variety of IC products at Jianyuan. He anticipated the factory might soon require automating electroplating to increase its business opportunities and productivity. He then sent his engineers to tour other IC packaging factories in Taiwan that included electroplating in their assembly line. Lo's friends in such factories agreed to show his engineers around, though it was not clear whether the foreign owners or investors of these IC packaging factories were aware of these visits. His engineers soon created a mechanical system to automate electroplating. Lo then gained leverage—first, by making IC samples to show potential customers (for example, other Philips' subsidiaries that made wafers) that his factory could do electroplating on IC, and, second, by urging the Philips headquarters to invest in the automation of electroplating in his factory.[71]

Y. C. Lo's biographer noted that he was able to prevent layoffs at Jianyuan when Philips called for them in the 1970s. In 1971 and 1972, an economic recession brought Jianyuan challenges in receiving enough orders from Philips headquarters, which decided that it was time to lay off personnel in Taiwan. Y. C. Lo wrote reports to headquarters to suggest an implementation of job trainings instead. Lo thought that it was a perfect time to offer training, since there were no jobs for the assembly line. He believed that his factory workers were highly skilled and needed to be retained to ensure the quality of products in the long run. He then utilized a rainy-day fund of profits that Jianyuan had set aside before the recession to get through the challenging times.[72]

As an expert in IC packaging, Y. C. Lo had been interested in IC wafer manufacturing for a long time. He had a strong belief that if his factory could make good IC packaging, then there should be no problems with wafer fabrication. As early as 1974, he proposed that Philips set up a wafer fabrication factory in Taiwan, but it was never approved. In 1985, when Morris Chang was raising money for the founding of TSMC in Taiwan, neither foreign nor domestic companies expressed much interest in investing in his new company. But Philips' headquarters decided to offer 27.5 percent of the start-up funds, and licensed patents to TSMC. As the chief manager of Jianyuan at that time, Y. C. Lo gladly took responsibility for persuading Philips headquarters to invest, as he had long believed that Taiwanese engineers and workers could do it. Several groups of TSMC engineers began their jobs by flying to the Netherlands to be trained at Philips. In the 1990s, the value of TSMC stock was one of the company's most profitable investments.[73]

ENTREPRENEURS FROM PHILCO-FORD AND BEYOND

In addition to Chew, who founded Unitron, at least two engineers (Zhengming Ou and Raymond Soong) left Philco-Ford's Kaohsiung Electronics and founded or joined two electronics manufacturing start-ups in Taiwan. Two Kaohsiung Electronics managerial staff (Stanley T. Wang and Frank Yih) left the company and founded two companies overseas. Three such "spin-offs," broadly defined, also began with IC packaging and testing as their major focus. Except for Unitron, the majority of the "spin-offs" remain active in global computer peripheral and semiconductor industries today.

Zhengming Ou was employed by Kaohsiung Electronics in 1966. He was born in Taiwan in 1944, and graduated with an electrical engineering BS degree from Cheng-Kung in 1962. In 1971, Ou and a few of his colleagues joined and helped with the initial operation of Orient Semiconductor Electronics (Huatai), one of the first domestic companies that fully focused on this field.[74] In 1984 he founded Universal Microelectronics (Huanlong keji) to make power-switching supplies. In the 1990s, his company made mostly internet-related devices. Ou's personal assets were estimated to be around $140 million US in 2000.[75] California-based Pantronix is an IC packaging company founded by a former Philco-Ford employee Stanley T. Wang (Dazhuang Wang). The company's semiconductors were used in Patriot missiles, F-22 fighters, and in the fields of health care and telecommunications.[76]

Frank Yih, Philco-Ford's MIT-trained and former Fairchild employee, was dubbed a "patriotic youth (aiguo qingnian)" by a Taiwanese newspaper. The major reason he left was because Philco-Ford's "Research and Development Department failed to bring out new and advanced IC products." Yih left for Silicon Valley to join the founding of Intersil. He contributed 18 percent of the company's start-up funds, while Fairchild Semiconductor's cofounder Jean Hoerni contributed 82 percent. Yih then relocated to Singapore to work as managing director for Intersil's Singapore facilities. But he soon returned to Taiwan in 1973 and founded his own company, Asionics. He considered himself the pioneer of Taiwan's electronic watch industry. In 1992, he returned to Shanghai to worked with a Chinese company to set up facilities to do IC assembly and later became a real estate developer and full-time philanthropist. While spending most of his time in Los Altos, California, and Shanghai, in 2019, at age eighty-five, he organized a reunion of hundreds of former Philco-Ford employees of all ranks to return to the Kaohsiung export processing zone.[77]

Born in Kaohsiung in 1942, Lite-On founder Raymond Soong (Gongyuan Song) worked at Philco-Ford's Kaohsiung Electronics in semiconductor packaging from 1966 to 1969. He also briefly worked as the chief engineer for Texas Instruments' Taiwan plant and worked on LED (light-emitting diodes) packaging. In 1974, affected by the oil crisis, Texas Instruments planned to lay off a thousand employees in Taipei and in Singapore, with Soong's production line slated to close. He and two other Texas Instruments engineers then raised funds to set up Lite-On in 1975. Lite-On bought machinery from Texas Instruments to begin its LED packaging business, and in 1983 his company was listed on Taiwan's local stock exchange, the first Taiwanese electronics company to do so. Later, Lite-On pioneered producing computer peripherals such as CD-ROM players, printers, and computer and game console casings.[78]

Besides those from Philco-Ford, two top-level professional managers and a women entrepreneur from prominent Taiwanese electronics companies worked as engineers at foreign-owned electronics manufacturers in the 1970s. All of them are alumni of the Taipei Institute of Technology. Fuxiang Song worked over two decades at Taiwan facilities of RCA, Ampex, PC manufacturer AST Research, and flash memory manufacturer Kingston Technology and became the CEO of Simplo Technology, which manufactured lithium batteries in 1998.[79] Dugong Cai worked for EMM's subsidiary in Taiwan as an engineer for nine years, for Ampex for five years, and he worked for AST Research and Kingston for years. Cai helped his former EMM colleague David Sun, who founded Kingston in California, to begin, expand, and manage Kingston's Far East manufacturing facilities. In 1999, he and Sun invested in Powertech Technology Inc. (Licheng), which focused on DRAM packaging and testing. He soon became the CEO of the company and shifted the company's focus from packaging and testing DRAM to a wide range of semiconductor products.[80] Suimei Chang was one of Taiwan's few women electronics entrepreneurs. Chang was born in 1953 and a graduate of the Taipei Institute of Technology around 1975; she was one of eight women in her class, which had over forty-eight students in total. After graduation, she was employed by a Japanese-owned electronics company, Taiwan Showa Electronics (Taizhao dianzi), to do research and development, and had participated in designing and manufacturing keyboards for IBM and Apple computers. She noted that during seven years at Taiwan Showa, she had learned not only design but also the nitty-gritty of manufacturing processes such as metal stamping and plastic injection molding. She then joined the Taiwanese subsidiary of a US electronics manufacturers, Arvin, which made satellite broadcasting devices in its Taiwan plants. In 1985, at thirty-three years old, Chang started her own company, Chyao Shiunn Electronic Industrial Ltd., to make connectors for TV sets, audio systems, and telephones, immediately winning contracts to supply connectors to two companies that made push-button telephones, which were replacing rotary dial telephones and in great demand. Her company continues supplying connectors for telecommunications and computer products.[81]

CONCLUSION

International electronics manufacturers set up factories in Taiwan beginning in mid-1966. Inexpensive Taiwanese labor was instrumental in helping manufacturers remain competitive globally. Their products included magnetic core memory units, transistors, rectifiers, capacitors, and ICs for TV and radio sets, calculators, and computers, though Taiwanese factories were primarily responsible for the packaging process of these semiconductor products instead of fabricating them from scratch.

The factories provided jobs for Taiwanese women and men, though the opportunities were highly gendered. Considered valuable employees, women workers in electronics factories gained economic independence, but they faced mandatory overtime, low wages, and occupational health and safety violations. Writing about women factory worker's lives in his fictional stories, Ching-chu Yang highlighted work injuries when he set his protagonists in electronics factories. Women's opportunities to advance from factory jobs were limited. They did not get promoted frequently. Many did not have specific plans for future careers, and others did not have education opportunities or financial resources to move upward from factory jobs. The scattered oral history interviews and historical documentation of their experiences, however, indicates that women workers' willingness to deliver skilled assembly work was critical in ensuring high quality— high yield—electronics manufacturing in Taiwan. Thomas J. Misa has discussed that the formulations of Moore's law were a result of coordinated efforts by the semiconductor industry. Along the same line, I argue that Taiwanese electronics factory women enabled the miniaturization of electronics, and mass production of them, in the late 1960s and 1970s. Their assembly of electronics contributed to the realization of Moore's law and made possible offshore production of highly sophisticated electronics for multinational corporations.[82]

Asked to create standardized products on assembly lines, women factory workers were required to engage in trial and error, tinker with electronics, and repeatedly practice their skills on a long list of tasks—die attachment, wire bonding, molding, electroplating, trimming, printing marks on the top of the chip, inspection, and so on. Women factory workers' employers and their supervisors or engineering colleagues, however, often recognized only the importance of eyesight for the assembly job, despite the presence of microscopes. The free milk, vitamins, and fish oil capsules, provided by employers to strengthen their women workers' health and eyesight, also diverted attention from women's skill in manipulating parts on a tiny transistor or magnetic core memory unit. Employers also did not recognize women workers' commitment to high-quality work— "no merest blemish," and "no tiny dent," as Xiang Ye's poem reveals. Although Fairchild's brochure highly racialized Navajo women factory workers' assembly skills by associating them with Navajo's traditional weaving culture, the brochure noticed women's skills. In contrast, Taiwan electronics factories' employers and supervisors attributed the high quality of women's assembly work to eyesight and failed to "see" women's skills at all.

Women worker's reflections on their jobs went beyond employers' limited understanding of the nature of the jobs. Many women workers in 1960s and 1970s Taiwan developed a unique perspective on computers. Acknowledging the extraordinary power of a computer's information storage, a quality-control worker at Unitron commented that factory shop employees could excel by emulating computers through memorization. Women electronics factory workers even noted that they expected to be called by their specific job titles, instead of the generic category of "women workers," which did little more than indicate their gender, social class, and low status in the hierarchy of a factory.

In contrast, better-educated engineer colleagues, mostly men, at the same factories witnessed the scaling up of electronics manufacturing and aspired to start their own businesses. The engineers in these overseas factories were critical in maintaining the operation of the factories with no immediate technical support from their foreign employers. They had to improvise technically to ensure the smooth operation of the production line. Y. C. Lo and Andrew Chew were examples of resourceful engineers who handled unexpected technical issues by tinkering with factory facilities and making devices in-house. Delta Electronics founder Bruce Cheng began his career as an aircraft maintenance engineer and shifted to the production of TV and computer parts. His experience working with local suppliers on the quality of source materials also constitutes the act of tinkering. He had to improve the production of TV and computer parts by procuring higher-quality source materials and product designs. Running the overseas factories, engineers gained access to the technical knowledge of manufacturing, valuable experience on the assembly line, and a savvy understanding of the semiconductor market. The experience granted the well-educated engineers, mostly men, the knowledge and confidence needed to start their own factories when opportunities arose. Several new companies were set up and run by their founders or top executives after leaving international electronics manufacturers. These domestic enterprises were founded from 1969 to 1999, but most opened by 1974. They offered similar services and products as the companies for which they had worked, ranging from TV parts to IC packaging and testing. Many became computer peripheral manufacturers, representative of Taiwan's 1980s networks of computer part suppliers.

The ten entrepreneurs and executives discussed in this chapter once worked for Philco-Ford, TRW, Texas Instruments, Wang Laboratories, Ampex, and others. Philco-Ford's Taiwan factory alone trained a group of engineers that set up or helped set up the first group of domestic electronics manufacturers, including Unitron in 1969 and Orient Semiconductor Electronics in 1971. Former TRW engineer Bruce Cheng became the founder of Delta Electronics. Former Philco-Ford and Texas Instruments engineer Raymond Soong became the founder of Lite-On. Both enterprises had been prominent computer peripheral suppliers since the 1980s. Delta Electronics, especially, worked with Acer closely in developing and manufacturing Acer's computers in the 1980s (chapters 8 and 9). While not an entrepreneur, Y. C. Lo, an engineer of one of Philips' most remote factories in the 1970s, remarkably advanced to the board of directors in the 1990s. Philips' long investment in its Taiwanese subsidiary also created an opportunity for it to join the founding of TSMC and enjoy a great financial return from it in the 1990s. In chapter 7, I discuss another group of entrepreneurs who made calculators in the first half of the 1970s and shifted to computer part suppliers in the 1980s.

7 MASS-PRODUCING CALCULATORS: SOLDERERS, ENGINEERS, AND ENTREPRENEURS

Chung Tung received his BSEE degree from National Taiwan University in 1961 and his MSEE degree from the University of California at Berkeley in 1964. As a member of HP Laboratories from 1965 to 1972, he was involved in the design of the 9100 Calculator and the HP-35 and HP-45 Pocket Calculators. . . . He is married, has a son and a daughter, and lives in Santa Clara, California. Music, reading, and table tennis are his principal modes of relaxation.

—Chung C. Tung, May 1974[1]

Having grown up in Taiwan, and with an undergraduate degree from there, Chung C. Tung joined Hewlett-Packard as a researcher in 1965 and was subsequently promoted to a manager around 1972 to lead development of the HP-65, the first programmable pocket calculator. Two years later, when the project he managed bore fruit, he penned "The 'Personal Computer': A Fully Programmable Pocket Calculator," in the *Hewlett-Packard Journal* (quoted above). Tung's article was one of the early appearances of the term "personal computer" in print. Three months later, in Taiwan, *Economic Daily News* wrote about Tung as an expatriate scientist leading the development of the state-of-the-art calculator. But no one in Taiwan really knew much about the guy who bravely took ownership of the term "personal computer." The article offered no new news; probably an overseas reader of the journal came across Tung's article and passed it to a journalist at *Economic Daily News* to showcase the accomplishments of engineers from the island.[2]

When Tung and other professionals participated in the development of and embraced the concept of "personal computers" in the United States, entrepreneurs in Taiwan searched for business opportunities to develop and assemble calculators. Several local Taiwanese entrepreneurs began mass-manufacturing calculators after the three student groups experimented in making computers and calculators on their campuses from 1968 to 1972 (chapter 5). The first of these locally made calculators went to market in April 1972. Leading engineers and entrepreneurs at the local calculator start-ups tended to have either participated in university-based minicomputer-building projects or were employed by international electronics manufacturers. In this chapter I present another emerging group of engineers-turned-entrepreneurs whose work on calculators in the first half of the 1970s differed from those who focused on electronics manufacturing (chapter 6). The calculators added to the abundant market of electronics products and parts manufactured by Taiwanese engineers and factory workers. More and more local entrepreneurs and engineers in the 1970s seized on business opportunities and eventually contributed to the mushrooming Taiwanese computer and computer parts industry of the 1980s.

Desk and portable calculators available on the market in the early 1970s occupy a unique place in the history of computing. In US society, portable calculators stimulated new forms of culture. With the aid of Texas Instruments and Hewlett-Packard, users of the TI-59 and HP-41CV calculators shared software in their respective communities in the late 1970s, and user communities evolved across more than thirty countries.[3] Historian Paul Ceruzzi, for one, chose a Hewlett-Packard portable calculator, from a list of new technologies he owned during his early career, to illustrate examples of Moore's law and its implications for technological determinism.[4] Specifically, his HP-25C calculator replaced his "beloved" slide rule.

In Fred Turner's research on the *Whole Earth Catalog*, a Hewlett-Packard 9100A Calculator, priced at $4,900 US ($40,474US in 2024), was handpicked by editor Stewart Brand to review in one of the issues. The *Catalog*, published from 1968 to 1971, appealed to followers of the back-to-the-land movement but eventually enjoyed a wider circulation beyond communes. At first glance, one would not consider an expensive desk calculator a necessary tool for New Communalism. The calculator was more expensive than an average car at that time. But Turner has demonstrated that the presence of the calculator in the *Catalog* spoke to New Communalists' avid interest in both cybernetics and small-scale technologies that could help attain their goals of transforming their minds. Their embrace of small-scale technologies can be juxtaposed with their rejection of mainframe computers, as the latter were deemed products of the military-industrial-academic complex. After the *Catalog* ceased to publish regularly, Stewart Brand wrote a 1972 *Rolling Stone* article that featured Xerox Palo Alto Research Center (PARC). In it appeared a photograph of a US $400 Hewlett-Packard pocket calculator. The calculator was comparable to other novel cultural artifacts, such as Spacewar computer game tournaments, the conceptualized Dynabook as a tablet computer for children, and the research center's beanbag chairs, which signified its cozy working environment. Brand included Xerox PARC researcher Peter Deutsch's remarks about portable calculators. "They'll reach millions when computer power becomes like telephone power. . . . I think it's important to bring computing to the people." The image of the portable calculator in the *Rolling Stone* article symbolizes the gradual distribution of computing power to more individuals in the early 1970s. The distribution also reached Taiwan around the same time, and Taiwanese businesspeople further committed to producing calculators for a global market.[5]

Japan's accomplishments in mass-manufacturing desk calculators began in the 1960s and came to dominate the global market since the late 1960s. Historians of Japan's computer industry consider their desk calculators a key part of that history. The Information Processing Society of Japan (IPSJ) certified Sharp's CS-10A electronic calculator, developed in 1964, as one of the artifacts of Japan's Information Processing Technology Heritage. It was one of the world's first all-transistorized desk calculators.[6] In Japan, the leading calculator makers in the 1970s included Sharp, Canon, Sony, Busicom, and Casio. The demand for better chips from the industry in Japan prompted the building of the first commercialized microprocessor, the Intel 4004, designed and fabricated by Japanese engineers at Busicom and their counterparts at Intel, as William Aspray's research has shown. Among the Japanese calculator companies, Sharp, Canon, and Sony continued to play a dominant role in the personal computer market in the 1980s.[7] In the Soviet Union, the mass-manufactured programmable calculators in the mid-1970s contributed to an "early digital" culture there.[8]

In the United States, several microcomputer and personal computer makers produced calculators in the early 1970s. For example, forerunners Commodore and Micro Instrumentation and Telemetry Systems (MITS) made calculators before they made microcomputers. Commodore sourced Texas Instruments chips to make calculators before it acquired Chuck Peddle's company and begin to make microcomputers. Peddle was the designer of MOS Technology 6502 microprocessors, which were used for the Apple I computers and the Commodore PET. Similarly, Ed Roberts, who founded MITS in 1969, began to sell programmable calculator kits in the early 1970s. The recession following the oil crisis and the competitive calculator market negatively affected MITS alongside many calculator makers. Ed Roberts then began his microcomputer business. Along with Commodore, several companies that were well known for their calculators in the 1970s, such as Hewlett-Packard and Texas Instruments, also became big players in the personal computer market in the 1980s.[9]

Taiwan's calculator makers, along with electronics manufacturers, succeeded in the 1970s in cultivating an industry that trained and employed both an increasing number of engineers and factory workforces. This chapter focuses on three major calculator start-ups in Taiwan in the early 1970s: Unitron, Qualitron, and Santron, which were also founded by engineers-turned-entrepreneurs, similar to many of those entrepreneurs in chapter 6. I begin with the history of Unitron, the first domestic electronics assembly factory in Taiwan and the first local manufacturer of calculators. Unitron's varied products and services reflected strategies that early Taiwanese electronics manufacturers adopted to compete with well-established international manufacturers. At Unitron and Qualitron, Acer Computer founder Stan Shih has his first and second jobs. In 1976, he founded Multitech, renamed Acer in 1987. Santron provided first jobs to the entrepreneurs of several computer manufacturers. For example, Barry Pak-Lee Lam, the founder of Quanta Computer, a leading notebook computer manufacturer today, began there.

UNITRON: A JOURNEY TO THE UNIVERSE OF ELECTRONICS

The first locally made calculator was assembled by Unitron, which was in business from 1969 to 1974 and one of the first domestic electronics manufacturers in Taiwan. The literal translation of the company's Chinese name, Huanyu, is "traveling around the universe." After 1974, Unitron was acquired by the North America–based International Telephone and Telegraph Company (ITT). Andrew Chew, whom we met in chapters 5 and 6, provided leadership in building, managing, and making Unitron profitable. Chew worked at Philco-Ford's subsidiary Kaohsiung Electronics till 1969 and founded Unitron with Simon Min Sze. Unitron's seed money came primarily from the investment of a family that owned a textile business. Unitron's business covered a wide range of products and services, which reflected Chew's strategy of diversifying the company's contracts. It began with assembling transistors, integrated circuits, rectifiers, magnetic core memory, digital alarm clocks, frequency counters, walkie-talkies, and, eventually, desk calculators. The list of products represented the labor-intensive nature of a series of technologies that Chew found Unitron could take advantage of. On Chew, Stan Shih noted in the foreword to Chew's biography, "Some might think of me as a pioneer, a member of the first generation that built Taiwan's electronics industry. But my senior Chew is of the zeroth generation." Barry Pak-Lee Lam referred to Chew as the "father of the semiconductor" in Taiwan. While it is an overstatement, it shows the respect Chew won in the industry.[10]

In 1968, Simon Min Sze (see figures 7.1, 7.2, and 7.3) took a leave from Bell Laboratories and returned to Taiwan to teach at Chiao-Tung as a visiting professor. During his one-year stint, he was introduced to Chew, who showed strong interest after hearing

FIGURE 7.1

Simon Sze, the second man from the right, the cofounder of Huanyu, giving a tour of Unitron (Huanyu) for Minster of Economic Affairs Yun-suan Sun (the fifth man from the right) and Vice President Chia-kan Yen (the third man from the right), in 1973. They were observing tasks conducted by women electronics assembly workers. *Source:* Figures 7.1, 7.2, and 7.3 are from pp. 4, 6, and 8 of "Fuzongtong Yan Jiagan canguan Huanyu dianzi gongye gufen youxian gongsi" [Vice president Chia-kan Yen visited Huanyu Electronics], "Minguo liushier nian Yan Jiagan fuzongtong huodong ji (shiyi)" [Vice President Chia-kan Yen events in 1973, part 11], Yan Jiagan zongtong wenwu 1966–1972 [President Chia-kan Yen files 1966–1972], archival document no. 006-030204-00021-004, July 10, 1973, AH. Reprinted with permission from Academia Historica, Taipei, Taiwan.

Sze's proposal about starting up an electronics manufacturing company in Taiwan to offer jobs to talented students trained by Chiao-Tung. The two connected through their Chiao-Tung network and then began to build a new company.

Sze moved from China to Taiwan in 1948 when he was a junior high school student. He studied at National Taiwan University for his undergraduate degree from 1953 to 1957 and obtained a master's from the University of Washington in Seattle and a PhD from Stanford University from 1959 to 1963. At Stanford, he worked with John L. Moll, who later advised Morris Chang and Chun-Yuan Duh (cofounder of Orient

FIGURE 7.2

Women electronics assembly laborers working at their desks at Unitron (Huanyu) in 1973. Yun-suan Sun standing in the back of the room, the second from the right, when visiting the company with Vice President Chia-kan Yen, the man in the middle with a black suit. Simon Sze, the tallest man in the back, briefing Yen. Reprinted with permission from Academia Historica, Taipei, Taiwan.

Semiconductor Electronics; see chapter 6). Sze then began his research career at Bell Labs. In 1967, when Sze was thirty-one years old, he invented floating-gate nonvolatile semiconductor memory, with D. Kahng, Sze's colleague at Bell Labs. Floating-gate memory contributed to the later development of flash memory. He wrote a popular textbook, *Physics of Semiconductor Devices*, the first edition of which was published by Wiley in 1969; the second, third, and fourth editions were published in 1981, 2007, and 2021, and at least 1.5 million copies have sold since its initial publication.[11] In 1991, he received the IEEE Electron Devices Society's J.J. Ebers Award for his "fundamental and pioneering contributions, and the authorship of widely-used technical text and [image resonance] books, in the field of electron devices."[12] Stan Shih lauded Sze for planting the seeds for Taiwan's high-tech industry. Shih believed that Sze "built a bridge" between Unitron and the US electronics industry. To Shih, Unitron was a

FIGURE 7.3
Women workers using microscopes in assembling electronics while Simon Min Sze (man on the right) gives a tour of Unitron (Huanyu) for Chia-kan Yen (second man from the right) and Yun-suan Sun (third man from the right) in 1973. Reprinted with permission from Academia Historica, Taipei, Taiwan.

cradle of talent, including himself and Jiahe Lin, who later became Acer's first director of R&D.[13]

Sze's visiting professorship at Chiao-Tung was endowed by Hong Kong–based freight mogul Chao-yung Tung. Sze met Tung in New York before he took the professorship. Tung told Sze that he was glad to offer support to Chiao-Tung, as many of his prominent employees were Chiao-Tung alumni. It might have been uncommon for an employed engineer to take a yearlong leave for a fellowship in 1968, but MIT professor and Chiao-Tung alumnus Lan Jen Chu phoned his friend, Bell Labs' president James Fisk, and persuaded Fisk to grant Sze a leave. Chu and Fisk knew each other, since they were graduate students together at MIT in the 1930s. Sze's visit to Taiwan was possible only because of overseas Chinese connections, as well as their enthusiasm about Chiao-Tung.[14]

At Chiao-Tung, Sze decided to found a company to offer job opportunities for Chiao-Tung's electrical engineering graduates. He noticed that many graduates were employed by foreign-owned electronics manufacturers, as there were not many local ones. Despite Chiao-Tung's excellent education, the university seemed to train engineers only for foreign-owned companies. His worry was consistent with that of the Abnormal Club (chapter 5), and Chew is a member of the club. Sze's professional identity as an electrical engineer might have factored into his decision to offer jobs for Chiao-Tung graduates. After getting to know Chew, Sze was introduced to Peiyuan Lin, a Lin family member in Lukang. They owned a large textile company established at the end of World War II and were interested in investing in a tech company. When Unitron was set up, the company's registered capital was $50 million NT (or $1.6 million US, the equivalent of $14 million US in 2024). Sze and Chew each invested 10 percent of the capital money, and the remaining financing was from the Lin family. Sze purchased land of about 192,000 square feet and built a three-story factory of about 71,000 square feet as well as a dormitory for up to four hundred workers. Sze and Chew recruited several Chiao-Tung master's students to work at Unitron. In the first few years, Sze arranged for the company to donate 5 percent of its profits to Chiao-Tung.[15]

For Chew in 1968, giving up his well-paid job was not an easy decision. He debated leaving Kaohsiung Electronics, knowing that his salary was going to shrink by half. But he was also keen to contribute to a company that was primarily locally owned. He aspired to make Unitron become Philips one day. To Chew, Philips was a respectable rival to Philco-Ford, since they were both among the earliest and largest international companies to set up electronics plants in Taiwan.[16]

After spending a year teaching and setting up a new company, Sze had to go back to Bell Labs in 1969 to continue his original full-time job. But acting as a consultant for Unitron, he passed along news about the latest semiconductor techniques to Chew, who ran the company's sales and R&D, though nominally the general manager was from the Lin family. By February 1971, Unitron had hired five hundred employees. At its peak, the company employed 1,200 workers, nearly a hundred engineers, and several dozen managers. Surviving photographs of the company show that most of the assemblers were women (figures 7.1, 7.2, and 7.3).[17] In 1971, the company spent $1 million NT for its R&D department. The R&D budget increased to $2 million NT at the end of 1972 ($64,000 US, or $490,000 US in 2024). Chew was proud that Unitron hired more engineers with a master's degree than many other state-owned enterprises in Taiwan at that time. A subdivision of the R&D department was responsible for looking for applications of MSI (medium-scale integration) chips. The chief director of the R&D department was Chiao-Tung alumnus Min-wen Kao. Most R&D employees were Chiao-Tung graduates.[18]

Stan Shih joined Unitron in 1971 after graduating from Chiao-Tung's master's program in electronics. He was born in 1944 in Lukang, Taiwan. Shih was also recruited

by Philips' Jianyuan Electronics, but he humbly stated in his biographies and oral history interviews that he declined a foreign-owned company because he thought that his
English was not good enough. Unitron, as a local company, beat Philips and won the
soon-to-be pioneer in computer manufacturing. Shih and Jiahe Lin joined Unitron's
R&D department at the same time; the two had been classmates at Chiao-Tung, and
they later founded Multitech. Stan Shih noted that, beyond the R&D department, some
technicians were extremely talented. He recalled that he met a technician who did not
go to college but was truly an inventor genius. Within a year, Stan Shih's $7,000 NT
monthly salary at Unitron had climbed to $10,000 NT. According to Simon Min Sze's
biographer, Shih's salary was only one-third of the average salary for a similar position
in a foreign-owned company. While the comparison may be overstated, foreign-owned
companies did pay better.[19]

DIVERSIFICATION OF PRODUCTS

Before Unitron's calculators were developed in 1972, the nascent company worked on
assembling or making different electronics. Its workers also assembled magnetic core
memory, as mentioned in chapter 6. Chew also took orders to assemble beepers and
walkie-talkies for Bell Canada. He noted that Unitron had to make more than a hundred components for the walkie-talkies and overwhelmingly relied on human labor to
put together the components. Challenging as it was, it enhanced the company's confidence and capability and led Chew to enter the calculator market.[20]

For 1972, the company expected to produce eighty million transistors, twenty million integrated circuits, fifteen million diodes, and twenty-five thousand magnetic core
memory planes.[21] One of Unitron's ambitious plans was to move from integrated circuit assembly to manufacture. By March 1972, Unitron was already able to produce
IC packaging with either plastic dual in-line or ceramic dual in-line packages. At that
time, Unitron had begun planning the fabrication of bipolar junction transistors from
scratch. Chew shared with the journalists of Economic Daily News that Unitron was
going to manufacture a variety of semiconductor electronics at full capacity, using pho-
tomask techniques to print the design on chips and assembling and testing the chips.
By then, Unitron had purchased machinery for these tasks.[22]

Chew strategically diversified Unitron's services and products, with the goal of lowering the risk in the fluctuating market of semiconductor products. With years of experience in the industry, he realized peak seasons and off seasons alternated. Unitron's
business was tightly connected to the global market because large international manufacturers commissioned Unitron to share a portion of production beyond their original

manufacturing capacity. To manage off seasons, Chew explored products that Unitron could utilize to ensure there were year-round jobs to lower off season downturns and financial risk.[23]

Chew succeeded in implementing a strategy of diversification by having companies ship the specific capital equipment needed for a product from overseas locations to Taiwan. This was an unusual arrangement, but Chew persuaded the companies. According to Chew, investors from the Lin family did not offer funds beyond their initial $10 million NT. Chew was short of funds to purchase machinery after the purchase of the land and the construction of the factory. When a US company phoned Chew to inquire about integrated circuit production or assembly, Chew directly told his customer that he did not have the necessary machinery to do it. He persuaded the US customer, however, to ship the required machinery to Taiwan, as the total production cost in Taiwan plus shipment would be less expensive than that in the United States. On another occasion, Chew also signed a contract to make rectifiers for a Japanese company by asking the company to ship the necessary machinery to Taiwan. The Japanese company even sent engineers to Chew's factory. They were to be stationed in Taiwan and were to transfer specific knowledge and techniques for the production. While the example illustrates that Chew was a resourceful engineer and acute businessman, and that engineers and laborers were versatile at Unitron, it was also an example of the insufficient funds available to local start-ups. As UN expert Skoumal emphasized in his observation of the domestic electronics industry in 1970, company owners tended not to allocate profits to better the companies. Unitron's textile-industry-based investors were unwilling to invest more. According to Chew, more troubling was that when Unitron needed funds, the Lin family decided to loan money to Unitron with an interest rate of 24 percent. As a result, although Unitron made a profit every year, it was entirely allocated to paying back interest.[24]

However resourceful Chew was, one of the technologies he did not find feasible to assemble in Taiwan was fax machines. But his judgment to stay away from the fax machine market also shows his thoughtful decision in reserving resources for projects that he believed would succeed. Two companies, among others, that specialized in fax machine manufacturing at Chew's time were Xerox and Stewart Warner. According to Chew, Stewart Warner maintained a long-term business relationship with Unitron by procuring Unitron's services to assemble integrated circuits. At some point, Stewart Warner consulted Chew about the possibility of making fax machines in Taiwan. The US company shipped two fax machines for Chew to disassemble and examine. But Chew and his colleagues carefully considered and decided that the required techniques and machinery were beyond what Unitron could offer at that time.[25] In contrast, at the end of 1972, when Unitron's calculator business was thriving, Chew was so ambitious that he told journalists that Unitron would be able to make computers in 1974, indicating that Chew shared an avid interest in manufacturing computers with Chiao-Tung members such as S. M. Lee, Gisson Chi-Chen Chien, Ching-Chun Hsieh, and Jong-Chuang Tsay (chapter 5).[26]

HUNTING FOR CHIPS FOR THE FIRST CALCULATOR—*BAILI* (BIG PROFIT)

At Unitron, Chew closely monitored news about the market for integrated circuits. For a long time, he was interested in the integrated circuits used by Sharp for its calculators. But Chew could not directly purchase these chips customized for Sharp. At some point in 1971, he discovered that integrated circuits for calculators were available from other companies and took an international trip to procure the chips. Chew did not specify in his biography which company he visited. But, according to Stan Shih, the integrated circuits for Unitron's first calculators were LSI chips from Mostek, one of the start-ups among US integrated circuit manufacturers of the late 1960s that capitalized on MOS techniques. Mostek customized a chip, MK6010, for Busicom's handheld calculator in 1971. It was likely that Chew procured subsequent Mostek chips, which became widely used by various companies for making calculators.[27] Soon after Chew secured the integrated circuits, he immediately left for Fifth Avenue in Manhattan to find potential customers for Unitron's calculators. He persuaded a client to place an order for the desk calculators he was going to make in Taiwan, using only a diagram of the calculator he sketched on-site.

It took Unitron engineers eight months of research and development before the company announced that they were ready to deliver four models of handheld and desk calculators in March 1972. Unitron's calculator became known in newspapers right after the completion of Cheng-Kung I calculator and right before the NTU minicomputer was announced (chapter 5). To advertise the calculator, Unitron solicited the public for submissions of calculator names. Eventually, the calculator was named Baili, whose two characters literately referred to "hundred" and "profit." A translation of Baili can be "big profit," with the connotation that businesspeople with the calculator could make a fortune. Baili was the first calculator made by a Taiwanese company. The sale price was $8,000 NT, equivalent to three months' salary for an average white-collar worker in Taiwan at that time, or $1,900 US in 2024. They were primarily for export. Chew originally estimated that Unitron should be able to make fifty thousand calculators per month, but it was all that Unitron could produce in 1972. The calculators added, subtracted, multiplied, divided, calculated square roots, and continued fractions up to twelve digits.[28]

Economic Daily News revealed that Baili calculators were developed by seven Chiao-Tung master's degree holders. Now we know that one of the engineers was Stan Shih. Shih identified a technical glitch in the initial batch of Baili desk calculators—key bounce. When a user pressed a number key, the key would sometimes repeat automatically. Shih fixed it with a debouncing mechanism. The numbers printed on the keys also wore out easily. Because the number keys were made by a company in Japan, Shih had no way to solve the problem directly. Another quality issue occurred on the printed circuit boards, which were made in Taiwan. They were fragile and could be damaged from physical impact. He aimed at improving these quality issues for the next calculator he was going to design. He did so—but only after leaving for another company, Qualitron.[29]

SELLING THE THRIVING COMPANY

In 1974, when its business was thriving, Unitron was sold to ITT. The deal was reached in part because major stockholders and managers did not get along with each other, which Sze and Chew both confirmed in their biographies. Peiyuan Lin, the major investor from the textile family and the general manager of Unitron, had his younger brother Sen Lin working as a vice-manager soon after Sen Lin returned to Taiwan with a master's degree in statistics and business management from the University of Memphis. The way the new vice-manager ran the company was at odds with the way Chew ran it. Chew once resigned from the company, likely in 1972, but the Lin family asked him to return within two months, as Unitron without Chew could not function. After the Lin brothers met with Chew and other shareholders, Sen Lin left with the calculator department and founded a new company called Rongtai, or Qualitron. *Rong* and *tai* stand for thriving and peaceful. Stan Shih decided to join Qualitron. His decision was in part based on the belief that the Lin family was going to withdraw its investment in Unitron. Qualitron was established in September 1972, with the Lin family investing $1 million NT as seed money, an equivalent of $245,000 US in 2024. In addition, to recruit Shih, Sen Lin offered him stock in the new company to encourage his technical contribution.[30]

While staying to run the company, Chew was disappointed that the department of calculators was taken away. The department was his brainchild, and it was also the company's most promising and profitable department. He lost his passion for securing contracts and looking for niche technical applications. When ITT, a long-term client of Unitron, approached Chew and expressed interest in acquiring the company, Chew suggested the board of the company sell the company. No one in the company absolutely objected to the idea, but the then minister of economic affairs of Taiwan tried to turn the tide. Minister Sun Yun-suan (Sun yunxuan) (see figures 7.1, 7.2, and 7.3) was shocked by the decision and did not wish to see a domestic company, and in fact the first electronics manufacturer, acquired by a foreign company. Sun then approached Chew, and Chew candidly explained why he favored selling the company. First, Chew believed that Unitron had reached a growth threshold. He saw no room for the company to make more profits or expand its scale. Second, the unpredictable fluctuation of subcontracting deals in the semiconductor industry at that time might put the company at risk in the near future. It was not an easy task to ensure the company win contracts, and hence profits, continuously. In his autobiography, Chew noted that he was probably more pessimistic about the future of the company than anyone else could have been. But he clearly knew that he had lost his motivation to run the company. After hearing from Chew, Sun nodded and asked, "How about arranging 20% of the stocks to remain in the hands of someone local? That makes the company still a ROC company."[31]

There was no lamenting its acquisition in newspaper coverage when journalists found that ITT had spent more than $2 million US to "invest" in Unitron. Neither did news articles directly bring up Unitron's lack of local capital investment. It is unclear whether Chew managed to have any ROC citizen hold 20 percent of the company stock. More likely, Sun dissuaded journalists from discussing the acquisition in detail. Chew signed a contract with ITT to stay in the company for another two years. After that, he left to begin his own electronics business, including manufacture and trade. By 1987, he was already a millionaire by running a semiconductor factory and a computer assembly factory for the GDR government in East Germany.[32]

For the acquisition, ITT paid $80 million NT (about $2.5 million US at that time, $17.4 million US in 2024). The worth of the company in 1974 was about eight times the capital that the Lin family had made available in 1969. Sze and Chew did not put to use the latest machinery for transistors and integrated circuit fabrication they had ordered. When Unitron was sold, Sze arranged to resell the machinery to the research institute of the Directorate General of Telecommunications.[33]

QUALITRON'S CALCULATORS, MADE ON A BALCONY (1972–1976)

Shih joined Qualitron in September 1972 and left the company in September 1976 (figure 7.4). He spent the first three months developing a handheld calculator with Texas Instruments' TMS 4-bit chip.[34] He also sought to improve the technical problems of the precious calculator he designed. To ensure the quality of the printed circuit

FIGURE 7.4
Women electronics assembly laborers working at their desks while Stan Shih gives a tour of Quali-tron for Provincial Chairperson Tung-min Hsieh. *Source:* "Taiwan sheng zhengfu zhuxi Xie Dong-min fangwen Taibei shi Rongtai dianzi gufen youxian gongsi" [Provincial Chairperson Tung-min Hsieh visited Qualitron (Rongtai electronics)], p. 1, archival document no. 009-030204-00058-085, December 24, 1975, "Xie Dongmin xiansheng shengzhengfu zhuxi shiqi zhao (wushiba)" [Provincial chairperson Tung-min Hsieh photographs], Xie Dongmin zongtong wenwu [Vice President Tung-min Hsieh files], AH. Reprinted with permission from Academia Historica, Taipei, Taiwan.

boards, he went to the Chung-Hua Arcade to purchase a board with a layer of cop-per foil and chemical etching materials to craft his own printed circuit board for the prototypes, and dried them on the balcony of the company's office space, which was located in Taipei (for a glimpse of his office, though with no view of the balcony, see figure 7.4).[35] He also spent time and energy finding ways to manufacture number keys locally to ensure the numbers would not quickly wear out. He closely monitored the production of the plastic cases of the calculators, as he believed that was decisive in forming the first impression a customer had of Qualitron's calculators. Shih pointed out that Sen Lin brought in new ideas that he had learned in the United States, especially those about industrial design and marketing. A marketing company proposed that Qualitron design a handgun-like calculator, which was not adopted.

The first model of Shih's handheld calculators was in a great demand, in part because it was less expensive than Japanese products. Shih also got into the market of scientific calculators. Qualitron immediately attracted the attention of Japanese and US calculator makers, and eventually earned profitable contracts from them. For example, US company Rockwell bought the calculators made by Qualitron and marketed them as the Rockwell 24k calculators. The other two popular products that Shih designed at Qualitron were digital watches and digital-clock-embedded ballpoint pens.[36]

The booming calculator business allowed Qualitron's capital to grow from $1 million to $10 million NT from 1972 to 1976. The company's revenue in 1976 was $20 million US ($108 million US in 2024). But Stan Shih left Qualitron and founded Multitech, the predecessor of Acer, in September 1976. The decision was prompted by a financial management problem at Qualitron, a problem similar to the one that had vexed Andrew Chew a few years earlier.[37]

The 1973 oil crisis had a great impact on the Lin family's textile business, which began to borrow funds from Qualitron to an extent that Qualitron ran out of working capital and began to incur debts. In April 1976, Shih was already aware of and worried about Qualitron's financial situation. He discussed the problem with Sen Lin directly, but Sen Lin was unable to say no to his family. Shih was especially concerned that, as one of the major shareholders and a member of the board, he had no control over this issue, and he might soon be burdened with debt. Shih then consulted Mao-pang Chen, the CEO of Sampo Corporation and the president of the Association of Electrical Appliance and Machine Manufacturers (Diangong qicai gonghui), who had earned great respect in the industry and served his tenure as president from the 1960s to the 1980s. After the meeting with Chen, who offered a pessimistic diagnosis of Qualitron's finances, Shih made up his mind to leave Qualitron, even though Lin reassured him that the financial situation would be improved soon. In September 1976, with $1 million NT ($32,310 US, or $181,000 US in 2024) as seed money, Shih set up his own company, Multitech, to focus on microprocessor applications, an area Shih had worked on with several engineers at Qualitron. The seed money was raised by Shih and his wife, Carolyn Chi Hua Yeh (Zihua Ye), and five partners: three previous Qualitron engineers, a Chiao-Tung alumnus Shih had known since his undergraduate time, and the alumnus's friend. Sen Lin shut down Qualitron not long after Shih left. The Lin family promised to repay all debts and did not get Shih involved. Multitech was renamed Acer in 1987. In the 1990s Acer grew to the seventh largest personal computer company in the world. As the CEO of such a large enterprise, Shih generously noted, "If Qualitron had not encountered financial mismanagement, what Acer has achieved today would have been achieved by Qualitron."[38]

THE GROWING MARKET FOR CALCULATORS

At the time Unitron developed its first calculators, Japanese calculators were also available in Taiwan. For example, Sampo, a local television and radio assembler and maker, also acted as authorized dealers of Japanese appliances. In May 1972, Sampo carried several models of calculator, from eight-digit to sixteen-digit, with prices ranging from $14,500 to $35,000 NT. Sampo's twelve-digit calculator cost a customer more than twice the price of Unitron's. Sampo also carried a programmable calculator model targeting educational and research institutions, priced at $65,000 NT. Sampo probably imported these calculators from Japan or did final assembly work in Taiwan. Toshiba, Sanyo, and Sony sold calculators ranging from $10,000 to $45,000 NT in Taiwan. At the same time, Wang Laboratories' 300 series were on the market too, costing $200,000 NT. However, models in the 300 series were able to offer more programmable features and connected to four terminals through a time-sharing system.[39]

Being the first calculator maker in Taiwan, Unitron's widely celebrated success invited many entrepreneurs to join this profitable business. In 1972, according to *Economic Daily News*, four domestic companies made calculators: Unitron, Qualitron, Santron, and Zhangcheng. But the number of calculator makers increased to forty-two companies the next year. The new calculator makers included both start-ups and domestic home appliance companies such as Tatung and TECO. The former had experience making TV sets and magnetic core memory, and the latter specialized in making electric motors. TECO's calculator business lasted roughly from 1973 to 1977. In the beginning, TECO's engineers purchased and researched the Japanese company Ricoh's calculators, and successfully developed their own calculators. While TECO's calculators were in great demand after 1973, TECO stopped its calculator business because the price of calculators dropped significantly. Furthermore, TECO's production focused on desk calculators with LED display instead of the much more popular handheld calculators with LCD display. But soon after, TECO began manufacturing monitors that could process and show Chinese characters and then supplied monitors to IBM in the early 1980s.[40]

In a 1974 news article, a journalist noted that the managers of these calculator manufacturers were all graduates from prestigious universities such as National Taiwan University, Chiao-Tung, Tsing Hua, and Cheng-Kung; this high percentage of university or master's degree holders was not seen in other industries. The number of calculators produced in Taiwan had grown significantly since 1972: in 1972, among the 55,600 calculators made in Taiwan, ten thousand were handheld calculators and the rest were desk calculators; by 1973, the number of calculators produced had reached two hundred thousand. In November 1973, it was estimated that production was going to double or triple in the following year. In September 1975, Taiwan expected to export a million calculators that year, a figure that had doubled from the previous year. In 1976, the second largest calculator manufacturer in Japan, Systech Corporation, went bankrupt. *Japan Times* saw that fierce competition from cheaper products from Hong Kong, Taiwan, and Singapore had greatly affected Japanese companies' US and European market share. In 1978, the six million calculators produced in Taiwan accounted for 20 percent of global market share, 90 percent of which were for the export market. The major suppliers of ICs at that time were Japanese companies that made CMOS chips.[41]

Some calculator start-ups were founded by veterans of the nascent electronics industry in Taiwan. An employee formerly hired by Unitron, Jinzhang Chen, set up Haozhou to make calculators in 1973. Unlike Stan Shih and other Unitron engineers with engineering degrees, Chen had attended an agricultural professional school but became interested in electronics. When fulfilling his mandatory military service at the turn of the 1960s, he was recruited to work as a radar operator in the navy. In the 1960s he worked for various electronics manufacturers and designed transistor radios and worked in Thailand to set up an electronics factory. He became an engineer at Unitron in 1971 and designed calculators. In October 1973, his company Haozhou had five models of calculators, and *Economic Daily News* portrayed him as a promising entrepreneur from a humble background. It is unclear, though, whether the company survived into the 1980s.[42]

Another early calculator maker in Taiwan was a company of Chiao-Tung alumni. Zhangcheng, literally translated as "the Great Wall," planned to make calculators in 1972. Zhangcheng was founded by Chao-Chih Yang (chapter 4) and several Chiao-Tung alumni in August 1968. *Economic Daily News* noted that the company was run by holders of four doctoral degrees, two master's, and a bachelor's. Their investors included businessman Zhihou Yin, who owned a textile company among his many other prosperous businesses.[43] The group of Chiao-Tung alumni first focused on manufacturing testing equipment for factories, vocational schools, and colleges of technology. The company also designed and set up laboratories for educational institutions. In terms of technical service, it offered maintenance service and technical consultancy for electronics factories and programming service for companies that would like to use computers. At that time, the company was located in Hsinchu and hired a couple of students at local vocational schools to help with soldering in the factory. By 1972, Zhangcheng expanded to supply digital circuit kits for educational institutions and made electronic kits for students to learn how to fix TV sets at one-tenth the price of imported kits. The company further developed testing equipment for transistors and targeted domestic factories as its main customers. For such transistor testing equipment, Zhangcheng meant to aid Taiwanese companies that had purchased large quantities of transistors but needed testing equipment to distinguish the workable transistors from inferior ones.

Soon after Unitron's calculators became available on the market, in May 1972, Zhangcheng also announced that it had developed a calculator specifically for teaching, though the details of the calculator were not revealed in newspapers. One of Zhangcheng's new investors, Zhitong Yin, told a journalist from *Economic Daily News* that the company was going to make "minicomputers" for schools, by which they probably meant programmable desk calculators similar to (and likely inspired by) Wang Laboratories products. Yin estimated that they would be able to deliver those "minicomputers" in summer, and the sale price would be about $50,000 to $60,000 NT. Each machine would require two or three months to deliver. Educational institutions or factories with more than one thousand employees would benefit from the adoption of this "minicomputer." However, the proposed minicomputer was never again brought up in the public media. Had the company successfully manufactured programmable desk calculators, it would have been celebrated widely. But Zhangcheng produced an office-use desk calculator in 1973, which sold for $6,000 NT. The company expanded greatly after getting into this booming industry. By November 1974, Zhangcheng had ninety employees, and its monthly revenue was $10 million NT. The company's major domestic customers in 1974 were college students, whereas its major export markets were in the United States, Europe, and South Africa. But both Zhangcheng and Haozhou did not seem to continue into the 1980s. It is likely that they made a handsome profit from the calculator market but could not develop future products or find an equally profitable new business after the market became saturated and more competitive.[44]

When calculators became inexpensive and popular, this development stimulated some cultural negotiations in Taiwan. The first negotiation was about women's uses of calculators. The second negotiation was about in what ways Taiwanese could make sense of the abacus, considering calculators were now available to everyone as an alternative and effective tool.

Qualitron saw a market for women consumers of calculators as early as 1973. As the company emphasized product design, it developed a calculator that obviously targeted women customers. The model, La Femme, had oval-shaped number keys, an ellipse-shaped LED screen, and a mirror on the back. The colors used for the calculator were shades of brown, which made the calculator look like an eye shadow palette (see figure 7.5).

The price drop of calculators in 1975 generated demand from housewives. *Economic Daily News* published a news article meant to offer guidance for housewives about how to choose a suitable calculator to manage their bookkeeping at home. The journalist urged housewives not to leave calculators anywhere close to television sets, refrigerators, and record players to avoid damage caused by heat, humidity, and magnetic fields. The news article presumed that a housewife might not be able to distinguish portable calculators from mainframe computers, as the terms computers and calculators are interchangeable in Chinese language. The journalist implied that calculators were new, affordable gadgets for housewives, but complex computers were not meant for housewives.[45]

FIGURE 7.5
Qualitron's La Femme portable calculator. Photography by Guy Ball. Reprinted with permission.

Electronic desk and portable calculators in early 1970s Taiwan were sometimes referred to as "electronic abacuses," indicating that calculators were viewed as an extension of an old technology. But on other occasions, calculators were viewed as a technology competing with the abacus. In the summer of 1973, *United Daily News* covered an abacus competition among Taiwanese, Japanese, and Korean participants, and attributed the relatively poor performance of Japanese players to the popularity of electronic calculators in Japan. In early December of that year, a retailer of office machines held an arithmetic competition among its four hundred employees to promote Sampo calculators, which were imported from Japan and sold with the brand name Sampo. A week later, a middle school held another arithmetic competition between abacus users and calculator users. The twenty-five abacus masters were elementary and junior high school students. But the fifteen calculator masters were students in their late teens from a vocational school. Abacus masters won the competitions in both multiplication and division. For multiplication, players were asked to multiply a five-digit number by a six-digit number. For division, players were asked to divide ten-digit numbers by five-digit numbers. After the event, the head of the association of abacus calculation told a journalist that he believed the two technologies were complementary.[46]

Economic Daily News, in 1975, translated a news article from the Associated Press about the trend that showed the younger generation in Singapore was less familiar with abacuses. But the Associated Press journalist noted that abacuses were salient in the central fish market. Between 3 p.m. and dawn, when the market was running auctions, more than two or three hundred bidders moved the beads of their abacuses. The journalist added, "The humid and salty air in the fish market at late night may make the delicate and expensive calculators fail to function after a few months of use. The wholesalers decided to stick with abacuses."[47]

Furthermore, in August 1975, Zhangcheng invited its consumers to stop by several malls in the three largest cities in Taiwan, where consumers could exchange a used abacus for a calculator by paying an additional $520 NT (an equivalent of $16 US in 1975, and $96 US in 2024). The sale price of a calculator in 1975 was several hundred NT dollars. Zhangcheng could still make some money from the sales. Interestingly, Zhangcheng did not entirely view abacuses as competing or obsolete products, as it announced that the company would donate the collected abacuses to students who need the devices. Sampo also ran a similar promotion in July 1976. Customers could exchange used abacuses, slide rules, and used calculators for new calculators, after paying an additional fee. Like Zhangcheng, Sampo planned to fix and donate the used items to the first-year college students in August.[48]

The production of calculators also supported domestic suppliers of calculator parts. As mentioned in chapter 6, Dee Van was a company founded by Xingdao Zheng in 1972 that wired magnetic core memory units for manufacturers. After giving up the business of assembling magnetic core memory units at the end of the year, Zheng shifted to making rectifiers and AC adapters to supply local start-up calculator manufacturers. By 1976, his company had sold 300,000 AC adapters to Qualitron, Zhangcheng, Santron, and companies in Europe and North America. Zheng was thirty-two years old in 1976. The average age of the company's 250 employees was younger than twenty-five, and the majority were women laborers. The beginning monthly salary for women assembly workers at Dee Van in 1973 was about $3,000 NT. Recall that Stan Shih's monthly salary at Unitron in 1971 was $7,000 NT. But Zheng believed that it was a pretty high salary for the Xindian area, a town close to Taipei city and the location of the earliest General Instrument (GI) factory in Taiwan. While Zheng's experiences at GI and Wang Laboratories had been instrumental in starting up his company, he revealed that he had disassembled imported Japanese calculators to learn how to make rectifiers for calculators. He often could make a similar product with a lower price three months after he had thoroughly explored a new Japanese product. In 1976, Zheng set up a subsidiary to make power supplies in Los Angles, tapping into the US market. Dee Van later supplied PC cases to the growing number of computer manufacturers and made hi-fi audio systems. In 2000, it was known as the largest Taiwanese manufacturer of power supplies in Shenzhen, China.[49]

SANTRON CALCULATORS, DESIGNED BY THE BEST SOLDERER (1972–1973)

A few weeks after Unitron released news about its first calculator product, in March 1972, National Taiwan University announced that Barry Pak-Lee Lam and Sayling Wen (Shiren Wen), along with a professor and fellow graduate students, would soon complete a minicomputer (chapter 5). Both Wen and Lam came from humble backgrounds. Wen was born in Taipei in 1948, a native of Taiwan. Wen had been working as a licensed plumber when he was an undergraduate student. He learned the trade from his father, whose customers include a hotel run by the Gao family, who had founded a company called Sande Construction. The Gao family learned about Wen's research from newspaper coverage and contacted Wen's father to recruit Wen to work for a start-up that was going to be run by the youngest son of the Gao family, Congfu Gao. Wen then invited Lam to join the company. The Gao family also recruited Guoyi Ye, a friend of one of Zongfu Gao's siblings, to manage the company's finances. The Gao family invested $5 million NT in the start-up. The name Santron was possibly shortened from Sande and electron.[50]

Either the Gao family or Wen came up with a preliminary idea that the start-up could consider calculator manufacturing. But the idea felt feasible after Lam fixed Congfu Gao's Casio calculator purchased during Gao's trip to Japan. Gao asked for Lam to take a look at the broken calculator. Lam opened its case, resoldered the joints where the wires had connected to circuits boards, and breathed life back into the calculator. Lam's magic touch over the malfunctioning luxury gadget from Japan impressed the Gao family and demonstrated Lam and Wen's expertise in calculators and computing devices. The group decided that their start-up should make calculators; the company was founded on April 4, 1972. The episode has been retold many times with variations, including in a news magazine article in 2002 and in Lam's oral history interview in 2011. But the earliest record of the story was narrated in 1973 in *Economic Daily News*. After speaking with Wen and Lam in 1973, a journalist stated that Santron engineers were interested in imitating (*fangzhi*) a desk calculator brought back from Japan by Gao. After they opened the case and examined the components, they believed that they could make such a calculator with procured parts. In three months, they made a calculator, the sale price of which was $9,000 NT. The original Japanese calculator was twice as expensive. Lam explained later that the action might not constitute "copying." In the 2011 oral history interview, historian Ling-Fei Lin asked Lam whether anyone did reverse-engineering for the first calculator Santron manufactured. Lam noted, "There really was nothing to reverse. All those ICs [and their] applications were already defined." Lam emphasized that the availability of integrated circuits with various technical applications in the early 1970s rendered multiple possibilities for developing commercial products. He explained that developments in the IC market granted Taiwanese companies opportunities to get into the calculator market.[51]

The calculators developed by Lam and Wen at Santron used the MK5013P and MK5014P chips from Mostek, the same manufacturer of the chips for Shih's first calculators. The latest Mostek chips significantly powered up Santron's calculator and made it competitive. More than a thousand calculators were sold in the first two months after they became available on the market. The release date of the calculator was sometime between April 1972 and March 1973; ten thousand calculators had been exported by March 1973. These calculators were priced at more than $10,000 NT each, slightly more expensive than Unitron's Baili. Ye, who oversaw finance, said that it was a lucrative business at the time. For the first batch of products, they made three hundred calculators, and the profit of that batch sustained the company for half a year.[52]

The scale of Santron was quite different from that of Unitron. Santron owned office space of only 1,780 square feet in a factory building made of asbestos, as opposed to Unitron's 71,000 square feet of industrial property. Santron employed, at most, 150 workers. While they might not all be working in the factory at the same time, the presence of half would make the workspace quite crowded. Wen and Lam told journalists that, as everyone was working on a table and sitting back-to-back with coworkers of another table, on hot and humid days one could immediately tell if others had started to sweat. To illustrate how economical they were in running the company, during mealtime, a manager would order inexpensive food in big pots, often rice with dried fish and vegetables, delivered to the office for all employees to share. Each person's meal cost only $1.50 NT at that time (worth $0.40 US in 2024). The size of Santron might be similar to that of Qualitron in 1973, though Qualitron grew to a company of a thousand employees in 1976. Surviving photos indicate that Qualitron's office had been renovated with workers occupying designated office desks and chairs (figure 7.4). Young workers, mostly women, wore navy blue jackets with the company's name embroidered on the top left of the front side. The young Stan Shih also wore the same kind of jacket, instead of a suit, when Taiwan's Vice President Tung-min Hsieh visited Qualitron in December 1975. In contrast, Santron's workers sat casually around a long table when soldering the printed circuit board in the office. The interior of the office seemed to show signs of wear.[53]

Lam might have joined the workers to weld the printed circuit boards and wires. Wen recalled that to meet the deadline to deliver a batch of calculators, the entire staff stayed up all night for three days to assemble the calculators. Lam must have been a member of the sleepless workforce. In the 2011 oral history interview, Lam proudly declared that he was the best in soldering among all the workers, and he self-depreciatingly made fun of his unimpressive grades when he majored in electrical engineering at National Taiwan University. He noted more than once during the interview that, precisely because he had done poorly in written tests, he knew that his talent must be in those hands-on skills. In contrast to high achievers able to apply for graduate programs in the United States, Lam found that as a mediocre student he could enjoy the ample opportunities available in Taiwan. Lam was certainly trying to play down the efforts he made, but his hands-on skills and his interest in building computers brought him opportunities to get into and excel in the global business of calculators, computers, and laptops.[54]

Santron's labor force had worked day and night. But that was not enough to succeed. A part imported from Japan did not arrive as expected. Fortunately, the crisis was resolved by a colleague bringing back the parts in person from Japan to Taiwan directly by taking a commercial flight. To conclude his story of the company's efforts to meet the delivery deadline, Wen stated that the calculator business at that time was not as easy as we may imagine now; many parts of the calculator, including screws, had to be imported. Andrew Chew at Unitron made a similar comment in 1973 in a newspaper interview, urging the government to simplify the paperwork and regulations on imports, especially for the electronics industry, which relied heavily on imported parts.[55]

Wen's and Lam's accomplishments at National Taiwan University and Santron in 1972 bought them an award in 1973, granted by the China Youth Corps (CYC), an organization with strong financial and political ties to the Nationalist Party. Every year CYC granted an award to ten young adults to commemorate the youth sacrificed during the 1911 revolution in China. Jong-Chuang Tsay (chapter 5) also received the award for building a programmable calculator the year before. But for the 1973 awardees, premier Chiang Ching-kuo met with Wen and Lam right before the award ceremony. Interviewed by journalists after the ceremony, Wen noted that he was appreciative of the opportunity to start the calculator business, whereas he did not think that other overseas Taiwanese would enjoy similar financial support in starting a company. He alluded to top students who had the resources to choose to study and then work in the United States. Wen also stated that the electronics industry in Taiwan should consider moving beyond assembling imported components into finished products, as it was merely shortsighted and opportunistic. It was crucial to the industry to make electronic components locally and "root" the industry in Taiwan.[56]

By the time they received the award, Wen and Lam had decided to leave Santron and join another start-up, Kinpo. Despite the profitable calculator business, Congfu Gao was an audiophile and wanted to focus the company's business on home audio systems. Lam realized that this was not his specialty. In the 2011 oral history interview, Lam stated that he was good at digital stuff, instead of analog things, though, in hindsight, soon the home audio systems became involved with digital techniques. Lam decided to continue his development of calculators somewhere else.

An online search of surviving vintage calculators reveals that Santron continued to make many models after Lam and Wen left. Gao apparently decided, despite developing home audio systems, not to give up the lucrative market in calculators. Some Santron calculators were sold to the European market.[57] Similar to Shih's generous remarks about Qualitron, Lam expressed thanks for gaining valuable experiences at Santron. In 2002, almost three decades after working at Santron, Lam and Wen met Gao again. The gathering was arranged as a press conference. Lam and Wen were leaders of Taiwan's vigorous tech industry, yet Gao had not maintained personal or business contacts with them for decades. He ended Santron in 1984 and emigrated to the United States. Lam and Gao finally met again in front of journalists and gave each other a long hug, with all sorts of feelings welling up in their minds. Lam emotionally told the journalists,

"Without Santron, I would not have accomplished so much today." He continued, "I would have been an unremarkable engineer in Hong Kong."[58]

KINPO (1973–1988), MAKING "CAL-COMPS"

After deciding to leave Santron, Wen and Lam, in their mid-twenties, spent a few months avidly looking for their next jobs. They were then introduced to an affluent family in Taoyuan, the Xu family. Chaoying Xu met with Lam, Wen, and a few more engineers, over a dinner, and decided to invest in their calculator business. Xu came up with the name of the company, Kinpo, during the three-hour dinner and toasted the birth of the company. Xu later recalled that all these NTU alumni seemed trustworthy and earnest, which helped him make the investment decision without hesitation. Xu raised $6 million NT in seed money from investors in his family's hometown. His son, Shengxiong Xu, also joined the start-up as the general manager and was in charge of sales. The literal translation of Kinpo is "golden treasure." The logo of Kinpo, however, consisted of two overlapping "C" letters. An engineer who followed Wen and Lam to work for Kinpo noted that the two letter Cs referred to the computer and calculator, which were proposed by Wen and Lam to reflect their strong belief in developing the two technologies. The English name shown on Kinpo's calculators were "Cal-Comp" or "Compex."[59]

Kinpo was founded in June 1973 and developed a handheld and a desk calculator that September. Lam indicated in his oral history interview that it was easy to design a new calculator at that time. Lam, twenty-five years old, was promoted to general manager of the factory in 1974. The chief engineer was twenty-four-year-old Chee-Chun Leung (Cizhen Liang), who graduated from NTU with a major in physics and joined Santron and then Kinpo with Lam. Both Lam and Leung grew up in Hong Kong and had become close friends when they studied in the same high school in Hong Kong and at NTU. When interviewed by journalists in 1974, Lam and Leung noted that they were looking forward to manufacturing computers soon. The dream of making computers held a special spot in Lam's heart. By April 1974, Kinpo was making three thousand desk and handheld calculators per month, the buyers of which were in Europe and Australia. The company had grown from twenty to one hundred employees in the first year. In September 1975, Kinpo was already making fifty thousand calculators per month. One of the Kinpo products was selected as the eleventh best design of calculators by a Japanese magazine after having been displayed in an expo of calculators in Osaka. Wen noted at that time that the numbers of calculators Taiwan exported was only one-tenth that of Japan. He hoped that the enhanced quality of calculators would allow Kinpo to compete with Japanese companies.[60]

Lam felt that Kinpo was able to succeed in the calculator business in part because the company made a swift shift when the display technology of calculators transitioned from light-emitting diodes (LED) to liquid crystal displays (LCD) around 1974. Kinpo expanded its market share when it provided LCD calculators. In addition to making Kinpo-branded calculators, the company also made calculators for US and European companies. Lam noted that he had to put tremendous effort into persuading Morris C. M. Chang at Texas Instruments that Kinpo could produce calculators for Texas Instruments in the 1970s. Lam especially recalled that Chang set a high standard for the final products. In 1975, the *Economic Daily News* covered Lam's visit to Hong Kong and Japan to research the emerging microprocessor market. In less than two years, Kinpo began to make microprocessor-cashier-registers.[61]

Lam continued believing that someday he would make computers. He soon left Kinpo to begin his own start-up, Quobao, to make arcade game consoles in 1979. But the business failed, and Lam returned to Kinpo.[62] In the 1980s, Kinpo and its subsidiary, Compal Electronics (Renbao), set up in 1983, focused on manufacturing computer terminals. Lam and Leung continued working at Kinpo in the 1980s, but Wen joined Inventec in 1980 and kept his focus on manufacturing calculators. Inventec was founded in 1975 by Guoyi Ye, a friend of the Gao family and the chief financial manager at Santron. In 1987, one of Compal's factories had a serious fire; Lam and Leung took responsibility and resigned. The next year, Lam founded Quanta to begin his business manufacturing laptops, with a large portion of investment coming from Wen and Ye. Both Quanta and Inventec worked on laptop manufacturing in the 1990s. In 2011, Taiwan's laptop manufacturers made two hundred million laptops, which accounted for 90 percent of the world market. They made laptops marketed with their own brands and also produced them for other companies; the four largest manufacturers have been Quanta, Compal, Wistron (a spin-off of Acer), and Inventec since the 2000s. Most of their factories are now located in China.[63]

CONCLUSION

Unitron, Qualitron, Santron, Kinpo, and other companies in Taiwan carved out a successful path manufacturing calculators in the early 1970s. Taiwan's calculator industry, as their US and Japanese counterparts, also produced a strong legacy in the subsequent microcomputer and personal computer industries. But the leading engineers of the Taiwanese calculator industry had unique experiences building their own minicomputers on campuses or working for electronics manufacturers. Before manufacturing calculators, Chew, Lam, and Wen participated in university-based minicomputer-building projects; Chew, Shih, and Haozhou's founder Jinzhang Chen were employed by electronics manufacturers. With experience in colleges or at factories, these engineers-turned-entrepreneurs kept abreast of the development of ICs and the latest calculator products from the United States and Japan. Chew, Shih, and Lam also exhibited enthusiasm for hardware tinkering, including crafting printed circuit boards, and a deep understanding of design and manufacturing process of calculators. They swiftly took advantage of the availability of chips that integrated calculators' many functions. Beyond the start-ups, Tatung and TECO, a home appliance maker and an electric motor maker that had previously assembled TV sets for international TV makers, also seized the lucrative opportunity and made calculators for the export market.

The Taiwanese companies' achievements were highly contingent on technological innovations in integrated circuits in the United States, but Taiwanese engineers made critical moves to program the integrated circuits and create their own commodities. The ingenuity of Taiwanese engineers is comparable to the earlier accomplishment of Japanese engineers. While Japanese portable calculators had great success globally in 1970, the majority the LSI chips in Japanese calculators were manufactured by US companies.[64] Witnessing Japanese calculator's enviable triumph, US calculator makers soon decided to go for single chips with greater functionality that were designed in the United States. Lewis H. Young, the editor-in-chief of *Business Week* wrote,

> The reason the U.S. has taken back a great chunk of the electronic calculator market from the Japanese is due to technological advances. . . . Japan's advantage was based on the low wage rates paid Japanese workers to connect the U.S.-made semiconductors and put them in a plastic case. With the introduction of the single chip calculator, using a technological process that engineers call large scale integration (LSI), the Japanese edge disappeared. As a result, calculator manufacturing has quickly moved back.[65]

The highly integrated circuit chips not only allowed the Americans to reclaim the market, but also invited Taiwanese engineers and companies outside the United States and Japan to take part in the game. The Taiwanese companies, with low-cost labor and highly educated engineering talent, eventually captured one-fifth of the global market for calculators in 1978.[66]

Major Taiwanese calculator makers and their engineers-turned-entrepreneurs initiated their own start-ups to make computer terminals, computers, and laptops in the 1980s. Acer and Quanta especially made their way to become leading companies in the global industry of computer manufacturing. The success of engineers-turned-entrepreneurs was not an accidental development. Instead, Chew, Lam, and Wen had aspired to make electronic computers since the 1970s. Attentive to the global electronics market, Shih and Lam had witnessed various microprocessor applications since the mid-1970s. As the chief engineer of Kinpo, Lam toured Japan and Hong Kong around November 1975 to gain knowledge of novel electronics products that utilized microprocessors.[67] In the same year, MiTAC was founded in Taiwan by Matthew F. C. Miau (Fengqiang Miau), an engineer who joined Intel in 1971 and participated in the design of the Intel 8080 and 8251 chips. Miao was born in China, moved to Taiwan around 1949, went to Hong Kong for his high school education, and attended University of California, Berkeley for his college education. MiTAC became the official sales agent for Intel products in Taiwan and later made personal computers and computer peripherals. MiTAC organized a talk by a Japanese Intel engineer in the summer of 1975 to promote the applications of Intel microprocessors at which attendees included representatives from Kinpo, Zhangcheng, TECO, Sampo and other companies and research institutes.[68]

Stan Shih had taken action a year earlier. In 1974, all the seats were taken at a microprocessor conference organized by Intel that he had intended to attend. Shih then scheduled a trip to Los Angeles for a one-week Parallel Processing System 4-bit conference, focusing on microprocessors, held by the US company Rockwell. After the trip, Shih told journalists that Qualitron was going to develop new products including microcomputers. Before Shih was able to realize the promise at Qualitron, he founded his own company, Multitech, in 1976. Multitech issued a free magazine available to customers, naming it *Voice from the Gardener*, as the company staff believed they were planting the seeds of microprocessors in Taiwan. The topics discussed in the magazine focused especially on programming microcomputers, and its circulation grew from two thousand to twenty thousand. The free magazine was later reissued as *0 and 1*. The circulation reached eight thousand issues, and its readers were primarily engineers, as Stan Shih recalled.[69]

Unlike Matthew Miau, who had worked for Intel, Andrew Chew, Stan Shih, Barry Lam, and Sayling Wen were trained at Chiao-Tung and National Taiwan University. Local engineers-turned-entrepreneurs gained their knowledge of manufacturing calculators as a by-product of their aspirations to make computer-like devices. They obtained experience by experimenting with and exploring the manufacturing process. Their enthusiasm for and entrepreneurship in manufacturing calculators and computers in Taiwan marked the beginning of an era, from the 1980s on, for making and designing personal computers in Taiwan. Stan Shih's idea of being a gardener who cultivated the island with numerous microprocessors would bear technological fruit.

8   INCOMPATIBLE COMPUTER DREAMS: CONTESTED COMPUTER EXPORTS TO THE UNITED STATES

As customs officials, computer companies, and the Taiwanese continue to crack down on computer clones, consumers should be wary of any computers coming to this country from Taiwan. Mr. Lu points out the Taiwanese make computer components. They don't manufacture any brand name computers of their own. So, if you see such a computer, don't buy it.

—*New Tech Times*, Wisconsin Public Television, circa 1983[1]

Mass-manufacturing computers in Taiwan had been a technological dream of many scientists, engineers, entrepreneurs, and computer users since the late 1950s. Roughly two decades later, Chiao-Tung alumnus, seasoned microprocessor engineer, and fledgling entrepreneur Stan Shih came to realize the dream (figure 7.4). Beginning in the late 1970s, entrepreneurs around the world witnessed the unprecedented opportunity to import microprocessors such as the Zilog Z80, MOS technology 6502, or Intel 8088 to make microcomputers domestically. To build computers, they could purchase or customize motherboards, memory, cases, and the remaining computer parts locally or overseas. The strong electronics industry in Taiwan brought advantages to computer makers there to procure most parts they needed domestically. Shih and a good number of Taiwanese computer manufacturers began to make Apple and IBM compatible computers in the early 1980s. Founded in 1976, Shih's company Multitech, renamed Acer in 1987, was a forerunner in microprocessor programming. Calling his company "the gardener of microprocessors," he employed a group of engineers equipped with the knowledge and skills to develop microprocessor-based products.

The technological dream shared by Shih and his fellow entrepreneurs was challenged in the 1980s by Apple and IBM, which held patents and copyrights for their

microcomputers. While some Taiwanese computer manufacturers were making counterfeits, Shih and his contemporaneous competitors focused on developing Apple and
IBM compatibles for both domestic and export markets. Their strategy was not unique.
US companies such as Franklin and Compaq as well as others worldwide were taking
advantage of the availability of microprocessors to make Apple and IBM compatible
computers. Apple took a firm position against compatible makers. In contrast, IBM was
lax and allowed manufacturers to make IBM compatibles.

Amid the legal battles between Franklin and Apple, Shih shipped a thousand Micro-
Professor II computers to the United States in early 1983. But these Apple II compatibles were detained in US Customs in San Francisco due to the questionable copyright
status of their operating system. Shih's Micro-Professor II was considered counterfeit
after an inspection in US Customs. In 1982, Apple had requested the US International
Trade Commission bar Apple counterfeits at the border because Apple asserted that
foreign manufactured counterfeits constituted unfair international trade practices. US
Customs began intercepting counterfeits around July of that year.[2]

This chapter begins with the fast-growing company Multitech's success in developing products, including the Micro-Professor II computer, then describes and explains
the company's initial failure in entering the US market in 1983. In the early 1980s, US
media and congressional hearings portrayed Multitech and all other Taiwanese computer manufacturers as invaders and counterfeiters. Apple's determination to exclude
any computers compatible with their systems was elaborated in a series of congressional hearings regarding counterfeit computers in the United States in the late summer
and fall of 1983. This chapter reveals that much about Multitech and other Taiwanese
companies was misrepresented in the hearings and media. I argue that the misrepresentations indicate that the US computer industry, media, and analysts were unable to
make sense of Shih's success in making less expensive microcomputers. They assumed
the only way Taiwanese computer makers could succeed was by counterfeiting. This discussion focuses on the incompatible viewpoints between Stan Shih and Apple, whereas
chapter 9 centers on the conflict between Multitech and IBM, and the subcontracting
business opportunities that Multitech earned during the second half of the 1980s.[3] For
both chapters, I use Shih's oral histories and biographies, and newspaper articles about
him to delineate Shih's responses to Apple, IBM, and the US Customs' actions.

MULTITECH'S EARLY ACCOMPLISHMENTS

Micro-Professor II computers enjoyed remarkable success in Taiwan's domestic market
from their debut in June 1982; Stan Shih rolled out the new product in the overseas

market in the fall. Micro-Professor II computers were especially popular in Taiwan for both their technical achievements in processing Chinese characters and their moderate compatibility with Apple II computers. By early 1982, unauthorized copies of Apple IIs made in Taiwan were commonly seen in the domestic market. Shih sensed the booming market and planned to develop a new computer to take advantage of the surging consumer demand. Shih believed that the new product had to be different from those inexpensive Apple II counterfeits but compatible with the popular Apple computers. Some of Shih's coworkers were pessimistic about whether customers would choose their product over cheap counterfeits.[4] The sale price of the Micro-Professor II computer turned out to be competitive. It cost from $275 to $340 US in Taiwan. In contrast, other Apple II compatibles or counterfeits made in Taiwan cost $250 US on average, though the price could range from $150 to $750 US. These locally made computers were less pricey than the authentic Apple II, which ranged from $1,490 to $1,700 US in Taiwan in December 1982.[5]

Multitech's technical success in making Micro-Professor II and its successful financial performance in the early 1980s could be credited to Shih's enthusiasm for making innovative and marketable products with microprocessors. Shih left Qualitron, where he had been primarily an engineer, in 1976, to found Multitech. In starting his own company, he had to focus on two major business opportunities. First, he worked with his engineers to design new microprocessor-based products. Before Micro-Professor II computers, Multitech engineers had designed an automatic bridge machine, a computer terminal that could display Chinese characters, and a microprocessor trainer, tools for users to learn the workings of microprocessors. Second, Shih traded microprocessors. He imported microprocessors from the United States and sold them to Taiwanese customers. Multitech became a franchised dealer for Zilog and Texas Instruments in the late 1970s. In 1977, Shih even established a branch in California to better manage the relationship with US clients. Microprocessor trading was especially profitable and made Multitech thrive. Owing to the mushrooming of arcade game console makers in Taiwan, the company had a fabulous year in 1980 selling both Zilog and Texas Instruments chips. Its employees grew from eleven in 1976 to over a hundred in 1980. Revenue in 1980 doubled from 1979, reaching $160 million NT, equivalent to $5.1 million US. In comparison, Microsoft's year-end revenue in 1980 was $7.5 million US, though the number of its employees was forty.[6]

Multitech's early collaboration with the arcade game console industry speaks to its complex relationship with counterfeiters. The late 1970s witnessed a growth in Taiwanese counterfeiters of primarily Japanese and US arcade game consoles and a few local arcade game developers. The industry made handsome profits when arcade games enjoyed a remarkable popularity among the youth of the island, which was still under martial law. Around 1978, Multitech became the exclusive franchiser of Texas Instruments in Taiwan and sold its transistor–transistor logic (TTL) chips to arcade game console makers and counterfeiters there. Multitech profited greatly from this business. Occasionally, upon buying a large quantity of microprocessors from Multitech, arcade game console counterfeiters would request Multitech engineers to help them copy the program logic array of a game console so they could make copies of arcade game programs. As time went by, Multitech engineers accumulated enough knowledge to design new arcade games for companies in Hong Kong.[7]

The Taiwanese government banned arcade game consoles in March 1982, intending to help youth "corrupted" by arcade games and loitering around the arcade-like public spaces. The government wanted them "back on the right track." The industry took a great hit, and some arcade game console makers pivoted to making illegal copies of Apple II computers. In an interview with a journalist about the ban, Stan Shih urged the government to reconsider the differences between gambling games and games of puzzles or brain teasers. He believed that the government should assist the arcade game industry one way or another. He did not state that he supported counterfeiting, but he urged the government to take the plight of struggling game designers and game console makers seriously.[8]

Shih's enthusiasm for making innovative and marketable microprocessor products prompted Multitech to develop several popular products before the Micro-Professor II. Multitech's first product after its founding was a machine that would allow bridge players to play the game in separate rooms to avoid cheating. The project was commissioned by a Chiao-Tung alumnus Chung Ching Charles Wei (Zhongqing Wei), a successful shipping industry tycoon based in New York and an extraordinary bridge player. Jiahe Lin, the first director of Multitech's R&D department, and an ex-employee of Unitron and Qualitron and a Chiao-Tung alumnus, delivered satisfactory prototypes to Wei, though the product was not commercialized. Lin was also tasked in 1977 with documenting and annotating Zilog 80 chip codes for Multitech customers. In addition to microprocessors, Multitech designed a cathode ray tube terminal for another Taiwanese company, Chengzhou, which Shih believed was the first Taiwanese company to do "original design manufacture" for a US company. The terminals were ordered by Hazeltine Corporation in the United States. Around the same time, Multitech completed the development of the Dragon Chinese Computer Terminal, in September 1980, which drew tremendous public attention.[9]

Starting in 1978, Multitech succeeded in distributing and selling a microprocessor trainer course in Taiwan. The trainer, EDU-80, was developed by another Taiwanese company, Quanya. The two companies worked together to offer microprocessor and

EDU-80 trainer courses around the country. These high-demand classes promoted the application of microprocessors, fitting into Shih's calling to be the gardener of microprocessors. Senior R&D engineers at Multitech taught clients or prospective consumers at night. Their students also included hobbyists, college students, full-time engineers, and military personnel. Multitech even had a series of classes solely for agencies of the navy, the military research institute CSIST (see chapter 4), and the government-sponsored Industrial Technology Research Institute (ITRI). Between 1978 and 1981, three thousand students attended classes. Many instructors, including bridge machine developer Jiahe Lin and Kun-Yao Lee (Kunyao Li), known in the industry as K. Y. Lee. Lee later oversaw Multitech's mass-manufacturing of IBM compatible computers (see chapter 9), and became a high-level Acer manager. He had teased Shih for "exploiting" Multitech engineers by asking them to teach in the evenings. Lee also mentioned that he was once sent to the United States to attend courses offered by prominent chip manufacturer AMD (Advanced Micro Devices). While an overseas trip would be considered an admirable experience, his trip was scheduled during the weekend, and he was expected to work overtime after he was back, to brief all engineers for days about what he had learned at AMD. Despite laughing about how busy they were in Multitech's early years, Lee pointed out that Taiwanese private enterprises at that time seriously delved into the research and applications of microprocessors.

Because the EDU-80 was a relatively pricey product for the Taiwanese market, Shih decided to roll out a microcomputer trainer that would be less expensive and more powerful than the EDU-80. In October 1981, Shih put on the market the Micro-Professor I, with the Zilog Z80 chip. He took the Micro-Professor I to electronics expos in the United States and Japan and won acclaim from consumers and the media. The German magazine *Chip*, for example, reviewed the Micro-Professor I and, in April 1982, nominated it one of the top ten computers of the year.[10]

Having designed calculators at Qualitron, Shih knew the global electronics market. Situated in a relatively small market, Shih and other calculator engineers and entrepreneurs recognized the importance of making their products exportable. In late 1981, Shih also met and worked with other Taiwanese microcomputer makers to brainstorm strategies to export Taiwanese products, including his Micro-Professor I computers.[11]

THE SUCCESSFUL ROLLOUT OF THE MICRO-PROFESSOR II COMPUTER

Owing to the company's years of experience selling microprocessors and holding classes about the new technology in Taiwan, Multitech engineers became familiar with the MOS 6502 chip used by Apple II computers. They began to develop the Micro-Professor

II with the same chip. The development was led by chief R&D engineer Jonney Chong-Tang Shih (Chongtang Shi). He also led the development of the Dragon Chinese Computer Terminal in 1980 and subsequent Multitech computers in the 1980s. Stan Shih recruited Jonney Shih after being introduced by a professor at Chiao-Tung with whom Shih was acquainted. Stan complimented Jonney, saying Jonney was much smarter than he was. He credited Multitech's technical development to Jonney's brilliant leadership. Jonney Shih's Multitech colleagues related that he was so preoccupied with engineering problems that he would stop at a red traffic light thinking about the problems, not realizing when the light turned green. Jonney Shih left Multitech and joined Asus in 1992, becoming the chairman of the board there.[12]

Stan Shih, Jonney Shih, and other engineers intended to make sure the final products of the Micro-Professor II computers would be compatible with the Apple II. But they were also aware that they should avoid directly copying from the Apple II computer to prevent copyright infringement. Eventually, the Micro-Professor II computers were only partially compatible with the Apple II. Using different computer accessories and parts from the Apple II, Multitech engineers also had to develop a new operating system to command the hardware configuration. Additionally, an outstanding feature of the new product was to display Chinese characters. Micro-Professor II users would use the computer language Basic to operate the computer. At the level of applications, Multitech ensured various Apple II applications could be run on the Micro-Professor II. But, because of partial compatibility with the Apple II, users would have to edit application codes on their own, which posed difficulties for novice users. For these applications, Stan Shih noted at that time that Multitech did not know many of the authors of these applications; to legally use these applications, Multitech contacted the American Institute in Taiwan (AIT, functioning after 1979 as a de facto US embassy and consulate, in lieu of an official diplomatic relation between the two countries) to set up an account to collect royalties derived from selling Micro-Professor II computers—although no one ever showed up to retrieve the royalties.[13]

Because Multitech's goal was to make Micro-Professor IIs compatible with Apple II, the company must have read and studied the original Autostart and Applesoft code. As Jonney Shih described in 2011, he "reverse engineered" several computer systems, including Apple II, in his career at Multitech. Stan Shih further pointed out that the Micro-Professor II design was executed before Shih's company and the software industry adopted cleanroom practices. Multitech adopted the method only when developing a 32-bit computer to be released in 1986. A cleanroom environment, where engineers develop software code without any exposure to the original software code of relevant products, could serve as a method to avoid copyright infringement.[14]

In terms of sourcing components, the Micro-Professor II was a product of the growing electronics design and manufacture industry in Taiwan. The manufacturing was done by two other Taiwanese companies. Many computer ICs were designed by a just-established IC design company in Taiwan, Syntek Semiconductor, a company founded by a participant in an RCA technology transfer project, in which more than thirty engineers were dispatched by ITRI to RCA to learn CMOS IC fabrication in the mid-1970s (more in chapter 10). The manufacturing of Micro-Professor II was contracted to another promising local company, Delta Electronics, which is presently a profitable large enterprise (see chapter 6).[15]

Micro-Professor II computers were soon sold in more than twenty countries around the world. Stan Shih had been ambitious about the US market since Multitech's beginning. He worked with Edward Chang (Guohua Zhang), his college friend and then an engineer at Hewlett Packard, to establish a US branch of Multitech in California in 1977. As president of the American branch, Chang maintained a good relationship with *Info-World* journalists and promoted the Micro-Professor II at electronics expos on the US West Coast.[16] *InfoWorld* covered the Micro-Professor I's speech synthesizer board and the Micro-Professor II's rollout in August 1982. When Multitech displayed the Micro-Professor II at the Mini/Micro 82 conference and exhibitions in Anaheim, California, in October 1982, its booth drew a big crowd from the expo's ten thousand visitors. A similarly popular booth noted by *InfoWorld* was one that showed Hewlett-Packard's H-75 portable computer. The juxtaposition was an indicator that the Micro-Professor II was as trendy as the new gadget developed by Hewlett-Packard, a prominent, admired technology company in Silicon Valley. In November, Multitech also participated in the Applefest Show in San Francisco to promote both the Micro-Professor II computer and the Dragon Chinese Computer Terminal.

At the Applefest Show, Multitech's domestic competitor MiTAC was also there to promote its Applemate, a floppy-disk drive for the Apple II computer. MiTAC's founder Matthew Miau, a former Intel engineer who participated in the design of Intel's 8080 and 8251 chips, also had his company develop products compatible with Apple and, subsequently, IBM computers. He was as ambitious about the US market as Shih. MiTAC's MIS-8000 series computer, a Chinese computer terminal co-developed with the government-sponsored research organization Electronics Research and Service Organization (ERSO) in Taiwan, appeared at another expo in June that year in the United States. ERSO was one of Industrial Technology Research Institute (ITRI)'s offices, and its goal was to promote the development of the computer industry.[17]

A Micro-Professor II computer was listed for $399 US in the United States in 1983, £200 British in the United Kingdom in October 1982, and $598 Australian in Australia in February 1983. The Micro-Professor II was widely available in Europe, as users from Finland and Spain left comments on their ownership or usage of the computer in their childhood in the Old Computer Online Museum, run by enthusiasts. British magazine *Your Computer* reviewed the Micro-Professor II in October 1982, along with three other microcomputers made in Japan, Hong Kong, and the United States (figure 8.1). On the cover of that issue, each computer was placed against the national flag of its origin country. Stan Shih was exuberant about his accomplishment: the Micro-Professor II review by *Your Computer* and being presented as a national icon.[18]

In February 1983, *Electronics Australia* included an advertisement for the Micro-Professor II, placed by recently founded Jaycar, which then primarily sold electronics kits in Sydney and is now a large electronics retailer in Australia. An excerpt of the advertisement:

> Computer Sensation!! . . . That is why we got so excited when we saw the "Micro Professor MkII." It is the closest thing that we have seen to be software compatible with the Apple. Yes, we know what you're thinking. It is NOT one of those cheap Taiwanese "Apple" copies which infringe Apple's copyright. The Micro Professor MkII is a completely new and unique design in its own right. It just so happens that most of the widely distributed Apple software will run on this machine. O.K. But why so excited? LOOK at THE PRICE! Check out the STANDARD FEATURES of this unit. Sit down. Think about it and COMPARE what you get with the Micro Professor MkII as STANDARD that are options on other machines!![19]

Micro-Professor II's sweet price, solid performance, and good compatibility were the reasons that it became popular in places beyond Taiwan. The above Australian advertisement began with a complaint about the "quite frankly expensive" home computers available on the market at that time. Jaycar had not been interested in selling these expensive home computers. But now Jaycar decided to carry Micro-Professor II computers, as they were mostly compatible with Apple, equipped with more computing power, and less pricey. Among these praises, Jaycar made clear to potential customers that Micro-Professor II was not the same as the better-known Apple clones from Taiwan.[20]

As of September 1983, the batch of unfortunate Micro-Professor IIs detained by US Customs were still stored somewhere in its warehouses. At the same time, Stan Shih had an opportunity to clarify that US Customs was "determining whether one of our products has been independently developed or infringes two copyrights held by a U.S. manufacturer." He was willing to "cooperate fully and openly with Customs" through Baker McKenzie lawyers hired by Multitech.[21] The batch of Micro-Professor IIs were never returned to Multitech. Apple Computer requested that Multitech recall all the Micro-Professor IIs sold in the United States and compensate Apple at the rate of $20

FIGURE 8.1

Micro-Professor II computer featured on the cover of the British magazine *Your Computer* in October 1982. Three other microcomputers, made in Japan, Hong Kong, and the United States, were reviewed in the same issue. *Source: Your Computer* 2, no. 10 (October 1982), cover image. Author's collection.

US per computer. But Multitech refused and never had to oblige Apple, which never filed a copyright infringement case against Multitech's Micro-Professor II in Taiwan.[22]

CONFLATING COUNTERFEITERS AND TAIWANESE COMPUTER MAKERS

The seizure of the Micro-Professor IIs in US Customs indicates that officials suspected the computers might be counterfeits. But the suspicion also reflected Apple Computer's efforts in expelling compatible computers from the market, as well as the widespread stereotypes that all Taiwanese computers were counterfeits. US journalists at that time tended to create an image in which all computers in Taiwan were counterfeits. They failed to notice the Taiwanese companies, including Multitech, that were producing noncounterfeit computers.

US newspaper correspondents in Asia were well aware of counterfeit Apple computers there. In June 1982, the *Wall Street Journal* covered the widely seen Apple clones in Hong Kong and Taiwan and decided that Taiwanese manufacturers were responsible for making these counterfeits and distributing them to Hong Kong, Singapore, and Indonesia. These computers were sometimes sold in kit form; "engineering students put them together for relatives and friends" in Taiwan, and a customer in Hong Kong could also "ask one of the eager students to do it, and you have an Apple II, Hong Kong style." Managers of Apple's regional distributor, Delta Communications Service, complained that no one was willing to buy genuine Apple computers in Taiwan. The market was "gone, finished, kaput!"[23]

From the perspective of Taiwanese computer enthusiasts, as early as January 1982, a reader in Taiwan could see advertisements for Focus-II and Green Apple computers in *0 & 1 Technology*, a free magazine produced by Multitech to share and promote knowledge of microprocessors and programming among its customers.[24] In March 1982, it was estimated that there were tens of thousands of microcomputers in Taiwan, mostly Apple II compatibles or counterfeits, built by enthusiasts, electronics shops, or entrepreneurs. By the end of 1982, a local journalist noted that Taiwan was home to at least a hundred companies manufacturing Apple II compatibles.[25]

Many of the counterfeits looked exactly like an Apple II computer and contained the same arrangement of ROM chips. An Apple II compatible needed the Autostart and Applesoft programs, stored in the ROM chips, to work in conjunction with an MOS 6502 chip. Autostart was firmware, similar to what we nowadays call BIOS, a term used in relation to IBM PC systems, and Applesoft was an operating system. To make properly functioning Apple II counterfeits, manufacturers had to copy the two programs. Apple attorney Gary Hecker explained in a congressional hearing in July 1983, and demonstrated in an episode of the US TV program *New Tech Times*, how he analyzed the ROM chips. He compared the ROM chips of counterfeits with the original code of Apple products and determined what percentage of the code was identical. Even if a counterfeit looked different from the outside, it might contain the same ROM code, essentially an unauthorized copy of an Apple II.[26]

An episode of *New Tech Times* aired in the United States around 1983 displayed computers with brand names Golden II, System-II, Orange+two, Pineapple, and Banana, intending to showcase US Customs–seized Apple counterfeits from Taiwan. But, in reality, Pineapple and Orange+two computers were likely made by companies in Hong Kong and the United States. Apolo-II (misspelled by the manufacturer), however, was certainly made by Taiwanese computer maker Sunrise, whose owner spoke openly to Taiwanese and US journalists and a private undercover investigator about copying Apple and his subsequent decision not to proceed.[27] These brand names, derived from "Apple," might well "hint that these machines are as good as the model they copy," as the host Mort Crim of *New Tech Times* said. But the names said more than that. First, the naming shows that some counterfeiters foolhardily and greedily grasped the market of the unprecedentedly novel gadgets.[28] Second, the names demonstrate that the companies might not expect Apple to sue them in Hong Kong or Taiwan, especially considering that software copyrights were a new idea, a gray area not fully recognized by the legal system in Taiwan or in many other countries in 1982.[29] Third, *New Tech Times* had several close-up shots of two Taiwanese computers—Multitech's Micro-Professor computers and MiTAC's Little Intelligent computers, which Apple Computer considered infringed its rights. However, the two companies took a different naming strategy and claimed that they had developed Apple compatibles on their own. Moreover, both computers could process and display Chinese characters.

APPLE'S RESPONSE TO COUNTERFEITERS AND COMPATIBLES

Around 1982, Apple began to notice the worldwide problem of counterfeits and compatibles. In March, Apple's vice president Albert Eisenstat returned to the United States from a trip to Taiwan and Southeast Asia. On his trip he discovered about half a dozen manufacturers that made Apple counterfeits in the region.[30] But Apple decided to focus on domestic issues first. In May, Apple filed a lawsuit in the United States against US company Franklin Computer Corporation for copyright infringement, including code stored in ROM chips. Franklin Computer believed that the company was making compatible computers with better hardware—though, later in court, one of Franklin's engineers did admit that it incorporated parts of Apple's code for Franklin's ACE 100 computers. Apple's legal battles against Franklin lasted from May 1982 to early 1984. While a district court found in favor of Franklin in 1982, the Court of Appeals decided to overturn the ruling in August 1983. Franklin had no luck in its appeals in fall and winter 1983 and finally settled with Apple in January 1984. Franklin was obliged to develop its ACE 500 computers independently.[31]

After bringing up a case against Franklin, Apple initiated several lawsuits in Taiwan to remove counterfeits from the Taiwanese market. In June 1982, it registered its trademark and the copyright for the Autostart and Applesoft programs in Taiwan, although, up to 1985, the implementation of copyright law in Taiwan did not include computer software at all. An Apple lawyer came to Taiwan, acquired an Apple II compatible, and brought it back to the United States in July 1982. Three months later, Apple brought a case against two Taiwanese companies, Sunrise and Guan-Haur, that manufactured unauthorized copies. Yet because Apple had not registered as a company in Taiwan, the Taiwanese courts declined to hear the case in February 1983. Eventually, Apple completed the registration, and in August 1983, a prosecutor charged six Taiwanese companies with copyright infringement of Apple computers. Concurrently, a different prosecutor charged another two companies.[32]

While Apple did not bring a case against Multitech in Taiwan, it worked intensely to stop Multitech from exporting Micro-Professors and other compatibles and counterfeits to the United States. By July 1983, though, Apple had taken legal actions to stop the export of questionable computers from Taiwan or other origins to several European countries, South Africa, Israel, Australia, and New Zealand. Sales of Micro-Professor IIs were affected. For example, in August 1983, Apple halted the sales of the Micro-professor II to Sirtel, a UK electronics company, and the Spectrum retail chain there recalled all Micro-Professor IIs from its dealers.[33]

To protect its domestic market, Apple testified at several hearings in 1983 for the Subcommittee on Oversight and Investigations of the Committee on Energy and Commerce of the US House of Representatives. The hearings were to assess the impact of "illegal and unfair trade practices on interstate commerce," especially in the steel and electronics industries.[34] The hearings, on July 27 and August 2, extensively discussed the negative impact of Apple clones and other counterfeits on US enterprises. In these hearings, Multitech and other Taiwanese computer makers were considered simply counterfeiters and invaders in the US market.

How the questionable computers entered the United States was a major concern in these hearings. Apple's attorney Gary Hecker believed that only one-tenth of 1 percent of the imported computers were examined, identified, and detained by US Customs. Only 5 to 10 percent of illegal copies of Apple computers were identified at customs.

Hence, from his perspective, it was a serious issue that required more intervention.[35] Participants in the July hearing discussed an "underground network" in which companies imported parts to the United States and made Apple counterfeits here. While the counterfeits in question had nothing to do with Multitech or MiTAC, the companies were mentioned a couple of times and vaguely associated with a criminal underground market. An affidavit from a special agent of the US Customs Service described the underground network. When a narcotics officer arranged an undercover operation in March 1983, he serendipitously found that Apple counterfeit parts had been shipped from Taiwan and assembled in the same apartment he visited. The officer then worked with the agent of US Customs to identify and prosecute six individuals and five companies involved in the case. The six individuals sold at least 350 computers and had an additional three hundred assembled computers. At least two men were sentenced to thirty days in prison and fined $15,000 each in May 1984. The case presented the first criminal charges for smuggling related to computer counterfeits or software piracy in the United States, according to the *Wall Street Journal*. One of multiple companies involved in the case was the Taiwan Machinery Trading Co.[36] However, only two of the six individuals had Chinese last names. The owner of Taiwan Machinery Trading Co. was a Filipino Chinese, according to Taiwanese news media, implying that the owner's nationality might not have been Taiwanese.[37] Regardless, the companies ordered parts from Taiwanese manufacturers. Hence, in bringing up this case in the hearing, Taiwan was portrayed as the center of criminally manufactured fake Apples. Moreover, the circulation of Taiwanese computers and parts infiltrated narcotics trafficking networks in the United States. Juxtaposed with the problematic criminal network at the hearing, Micro-Professor computers and Little Intelligent computers were referred to several times and labeled counterfeits.

The counterfeit problem in the hearings was framed as another illegal act—dumping—in which a company sells low-priced products, lower than the usual manufacturing costs, to increase market share. In the context of international trade, dumping was a way to orchestrate unfair competition. The July hearing's participants were deeply concerned with the scale of counterfeits from Taiwan. Stephen F. Sims, the special assistant to the subcommittee, asked Steve Waterson, an import specialist in the US Customs Service, to estimate Multitech's manufacturing capacity. Waterson believed that Multitech could make five thousand computers per month. Sims continued to ask if counterfeit production would reach fifteen thousand computers per month. Waterson then answered that he would not be surprised if Taiwanese companies made and exported 300,000 counterfeits to the United States. Waterson's and Sims' initial estimate of Multitech's monthly production capacity was close to what the company revealed to Taiwanese journalists in late 1982. But Waterson did not explain from where his estimate of 15,000 and 300,000 Taiwanese computers came. Seeing overseas mass-manufactured computers as a threat, estimates grew higher and higher as the conversation went on. Waterson's and Sims' estimate was unreasonably high, if one considers that IBM sold 75,000 personal computers and Compaq 53,000 in 1983. Furthermore, the potential buyers for Apple counterfeits in the United States were limited. In March 1983, Apple's associate counsel, Dan Wendin, noted that their major buyers in the United States continued to be computer hobbyists.[38]

Sims was startled by the manufacturing capacity of a single Taiwanese company. In the same vein, foreign businesspeople at that time were likewise impressed. Writing in October 1982, a *Los Angeles Times* journalist observed, "The only solution, some Western businessmen and bankers here say, is forming partnerships with Taiwan man- ufacturers." The partnership, what we call outsourcing today, soon happened—US per- sonal computer companies gradually contracted with Multitech, MiTAC, and other companies to manufacture their computers.[39]

MULTITECH OR MULTITEC, STAN SHIH OR JONNEY SHIH

In the hearings, facts about Multitech, MiTAC, and other Taiwanese computer mak- ers were often misrepresented, or misconstrued with inaccurate information. The igno- rance reflected misrepresentations—sometimes unintended, sometimes intended—of Multitech and MiTAC as counterfeiters. In the July 27 hearing, Gerry Sikorski, a subcom- mittee member from Minnesota, brought to the hearing the latest issue of *Business Week* that had "just hit the newsstands." A news article titled "High-Tech Entrepreneurs Cre- ate a Silicon Valley in Taiwan" drew his attention.[40] Sikorski went over each of the com- panies covered in the news article, including Multitech and MiTAC, and asked Apple's attorney Hecker if he knew whether any were "producers or importers of piratical copies of Apple Computer in the U.S."[41] Regarding MiTAC, Hecker answered that he had a MiTAC Little Intelligent computer and noted, "We have tested it, and it infringes our copyrights 100 percent." Sikorski and Hecker then checked on Stan Shih's Multitech:

Mr. SIKORSKI. Multitec [*sic*]?

Mr. HECKER. Multitec [*sic*] we are quite familiar with. In fact, we have here today two of their computers. This is the first Multiteck [*sic*] computer which we identified as being a counterfeit—100 percent infringement of Apple's copyrights. This computer is known as the Microprocessor II [*sic*]. Many of these have been seized at U.S. Customs. There is another computer here which is considered their next generation of a computer. It doesn't look anything like the Apple. It has a detachable keyboard, but same exact circuitry and software—100 percent copyright infringement—and that computer, I believe, is called the Microprocessor III and there is a Microprocessor IV [*sic*].[42]

The transcriber of the hearing probably did not have access to the news article, and thus misspelled Multitech three times in a row. Hecker also misnamed the Micro-Professor series as the Microprocessor series. The transcriber's misspelling and Hecker's mistaking, however, effectively signaled a reductionist portrait of Multitech in the hearing. Indeed, Multitech can be a confusing name, and naturally and easily invites minor spelling errors. Multitech was renamed to Acer precisely because it confused customers many times. For example, in April 1982, Multitech's Micro-Professor I was nominated as one of the top ten computers by German magazine *Chip*; but in that *Chip* issue, Multitech was misspelled as Microtek, which is a scanner manufacturer set up in Taiwan in 1980. Bo-bo Wang, Microtek's founder, a former Xerox engineer, and an acquaintance of Shih's, received and transferred *Chip* readers' and interested buyers' inquiries to Multitech. In 1986, when Stan Shih was informed that Multitech has been used by a US modem manufacturer, he then decided to rename his company.[43]

By reductionist, however, I mean that participants in the hearings lumped Micro-Professor II with the group of Apple counterfeits. Hecker's statement conflicted with Stan Shih's later statement, insisting that his company had redesigned the computer to make it compatible with Apple IIs. If Shih were to be honest, Hecker's notion of "100 percent copyright infringement" might not be precise enough to describe how many lines of Multitech's code were like Apple Computer's Applesoft and Autostart programs. Robert King, journalist at the *Wall Street Journal*, remarked that in June 1983, the US Customs Service was unclear about "how much the software in these machines [computers] has to resemble Apple's before they can be confiscated."[44] In addition, as a review in *InfoWorld* pointed out, the arrangement of bus structure and other hardware of the Micro-Professor II computer was intentionally different from the Apple II, which would make some differences in "circuitry" in Hecker's notion.[45] As I have shown earlier, Stan Shih was aware of counterfeit problems and planned to avoid being identified as a counterfeiter by his customers. He also wanted the new computer to process Chinese and be compatible with the Apple II. To achieve compatibility, it was likely that the Micro-Professor II computer could not be a brand-new invention. A compatible computer had to use some Apple II code, as the Franklin model had.

The third congressional hearing, on August 2, 1983, finally got Stan Shih's attention and response, though only through a letter. A witness testifying at the hearing stated that Multitech was not only a counterfeiter but also sold computers to China. Given

Taiwan was then under martial law, the accusation could potentially get Shih in a big trouble. The witness's statement prompted Shih to send a letter, on September 12, 1983, to Michigan Congressman John D. Dingell, the chair of the subcommittee.

The witness, James Tunnell, had been hired by the Osborne Computer Corporation in 1982 to investigate counterfeiters of Osborne's portable computer. For that reason, Tunnell was paid to travel to Hong Kong, Taiwan, Tokyo, Singapore, and perhaps other parts of Asia. He also collected information about counterfeit computers other than Osborne's on his own time. In his testimony, he listed MiTAC and Multitech, along with many Hong Kong, Taiwanese, and a few Singapore companies in his thirteen-page "profile list of infringers." Among those Taiwanese companies, the government-sponsored research institute ERSO, and two other corporations—Microtek and Tatung—were all mistakenly listed there as counterfeiters, despite not making Apple or IBM compatibles in 1982. Much of the information Tunnell gathered about MiTAC, Multitech, and the other three entities was full of minor and major factual mistakes. Or, like today's notion of misinformation or fake news, some inaccurate information was placed with some truthful information in Tunnell's report. Tunnell might not have intentionally created misinformation. But, the inaccuracy he presented reflects, first, the unclear definitions of counterfeit and compatible computers at that time and, second, a pre-occupation with and ingrained understanding of Taiwanese computer makers as counterfeiters. Tunnell conflated all Taiwanese computer companies.[46]

Tunnell first stated that Multitech was now "in hot water with Apple." He then noted that the owner of the company was "Jonney (John) Shih, U.S. educated and trained." In fact, Stan Shih never gained a US degree—he is famous for being successful without a US degree. Jonney Shih was at that time the chief R&D engineer at Multitech and happened to have the same last name as Stan. Both Shihs hailed from the historic port city of Lukang, where Shih is a common surname. In the report, Tunnell seemed to develop a pattern: he mistook a top managerial-level person for the owner or the CEO of a company, perhaps because of incapable or unreliable informants or misunderstandings in his field visits. For MiTAC, he misidentified the founder, Matthew Miau, a former Intel engineer, as a Mr. Chu. He then explained that Mr. Chu's first name was unknown because Mr. Chu "never used [it] when he signs documents." But Matthew Miau's full name had appeared in Taiwanese newspaper articles a couple of times. Moreover, thanks to Miau's family reputation, Miau was already a celebrity in the industry.[47]

Stan Shih's letter to Dingell and the subcommittee was completed on September 12, more than a month after Tunnell's testimony. The letter may have arrived at its destination before the fourth hearing, scheduled on September 21, but subsequent hearings never discussed it, perhaps because it was not made under oath. In the beginning of the letter, he noted that his company imported over $30 million worth of electronics from the United States to Taiwan because at the time Multitech was a franchiser or exclusive franchiser of many US companies, including Zilog, Texas Instruments, and Intel. Shih's strategy was to reaffirm that he had no intention of creating a dumping issue, that he was not an outlaw, and that he had never wanted to be one because he wanted to maintain his US electronics manufacturer franchises.[48]

Stan Shih then stressed that his company did not make counterfeit computers. Instead, his company intended to make computers to be "compatible with the best-selling" products on the market.[49] He explained, "Multitech shares the Subcommittee's concern over the growing trade in truly counterfeit products. It is also concerned, however, that in the effort to stop this trade, the distinction between counterfeit merchandise and legitimate electronics and computer products independently developed to be compatible with manufacturers' products will be lost."[50] He emphasized that Multitech's computers were totally different from "illegal" and "blatant counterfeits," and he did it repeatedly in the letter. For example, he continued,

> The distinction between illegal counterfeits and independently developed products is not always clear, of course. Computer software and firmware are presenting a range of difficult and unique questions under the copyright laws of the United States and other countries. These questions will more often be raised in disputes between companies acting in good faith rather than in actions against blatant counterfeiters.[51]

Apple Computer's answer to these difficult and unique questions was clear: it decided not to allow copyright infringement of its operating system code and would exclude any compatible computers. In contrast, Stan Shih noted that counterfeiters should be discouraged, whereas competition between "legitimate high-technology" manufacturers, especially US, Taiwanese, and other Asian companies, "will benefit" all. "None of us can afford to ignore the legitimacy or potential competitiveness" of newcomers in the industry, he remarked. He suggested that his company was there to contribute to healthy competition in the industry. Shih's use of the term "legitimate" indicates his belief that his company, "like many US companies," was "independently developing computer products."[52]

Stan Shih and his legal team might have noticed the ongoing lawsuit that Apple brought against Franklin. He might have expected Franklin to continue challenging the court's ruling in favor of Apple, which was announced in late August—he drafted the letter only two weeks after the ruling. Other computer counterfeiters and compatible makers also hoped the Franklin case would not be settled soon.[53] That way, producing Apple compatibles remained a not-entirely-outlawed industry before a clear line could be drawn when the case was closed.

In his later biographies and interviews, Shih also emphasized that Multitech developed Micro-Professor IIs independently. As mentioned earlier, he decided to develop the Micro-Professor II in late 1981 or early 1982, when bogus Apples emerged. In this context, he intended to differentiate Multitech from counterfeiters. Edward Chang, president of Multitech's branch in California, assured *InfoWorld* journalists that the Micro-Professor II was a "completely different machine" from fake Apple computers. *InfoWorld* noted in August 1982 that Chang "anticipates no legal problems with Apple computer." Nevertheless, since the cleanroom practice had not been implemented, the act of reverse engineering probably led to an unknown level of similarity between Multitech's and Apple's code, allowing US Customs to stop the Micro-Professor II from entering the US market.[54]

In later biographies and interviews, Shih noted, "It was difficult for Apple Computer to substantiate the accusation of Multitech's [1983] copyright infringement." "However," he continued, "our English manual was founded problematic." To help novices learn how to use the computer, Multitech produced an over two-hundred-page user's manual in the form of a comic book. Due to lax supervision of a translator hired in Taiwan, portions of Multitech's manual were copied word-for-word from the Apple II manual. With this, Apple pressed Multitech to pay $20 US for each Micro-Professor II sold. But, in the end, Multitech did not oblige them. When recalling these disputes, the mild-mannered Shih put it this way: the trouble with US Customs and Apple was an opportunity for Multitech to learn and strengthen the company's copyright literacy. After the incident, Shih immediately promoted training courses on computer-related copyrights, patents, and trademarks in Taiwan, by working with both the Taiwanese government and associations of computer manufacturers. In 1984, Shih told journalist that his company spent over $10 million NT, an equivalent of $323,100 US at that time, to cover all expenses hiring attorneys to negotiate with Apple Computer in the United States and the US Customs Service.[55]

The rest of Shih's letter to Dingell and his subcommittee was a repudiation of Tunnell's statement about Multitech's trade with China. For Multitech's profile, Tunnell wrote,

> It is reported that MultiTech [sic] . . . has exported three different computer designs (The Dragon Chinese)—a Chinese high-level language Computer (TDC), and the MCZ-1 and ZDS-1 systems[56] to Mainland China. All three systems are known to be used in military applications. MultiTech International has developed an IBM PC-compatible computer product and is prepared to export that product to established dealers in Europe and the United States. MultiTech International has also produced copies of the U.S. made Digital (Digital Equipment Corp.) computer hardware for export to Mainland China. Exact model duplication is unknown, as products copied have been variations of the original, although two versions of the counterfeit Digital PDP-11 have been seen.[57]

In repudiating the allegations, Shih clarified that his company did not copy Digital Equipment Corporation's computers and, therefore, did not sell any DEC computers to Chinese customers or agents. His company would not sell any other computers to a customer or distributor who lived in mainland China. Because the Nationalist Party–led authoritarian government did not allow trade with China at that time, Shih told Dingell, "As you can well appreciate, these [conducting trade with China] are grave charges in Taiwan." Shih pointed out that he could not stop any of his customers from bringing or reselling Multitech products to China. Reselling Multitech products seemed possible but must have been through individuals or middlepersons before the summer of 1985. Only after that summer did the steadfastly anticommunist government became more lax about trade with China. It then allowed Taiwanese companies to sell products to China if transactions were completed through intermediaries in Hong Kong, Japan, or other places. According to a *Wall Street Journal* article, a Taiwanese businessman declared in 1985 that his company made seven thousand personal computers in 1984, and 2,500 were delivered into the hands of the PRC. Tunnell's misinformation about Multitech's involvement in trade with China might again be from his unreliable informants or his inability to verify.[58]

In the very last part of Stan Shih's letter, he finally clarified that, instead of Jonney Shih, he, the undersigned, was the president and the chairman of the board of Multitech.[59]

In addition to Multitech, Tunnell devoted space in his report to intelligence about MiTAC, which he deemed a counterfeiter. MiTAC, he noted, was developing an IBM PC and Apple IIe compatibles. He suggested that both were counterfeits and emphasized that MiTAC would be able to make 150,000 Apple IIe compatibles for the US market.[60]

MiTAC's founder Miau responded to the accusation in a press conference in Taiwan in September, and it was likely that the company had spoken to Apple Computer or the subcommittee, as Multitech did. Miau clarified that the design of its latest computer was going to be different from an IBM PC. Miau, a former Intel engineer, noted that he was clearly aware of patents and copyrights held by other companies. His company had contacted Microsoft and contracted ERSO to get authorization from Microsoft for the new computer under development. He stated that Tunnell was totally confused by and ignorant about the difference between counterfeits and compatibles. Compatibles were going to be the future of the computer industry, as many large US companies were now making IBM compatibles. Miau did not seem to openly comment on MiTAC's Apple compatibles in the 1980s. But in 2021, MiTAC's webpage frankly revealed that MiTAC gave up developing Apple compatibles after eighteen months of attempting to avoid violating Apple's patents, despite that MiTAC's Little Intelligent computer was available in the Taiwanese market in mid-1982, earlier than the rollout of the Micro-Professor II.[61]

Tunnell's comments on the research institute ERSO were inaccurate as well. In his report, he claimed that ERSO's more than 1,200 employees were downsized to eighteen staff members, which could not be true. From the late 1970s to 1981, ERSO poured resources into building a computer to simulate the PDP-11 minicomputer. Tunnell might have mistaken this project as a Multitech project. The building of a copy of a PDP-11 was like Chiao-Tung's minicomputer-building project described in chapter 5, aimed at learning the essence of computer manufacturing techniques instead of counterfeiting. According to Ding-Yuan Yang, one of the leaders of the ERSO project, he acquired technical documents from Hewlett-Packard and DEC and had ERSO researchers study those documents and build a computer. The PDP copy was renamed the EM-1104 computer and was the only copy ERSO built.[62]

Tunnell also mistook the rising scanner manufacturer Microtek, founded by a former Xerox engineer, for a counterfeit computer maker. Furthermore, Tatung Corporation was on Tunnell's counterfeiter list, too, but Tatung made the company's first computer, the Einstein, only in June 1984. It was a unique computer system, neither a counterfeit nor compatible, and it was made in England. Tatung had made magnetic core memory for IBM since the late 1960s (chapter 6) and began to make CRTs for their own brand names and US companies in the late 1970s. As of August 1983, when Tunnell testified, Tatung did not make computers. Bogus Apple II makers, however, might have purchased Tatung monitors, which could have prompted Tunnell to include the company in his report. Tatung, in May 1981, set up a subsidiary in England, and had hired engineer Roy Clarke and his team in Bradford, England, to develop the Einstein computer. Clarke emphasized, "It was all done here. None of it was done in Taiwan." The prototype took about nine months to develop. At the same time, a 41-acre production complex in Telford was arranged, which cost £10 million British. The first batch of Einstein computers was delivered in June 1984, and the company was scheduled to make fifty thousand copies in the second half of 1984. A basic model with no monitor cost £499 British. Einstein computer enthusiasts published an independent users' quarterly magazine starting in November 1984 to exchange knowledge about the computer. Tatung's owner, T. S. Lin, provided a greeting overleaf in the first issue of the magazine, which continued publishing till early 1991.[63]

Three paragraphs in Tunnell's report were almost identical to the text of a *Wall Street Journal* news article. It is unclear whether Tunnell simply mispresented the work as his own writing. In one paragraph, he reported that US customers of fake Apple II computers would be able to easily swap the standard 220-volt power supply in the Sham

Shui Po district of Hong Kong with a 110-volt power supply, as if he had witnessed it in Hong Kong. In another paragraph, he depicted a seller in Hong Kong who tested the computer and had it show "a folksy 'Looks good!'" on the video screen. Moreover, in another paragraph, he noted that customers often would ask for a discount, and the shop owners would agree to a 5 percent discount. He stated that "the final ritual isn't something designed in California's Silicon Valley, it's purely Hong Kong." For unknown reasons, the three paragraphs were almost the same as sentences from Anthony Spaeth's October 1982 *Wall Street Journal* news article. Stan Shih, in his letter to Dingell, alluding to Tunnell's lack of credibility, pointed out that some of Tunnell's written testimony seemed to be plagiarized from the *Wall Street Journal*.[64]

Several months after the congressional hearings, a high-profile US magazine, *Life*, inaccurately portrayed Taiwanese products as counterfeits when, in March 1984, several *Life* magazine journalists and photographers traveled to Taiwan and claimed that they would like to visit a keyboard manufacturer. There, they took a photograph of three staff members casually looking at fourteen keyboards in a factory. In the September issue of the magazine, the photograph appeared, indicating a booming counterfeiting business in Taiwan that included copies of Cabbage Patch Kids, Ray-Ban sunglasses, and truck brake shoes. The photo caption of the keyboards announced, "Sham Apple keyboards are readied for export." The engineer who designed the keyboard, T. Y. Chen (Zongyuan Chen), was outraged when he discovered the true intention of the journalists. He told Taiwanese media that he designed the keyboard and had "design by T. Y. CHEN" everywhere on the keyboard's printed circuit board. He emphasized that the keyboard contained eleven ICs. In contrast, Apple's keyboards contained no printed circuit boards at all. Chen also noted that had he been a counterfeiter, he would not have had journalists visit his factory. Quality control staff member Mingzhi Kao, who was shown in the photograph, challenged the *Life* caption too, responding that if he were readying the keyboards for export, as the caption claimed, he would have faced the keyboard toward himself instead of toward the camera. Kao indirectly suggested that the writer of the caption knew nothing about manufacturing keyboards. Chen and Kao believed the company was framed.[65]

In the same *Life* article, the photograph of Kao accompanied two other large photographs that featured an aged man and woman, suggesting that counterfeiting in Taiwan represented the country's stagnant technological development. The first photograph, across two pages, was an eighty-four-year-old woman eating food from a bowl with chopsticks, on a bed fully covered by at least twenty fake Cabbage Patch Kids and a little real human baby, barely the size of the doll. The live baby was difficult to spot at first glance, which was meant to emphasize "bogus babies" around her. The old woman's dress was in an early twentieth-century style. The caption says she was watching her granddaughter but did not specify whether she was also making these counterfeit Cabbage Patch Kids or someone else in the family was doing it. The elderly woman and the unnoticed human baby were deployed to stress the characteristics of the phony Cabbage Patch Kids—unauthentic and inert, the opposite of US ingenuity, protected by trademarks and possibly patents and demonstrating intellectual energy. Hence, the old lady and the little baby had no ingenuity. This image, read with the case of "sham Apple keyboards," also suggested that babies, dolls, and computer peripherals are treated as immature playthings. As such, even if the keyboard was not a counterfeit, it was nothing genius. In a similar vein, the other across-two-page photograph featured an aged technician, grinning with his five silver teeth. He allegedly made counterfeit truck brake shoes not as durable as "their made-in-U.S.A counterparts." The man, face full of wrinkles, was again represented as "Oriental," old-fashioned, and early twentieth-century-like, hence the epitome of the stagnant technological progress of Taiwanese society.[66]

CONCLUSION

Stan Shih and other Taiwanese computer makers embraced the role of entrepreneurs, sought technical improvements, and attempted to keep themselves away from copyright infringements, despite that some US competitors and analysts viewed them as copycats and invaders. Shih was aware of the rampant counterfeits in Taiwan when his engineers began to develop the Micro-Professor II computer. He intended to avoid being identified as a counterfeiter by his customers. But his manufacturing of Micro-Professor II computers, an Apple II compatible system, was not acceptable to Apple. After Apple acted to stop questionable machines being sold in the United States, Shih's initial success in exporting Micro-Professor II computers to Europe and Australia became limited, and the Micro-Professor II computer could not enter the US market. The first batch of Micro-Professor II computers were detained in US Customs in early 1983. In chapter 9, I focus on Shih's success in developing and selling IBM compatibles, in the context of IBM's relatively lax attitude toward compatible systems before 1987.

Despite Shih and Miau's intention to move beyond copying existing commodities, there was a discrepancy between their and Apple's attitudes toward compatible computers. The discrepancy led to a series of misrepresentations of Multitech and rising Taiwanese computer manufacturers to the public in the United States. Apple's rejection of counterfeits and any compatible computer systems became the focus of a series of congressional hearings in the United States from June to August 1983. The information gathered in the hearings and some US media, such as *New Tech Times*, presented

Multitech and other Taiwanese computer manufacturers as counterfeiters only. The image of counterfeiters was a departure from the "high-tech entrepreneurs," portrayed in a 1983 *BusinessWeek* article complimenting Multitech and MiTAC on their contribution to creating "a Silicon Valley in Taiwan." Participants in the hearings, including several congressmen, Apple's legal team, and a private investigator, painted a threatening image of Taiwan's manufacturing capacity swallowing whole the US microcomputer market.

There may have been many reasons behind hearing participants mistaking all Taiwanese computer makers as counterfeiters. Some might have had no knowledge of and hence ignored the technical capacity of Taiwanese computer manufacturers. The ignorance reflects the presumption of a lack of innovation in the fledging computer industry in Taiwan. The endurance of such a presumption has been elaborated in postcolonial studies. Edward Said has argued, in *Orientalism*, that Western scholars had maintained an unchanging understanding of the "Orient"—demonic, defeated, grotesque, illogical—from ancient times through the Middle Ages and early Renaissance to the twentieth century. Before the invasion of Egypt by Napoleon in 1798, Western understandings of "the Orient" had circulated in textual works primarily. But after the invasion, Western visits to and colonization of the area did not change previous understanding. "Orientalism" precisely describes ingrained understandings of the Orient that lasted for centuries, before and after the increased in-person interactions between the Orient and the Occident. In this chapter, Congressman Sikorski, special assistant to the subcommittee Sims, and the producer of *New Tech Times* were developing their understanding of the "facts" about computer counterfeits manufactured in Taiwan through their reading of various textual materials, such as testimonies and news articles. Many of the sources they read, and the reports they produced, were Orientalist. These textual materials simply reinforced the classification of all Taiwanese computer makers as invaders, counterfeiters, and criminals. It was especially astonishing to see that, even after making trips to Taiwan, private detective Tunnell and *Life* magazine journalists and photographers simply utilized what they knew before their trips to represent what they "saw" during the trips. Their actual trips to Taiwan did not matter much, as they did not serve as an investigation. Instead, the trips were a reification, to confirm their impression of the backward island, the Orient, or the cunning copycats. Some of these paradoxical "facts" that they crafted for their audience were fact-like misinformation.[67]

The Orientalist representations of Taiwanese computer manufacturers can be self-contradictory. On the one hand, when it comes to the manufacturing capacity of Taiwanese computer makers, participants in the congressional hearings exaggerated the numbers of counterfeit computers Taiwanese companies dumped in the United States.

On the other hand, more and more US computer manufacturers gradually recognized the manufacturing capacity of Taiwanese companies and subcontracted them to make computers at lower cost for the expanding US and global personal computer market. The contradictory understandings of Taiwanese computer makers bear similarities to the discourse about science in the context of the imperialism, elaborated in historians Gyan Prakash's and Warwick Anderson's works. British imperialism was deeply tied to the empire's economic expansion. But British imperialists claimed to embark on a civilizing mission in their colonies. The mission was accompanied by an assumption that Indians and other colonized populations would never be fully civilized. A commitment to a civilizing mission could also be found in US imperialism in the Philippines. A constructed image of the colonized, Oriental populations' undeveloped minds and undisciplined bodies further reinforced colonizers' assertion of legitimacy. Similarly, at the level of representation, Taiwanese computer manufacturers were portrayed in the hearings and the media as having no capacity to innovate and acting like mobs disturbing the US market. The portrait was inconsistent with an emerging trend developing in Silicon Valley—Taiwanese companies soon became crucial in assisting some US computer companies in lowering their manufacturing costs.[68]

Under the logic of capitalist competition, profit-driven innovators would unavoidably encounter counterfeiters, as their goal was to draw the line between the original and the imitator. Defending their own commercial interests, Apple's complete rejection of compatible models and IBM's requests for royalties from PC clone makers after 1987 (described in chapter 9) imposed a "copyright modernity" by taking aggressive legal actions against companies that made Apple and IBM compatibles. Historian Ruth Rogaski has pointed out that, in late nineteenth-century northeast China, Western and Japanese officials promoted "hygiene modernity," but "progress" in public health only arrived "on the heels of destruction and was established at gunpoint." Anderson has also revealed that, in the colonial Philippines, science was used to highlight the difference between races. Laboratory-based assays of cells of the colonizers and colonized served this purpose. The similarity tests of ROM chips at US Customs was also a quest to distinguish artifacts created by *different* groups. A full implementation of copyright protection was important for a country to enter the world of modernity. After all, to make progress in achieving technological modernity, one would be expected to practice codes of "hygiene" in a "cleanroom" environment when developing new code for personal computers.[69]

The Orientalist image of a swarm of copycats also further limited the available thought space for discussing innovation demonstrated by Taiwanese computer makers: they strove to find ways to allow Apple compatibles to process Chinese characters;

they also offered technical support to users of Apple compatibles. Even in the contexts of "pirating" and "copyright infringement," one had to acquire a degree of knowledge of computers before one could copy the computer code. Similar to the legal action that stopped Micro-Professor IIs from entering the US market, the representations of the Other could have some material effects in reality. In this case, the Oriental image persuaded Western viewers that Taiwan had no innovation.

Right after sending his repudiations to the chair of the subcommittee in September 1983, Shih proposed "number-two-ism," which described his philosophy of being confident in being the second best in the industry. Being the first could cost much, he said. If his company was not yet ready to be the best, he would accept and be content with the status quo. If an opportunity arose for him to take the first position, he would not hesitate to take it. In 1983, Shih was probably referring to Multitech, MiTAC, and other Taiwanese companies in their tie for first place among the nation's high-tech companies. Soon, Shih's competition became international. After resourceful US IBM clone makers recognized the manufacturing capacities of Taiwanese computer makers, Multitech was able to leverage subcontracting opportunities into making profits and growing bigger. In 1986, when a journalist asked the leading R&D engineer Jiahe Lin about Shih's "number-two-ism," Lin answered that IBM was the best, and Multitech was the second best. Stan Shih, Lin, and their coworkers such as Jonney Shih were aiming to catch up with the first in the industry by developing faster and cheaper IBM compatible computers.[70] However, the competition between Multitech and IBM, from Shih and Lin's perspective in 1986, was more about a competition of technical achievements than of market share. Shih's company eventually made it to the fifth largest PC maker in the world in 1995. In the big picture, Taiwanese computer manufacturers altogether attained a dominant market share in the personal computer industry in the 1990s. The fast-growing number of computers and components developed and made by Taiwanese companies eventually persuaded the US news media to dub Taiwan a "PC Powerhouse" through "a well-worn road to dominance," in 1988, and "the Republic of Computers" in 1995. In chapter 9, I discuss Multitech's success and challenges in manufacturing IBM clones and the company's leadership in making Taiwan "the Republic of Computers."[71]

9 THE REPUBLIC OF COMPUTERS: FROM SCOUNDRELS TO ASIAN HEROES

Thanks in no small part to Shih's pioneering example, the integrated-supply-chain model is now the hallmark of PC contract manufacturing worldwide. In other words, he's a big reason why your PC costs $1,000, not $10,000. . . . Instead of merely manufacturing cheap capacitors, radios and the like, Taiwan's PC industry grew far more ambitious under Shih's mentorship.

—Paul S. Otellini (then CEO of Intel), November 13, 2006, *Time*

In 1983 and 1984, Stan Shih's company, Multitech, stumbled twice as US Customs detained its computers from entering the US market. The first occurrence was in early 1983, and the computers were Apple II compatibles. The second was in early 1984, and the computers were IBM PC/XT compatibles. While chapter 8 discusses the disputes over Taiwan-produced Apple II compatibles, this chapter focuses on Shih and his fellow Taiwanese computer makers, such as MiTAC, who began to manufacture IBM PC clones, and eventually became indispensable subcontractors for US computer makers. Acer, Multitech's new name after 1986, became one of the world's top ten computer makers in the mid-1990s.

Shih's success can be attributed to several achievements he and his company attained in the 1980s. Similar to other fast-growing US PC clone makers, Shih increased his employment of R&D engineers over time to develop computers with stronger technical performance. Benefiting from the infrastructure of Taiwan's existing electronics industry, Shih's engineers, factory workers, and suppliers readily supported his plan for mass-manufacturing computers in Taiwan. Shih strove to sell computers from the island to around the world and, with his fellow Taiwanese competitors, eventually set up factories in overseas locations to ensure their products' low cost and swift delivery.

To catch up with the advances in computing technology, Shih started new companies and acquired Silicon Valley enterprises. In 1986, his engineers leveraged the resources they had in Silicon Valley to work closely with Intel to develop a 32-bit microcomputer, the Intel 80386-based computer known as the Multitech 1100, one of the earliest 32-bit microcomputers. It appeared only a few months after the fast-rising US company Compaq developed its 32-bit microcomputers, the world's first. The success of the 32-bit microcomputer helped Acer become so well-known that it won contracts to make PC clones for top computer manufacturers in the United States, Japan, and Europe. Beyond subcontracting, Acer has continued to make, market, and sell its own computers directly to consumers globally.

As this chapter shows, Shih could not avoid the challenges posed by IBM, the computer industry giant of decades. After his IBM PC/XT compatibles were detained at US Customs in early 1984, Shih had his engineers swiftly redesign the computer. In 1986, Shih began to promote its IBM PC/AT compatibles and then 32-bit IBM compatible computers. Both were well-received in the market and allowed the company to grow significantly. In 1987, however, IBM requested PC clone makers across the globe, including Acer, Compaq, Dell, and others, to pay royalties, dating back to 1983 or earlier, on IBM PC/XT and AT compatible computers they manufactured. Big Blue believed that, to make IBM compatible computers, these companies must have used thousands of patents that IBM held for personal computers. Therefore, they owed royalties to IBM. Furthermore, in 1989, IBM informed Acer of copyright infringements in its IBM PC/AT compatible computers. After protracted negotiation, Acer reached a $9 million US settlement with IBM. Despite the considerable settlement and the royalties requested by IBM, Acer continued to profit and expand its business globally in the 1990s.

At US congressional hearings in 1983, Shih's company Multitech was viewed as an invader. In this chapter, the ecology of the industry in the 1980s, however, made Multitech a sought-after subcontractor for prestigious computer makers around the world. US computer makers were keen to establish their share in the thriving personal computer market, and they found that Acer and other Taiwanese computer manufacturers had qualified subcontractors that could supply them inexpensive and profitable computers for that market. Acer's commercial and technical success, nevertheless, did not change how Acer's competitors would view it. Despite that IBM did not openly describe Acer and MiTAC as counterfeiters, the media and IBM employees viewed Taiwanese computer makers and other PC clone makers as flies and scoundrels that stole IBM's technology. In a recent documentary, Compaq founders were portrayed as "silicon cowboys," who dared to challenge Big Blue.[1] Stan Shih and his engineers probably did not think of themselves as cowboys, but they did share the same spirited entrepreneurship that Compaq founders exhibited—why can't we make inexpensive and even more powerful personal computers to run software applications designed for IBM machines?

SHIH'S SWITCH FROM APPLE COMPATIBLES TO IBM COMPATIBLES

In fall 1982, Stan Shih attended the Computer Distributors Exhibition (Comdex), an annual microcomputer expo, held that year in Las Vegas. In no time, he found that everybody was talking about Compaq's new IBM compatible computer. He was amazed by Compaq's technical achievement and business strategy in proposing something to be compatible with the products of a computer industry leader, the same strategy that Shih had believed in since he began making Apple compatibles. Given his company had rolled out their Apple II compatible computer—the Micro-Professor II—earlier that year, he thought about developing an IBM compatible after witnessing Compaq's success. Because Multitech's engineers were swamped with the development of the Micro-Professor III, Shih commissioned government-sponsored research organization ERSO to develop an IBM-compatible personal computer. The development would cost Multitech $15 million NT ($484,650 US in 1982). Shih also immediately found a buyer for the IBM compatibles, the National Cash Register (NCR) subsidiary Applied Digital Data Systems (ADDS). ADDS was going to sell Multitech's computers as ADDS computers in the United States.

Micro-Professor II computers' unfortunate encounter with US Customs in early 1983 did not discourage Shih from developing an IBM compatible. But in the summer of 1983, ERSO wanted to modify Shih's project. ERSO notified Shih that it intended to include four additional computer companies in the research and development project for an IBM compatible personal computer. The computer would be compatible with the IBM PC/XT. ERSO staff felt pressure to modify the project with some officials believing that ERSO, as a semigovernmental organization, should serve the local computer industry instead of working for a single company.

ERSO returned four-fifths of the original Multitech funds and replaced them with additional monies from the four companies. According to Ding-Yuan Yang, a former Harris Semiconductor engineer and now the deputy director of ERSO, Stan Shih did not like the new plan but relented. Decades later, Shih got over it and humorously noted that he was happy to use only $3 million NT for the design of an IBM XT compatible. "It did not matter too much, retrospectively," Shih reflected. ERSO's arbitrary decision in 1983 did not affect Shih's business for two reasons. First, the four companies stopped being Multitech's competitors within ten years, as Shih's company became the largest computer manufacturer in Taiwan and one of the top companies in the world. Second, US Customs unexpectedly stopped the batch of IBM XT compatibles that ERSO developed for Multitech from entering the United States.[2]

In February 1984, the first batch of Multitech-manufactured IBM XT compatibles for ADDS was detained by US Customs in Seattle, suffering the same fate as the Micro-Professor II computers in early 1983. The seizure was primarily due to the high degree of similarity between ERSO's and IBM's basic input/output system (BIOS). US Customs detected the similarity because IBM supplied US Customs with programs that compared IBM code with other computers on their way to the US market. Earlier that year, IBM had taken legal action to stop manufacturers including Handwell Computer and Corona Data Systems from infringing on its BIOS copyright. The seizure also occurred shortly after Franklin settled with Apple. By the end of July 1984, US Customs had detained at least seven hundred Taiwanese-manufactured computers with ERSO design.

Like Apple, IBM initially pushed customs to stop PC clones from entering the US market but eventually enacted a different strategy. In 1987, it pivoted and asked Compaq and other manufacturers to pay royalties for making PC clones. Though, in 1984, IBM approached ERSO to investigate copyright infringement. It then offered ERSO a chance to revise its BIOS program in question. Seeing that ERSO was not a for-profit company, US Customs took the unusual step of returning the seized computers to Multitech.

Despite that the computers were detained, ADDS' president David G. Laws expressed faith in Stan Shih's product. Shih, however, was not going to wait for the revised version of the BIOS program, as he felt pressure to meet client deadlines. ADDS' contract was especially important, as it would bring a revenue of $20 million US. Shih decided to abandon ERSO's BIOS, opting to license a US company's workable version.[3]

Furthermore, Shih immediately decided to pay $1 million US to Digital Research for authorization to use Concurrent CP/M to replace the suspect ERSO operating system. Although Shih had another option—getting the licensing of MS-DOS from Microsoft—he did not choose that route.[4] Concurrent CP/M was almost entirely but not 100 percent compatible with MS-DOS. Shih regretted his choice when he observed IBM XT compatibility issues in Multitech's final products. Eventually, Multitech shipped the promised computers to the United States and delivered them to ADDS in late July with no hassle. To facilitate the customs inspection, Shih even arranged a special cargo flight of three hundred computers so that US Customs could do a thorough inspection in the presence of Multitech's attorneys. Subsequently, Multitech's reputation benefited from being one of the first manufacturers to make IBM-compatible PCs.[5]

IBM GENTLEMEN SPARED ERSO

In contrast to Shih's quick resolution of the BIOS issue, ERSO took a longer time to respond to IBM's accusation. Detaining Multitech's computers in February 1984 distressed Taiwanese officials and ERSO staff. K. T. Li, then a senior adviser to the president of Taiwan, decided to reach out to IBM to find a solution. Ding-Yuan Yang, who led the development project for the IBM XT compatible computers at ERSO, recalled that IBM immediately dispatched engineers to Taiwan to investigate the issue.[6] Interviewed in a 1984 issue of *BusinessWeek*, Yang noted, "The lawyers always say that independent development work is the most important condition." He added, "Now we know that is not enough. We cannot have too much similarity because of the [US] Customs test."[7] IBM engineers interviewed ERSO engineers after they arrived in Taiwan and decided that Taiwanese engineers were ignorant about copyrights, but that they had not maliciously copied IBM's code. Instead, engineers copied code that they deemed simple, and therefore saw "no point to re-write." During the investigation, IBM engineers even complimented ERSO's design of the driver for the hard disk, according to Yang. IBM then informed ERSO that the team should rewrite the code and hire Arthur D. Little consultants to arbitrate. IBM would let go of the case, if Arthur D. Little decided that ERSO's revised BIOS involved no infringement of IBM's copyright.[8]

Yang remarked that IBM was relatively "gentlemanly" throughout the process. Yang secured a housing facility at the Industrial Technology Research Institute (ITRI) in Hsinchu, as ERSO was an office under ITRI, and asked all his engineers to move there to rewrite the BIOS program day and night. It took a few weeks to come up with the new BIOS, but Arthur D. Little spent three months verifying that ERSO's new BIOS did not infringe IBM's copyright. In July, IBM formally informed ERSO of the result and subsequently helped some ERSO engineers gain training in the United States. Even before Arthur D. Little released the arbitration decision, IBM had set up a research and development company, New Development Corporation (known as Qianzhan in Chinese, meaning "visionary"), in Taiwan in April with investment from ITRI and other semigovernmental funds. The first CEO of the company was Paul Wang (Boyuan Wang), a former US IBM employee, who grew up in Taiwan before pursuing a PhD in physics at Carnegie Mellon University. The new company employed eight personnel from the United States and aimed to recruit a hundred engineers in Taiwan. Many ERSO staff members were invited to join the new company, which generated concerns about causing an ERSO brain drain. Journalists questioned why talented engineers at ERSO employed with state funds would become inexpensive programming labor power for a US tech giant. But Yang noted that the turnover rate of ERSO engineers had averaged 15 percent every year. Therefore, the new company, Qianzhan, did not pose a threat to ERSO.[9]

In contrast to Yang's description of IBM's "gentlemanly" approach, *BusinessWeek* journalists had a different take. According to a September 1984 *BusinessWeek* report, US companies were "worried that Taiwan could make a dent in the US market," and "the Taiwanese had initially planned their assault" in early 1984. The assault referred to the first batch of ERSO computers, made by Multitech for ADDS and detained by US Customs in February. Now, "the Taiwanese seem[ed] to have missed the boat," after the delay in delivering those computers to the United States. Between February and September, Big Blue lowered the price of its PC/XT and planned the release of a new model IBM PC/AT. The low labor cost for making computers in Taiwan was suddenly less competitive than it had been a few months earlier. IBM Taiwan's general manager, Barry B. Lennon, pointed out that it was wary of the mushrooming growth of new computer makers. "It's not like swatting at a single fly" he noted. "There's a swarm of flies." From *BusinessWeek*'s vantage point, neither Taiwanese companies nor IBM were "gentle" in any sense in this rivalry. Unnamed Taiwanese companies suspected that IBM had deliberately deployed delaying tactics to stop their computers from entering the US market. *BusinessWeek* echoed the suspicion. A subsequent Taiwanese news report refuted the claim that Taiwan missed the boat.[10]

MULTITECH SCALES UP PRODUCTION

In addition to the ADDS contract, as *BusinessWeek* revealed in September 1984, Multitech began building forty thousand IBM PC compatibles for another US company, ITT. ITT wanted to roll out the ITT Xtra computer, compatible with the IBM XT. ITT had focused on manufacturing ICs since the 1970s and had acquired Unitron in Taiwan in 1974 (chapter 7). Multitech was going to make ITT's first microcomputer products. But, in addition to Multitech, ITT also awarded MiTAC a similar contract, possibly to ensure that both companies would supply enough computers for the US market. Both Shih and Miau acknowledged that they won contracts from ITT because of their connections with David S. Lee (Xinlin Li). Lee had founded printer manufacturer Qume in 1973 in California and sold the company to ITT in 1978. As an employee of ITT, he then built a factory for ITT in the newly developed Hsinchu Science Park in 1982 and made floppy disk drives in 1983. Miau got to know Lee in Silicon Valley when Miau was at Intel. Shih met Lee when Shih set up a factory in Hsinchu Science Park in 1981. When the number of start-up companies in Hsinchu Science Park was small, they tended to know each other well. Lee would be a key person for ITT executives to consult about whether ITT should contract with Multitech and MiTAC to make computers.[11]

The contract from ITT was a turning point for Shih's company. Shih was now making two computers for two US companies. He set up a new company, Continental Systems (Mingji, later renamed as BenQ), to focus on ITT's contracts, a measure to ensure ADDS that Multitech would keep all the technical details of ADDS computers from ITT and vice versa. Continental Systems' start-up capital came from Taiwan's first venture capital company, cofounded by construction tycoon Zhihao Yin, Shih, and others in 1984.[12]

ITT's order of forty thousand computers further prompted Multitech to improve its productivity. Shih assigned Kun-Yao Lee (K. Y. Lee), one of the early employees of Multitech, to manage the new company Continental Systems. He and several engineers visited ITT's San Jose, California, facility for training. ITT provided support, ranging from setting up product standards to ensuring production quality in Taiwan. Lee noticed, however, that ITT, hoping to retain its technical advantage, failed to reveal to him every technical detail at the beginning of the training.[13]

One of the major factors that made Continental Systems' project successful was the experienced employees hired and trained since the late 1960s by Taiwan's flourishing electronics industry. To scale up production to forty thousand computers for ITT, Lee had to recruit more employees, whom he hoped would be capable of assembly and management in a factory setting. He recalled recruiting a new group of women workers for the assembly line from a region where many electronics factories were located. He hired two former quality control engineers from Digital Equipment Corporation's (DEC) Taiwanese factory, which began operations in 1972. Shih noted that he also made efforts to recruit an electronics industry veteran, Zhengtang Chen, who had worked for Zenith Electronics' Taiwanese facilities in the 1960s and then joined Conic, an electronics manufacturer in Hong Kong in the 1970s. Chen was also a Chiao-Tung alumnus, a college friend of Shih's; Chen and Shih had been members of Chiao-Tung's volleyball team. Chen's management experience tremendously helped ensure Continental Systems' high productivity.[14]

Besides employees, Lee also worked closely with local suppliers to source parts for ITT computers, pointing out that the ITT contract pushed suppliers' manufacturing capability further. Initially, it was not easy to find suppliers that could make printed circuit boards, cases, or cables to meet ITT's specifications and exacting standards. Together, it would be up to Lee and suppliers to satisfy ITT's requirements. Delta Electronics, which made Micro-Professor computers, was now developing power supplies for ITT's computers. Its founder Bruce C. H. Cheng similarly worked with local suppliers to procure source materials for his then-employer TRW in the late 1960s (chapter 6). Now, Lee worked with Cheng to procure power supplies for ITT. Lee might have been slightly more fortunate than Cheng, as the industry of 1980s Taiwan was more resourceful than that of 1960s Taiwan. Lee believed that, starting in the 1960s, international electronics manufacturers such as Philco-Ford trained a good number of seasoned engineers. Furthermore, the late 1970s arcade game console industry and hi-fi audio systems manufacturers gave rise to a group of suppliers, some of whom swiftly redesigned their products to meet the demand of the growing computer industry in the early 1980s.[15]

Without being involved in the technical design of the computer, Continental Systems originally provided electronics manufacturing service (EMS) to ITT. But Lee and his employees frequently offered feedback to ITT, and gradually became more and more crucial in leading the development of products. When ITT's computer sales in the US market stagnated due to its high price, Lee lowered manufacturing costs so that ITT could compete with other computer makers. Their collaboration allowed Continental Systems to transform itself from an EMS company to an original design manufacturer of ITT, which demonstrated Continental Systems' technical capability in developing microcomputers.[16]

ITT's contract marked the beginning of the mass-manufacturing of computers in Taiwan, but Taiwan's potential had been recognized by others a few years ago. Shih's competitor, Matthew Miau, revealed that the first CEO of Compaq, Rod Canion, and his wife had visited Taiwan and considered giving MiTAC a contract to make Compaq computers. Canion's trip took place even before ITT had awarded Shih and Miau contracts. Compaq did not award Miau a contract until 1995, and the CEO that made the offer, after Canion's ouster in 1991, was Eckhard Pfeiffer. But Canion's visit in the early 1980s demonstrated US microcomputer makers' growing interest in outsourcing their manufacturing to Taiwan.[17]

OUTSOURCING DEMANDS FROM THE UNITED STATES

"The small companies in the U. S. will go out of business, and the larger ones will then have to subcontract to us." This prediction from Shih appeared in *BusinessWeek* in September 1984. *BusinessWeek* might have polished Shih's wording to make the statement sound more assertive than he usually sounded—he tended to describe Multitech in humbler ways—but he had faith in his business judgement and strategies. Whichever way Shih told *BusinessWeek*, he clearly spotted a niche for Multitech in the heated competition among IBM clone makers.[18]

Shih's prediction was identical with US industry analysts' observations. In summer 1983, *Computerworld* covered investment analyst Mike Murphy's enthusiasm for offshore manufacturing. He pointed out that offshore manufacturing was going to bring US companies not only lower material and assembly costs, but also prime quality control. "Quality control is still a labor-intensive operation," he noted. Computer products or parts required a higher degree of quality control in the production process. Therefore, companies in Taiwan, Singapore, and South Korea could employ a large number of workers to test and ensure quality. The labor force for quality control constituted their competitive advantage, according to Murphy. Another analyst, Barry Rogstad, pointed out that computer equipment imports from Japan, South Korea, Taiwan, and Singapore to the United States soared in 1983; for each country, the growth ranged from 53 to 113 percent.[19] Analysts were clearly aware that many US companies moved their production overseas to save costs. Atari's announcement that it was laying off 1,700 employees, in February 1983, alarmed the industry. The game console and computer maker had decided to move its production from Sunnyvale, California, to Taiwan and Hong Kong, and its announcement generated anxiety among factory and office workers as well as high-level managers and executives. Workers worried about job security, while executives pondered whether it was time to make their products offshore.[20]

Even if a US computer maker did not move its entire production line overseas, it might source parts from overseas companies. It was common for US manufacturers to make, outsource, or purchase electronics parts overseas as early as the late 1960s (as shown in chapter 6). In 1986, *Wall Street Journal* journalist Patricia B. Gray pointed out that "only half in jest, analysts call the IBM Personal Computer an import because many of its components are made overseas." While not specified in the news and perhaps unknown by US analysts, Tatung, which had made magnetic core memory units for IBM since the late 1960s, turned out to be IBM's major supplier of computer monitors in the 1980s and claimed to be the largest supplier in the world. IBM purchased $300 million US worth of computer parts in Taiwan in 1986, one-fifth of the island's computer parts exports. A Tatung manager noted that IBM's greatest influence on his company was Big Blue's quality control enhancement techniques, which had yields of 99.5 percent. The same *Wall Street Journal* article also revealed that notable companies such as Sperry, Tandy, and ITT were now contracting overseas manufacturers to fully assemble IBM clones. Clones from prestigious Japanese companies like Sharp and Epson were in high demand in the US market. Korean company Daewoo was making computers for Leading Edge Hardware Products Inc., which was lauded as one of the high-quality clones in the US market.[21]

CATCHING UP WITH COMPAQ, OUTPACING IBM

Multitech rebranded itself Acer in late 1986, just as US PC makers became aware of its importance. The new name was necessary because a US modem maker had registered the same name in the United States and other countries earlier than Shih had. In the fiercely competitive industry of IBM PC clones, Acer accomplished two technical feats that year and invited increasing attention and business opportunities from mature computer manufacturers that contracted with Acer to make computers. First, in 1986, Multitech successfully developed a clone of the IBM PC/AT, two years after IBM's introduction of the model, which used the Intel 80286 chip. The IBM PC/AT clones brought Acer strong results in the global market. In 1986, Multitech's revenues were $258 million US, almost double the previous year. Additionally, taking advantage of the geopolitical situation, Taiwan-made computers became popular in the Soviet Bloc, as US companies had to go through a Commerce Department and Pentagon license examination to export more than ten units.[22]

Second, in October 1986, Multitech successfully developed a 32-bit microcomputer, only a few months behind Compaq's development of its 32-bit microcomputer. Multitech claimed that the company stood to be second company in the global computer industry to do so, slightly ahead of a dozen other computer makers in the United States. IBM's 32-bit microcomputer was subsequently introduced in 1987. Acer thus won contracts to make PC clones for top companies, including Unisys (a merger of Burroughs and Sperry), Texas Instruments, the British company ICL, Japan's Canon, and European firms Siemens and Philips. Unisys even went ahead and licensed Acer's products.[23]

Acer's chief R&D engineer, Jonney Shih (who, as we have learned, is not related to Stan Shih), greatly contributed to Acer's 32-bit computer development. His short stints in Silicon Valley beginning at the end of 1984 were crucial to the development of this new product. During his time in Silicon Valley, he was based in Stan Shih's research firm, Suntek. Several Taiwanese engineers employed by National Semiconductor and other US companies reached out to Stan Shih in the early 1980s to urge him to start a company in Silicon Valley. In 1984, he founded Suntek, which he meant to carry out research and development for the Multitech's computer manufacturing business. Suntek engineers began to work on developing a new workstation computer with Motorola's 68000 chip. But operating costs rose beyond Stan Shih's estimates. He decided to dispatch Jonney Shih and a couple of engineers from Taiwan to California to lower the R&D cost. At the same time, he also expected Jonney Shih and other engineers to learn from their Silicon Valley colleagues, especially about the design of a workstation computer as well as the application of Motorola's 68000 chip. Jonney Shih then began to plan on using another Motorola product, the 68020 chip, to develop a 32-bit computer. Clearly, Stan Shih's strategy worked, as Jonney Shih was inspired to design a new computer product for Multitech.

Stan Shih intervened and decided to use the Intel 80386 chip for the 32-bit computer Jonney Shih was going to build. Stan asked Jonney to stay in Silicon Valley for another six months to stay close to Intel for Acer's 32-bit computer R&D. Why Stan Shih chose Intel over Motorola was unknown. But Intel was eager to promote its new 80386 chip at that time. As IBM was not enthusiastic about it in the first place, Intel might have reached out to other computer manufacturers. Compaq would have been one of them, and the company seized the opportunity to work with Intel for its 32-bit computer, the first in the world. Stan Shih probably was also approached by Intel or aware of Intel's plan and decided to take the opportunity to work with Intel. Multitech had begun to develop the 32-bit computer in August 1986, but Compaq was ahead—it completed the development of a 32-bit machine in September. In that fall, at least a dozen computer makers were working with Intel to develop their respective 32-bit computers. After the November Comdex began, it was clear that Zenith, Tandon, Advanced Logic Research, Kaypro, GoldStar (a Korean company), and PCs Limited (the predecessor of Dell Computer) also announced or demonstrated their new computer models with the latest Intel 80386 chips. Except for Compaq, the mass-manufacture and actual delivery of these companies' 32-bit computers did not happen until spring 1987. Despite many computer makers demonstrating their completed 32-bit computers, only Compaq, Acer, and Zenith could steadily deliver the new gizmos in 1987. Thereafter, Acer claimed that it was the second company in the world to roll out the 32-bit computers.[24]

When, right before Comdex, Jonney Shih finally completed the development of the new model with Intel's 80386 chip, he shared his excitement with Stan Shih immediately, especially that his computer was faster than Compaq's. He recalled, in an oral history interview,

> We used a very a special memory design technique, and I think we out-performed Compaq by around 8%. . . . I remember that very early in the morning, I went to see Stan Shih and told him, this time we could beat all our competitors. And I think that year, at Comdex in Las Vegas, at that time still the biggest computer show in the world, I brought my machine to compete with all the other competitors. We really beat everyone except one company that was called PC's Limited, it's the old name of Dell. Only Dell beat us. But I think they used the whole static memory, cache memory, no D-RAM, [it was] too expensive so they could not really do mass production. So we still won the best product of the year. And that's the year that Multitech became more internationally well-known.[25]

Despite Jonney Shih's pride in the technical performance of his computer, an in-depth news report on Comdex that year from the *Washington Post* failed to mention Multitech at all, although, in it, readers found the following sentence: "There were machines from fabled clone-mills of Taiwan, from France, Germany and Korea." Nor was best product prize that Jonney Shih referred to in his oral history interview covered in any news reports from the 1986 Comdex. Another possibility is that Jonney Shih might be referring to the prize Multitech won in 1987 in Cebit, an annual computer expo in Hanover, Germany. While there is no way to verify the details of the prize, we know that Jonney Shih's 32-bit computer was well-received in the industry.

Stan Shih's insistence in founding a new company in Silicon Valley and sending his Taiwan-based engineers there left a remarkable legacy. During one of Jonney Shih's presentations at Intel on the design of his 32-bit computer, Intel executives remarked that his meticulous attention to the engineering design made him an unusual vice president in the industry. He impressed the Intel team, which included Andy Grove, Albert Yu, Sean Maloney, and Paul Otellini. In the following decades, these Intel exec-utives would schedule a visit to meet Jonney Shih whenever they were in Taiwan. Beyond their collaborative experience in Silicon Valley, another reason they met with Jonney Shih in Taiwan was Intel's role as a major supplier, since 1989, for Asus. Jonney Shih joined Asus in 1992, though Intel also became an Asus rival during the years Intel also made motherboards.[26]

Moreover, the founding of Asus should be credited to the development of Multi-tech's 32-bit computers. When Jonney Shih led the remarkable project that outpaced IBM and others in the 32-bit computer race, several engineers among those sent to work briefly at Suntek in Silicon Valley decided to start their own companies after they returned to Taiwan. Former Acer engineers founded Elite Computer Systems and Asus in 1987 and 1989, respectively. Both companies focused on the design and manu-facture of motherboards. Both companies maintain a dominant market share in the worldwide motherboard industry today. Asus is also a household brand for US tech-savvy computer consumers. Asus founders, closely working with Jonney Shih on the 32-bit computer, gained a tremendous amount of experience integrating different indi-vidual ICs in a computer, before pre-integrated chipsets became common in the per-sonal computer industry. The experience led them to mastering motherboard design and turning it into a technical advantage for their new companies. In 1989, Jonney Shih considered joining four young Asus founders and leading the new company, but he decided not to leave Acer in 1989, after Stan Shih's sincere entreaties that Jonney Shih stay with him. Jonney Shih, however, contributed two-thirds of the start-up capi-tal Asus needed. He eventually left Acer in 1992.[27]

Ding-Yuan Yang commented that Taiwanese companies created the motherboard industry for IBM PC clones. He noted that, starting in the mid-1980s, Taiwanese companies made and sold motherboards, cases, and power supplies to the United States, and US IBM PC clone makers would have to procure only hard disk drives and CPUs to assemble them into a computer. He did not mean to claim that the first motherboard company was a Taiwanese company or that it was an idea from Taiwan—Nara technologies in California and Hauppauge Computer Works in New York were two early US-based motherboard vendors that readers of *InfoWorld* would have heard of in 1985 and 1987, respectively. Nevertheless, Taiwanese motherboard makers were among the forerunners of the motherboard industry and dominated the market starting at the beginning of the 1990s. In 1996, Taiwanese companies made 74.2 percent of the motherboards in the industry.[28]

Shih viewed the success of the 32-bit microcomputer as a meaningful milestone. He believed that his engineers now had attained the second-best position, as his principle of number-two-ism called for. Their achievement was "matching their Western counterparts when it comes to design," he told *Economic Magazine*, and added,

> Frankly speaking, we were far behind the world leaders in developing information products. I used to promote a "We try harder" philosophy in an effort to push our engineers to improve our product design capability from the number three or four position in terms of our entry position into a product's life cycle. By proving they've now reached a "number two" slot, Taiwan's engineers have shown that they can match their Western counterparts when it comes to design. Now we can work on moving into the leading position in specific areas.[29]

In 1985 and 1986, Shih boldly planned on making his company a multinational like those "number ones." After establishing Suntek in Silicon Valley in 1984, he toured Southeast Asia to station his staff alongside local partners who would promote Multitech computers the following year. In 1986, Multitech's revenue was $200 million US, and Shih announced Dragon Flying Decade (*longteng shinian*) in May, an ambitious plan that set goals for the company's revenue for the next ten years. Shih aimed to make Multitech reach revenues of $1 billion US in 1991 and $3 billion US in 1996, a plan he had even before he knew that his engineers would successfully develop the 32-bit microcomputer in fall 1986. A core idea in achieving the Dragon Flying Decade plan was to sell computers internationally, beyond the limited consumer market of Taiwan. He soon set up additional branches in Japan and Europe in 1986 and recruited experienced executives from other multinational corporations, including IBM, in the second half of the 1980s. Despite that Taiwanese business executives deemed his revenue projections unrealistic, Shih fulfilled them, and earlier than he had planned. In 1994, Acer's revenue had already reached $3.2 billion, and increased to $6 billion in the years to follow.[30]

IBM, TAKING ACTION AGAINST "SCOUNDRELS"

In 1987, IBM decided to ask PC clone makers worldwide to pay for copyright and patent infringement of the compatible computers made in the preceding four years. IBM sent notices to Taiwanese companies in March and a couple of US companies including Compaq, Dell, and AST, likely on later dates. Big Blue requested they pay royalties for manufacturing IBM PC XT and AT clones. It believed these clones, which could date back to 1983 or earlier, used patents held by IBM. According to Shih's biographer, IBM claimed that thirty thousand patents were breached in manufacturing the clones. The number might be exaggerated, but in the 2016 documentary *Silicon Cowboys*, which portrayed Compaq's history in the 1980s, Bob Jackson of IBM's technology licensing department noted that IBM owned at least nine thousand patents that IBM compatible computers required. The retroactive fee asked by IBM could range from 1 to 5 percent of the sale price of the clones. IBM also required these companies to pay the retroactive fees before IBM would license them new patents from its upcoming PS/2 model. Compaq CEO Rod Canion noted, "We think that the range IBM is talking about charging for its patents is out of line."[31]

To cope with the unexpected request, Stan Shih and Matthew Miau decided to ally with three large Taiwan computer companies for negotiations with IBM. An Acer staff member interviewed by the *Wall Street Journal*'s James McGregor charged IBM with "economic colonialism." Attorney Utz Toepke, who formerly worked for IBM but now represented Acer, commented, "They act as if they are emperor of the world—they march in here and say, 'Pay us!'" While McGregor interviewed Toepke and presented his "emperor" analogy, he also noted that "IBM's sympathizers" would view "Taiwanese computer makers [as] scoundrels who stole IBM's ideas, tacked on a few bells and whistles and grew rich by undercutting the U.S. company's prices." The sympathizers McGregor quoted left out US clone makers among the scoundrels they identified. In fall 1988, IBM's Asian patent specialist announced his team was going to take action in South Korea, Hong Kong, and Singapore. He bluntly noted, "We're getting more aggressive." The tone of his comment was similar to that of Bob Jackson, the former IBM technology licensing team member in *Silicon Cowboys*. Despite joining Compaq later in his career, Jackson described IBM's standpoint this way: "If they [IBM's competitors] misbehaved—loosely used term. But if they misbehaved, we had the howitzers lined up."[32]

When McGregor's news article was published in September 1988, Dell Computer had already paid IBM the retroactive royalties, totaling 3 percent of the sale price of the clones Dell had previously sold. Dell was first, followed by several US companies. In Taiwan, smaller companies also gave in early to survive. In 1988, Michael Chang, an owner of a 160-employee computer maker, was the first company in Taiwan to pay the retroactive royalties. "We were very scared of IBM," Chang admitted. In 1984, his company paid IBM $25,000 for copyright infringement for clones it manufactured. In 1988, his facilities' production rate was seven thousand computers per month. To ensure his company ran smoothly, he sought a deal with IBM sooner than later. He then decided to pay the retroactive royalties and negotiated a lower fee, which he was not allowed to reveal to McGregor. Chang only shared with McGregor, "I got a big discount."[33]

Shih's alliance took more than eighteen months to negotiate with IBM. His alliance also gained support from the Ministry of Economic Affairs in Taiwan. The number of companies that joined the alliance increased during the first year of the negotiation, as they believed that solidarity would help them reach a good deal and reduce the royalty percentage they paid IBM. But the alliance gradually failed when MiTAC and other members signed individual agreements with IBM. Acer reached an agreement with IBM in January 1989, paying IBM over $2.5 million US in retroactive royalties, which accounted for about 0.7 percent of Acer's revenue that year. What percentage of the sale price Acer paid per computer was never precisely disclosed. Nevertheless, Taiwanese companies smaller than Acer and MiTAC paid a 1 percent royalty. In the United States, the most prestigious IBM clone maker, Compaq, reached an agreement with IBM on July 17, 1989, but the amount of the money and percentage of sale price per computer remained undisclosed at that time.[34] Compaq general counsel Bill Fargo revealed in the 2016 documentary *Silicon Cowboys* that the company paid $130 million.

In the late 1980s, most Taiwanese computer makers had gross profits ranging from 5 to 10 percent of sales. The gross profit margin for US PC clone makers would be higher, but IBM raised the percentage of royalties in 1991 and again in 1993. In 1993, Big Blue sought up to 4 percent of the retail price of each computer. Taiwanese companies commented that it would not be wise for IBM to kill the geese that lay the golden eggs.[35]

IBM, probably unsatisfied with royalty negotiations with Acer in early 1989, moved forward to press Acer on another copyright infringement case, this time for an outdated computer. On May 12, 1989, Stan Shih was going to attend a press conference to announce his new company TI-Acer Semiconductor (Deji). A phone call he received the day before the press conference delivered unexpected news. Over the phone, Shih was asked to meet IBM's legal team for breakfast at the Sheraton hotel in Taipei the next morning. He was told that the BIOS of Acer's 80286 computer infringed IBM's copyrights; the model was compatible with the IBM PC/AT. Acer had released it in 1986, and was going to discontinue the model soon. The phone call threw cold water on Shih's plan to announce his ambitious investment, a joint venture of Acer and Texas

Instruments to manufacture DRAM (dynamic random-access memory) chips in Taiwan. DRAM chips were necessary components for microcomputers (and more recently, products such as smart phones). For years, Shih's dissatisfaction with the fluctuating market of DRAM, in which both supply and price proved unstable, grew. By setting up his own DRAM manufacturing facilities, Shih sought to secure a supply of DRAM for his computer products.

Stan Shih told IBM's legal team in the breakfast meeting on May 12, 1989, that the BIOS of the IBM PC/AT compatible computer was developed by a US company, Award Software. Stan Shih paid $50,000 US for the BIOS in 1986, and his engineers carefully reviewed and revised the BIOS to fit Acer's hardware. But the IBM legal team insisted that, after comparing the code of the two BIOS systems, Acer's BIOS constituted copyright infringement and the specific infringement was about the code that drive the keyboards. Following the breakfast, the two companies entered a yearlong negotiation. At the end, Acer reached a settlement with IBM for $9 million US. Industry rumors indicated that IBM initially requested $30 million US but settled for $9 million over five years. In 1990 alone, Acer paid IBM $1.25 million US for the infringement settlement. It accounted for about a third of Acer's net profits for the first half of 1990, and did not include the newly incurred royalties on new IBM-compatible computers that Acer owed IBM.[36] Facing Big Blue's challenges, Acer's well-established research and development structure continued to offer state-of-the-art products in the microcomputer market. Earning subcontracts from prominent computer manufacturers from Japan, Europe, and the United States, Acer's strong yearly revenue compensated for the royalties and settlement owed IBM.

Acer was already a global computer seller by the late 1980s, selling its computers through subsidiaries or agents in Europe, Australia, Southeast Asia, and the like. To connect better to the US market and Silicon Valley, Shih acquired at least two US companies, minicomputer maker Counter-Point and UNIX network computer company Altos, at the turn of the 1990s. While Acer suffered financial loss due to overexpansion globally in 1991, Shih was able to lead Acer back to a profitable track again by the mid-1990s. In terms of manufacturing, Acer began to set up facilities in Malaysia in 1989, in addition to its facilities in Taiwan and the United States. Acer started to expand its manufacturing facilities to China only in 1995. By 1996, Acer had over thirty-four assembly facilities around the world. As for sales, by 1995, Acer had formed joint ventures for at least twelve regions, including Mexico, Brazil, Chile, Argentina, South Africa, India, Thailand, and Indonesia. In 1998, Acer had fifteen thousand employees working across eighty offices in thirty-eight countries. Acer's products could be found in more than a hundred countries.[37]

By the mid-1990s, Acer became one of the top PC makers in the world, at its height the fifth largest. In the United States, Acer was the largest supplier of Best Buy, and Best Buy was Acer's largest US customer in the mid-1990s. Acer's market share was ahead of Dell, Toshiba, and Hewlett-Packard in 1995. In that year, Compaq anticipated selling six million computers, while Acer sold four million. Around 2009, Acer climbed to the second largest personal computer maker, trailing Hewlett-Packard. At the time of writing, Acer remained one of the top six PC makers. In addition to desktop computers, Acer also made laptops for well-known enterprises; among the 72,000 laptop computers Acer built each month in 1995, 70 percent were for Apple, Texas Instruments, Mitsubishi, Canon, and other firms.[38] Overall, Taiwan became the largest supplier of notebooks in 1994. In 2011 alone, Taiwan's laptop manufacturers made two hundred million laptops, which accounted for 90 percent of the world market. The top four largest Taiwanese manufacturers included Quanta, Compal, Wistron, and Inventec. Wistron was an Acer spin-off, and Quanta, Compal, and Inventec were related to calculator makers Santron and Kinpo (discussed in chapter 7), one way or another.[39]

Taiwanese corporations also fiercely competed for subcontracts from resourceful US computer makers, the majority of which recognized that assigning manufacturing contracts to Taiwanese computer makers saved their manufacturing cost. For Apple specifically, Acer began making two models of its PowerBook notebooks in 1993. Apple also contracted both Acer and MiTAC to build motherboards in July 1993, after Apple announced its intention to lay off 16 percent of employees, primarily in the United States. By 1995, Apple, IBM, Compaq, and Dell all had their own subcontractors in Taiwan making personal computers. To lower the manufacturing cost, Taiwanese companies sourced parts from various locations to put together computers. The *Los Angeles Times* covered MiTAC's offshore production for Compaq in 1995: "The Taiwan-produced Compaq Presario, for example, features a case and a power supply made in China; a keyboard made in Malaysia; a monitor made in South Korea; and a motherboard made in Taiwan. These are all assembled by Mitac Corp. in Fremont, Calif.—a move that enables the Taiwanese company to take advantage of last-minute price fluctuations in U.S."[40]

Beyond computers and laptops, Taiwanese companies held over half the world market share of most computer components. The market share percent of monitors made or developed in Taiwan was 53.4, scanners 61.2, mouses 65, keyboards 61, sound cards 55, video cards 50, motherboards 74.2, and switching power supplies 55.3 percent. By 1999, TSMC and UMC also supplied 10 percent of the worldwide finished semiconductor chips. Recall a 7.3 Richter scale earthquake hit Taiwan on September 21, 1999, and affected power and water supply throughout the island. US computer manufacturers, including Dell, Compaq, and Hewlett-Packard, which relied on Taiwanese subcontractors and suppliers, experienced disappointing earnings in that fall. The earthquake served as "a jolting reminder to the [US] computer industry on its reliance on this island nation," as *New York Times* journalist Mark Landler pointed out.[41]

CONCLUSION

Inspired by Compaq's success in 1982, Stan Shih embarked on manufacturing IBM compatible computers. Other Taiwanese computer makers and the R&D institute ERSO took the same path to explore the potential of mass-manufacturing sophisticated compatible computers. In addition to making their own products, Taiwanese computer manufacturers contracted with US computer makers that had decided to outsource their manufacturing tasks. Acer's technical achievement in developing the 32-bit microcomputer right after Compaq made Acer a sought-after subcontractor after 1986. The prosperous computer market encountered a drastic change after IBM decided to take legal actions against clone makers in 1987. Compared to Apple's longstanding and outright rejection of compatible systems,[42] IBM targeted counterfeiters before 1987 but eschewed extensive legal actions against compatible computer manufacturers. After 1987, IBM requested, retrospectively back to 1983, royalties from global PC clone makers. Compared to the framing of Acer and MiTAC as counterfeiters and invaders, the media and IBM employees viewed Taiwanese computer makers and other PC clone makers as scoundrels that stole IBM's technology and as flies that would not go away.

Counterfeiters, invaders, scoundrels, and flies represent those that inhabit the periphery as opposed to the center; they are secondary to the original. Their presence messes with the existing order, but their presence is necessary to complement the center or the original, in this case, Apple and IBM. But these representations also bring to mind the "number-two-ism" proposed by Stan Shih. Shih brought up "number-two-ism" in October 1983, after he wrote to John D. Dingell stating that "the distinction between illegal counterfeits and independently developed products is not always clear."[43] In 1983, his "number two" was alluding to the competition between Multitech and other Taiwanese companies. He stated that he was content with the status quo but would take the first position if an opportunity arose. In 1986, Shih's company Multitech announced that it stood to be the second company in the world to develop a 32-bit microcomputer. It was before IBM, but after Compaq. Shih believed that his engineers have matched their "Western counterparts."[44] After 1986, Shih began to boldly expand its global market share and set up ambitious plans for revenue growth. It looks like he would like to be the number one. But, at the same time, Shih and other Taiwanese computer manufacturers strove to be subcontractors of US, European, and Japanese companies. Under the logic of capitalist competition, top computer companies would require access to subcontractors, who would be able to take advantage of undercompensated women factory workers. Many Taiwanese electronics manufacturers also focused on supplying computer peripherals or components. The strategy taken by Shih and other Taiwanese tech executives, the "niche" they found, or the spot they were able to occupy, was still secondary to the center or the original.

In 2006, Intel's CEO, Paul S. Otellini, praised Stan Shih's integrated supply chain for making inexpensive PCs available to more consumers than ever. He noted, "He's a big reason why your PC costs $1,000, not $10,000." Shih's leadership in Acer realized the technological dream held by engineers and technocrats in Taiwan for decades: he not only made computers but also made sure those computers were IBM compatible and less expensive. While Compaq's silicon cowboy identity helped challenge IBM's PC dominance, it was not alone. Shih, too, was one such rebel. Also, recall Canion emphasized, in the prologue of his book about Compaq, that "an open industry standard matters," which Shih would echo. Shih and his fellow Taiwanese computer makers fought against the images of copycats, invaders, flies, and scoundrels in the 1980s and eventually became key players in the international computer industry. Acer named its post-1995 computer models "Aspire," which spoke precisely to Shih's ambition of transforming himself from an engineer to the builder of a global computer manufacturer, a role that many of his fellow Taiwanese engineers had likewise aspired to since the 1950s and a role that none of his fellow Taiwanese engineers had assumed earlier than he had in the 1980s.[45]

Subcontracting remains a profitable business model for most Taiwanese computer makers. The island's subcontractors in the industry of computer chip making, notably TSMC, also has led a niche market since the 2010s. In chapter 10, I discuss the entanglement of the history of semiconductor fabrication with the history of the computer industry in Taiwan, especially the experimenting with production conditions in factories and the role of technologies in geopolitics.

10 TSMC AND THE NEW GEOPOLITICS OF THE TWENTY-FIRST CENTURY

South Korean conglomerate Samsung Group's chairman, Lee Kun-hee, visited Taiwan at the end of January 1989. During this trip, Lee met with Acer's CEO Stan Shih, TSMC's founder Morris Chang, and many other computer manufacturing companies' executives and government officials. Lee extended an invitation to Shih and Chang to tour his facilities in South Korea, the only facilities outside the United States and Japan to fabricate DRAM chips at that time. The purpose of Lee's trip to Taiwan was likely twofold. First, Lee intended to encourage his customers—Taiwanese computer manufacturers—to continue ordering DRAM chips from Samsung. Second, he might have been concerned that Shih had developed an interest in manufacturing DRAM chips in Taiwan, and that Morris Chang, who had just set up TSMC in Taiwan in 1987, would consider entering the manufacturing business of DRAM chips. Lee sought to deter Shih and Chang from becoming formidable competitors.

Shih, a seasoned businessman, and Chang, an experienced corporate executive, clearly sensed Lee's implicit agenda of dissuading potential competitors from joining the game with his show of Samsung's accomplishments. Lee had even attempted to recruit Chang to work for Samsung. The dissuasion was not effective, however. First, Chang had planned to focus on manufacturing made-to-order VLSI (very-large-scale-integration) chips. He had never given a thought to making DRAM chips in Taiwan, as he knew that market had been highly competitive and volatile, based on lessons learned in his nearly three-decade career at Texas Instruments, whose dominance in the DRAM market was challenged by Japanese companies in the 1980s. Chang had left Texas Instruments in 1984. Shih, in contrast, was determined to source a steady supply of DRAM chips for his computer manufacturing business. His computer products needed DRAM chips, but these chips sometimes could be hard to obtain. In May 1989, a month after Shih paid a trip to Samsung's facilities in South Korea, he announced a joint venture with Texas Instruments, TI-Acer Semiconductor, known as Deji in Chinese, to manufacture DRAM chips in Taiwan (as described in chapter 9). Chang generously offered his assistance to Shih on this deal. He traveled with Shih to Texas when Shih negotiated details with Texas Instruments to transfer some production lines to Acer's Taiwan's facilities.[1]

TI-Acer Semiconductor's production began in 1990 and reached a steady yield in 1992, but the company suffered financial deficits after 1996. Shih attributed the decline of the company to two factors. First, Samsung already dominated the global DRAM market, leaving little room for most US, Japanese, and Taiwanese companies. Second, Texas Instruments, which supplied the technical know-how to the joint venture, did not come up with profit-making DRAM products. Texas Instruments eventually sold its DRAM business to Micron Technology in 1998. Industry giant IBM also exited the DRAM market in 1999. Along similar lines, Shih sold TI-Acer Semiconductor to TSMC in 1999. Chang converted the newly acquired facility to one of TSMC's ten fabrication factories (or fabs) to fabricate customized chips. Chang also immediately acquired another fab in Taiwan and rose to be the largest chip maker there. The expansion helped Chang's TSMC surpass his rival in Taiwan, Robert H. C. Tsao (Hsing-cheng Tsao, Xingcheng Cao), who led United Microelectronics Corporation (UMC).[2]

Yet, for Shih, his decision to sell TI-Acer semiconductor to TSMC paid off in later years. As part of the sale, the Acer Group received some TSMC stocks, the price of which surged, as TSMC dominated the market share of the made-to-order chips across the globe in the early 2000s. Among all semiconductor manufacturers, TSMC rose to be the eighth largest in 2005. Shih was invited to serve as a board member of TSMC in 2000, and his tenure lasted for more than two decades.[3]

The history of semiconductor fabrication in Taiwan is intertwined with the history of the computer industry in Taiwan. The engineers, entrepreneurs, universities, laborers, and government agencies that contributed to the maturing of the computer industry in Taiwan also fostered the rise of TSMC and UMC, the leading dedicated semiconductor foundries in the world. Both industries bear similarities. Chiao-Tung alumni continued to play a critical role in cultivating the semiconductor industry. Like the computer industry, chip manufacturing in Taiwan required engineers, researchers, and technicians' experimenting with various types of production conditions—an act of tinkering. The dedicated foundry model originated from Morris Chang's and Robert Tsao's similar understandings of Taiwan's successful computer manufacturing industry. Furthermore, semiconductor engineers and entrepreneurs, as their counterparts in the computer industry, have sought to make their technological capacity relevant to the business and diplomatic relationships between Taiwan and the United States. Just as Taiwanese engineers and scientists in the 1960s viewed computers as critical to aiding Taiwan's Cold War–affected national defense and industrial development, various parties in the United States and Taiwan in 2022 deemed TSMC's high-end chips valuable resources for cultivating international support to help Taiwan in deterring a potential Chinese incursion.

ATTEMPTS TO FABRICATE CHIPS

TSMC and UMC's dominance of the global market of made-to-order chips can be attributed to Industrial Technology Research Institute (ITRI)'s plan to cultivate a profitable, high-tech industry in the mid-1970s in Taiwan. ITRI was established as a research and development institute and supported by the Ministry of Economic Affairs in 1973. UMC was set up in 1980 by engineers selected and sent by ITRI to learn how to fabricate chips at RCA in the United States. Morris Chang moved to Taiwan and directed ITRI in 1985, and established TSMC in 1987.

But ITRI's vision was also shared by participants in the electronics industry in Taiwan. ITRI was not the first group to consider fabricating chips in Taiwan. Before ITRI carried out the RCA technology transfer project, several entrepreneurs, including those covered in the previous chapters, had recognized the urgency of fabricating transistors and chips locally. Multinational electronics manufacturers in late 1960s Taiwan prompted the rise of the domestic electronics assembly industry (as described in chapter 6). They also inspired entrepreneurs to consider making transistors and ICs from scratch there in the early 1970s, allowing Taiwanese workers to take up more profitable manufacturing tasks beyond assembly and packaging—the final stages of electronics production. Philips' manager Y. C. Lo, Unitron's Andrew Zaixing Chew and Simon Min Sze, and another engineer-turned-entrepreneur, Chun-yen Chang, drew up plans to make silicon wafers in Taiwan to fabricate transistors and ICs. But their projects did not succeed for different reasons in the early 1970s. In contrast, UMC and TSMC were founded in the 1980s with better financial backing, increased demand of chips from the domestic market, prearranged technical transfers from chip makers such as RCA and Philips, and well-trained engineers ready for the novel industry.

Chiao-Tung graduate Chun-yen Chang was the first person to be awarded a PhD in engineering by a Taiwanese university after 1945. In 1965 and 1966, on Chiao-Tung campus, he had experimented with making transistors and ICs from scratch, with instructions provided by visiting professors. These short-term visitors were employees of US semiconductor manufacturers who came to Taiwan to share what they knew with local students.[4] After Chun-yen Chang's graduation in 1970, he participated in the founding of two companies that fabricated transistors. First, Chang offered his technical expertise to a steel company to found Wanbang Electronics (Fine Product Microelectronics Corporation) in 1970, manufacturing light-emitting diode (LED) transistors. Wanbang's poor management and financial performance caused Chang to end his consultancy there in 1974, after which he started another company, Jicheng Electronics, focusing on making bipolar junction transistors. The founding of Wanbang and Jicheng were ambitious moves, but unfortunately they were latecomers in the industry of transistor fabrication, a market dominated by US and Japanese companies. Nevertheless, several engineers beginning their careers at Wanbang and Jicheng later joined UMC and TSMC in the 1980s.[5]

In academia, Chiao-Tung alumni worked together to arrange talks for engineering researchers in Taiwan to learn about semiconductor fabrication. These talks were delivered by overseas professionals keen to share what they knew about fabrication of ICs during their brief visits to Taiwan. A former RCA and Westinghouse engineer and Chiao-Tung alumnus, Hung Chang Jimmy Lin (Hongzhang Lin),[6] gave a talk, "Fabrication of Transistors and Integrated Circuits," for the Modern Engineering and Technology Seminars (METS) in 1966.[7] At the same event, James D. Meindl, the director of the Integrated Electronics Division of the US Army Electronics Laboratories in Fort Monmouth, New Jersey, lectured on the design of IC.[8]

Unitron and Philips, two well-established firms covered in earlier chapters, were also ready to fabricate transistors in Taiwan. In chapter 7, Unitron's Andrew Zaixing Chew and Simon Min Sze purchased machinery and planned to make transistor wafers to fabricate bipolar junction transistors in 1972. But Unitron was sold before it began the production. In chapter 6, we learned that, as early as 1974, Philips' chief manager in Taiwan, Y. C. Lo, attempted to persuade headquarters to set up a wafer fabrication factory in Taiwan, but it was never approved. He believed that if his factory employees could perform well in assembling transistors and IC packaging, they would certainly be able to do well in the fabrication of transistors and ICs.

A widely respected entrepreneur also urged the government to assist in developing the capacity to fabricate ICs and transistors domestically in the early 1970s. Recall Maopang Chen, Sampo's owner, whom Stan Shih consulted when he considered leaving the ill-managed Rongtai. In 1973, Chen urged the government to investigate plans to ensure that ICs and transistors would be fabricated locally for supplying the booming domestic calculator industry in 1973.[9] It was especially a pressing issue because the oil crisis had made imported semiconductor products more expensive than usual. As a leading appliance maker in Taiwan, Chen proposed the same thing that the appliance maker Samsung Electronics planned to do in 1973. The Samsung Group's founder, Lee Byung-chul, and his son Lee Kun-hee began to invest in fabricating transistors at the same time.[10]

TRANSFERRING CMOS IC FABRICATION FROM RCA TO TAIWAN, 1975

The industry's consensus on the urgency of fabricating IC domestically might have prompted Hua Fei, the secretary-general of the Executive Yuan (in office from 1972 to 1976), to initiate a technology transfer project to help Taiwanese engineers gain licensing and training in the semiconductor manufacturing industry. In 1974, he asked Wen Yuan Pan, an RCA engineer and his college friend, to share his insight on Taiwan's opportunity in the industry. Fei arranged Pan to meet with Minister of Economic Affairs Yun-suan Sun, who had just failed in persuading Andrew Zaixing Chew to rescue Unitron from being sold to ITT and turning from a model domestic company to a foreign-owned corporation (see chapter 7). On February 7, 1974, Fei, Sun, Pan, and four additional technocrats met for breakfast over soy milk, flatbread, and steamed buns. Five out of the seven attendees held engineering degrees awarded by the pre–World War II Chiao-Tung University. In the meeting, Pan briefed that it would probably require four years for a group of engineers to learn and transfer IC wafer manufacturing from a US company to Taiwan, and the cost would be about $10 million US. The breakfast meeting attendees later designated ITRI to execute such a technology transfer project. Morris Chang also played a part in the technology transfer project. In 1975, Morris Chang, then a vice president of the semiconductor group at Texas Instruments, was invited to offer his insights when Sun, Fei, and other technocrats were debating which US company ITRI should work with.[11]

In July 1974, Pan completed a proposal that budgeted $12 million US. Pan then organized the Technical Advisory Committee (TAC), consisting of seven US-based academics and engineers, to help contact candidate companies. After reaching out to over thirty manufacturers, TAC compiled a list of fourteen companies in February 1975. Then, the Ministry of Economic Affairs sent a formal invitation to each company to submit proposals for transferring IC wafer manufacturing technology to ITRI. Seven companies replied with interest. TAC reviewed and came up with a short list of viable proposals from GI, Fairchild, RCA, and Hughes Aircraft Company.[12] Fairchild's proposal was ruled out subsequently as it exceeded ITRI's budget. GI's proposal was set aside, as GI did not possess CMOS techniques, which was regarded as the most promising technology by Pan. CMOS (complementary metal-oxide semiconductor) chips—as opposed to NMOS (N-type metal-oxide semiconductor), PMOS (p-channel metal-oxide semiconductor), and bipolar chips—use less energy and do not easily get interfered with by other signals. Thus, they were ideal for space communications products, electronics watches, and clocks. However, it was not a mature technology at that time. Even for some US factories dedicated to making CMOS chips, yields were not satisfactory. Therefore, some TAC consultants were concerned about whether Taiwan could succeed in a product line that even companies like IBM were struggling with.

In the final stage of selection, RCA and Hughes won equal numbers of supporters among the panel of technocrats and TAC consultants. Xianqi Fang, the director of Telecommunications Research Institute and a Chiao-Tung alumnus, remained skeptical of the oddly low budget proposed by Hughes. He then got authorization from Sun to delegate Pan to carry out two tasks immediately. First, Pan, who was based in New Jersey, would visit the Hughes Aircraft Company in person and discuss with representatives there the details of the budget. Second, Pan would fly to Texas to meet Morris Chang and solicit his comments on both proposals. Pan faithfully executed the assignments. Hughes acknowledged that the company underbudgeted the expenses, especially the additional production machinery needed to train engineers from Taiwan. Morris Chang confirmed that RCA's proposal was reasonable. With these facts clarified, ITRI then signed a contract with RCA on March 5, 1976.[13]

The first group of Taiwanese engineers recruited and dispatched to RCA numbered thirteen, leaving for the United States in April 1976. In all, a total of thirty-seven Taiwanese engineers were trained at four different RCA sites. The IC design group went to RCA's facilities in Somerville, New Jersey; the testing group was dispatched to a different town in New Jersey. The CMOS production group headed to Findlay, Ohio. The NMOS production group visited Palm Beach Gardens, Florida. Six trainees were former employees of Jicheng and Wanbang, the two transistor fabrication manufacturers formed in the early 1970s, showing that the two companies, despite their lack of extraordinary performance, offered opportunities for engineers to accumulate experience in transistor fabrication.[14]

ROBERT TSAO AND IC FABRICATION AT UMC, THE 1980s

The returning trainees from various RCA facilities joined a pilot factory to begin IC fabrication in Taiwan. The factory was completed in 1977 and soon began to make three-inch 7.6 micrometer wafers. After six months of operation, the yield was stable at 70 percent, much higher than the 17 percent RCA originally guaranteed in its contract. The high yield brought the pilot factory a high gross profit, which could reach to 20 percent of the sale price or higher. One of the contracts the pilot factory received was to fabricate ICs for Japanese company CASIO's electronic watches. The government-sponsored research institute ITRI then helped raise capital to establish a private company, UMC, in 1980, and transferred the CMOS production techniques to UMC.[15]

Many of UMC's leading engineers and managers had been RCA trainees. Robert H. C. Tsao, one of the thirty-seven trainees, became an assistant manager, chief manager, and eventually the chairman of the board of UMC from 1981 to 2006. Born in 1947, Tsao holds a BS in electrical engineering from National Taiwan University and a master's degree in management science from Chiao-Tung. The subjects of the training he received at RCA were productivity management, material procurement, and warehouse management. He returned to Taiwan after a two-month training program in Findlay, Ohio, and participated in the preparation, especially the procurement of machinery, for the pilot factory and the founding of UMC. The most profitable product in its early company history were ICs for phone dialers. UMC exported 2.4 billion such chips to the United States beginning in 1982. The booming demand for new telephones in the United States brought UMC handsome profits and additional business contracts.[16]

With the thriving business, Tsao began to deliberate strategies to ensure the company's growth and resilience in the fluctuating chip market. UMC not only designed and manufactured ICs for its corporate customers but also produced ICs for other manufacturers that needed extra production capacity. The latter accounted for 25 percent of its revenue at that time.[17] Both businesses—the integrated service and the contracted manufacturing—were equally successful, but Tsao started wondering whether UMC should specialize in either IC design or IC fabrication, instead of both. His intention was to lower the business risk by making UMC focus on only one thing. If UMC were to become an IC design company or an IC fabrication company, ITRI could then work on supporting the establishment of another company to take care of the other process.

The emergence of a new type of business opportunity also prompted Tsao to deliberate the future of the company: manufacturing DRAM chips for IC design companies. Three IC design start-ups were established in 1984 in Taiwan by US-university-educated Taiwanese. They had designed DRAM chips but were struggling with setting up fabrication facilities. They all reached out to Tsao, but UMC was able to take up only a small batch of jobs for one of the three start-ups. The companies eventually had to manufacture their chips in Japan and South Korea. Similarly, in Silicon Valley, many overseas Taiwanese participated in newly founded IC design houses; some got in touch with Tsao. Witnessing IC design companies' lack of manufacturing facilities, Tsao concluded that it would be worth it to expand UMC's manufacturing business.[18]

Tsao proposed to Minister of Economic Affairs Li-teh Hsu that UMC should concentrate on manufacturing ICs for IC design companies, especially companies founded by Taiwanese in Taiwan and overseas in Silicon Valley. UMC could also continue to make ICs for other manufacturers that needed more production capacity than they had. Tsao's idea was similar to what we now call "dedicated foundry" or "pure-play foundry," as opposed to integrated device manufacturers (IDM) such as Intel, Samsung, Texas Instruments, IBM, and Toshiba. The latter category comprises a limited number of firms that control the capital and factories to develop chips in-house and manufacture chips in their own fabrication facilities. The former, however, only manufacture customized chips for clients that make cell phones, networking devices, computers, game consoles, and other electronic devices.[19]

Aware that Morris Chang had been consulted by Taiwanese officials in the past, Hsu asked Robert Tsao to visit Chang in the United States, have him review the plan, and solicit his comments. Chang, now CEO of General Instrument, did not think the plan was feasible. Chang's concerns included that the investment required to focus on manufacturing would be enormous, while the estimated financial return seemed too optimistic.[20]

Without Chang's endorsement, Tsao continued running UMC in its conventional manner. In 1987, however, Chang returned to Taiwan and established TSMC to manufacture chips for IC design houses and manufacturers that needed extra production capacity, known as the beginning of the dedicated foundry model. For many years, Tsao believed that Chang appropriated his idea. He had good reason to believe so, since TSMC has competed with UMC on the made-to-order chips market since the 1990s. Tsao and Chang did not appear at the same occasion publicly for two decades. Only in 2020 did seventy-three-year-old Robert Tsao reconcile with eighty-nine-year-old Morris Chang. In a ceremony to celebrate Hsinchu Science Park's fortieth anniversary, Tsao approached Chang, and the two shook hands.[21]

Robert Tsao's conceptualization of what we now call the "dedicated foundry" model was partially inspired by witnessing of the rise of Taiwanese-founded IC design companies in Taiwan and Silicon Valley and their demand for manufacturing capacity. Two additional trends also contributed to the idea. First, he must have been familiar with international electronics manufacturers, including Philips and General Instrument, that offshored the IC assembly tasks to Taiwan. It would have been possible for Taiwanese employees and Tsao to consider executing additional tasks that had not been offshored, especially the production of silicon wafers, a manufacturing stage before the IC assembly stage. Second, he also would have been aware of the business opportunities Taiwanese computer manufacturers such as Multitech, the predecessor of Acer, and MiTAC had seized to manufacture computers for US PC companies beginning in 1984.

Since Taiwanese companies now could manufacture sophisticated personal computers, it would have been possible to manufacture other types of electronics.

MORRIS CHANG'S FOUNDRY MODEL AT TSMC

Morris Chang, known to be the first entrepreneur to execute the "dedicated foundry" model, was born in Zhejiang, China, in 1931. After World War II, Chang moved from Hong Kong to Massachusetts, where one of his uncles had recently relocated. Chang applied for and was enrolled into Harvard University in 1949. He then transferred to MIT in his sophomore year in 1950. His parents also moved to the United States, as they were worried that Chinese communists would take over Hong Kong.[22]

At MIT, Chang planned to pursue a PhD in mechanical engineering. But after two failed qualifying exams, he left with a master's degree and began to work for Sylvania in 1955, to manage the manufacturing of germanium transistors. In 1958, he joined Texas Instruments. His first assignment there was to manage the production of a NPN transistor for IBM. Since he was in Sylvania, Chang had worked with technicians on the production line about the actual procedures to fabricate transistors. At Texas Instruments, in Chang's own words, he and his coworkers were trying different "recipes" for the production process. Here, "recipes" referred to "testing yet another combination of process parameters, changing the temperature and the time in the furnace and experimenting with different alloys and diffusion materials," as described by Tekla S. Perry in her interview of Morris Chang.[23] For the first few months at Texas Instruments, the yield Chang's team were able to achieve was "stably zero." Because IBM's in-house facilities' yield was "unstably low," Chang and his colleagues joked that they have achieved a better stability than IBM. After months of experiments, his team attained an outstanding yield, 30 percent. His success for this assignment allowed him to become known by many colleagues at Texas Instruments, including Jack Kilby, who was about to patent his "miniaturized electronic circuit," one of the first integrated circuits in the world. Chang and Kilby chatted about Kilby's ongoing development of the miniaturized electronic circuit. They two were especially familiar with each other because they joined Texas Instruments around the same time.[24]

Chang had a splendid career at Texas Instruments, witnessing the changes in the semiconductor industry for decades and obtaining a deep understanding of the volatile market. Taiwanese technocrats developed a strong trust with Chang after getting to know him through their overseas social network or Texas Instruments' investment in Taiwan. Chang was one of several vice presidents for the new Components Group in 1967. When the Semiconductor Group became independent from the Components Group, Chang was promoted to group vice president of the new group in 1972.[25] In Taiwan, Texas Instruments announced in 1970 that it would build facilities to fabricate transistor–transistor logic (TTL) ICs and MOS chips there. At that time, Chang was already a top-level manager at Texas Instruments and would have weighed in on these decisions, especially when the CEO once considered setting up facilities in Japan instead of Taiwan. He had visited Taiwan several times in the 1970s, and became more connected to Taiwanese technologists.[26] In 1974, Taiwanese technocrats dispatched Pan to consult Chang when they were unsure about whether RCA would be a suitable manufacturer for the IC fabrication technology transfer project. At some point in the 1970s, Barry Lam also contacted Chang and earned contracts to make calculators for Texas Instruments, as we saw in chapter 7. Although Chang had not lived in Taiwan to this point, he had shaped Taiwan's electronics industry in a substantive manner; he offered his insights to the Taiwanese government-led technology transfer project, and his employer was a critical investor in Taiwan, subcontracting work to companies and owning manufacturing facilities there.

Chang soon wearied of Texas Instruments' corporate ladder. Chang became senior vice president at Texas Instruments in 1981, the fifth officer listed on the company's annual report, and senior vice president for corporate quality, training, and productivity in 1982. Meanwhile, Texas Instruments gradually shifted the company's focus from the research and development of semiconductor chips to consumer electronics in the late 1970s. Chang's expertise was in the former field. He decided to leave Texas Instruments in 1983 and was selected to become president and chief operating officer of General Instrument in 1984, when he was consulted by a Taiwanese minister and Robert Tsao on the business strategies of UMC. However, at General Instrument, Chang did not feel that he was able to use his professional knowledge, as other executives and the board were more interested in using mergers and acquisitions to expand the company. He accepted an offer from Taiwan to become director of ITRI. After Chang's move, K. T. Li, now a minister without portfolio, invited Chang to propose a business plan to set up a new technology company to aid the development of the semiconductor industry in Taiwan. Li assured Chang that he and the premier would be able to offer whatever financial support Chang might need. Chang then proposed to establish a "foundry" that would focus on "wafer fabrication" exclusively to custom-make ICs for clients and would not design any ICs. Li unwaveringly supported Chang by persuading banks and enterprises in Taiwan to invest in the new company, named Taiwan Semiconductor Manufacturing Company.[27]

Chang's dedicated foundry idea also had several origins. First, in the mid-1970s, he had predicted that the demand for ICs would grow and thus non-semiconductor manufacturers might ask conventional semiconductor manufacturers to make ICs for their products. In 1976, as vice president of the Semiconductor Group, he had mentioned this idea in an executive meeting. He believed that soon, companies that did not have "expertise to produce semiconductors" but would like to make goods equipped with semiconductors, such as phones, cars, and dishwashers, would want to "outsource fabrication to a specialist." In this new business environment, larger-scale manufacturers with a higher production capacity would make more profits than smaller manufacturers. But Texas Instruments did not seriously consider Chang's plan at that time.[28] When moving to General Instrument, Chang also proposed a new business model to customize ICs for corporate clients, including design and manufacturing. But after careful evaluation, General Instrument did not adopt this model because it did not own a sufficient level of IC design capacity and it did not have the necessary marketing experience to make this business model succeed.[29]

Morris Chang noted that his dedicated foundry model was also inspired by his reading of a popular textbook, *Introduction to VLSI System*, written by Caltech professor Carver Mead and Xerox PARC computer scientist Lynn Conway. In the late 1970s, Mead, Conway, and other instructors used the manuscript of the textbook to teach undergraduate and graduate students to develop and automate the design of VLSI chips, as the circuits of such chips were unprecedently sophisticated. At the time, the design and manufacture of chips was conducted only in semiconductor firms. But Mead and Conway demonstrated that the design of VLSI chips could be done by trained professionals who did not work for IBM, Intel, and other major IDMs. The book gave rise to IC design industry and subsequently prompted the division between IC design and IC manufacture.[30]

While the 1980s witnessed the emergence of a few IC design companies, IC manufacturing was still something that could be done only in well-established semiconductor firms' fabrication facilities. Morris Chang noted that at Texas Instruments and General Instrument, many IC design engineers were interested in starting up their own companies. But they chose not to become entrepreneurs because they did not have enough financial backing to build facilities to fabricate IC products. It was nearly impossible for someone to design an IC and outsource the fabrication of the ICs to another company. In 1985, when Chang proposed establishing TSMC, he was betting that more IC design start-ups would emerge in the coming years, and his new company would be able to earn manufacturing contracts from these new IC design houses, but Chang's former colleagues, friends, and competitors in the semiconductor industry in the United States considered Chang's business idea unfeasible. When Chang attempted to raise seed money in the United States, only Intel and Texas Instruments welcomed him to present his ideas, but they eventually did not invest a dime.[31]

Third, the foundry idea was "born of an appreciation of both the virtues and limitations of Taiwan," according to *New York Times* journalist Mark Landler in 2000. When presenting the business plan to K. T. Li in 1985, Morris Chang carefully considered every step of the development of IC and decided that Taiwan's "highly trained, skilled line operators, technicians, and production engineers" would do well in manufacturing. Taiwan's talent for R&D, circuit design, IC product design, sales, marketing, and intellectual property, however, seemed relatively limited. Chang, in 1985 thus decided that his new company should simply focus on IC fabrication for manufacturers and IC design houses, and not pursue these other facets.[32]

To found TSMC, Morris Chang needed $200 million US in 1987, in addition to existing loans. At least two sources of the capital raised for TSMC came from the well-performing electronics industry in Taiwan. First, MiTAC, the rising computer manufacturer founded by a former Intel engineer Matthew Miau (see chapter 8), likely offered his insight for his business tycoon father, Yuxiu Miau, who invested in TSMC. One of Yuxiu Miau's firms subsequently supplied industrial gases to TSMC. Second, Philips' Dutch headquarters offered 27.5 percent of the start-up funds Morris Chang needed and licensed patents to TSMC. Philips' chief manager in Taiwan, Y. C. Lo (see chapter 6), who had been interested in fabricating IC wafers in Taiwan since the mid-1970s, convinced headquarters to invest in TSMC. The licenses granted by Philips to TSMC were only slightly behind industry leaders such as Intel at that time and were crucial to helping TSMC advance in manufacturing smaller and smarter chips in the years to come.[33]

Morris Chang recalled that, during the first few years of the company's history, TSMC was able to earn only "leftover" contracts from well-established chip manufacturers such as Intel and Philips. But "leftover" contracts also indicated that TSMC was recognized by these industry leaders. In winter 1987, Intel's CEO, Andrew Grove, toured TSMC's facilities and decided to send personnel to improve TSMC's manufacturing processes. Over a year's time, Intel personnel proposed hundreds of changes in operational procedures to make TSMC a qualified contractor for Intel. Thereafter, when Intel did not have enough capacity to make chips for their customers, it subcontracted jobs to TSMC. Only in the 1990s did burgeoning new IC design houses, such as rising stars Altera and Nvidia, become a leading group of clients for TSMC; fab-less IC design firms (those without fabrication factories) reached 60 percent of TSMC's sales in 1995. Nvidia's cofounder Jensen Huang established, in his description, a "long-term win-win partnership" with Chang to manufacture 3D graphic chips for games. A significant portion of the demand for TSMC chips was also local. In mid-1995, with its strong computer manufacturing industry, Taiwan surpassed Germany to become the third-largest chip market in the world.[34]

TSMC and UMC carved out the path of the dedicated foundry and attracted imitators such as Chartered Semiconductor in Singapore and Semiconductor Manufacturing International Corporation in China. In 1998, Taiwanese foundries accounted for almost 75 percent of the market share of the make-to-order chips. When this new industry expanded drastically in the twenty-first century, Taiwanese foundries owned two-thirds of the market—TSMC had 44 percent of the share and UMC had 14.5 percent in 2007. UMC remained the second- or third-largest dedicated foundry in the world from the mid-2000s until 2017. TSMC has held over 80 percent of the market for smaller and faster high-end chips since it began to make chips for iPhones in 2014. In 2020, in the midst of the COVID-19 pandemic, chip shortages and TSMC made headlines in international media; the public realized that 54 percent of global foundry revenue was going to TSMC. This was old news, however; TSMC had been leading the industry for more than a decade.[35]

Regardless of whether Morris Chang or Robert Tsao first conceptualized the idea of a dedicated semiconductor foundry, the business models of contract manufacturing adopted by Taiwan's electronics assembly factories and computer manufacturers inspired the two engineers-turned-entrepreneurs when they contemplated the future of their companies. In 1998, *Forbes* commented on the similar strategies adopted by the two industries: "Similar to Taiwan's contract-PC and peripheral equipment manufacturers, the foundries' strategy is to complement, rather than compete with, customers in the U.S. and elsewhere. They custom-make chips to order. Flexibility, speed and high yields are crucial. TSMC churns out hundreds of different products for over 300 different clients."[36] It was an innovative and successful strategy at the turn of the century.

DEFENSE, DIPLOMACY, AND SEMICONDUCTORS, 2019–2022

In the 1960s, Taiwanese technologists sought to enhance the technological capacity in the island, including getting access to computers and gaining knowledge to manufacture them. They believed that improved capacity could create advantages in the military confrontation and economic competition between Taiwan and China. Six decades later, TSMC's founder and executives similarly aimed to make their technological capacity relevant to the relationships among Taiwan, the United States, and China.

Under the Trump administration, the Pentagon became concerned that a portion of high-end chips used in weaponry were not made in the United States. For example, some chips used in the F-35 jet fighters were developed by IC design firm Xilinx, which outsourced the manufacturing to TSMC. As a dedicated foundry, TSMC had adopted a management system to protect its clients' IC design from espionage since the beginning of its company's history. But in 2019, the Trump administration raised two new national security issues associated with TSMC. First, TSMC also made chips for Huawei, the Chinese tech giant that the Trump administration was wary of. Second, they considered a military conflict between Taiwan and China imminent, possibly jeopardizing the supply of the chips used in US weaponry. TSMC's chairman of the board Mark Liu (Deyin Liu) and CEO C. C. Wei (Zhejia Wei) strove to meet the Pentagon's requests on security issues and reiterated that it would be too costly to make chips in the United States. Because TSMC has owned facilities in Camas, Washington, since 1998, Liu and Wei knew that making chips in the United States would be more expensive than in Taiwan. They assented, nevertheless, to the Trump administration's request to construct fabrication facilities in Arizona in May 2020. Building another factory in the United States was viable only with federal or state subsidies.[37]

The global chip shortage arising from the COVID-19 pandemic in March 2020 allowed TSMC to receive significantly more attention than in pre-pandemic times. During the first few months of the pandemic, the demand for automobiles dropped but the demand for computer devices rose, due to evolving work-from-home policies. Chip makers had no choice but to reduce their production to reflect changes in demand. But, when the demand for automobiles bounced back in 2021, the production of semiconductor chips, which required months of development and manufacturing, could not satisfy market needs. Business news website Bloomberg published an article on January 27, 2021, titled "The World Is Dangerously Dependent on Taiwan for Semiconductors," one of the earliest major media outlets to discuss the geopolitical implications of chip shortage. US President Joe Biden, after his inauguration on January 20, 2021, immediately ordered a review of the nation's supply chains to prevent shortages of various products and components that US manufacturers struggled to obtain, ranging from masks to semiconductor chips.[38]

In March 2021, the National Security Commission on Artificial Intelligence (NSCAI), formed in 2018 to make recommendations to the president and Congress, published its final report on the status quo of artificial intelligence in the United States in relation to China. The committee was chaired by Google's former CEO Eric Schmidt. Among the many recommendations for bolstering the AI industry in the United States, one was to "revitalize domestic microelectronics fabrication" by offering financial incentives to support US chip manufacturers, and encourage TSMC and Samsung to manufacture chips in the United States.[39] The report stated that US dependency on TSMC made the nation vulnerable to "adverse foreign government action," namely the risk of China's incursion on Taiwan. The report commented, "The dependency of the United States on semiconductor imports, particularly from Taiwan, creates a strategic vulnerability for both its economy and military to adverse foreign government action, natural disaster, and other events that can disrupt the supply chains for electronics."[40]

In June 2021, the supply chain review conducted by the White House also arrived at a similar conclusion. Their report, "Building Resilient Supply Chains," noted that the United States "lacks semiconductor production capability at the most advanced semiconductor process node," posing a national security problem. At this time, only TSMC and Samsung could manufacture the most advanced semiconductors.[41] With the presentation of the two reports, a bill was introduced in the House to offer financial support to the semiconductor manufacturing industry in the United States in July 2021, which evolved into the Chips and Science Act, which was passed in the Senate in July 2022.

Meanwhile, Pat Gelsinger, an Intel veteran, returned to become CEO of the company in early 2021. He was determined to lead Intel to surpass TSMC and Samsung in advanced chip manufacturing. After the publication of the NSCAI's final report and the White House report on supply chains, it was clear that the US government would be willing to offer financial incentives, about $52 billion, to chip makers and other relevant entities, including research institutes, to keep US chip makers ahead of Chinese chip makers. As Intel's CEO, he might have been worried that TSMC would compete for the $52 billion. On June 24, 2021, he wrote an op-ed for Politico.com, a political news site, to advocate that the federal incentives should be paid to US companies. He said, "By contrast, foreign chipmakers vying for U.S. subsidies will keep their valuable intellectual property on their own shores."[42] In an event hosted by *Fortune* on December 1, 2021, Fortune Media's CEO Alan Murray interviewed Pat Gelsinger and asked Gelsinger to explain why US taxpayers' money should be used to support domestic companies, as opposed to TSMC's and Samsung's plants under construction in Arizona and Texas. Murray pointed at Taiwan's vulnerability, consistently under military threat from China. Gelsinger then noted the twenty-seven sorties from China that had flown into Taiwan's air defense identification zone two days earlier. "Does that make you feel more comfortable or less," he questioned. TSMC was explicitly or implicitly mentioned whenever Pat Gelsinger sought to promote Intel in the competition for federal funds for chip manufacturing.[43]

In spring 2021, TSMC had begun the construction of its facilities in Arizona with an expectation of receiving subsidies, tax reductions, or similar measures from the US federal or state governments. Gelsinger's multiple attempts to advocate against financial support for TSMC disturbed Morris Chang. Chang had retired in 2018 at age eighty-seven and no longer held any official position at the company. After NSCAI's report dropped, Chang defended TSMC's operations in Taiwan, including its supply of excellent engineers, technicians, and operators, a fact he had shared with *New York Times* journalist Mark Landler years ago. These employees, he continued, possessed a strong work ethic, implying that they would be willing to work more overtime or longer shifts than US employees. While the United States possesses abundant land, electric power, and water resources, the overall costs to manufacture chips in the United States is much higher than in Taiwan.[44] In October 2021, Chang criticized Gelsinger's critiques of TSMC. Chang reiterated that it would be impossible to reestablish a complete semiconductor supply chain in the United States, even with hundreds of billions of dollars of funding. It would be too challenging for the United States to return to the era of the 1990s when it manufactured 42 percent of semiconductors in the world. When Gelsinger brought up the potential Chinese military attack on Taiwan and expressed his disapproval of the subsidies for TSMC's Arizona facilities in December 2021, Mark Liu responded that TSMC does not criticize its peers, implying that Gelsinger had frequently criticized TSMC. He reiterated TSMC's commitment to make chips in Phoenix, Arizona.[45]

Mark Liu, like many of the other figures under discussion, had relevant experience navigating these matters in both Taiwan and the United States. He holds a BS in electrical engineering from National Taiwan University, and a PhD in electrical engineering and computer science from the University of California, Berkeley. He has been in Taiwanese news headlines frequently since the summer of 2021 because of a widely discussed chip-for-vaccine proposal. Generally speaking, the arrival of COVID-19 vaccines in Taiwan was slower than in the United States and Europe. Some Taiwanese critics thus proposed that TSMC should prioritize its delivery of products to countries that would be willing to facilitate the delivery of vaccines to Taiwan. No such deals have been formally discussed or made between Taiwan and other governments, but the vaccine shortage in Taiwan and its association with the wishful plan made TSMC even more cherished by many Taiwanese. In 2020, the Taiwanese government had secured COVID-19 vaccines orders for Moderna and AstraZeneca when those vaccines were still in development. It could not, however, obtain Pfizer-BioNTech vaccines from a Shanghai-based company Fosun Pharma that had won the exclusive rights to sell the vaccine in China, Hong Kong, Macau, and Taiwan. Negotiations between Pfizer-BioNTech and the Taiwanese government stalled, as the two parties could not determine how Taiwan should be referred to in the contract. In June 2021, Foxconn's founder and self-made billionaire Terry Guo (Taiming Guo) requested that Taiwanese President Tsai Ing-wen authorize him to leverage his business connections to resolve the deadlock on Taiwan's negotiation with BioNTech. Guo's request posed a thorny political dilemma for Tsai because he had run in the rival Chinese Nationalist Party

(Kuomintang KMT)'s presidential primary in 2019.[46] Guo posted his request on a late-night Facebook page posting on June 17, testing Tsai's crisis management. The next morning Tsai agreed to meet with Guo. Media outlets noticed with surprise in the afternoon that Guo's meeting at the presidential office with Tsai also included Mark Liu (figure 10.1); Tsai had likely contacted Liu earlier that day. By late afternoon, she was able to announce that she authorized both Guo's nonprofit organization and TSMC to purchase BioNTech vaccines on behalf of the Taiwanese government. In doing so, BioNTech could sidestep an acknowledgment of Tsai's government and other geopolitical complications with China as it made a deal with Guo and Liu. By looping in Liu and TSMC, Tsai likely sought to ensure that Guo alone could not dictate national policy in the distribution of vaccines he helped acquire. TSMC made it clear that the company would not reserve any vaccines for its employees, instead allowing the government to distribute them as it saw fit. The two companies signed a contract with BioNTech in late July, ordering ten million doses worth $350 million US. While it

FIGURE 10.1
Taiwanese President Tsai Ing-wen (right) meeting with TSMC chairman Mark Liu (left) and Foxconn founder Terry Gou (second left) on negotiating with BioNTech to procure COVID-19 vaccines, on June 18, 2021. *Source*: Presidential Office, Taiwan

was not the chip-for-vaccine deal some had envisioned in early 2021, TSMC did play a predominant role in pressing BioNTech to sell vaccines to Taiwan. Mark Liu greatly assisted Tsai in easing this political crisis.[47]

The next summer, Liu was no less busy. On July 18, 2022, *Financial Times* released the exclusive news that House Speaker Nancy Pelosi planned to visit Taiwan in August. Pelosi's plan gravely agitated China's President Xi Jinping. Some politicians in Taiwan and in the United States viewed Pelosi as a steadfast critic of China's repressive government. Others considered her decision disruptive to US-China relations, which had been slowly improving. Some warned that it would prompt China's retaliation against Taiwan. Others were worried about another impending chip shortage if China were to attack Taiwan, considering Russia's invasion of Ukraine in February that year.[48] Fareed Zakaria, host of *Fareed Zakaria GPS* (Global Public Square), a news program aired on CNN on Sundays, interviewed Mark Liu in advance of Pelosi's visit. Zakaria asked Liu to evaluate the impact of a Chinese invasion of Taiwan. Liu answered, "The war brings no winners. Everybody is losers." He warned that Taiwan's hard-earned democracy would be gone. Zakaria followed up and asked Liu to specify what would happen if the PRC government controlled TSMC after an invasion. Liu answered,

> Nobody can control TSMC by force. If you take a military force or invasion you will render TSMC's factory non-operable because this is such a sophisticated manufacturing facility. It depends on the real time correction with the outside world, with Europe, with Japan, with the U.S., from materials to chemicals to spare parts to engineering software diagnosis. And it's everybody's effort to make this factory operable. So if you take it over by force you can no longer make it operable.[49]

After her trips to Singapore and Malaysia, Pelosi and a delegation of five congressmen and congresswomen boarded a US military aircraft for Taiwan, taking a three-hour detour to avoid the South China Sea. It was a brief visit: Pelosi arrived in Taipei late on the night of August 2 and left for South Korea before evening the next day. Morris Chang and Mark Liu were invited to have lunch with Pelosi and President Tsai. Rumors indicated that over lunch Chang contended again that it would be difficult to bring chip manufacturing back to the United States. Immediately after Pelosi's departure, the PRC announced that it would begin live-fire exercises surrounding Taiwan on August 6. A series of unprecedentedly intense military exercises in the region lasted until mid-August. It has been known as the Fourth Taiwan Strait Crisis.[50] In late August, Stan Shih noted in an interview that a blockade of Taiwan would be apocalyptic, because chips are now essential to every aspect of societies around the world. Chang also reiterated what Mark Liu and Stan Shih had said. Lesley Stahl, the correspondent of CBS *60 Minutes*, went to Taiwan with her crew for an episode aired on October 9, covering how people of the island lived with the persistent military threat from China. She interviewed Morris Chang and asked him about what if PRC nationalizes TSMC. Chang answered, "If there's a war, I mean, it would be destroyed. Everything will be destroyed." The Fourth Taiwan Strait Crisis gave TSMC founder and executives an opportunity to express their concerns over a war, and the devastating effects of a war on their company. They sent their protests to PRC's military exercises in a subtle way. [51]

A month later, in November 2022, more than a hundred TSMC engineers and their families left Taiwan for Arizona to work at the company's new facilities in the United States. Taiwan's *Business Weekly* estimated that an additional six flights have been arranged for the months to come, to bring a thousand engineers and their families to the same destination.[52] Taiwanese Minister of Economic Affairs Mei-hua Wang had to explain on various public occasions that TSMC's Arizona facilities will not affect the company's continued investment in Taiwan, especially research and development for the manufacturing of high-end chips. Similarly, TSMC's CEO, C. C. Wei, who holds a

FIGURE 10.2
US President Joe Biden (middle), TSMC chairman Mark Liu (right), and CEO C. C. Wei (left), at TSMC's "first tool-in" ceremony in Arizona on December 6, 2022. Photographed by Jonathan Ernst. Reprinted with permission from the Reuters News & Media INC.

BS and a PhD in electrical engineering from Chiao-Tung University and Yale University, respectively, has had to reassure Taiwanese that TSMC's operation in Arizona will not cause brain drain from Taiwan and will not undermine TSMC's growth on the island. He addressed the brain drain issue in a talk at NCTU on December 3, 2022. In this talk, he also noted that if there had been a Chiao-Tung University in the United States, it would have been effortless for him to recruit engineers to work for TSMC's Arizona facilities. His remarks should be read as a compliment to Chiao-Tung rather than as a disparagement of US engineering education.[53]

On December 6, 2022, the first batch of TSMC machinery shipped from Taiwan arrived in Phoenix. TSMC held a "first tool-in" ceremony to celebrate the installation of the machinery, a ceremony similar to a groundbreaking or inauguration but unique to Taiwan's high-tech industry. US President Joe Biden, along with tech CEOs including Apple's Tim Cook, Nvidia's Jensen Huang, Micron's Sanjay Mehrotra, and AMD's Lisa Su, attended the ceremony (figure 10.2). Morris Chang congratulated the launch of TSMC's Arizona operation. He shared with the audience that he had dreamed of building a fab in the United States for almost four decades. Under his leadership, TSMC had previously invested in a fab in Camas, Washington, but the dream soon became a nightmare because of financial deficits. Now, TSMC was much more prepared, and he was glad to witness that Mark Liu had realized this dream and built a fab in Arizona. Welcoming President Biden, Mark Liu asserted that, following the passage of the Chips and Science Act, TSMC was also taking "a giant step forward to help build a vibrant semiconductor ecosystem in the United States." Taking the stage, Biden warmly noted, "I owe an awful lot to this company." He thanked everyone at TSMC, especially Morris Chang. He also mentioned Chang's wife, Sophie, who lived in Delaware in 1972 and worked on Biden's first senatorial campaign, though they had not stayed in contact. After acknowledging Apple's Tim Cook and Micron's Sanjay Mehrotra in the audience, Biden declared, "American manufacturing is back," with attendees applauding enthusiastically. Nonetheless, on the same day, two financial online news columnists wrote that TSMC's Arizona fabs will produce only a small portion of chips made in the United States and will not change much in terms of revitalizing US chip manufacturing.[54]

Two days later, China's *Global Times* published an editorial denouncing the United States' forcing of TSMC, "developing well in China's Taiwan island," to move to its territory and accusing the United States of draining the resources of "our Taiwan region."[55] Despite being at the eye of the storm, the US-educated electrical engineering executives Mark Liu, C. C. Wei, and Morris Chang will likely continue to ensure that TSMC leads in manufacturing chips, which will be critical in maintaining international support for Taiwan for years to come.

EPILOGUE

ASPIRATIONAL TINKERERS

Resettling on the war-threatened island, Taiwan-based technologists were aspirational and ambitious for the industrial future of the island at the beginning of the Cold War. Before the first mainframe computer arrived in Taiwan, at least two prominent Chiao-Tung alumni had anticipated that computers could offer hands-on experience for engineering students. Telecommunications bureaucrat Gisson Chi-Chen Chien and director of graduate studies S. M. Lee had pictured graduate students on the NCTU campus soon manufacturing computers, after they heard that Taiwan Sugar Corporation had rented accounting machinery from IBM in 1958. They had gained more confidence when they knew the United Nations approved Chiao-Tung University's technical-aid program in 1961. Chien and Lee believed that graduate students would figure out how to manufacture a computer after studying an on-site electronic digital computer. Their understanding of manufacturing might have seemed naïve, or they might have decided to deliver a rosy picture to boost the morale of Chiao-Tung alumni that resettled in Taiwan, the United States, and other places. In any case, developing Chiao-Tung students' engineering expertise might make the island of Taiwan, their new home, a prosperous place and win support from allies when artillery shells still fell on Quemoy at least every other day. But Chien and Lee's vision also showed their enthusiasm for computers and opportunities for tinkering with this novel artifact. They believed that with a computer on campus, students would obtain hands-on experiences with the technology, gain a better understanding of it, and explore related manufacturing endeavors.

NCTU faculty, staff, and students, mostly men, probed and maintained the computers after the IBM 650 and 1620 computers were installed on campus in 1962 and 1964, respectively. Several women computer operators began their careers there at the same time. The machines were especially precious, as the shipments took months traveling from the United States to the island. They were also precious as the only way for interested Taiwanese to learn the workings of a computer. NCTU members took an audacious move: they went ahead and declined IBM's service contracts. For the IBM 1620 specifically, university administrators and instructors substituted an in-house technician for IBM's official service support. While self-servicing was also a strategy to lower costs, Chiao-Tung folks signaled their confidence in maintaining the computers independently of IBM. They even tinkered with them by attempting to connect the IBM 650's peripherals to the 1620. The experience of exploring, using, and maintaining the two mainframe computers offered the technologists a realistic view of what manufacturing a sophisticated technology would encompass.

In a May 1964 technical aid application to the UN, Lee and Chien reenvisioned the Chiao-Tung campus from a computer manufacturing start-up to a laboratory to test run transistor manufacturing. The application was not granted. However, unseen by Lee and Chien in the early 1960s, an electronics manufacturing industry had already been budding in Taiwan since the mid-1960s. The industry was primarily initiated by US, Japanese, and European investors, constructing assembly factories to make transistors, integrated circuits, semiconductors, computer parts, and transistorized radios and TV sets. Investing in facilities 6,000 miles away from their headquarters, these overseas companies hired locally—assembly workers, mostly women, and engineers, mostly men—to operate the factories. To execute their jobs, Taiwanese employees experimented and identified best practices for fabricating electronics products and parts. And when headquarters frequently failed to offer timely technical support, local folks improvised to keep the offshoring factories functional.

Island technologists in the 1960s and 1970s not only welcomed opportunities to take care of technologies and machinery from afar but also continued to look for opportunities to build technologies locally. In parallel with the assembly factories' menial tasks, in the late 1960s, three groups of college and graduate students developed an avid interest in getting hands-on experience building minicomputers and calculators. They wanted to put together a minicomputer or calculator by sourcing components ranging from recycled electronics parts to items customized at multinational-owned electronics factories. They believed this was the path that would lead them to acquiring mass-manufacturing technology. NTU graduate student and minicomputer builder Barry Pak-Lee Lam became a pioneer of calculator manufacturing in the early 1970s and founded Quanta Computer for laptop manufacturing in the late 1980s. Chiao-Tung minicomputer building project leader Jong-Chuang Tsay became a faculty member in the Department of Computer Engineering at Chiao-Tung, after the completion of the computer and a two-year visiting fellowship at the University of Florida. National Cheng-Kung University's undergraduate Wang Tang, who built a calculator from scratch, left Taiwan to pursue a PhD in electrical engineering in the United States. He then stayed to work and retire in the United States. Engineering students who grew up with the nascent electronics manufacturing industry in Taiwan became the talent for both Taiwanese and US industries and academia.

Beyond factories and universities, calculator engineers tweaked their products in their offices. Stan Shih and Barry Pak-Lee Lam, both with master's degrees in electrical engineering, were attentive to manufacturing products for the domestic electronics market. Starting in 1972, they strove to tinker with and commodify calculators for both domestic and export markets. Working for a start-up company called Qualitron, Shih prototyped printed circuit boards on the balcony of his office by sourcing materials from the Chung-Hua Arcade, where customers could look for new or used electronics parts. At another start-up, Santron, Lam, an excellent solderer, would join employees in assembling calculators to meet delivery deadlines. Shih and Lam closely monitored the manufacturing of parts that went into the calculators they designed. As Taiwan's domestic market was small, they also looked for business opportunities to sell their products to other parts of the world. Their enthusiasm for microprocessors, hands-on experience, engineering knowledge, and entrepreneurial skill in manufacturing calculators and selling them overseas soon contributed to their success in manufacturing and exporting personal computers and laptops in 1980s Taiwan.

Island technologists' strategies of emulation and their embrace of innovation were not always appreciated by continental enterprises. Shih and Taiwanese computer makers, such as MiTAC, began developing and manufacturing Apple compatibles and IBM PC clones in the 1980s. Their design and production of personal computers benefited from the engineers, suppliers, managers, and women factory workers who had participated in the local electronics manufacturing industry since the 1970s. Apple and IBM viewed Shih's start-up company Acer and other Taiwanese companies as copycats, invaders, flies, and scoundrels, exporting compatible computers to compete for US and global market share. But Shih, Lam, and many other Taiwanese computer makers eventually leveraged their manufacturing capacity to work with well-established tech companies. They became indispensable subcontractors for US, Japanese, and European computer companies during the second half of the 1980s. Acer also sold its own computers globally and ascended to one of the world's top ten computer makers in the mid-1990s and one of the top five from 2004 to 2014. "Aspire," the name for Acer's post-1995 computer models, echoed the aspirations shared, since the late 1950s, by

Lee, Chien, Lam, and other technologists toward the manufacture of sophisticated computing technologies on the prosperous but precarious island.

Evolving alongside Taiwan's aspirations toward computer manufacturing was Morris Chang's TSMC and his foundry model. Chang had extensive knowledge of the global semiconductor assembly industry. In 1955, he began to work on fabricating transistors for Sylvania and then Texas Instruments in the United States. In the beginning, he had to teach himself how to manufacture transistors—his MIT degrees were in mechanical engineering. With his career on track, he and his colleagues tweaked what he termed "recipes," or the various parameters in the fabrication process, for increasing the yield of transistors. When Chang rose to a top-level manager at Texas Instruments in the early 1970s, he weighed in on the company's decision to set up an offshoring transistor assembly factory in Taiwan. Like Chang, Taiwanese engineers tinkered with assembly-line processes at various factories. He also had subcontracted calculator manufacturing to Lam in the 1970s when Lam worked for Kinpo. After Chang relocated to Taiwan in 1985, his creation of a dedicated semiconductor foundry model reflected his evaluation of "both the virtues and limitations" of Taiwanese manufacturing talent at that time. While he is better known for his wisdom regarding the global semiconductor market, his knowledge of Taiwan's strengths in manufacturing maximized his custom chip fabrication business.[1]

Hands-on experiences with computing technology also became popular among Taiwanese computer users beginning in the 1980s. Computer users built their own computers by putting together parts purchased from retail shops. These homebrew personal computers have enjoyed a significant market share in Taiwan ever since.[2] Nvidia's CEO Jensen Huang, a Taiwanese American, praised the homebrew computer culture in 2023. Nvidia's market capitalization crossed the $1 trillion mark on May 30, 2023, thanks to the rising demand for Nvidia chips to be used in AI systems. Three days before its trillion-dollar valuation, Huang gave a speech at NTU's commencement ceremony, recalling his first visit to NTU over a decade ago. Invited to tour chemical engineering professor Wen-chang Chen's lab, he witnessed Nvdia's graphics processing cards computing quantum physics simulations. Chen, who sat behind Huang during the commencement ceremony, had become the university's president in early 2023. During Huang's visit over a decade ago, Huang saw that "Nvidia's GeForce gaming cards filled the room, plugged into open PC motherboards, and sitting on metal shelves in the aisles were oscillating Tatung fans." He then noted, "Dr. Chen had built a homemade supercomputer, the Taiwanese Way, out of gaming graphics cards."[3] It is unclear whether Huang was familiar with the history of homebrew computer culture in Taiwan, but he recognized a unique "Taiwanese way," a style that he associated with Taiwan, where computer users build their computers and make them workable.

Morris Chang's and Jensen Huang's career paths and business strategies also demonstrate the ties between the industries in Taiwan and the United States, a product of the transnational network built by Taiwanese technologists and their overseas counterparts. Since the Cold War, island tinkerers have cultivated their relations with the United States as one of the superpowers. The relations between this superpower and its client state were not always reciprocal, and required the technologists of the client state to consistently advocate for their interests. Taiwanese military researchers struggled to find a way to support their projects financially and technically; women assembly workers made less money than their counterparts in the United States; and microcomputer manufacturers providing less expensive products to consumers were not greeted favorably by US enterprises in the early 1980s. Nonetheless, during the decade prior to the dissolution of the Soviet Union, Taiwanese computers and electronics made their way to and were welcomed in the Eastern Bloc, including the Soviet Union, East Germany and China through formal trade and smuggling routes. Eventually, both Taiwanese and US entrepreneurs eagerly coordinated investments in China during the late 1990s.[4] But relations between Taiwan and the United States have become indispensable for both sides in the 2020s: TSMC strives to complete its fab in Arizona, following the Chips and Science Act; Taiwanese subcontractors remain critical of US tech companies; and the military alliance of the two countries has grown stronger since the Russia-Ukraine War broke out.

NEW CHALLENGES TO SUBCONTRACTORS

Since the early 2000s, Acer has been moving away from manufacturing and toward marketing. It no longer subcontracts for accomplished US, European, and Japanese companies. Instead, its spin-off companies, including Wistron, focus on fulfilling subcontracting orders and manufactured Acer's products. Acer, however, suffered losses of an unprecedented scale from 2011 to 2014, in part due to a drop in demand for PCs that accompanied the rise of smartphones and tablets, especially Apple products. Furthermore, Acer failed to come up with versatile lines of PCs, tablets, and laptops, unlike Jonney Shih's Asus, which in 2015 surpassed Acer to be the world's fourth largest PC maker. Acer hired a new CEO, Jason Chen (Junsheng Chen), to lead the company at the beginning of 2014. Chen had worked for Intel for fourteen years and had become a senior vice president before he left Intel and joined TSMC in 2005. In 2013,

Stan Shih recruited Chen immediately after he heard that Chen was considering leaving TSMC. To revive Acer, Chen's strategy focused on high-end gaming laptops, lightweight laptops, virtual reality (VR) headsets, and other products with higher profit margins. Acer became profitable again in a few years. After the outbreak of the COVID-19 pandemic, which dislodged supply chains, Acer swiftly sourced devices to meet the surge in demand for laptops and computers, returning to the world's fourth largest PC maker in 2020. In 2021, Acer's revenue was $11.42 billion US. But Asus' revenues were almost twice Acer's, and Quanta earned revenues almost four times Acer's.[5]

In contrast, Quanta remains focused on manufacturing with low margin—lower than US computer makers; since the mid-2010s, it has enjoyed financial success generated by its server business. The company had been a seasoned contract manufacturer for Apple, Dell, and other notable US computer makers before it climbed to the world's largest laptop maker in 2000. At the turn of the century, Lam made two decisions that laid the foundation for Quanta's continuous success: he began to make servers in Asia for US computer makers, and he invested in facilities in Fremont, California, where Quanta based its US operations. The Fremont facilities have served at least two goals. First, after Apple's CEO Tim Cook committed to move Apple's manufacturing tasks back to the United States in 2012, Quanta made its Fremont facilities a final assembly plant for Apple—receiving components of Apple products made overseas and putting together the components for the US market. Quanta also began to assemble servers in Fremont as early as 2009. By 2014, Quanta had become the world's third largest server manufacturer, with a 17 percent market share, trailing Hewlett-Packard and Dell. Quanta had its Fremont facilities and a Nashville factory to assemble and supply servers directly to Amazon, Microsoft, Facebook, and other companies that owned data centers, though executives of Hewlett-Packard and Dell claimed that Quanta's margin was low and would be unable to offer comparable technical support and services.[6]

Acer, Quanta, and TSMC continue to look for the best strategies to push their companies to grow and be profitable, whether it is marketing-intensive, low margin, or generous investment in R&D. Founders and executives of these companies have been dynamic in terms of dealing with the intense global business competition and ongoing geopolitical challenges. Popular narratives about the success of the Taiwanese computer industry, however, tend to boil down to the companies' subcontracting strategy. Acer, Quanta, TSMC, and many successful Taiwanese electronics manufacturers, including Foxconn, used to be, or still are, known as successful subcontractors.[7]

Subcontracting, contract manufacturing, or *daigong* in Mandarin, as a business strategy seems to be deeply ingrained in the mentality of some Taiwanese. The internalized subcontracting image can result in a conservative approach to invest in innovation.

In summer 2021, rival political parties attacked President Tsai's administration for COVID-19 vaccine scarcity, even though many external factors contributed to the problem, including geopolitical issues, Taiwan's low level of COVID-19 cases, and the distribution policy of COVAX (COVID-19 Vaccines Global Access, a project backed by international organizations to ensure fair distribution of vaccines among participant countries). The then mayor of Taipei, Wen-je Ko, who is a surgeon, blamed the Tsai administration for not taking opportunities in the first place to become a subcontractor for vaccine developers such as AstraZeneca and Moderna. He stated in his daily press conferences that if he had been in charge, he would have found subcontracting opportunities to ensure the availability of vaccines in Taiwan. He believed that *daigong* is Taiwan's strength. In contrast, the Tsai administration allegedly placed all its eggs in a single basket by supporting two Taiwanese companies to develop protein subunit COVID-19 vaccines. He thought that the Tsai administration should not have pursued the breakthrough in developing new vaccines. Note that he clearly knew that Novavax worked on the same type of protein subunit COVID-19 vaccine in the United States, as did other companies in Japan, India, and China, and that protein subunit vaccines are commonly used for many diseases.[8] Without considering the difference between making vaccines and making computers, Ko extrapolated subcontracting to be the most effective strategy for Taiwan's pharmaceutical industry.

Leaving aside the thought process and intention of the mayor's criticism, what stood out was how his rhetoric reflected an acknowledgment of the accompanying role of contract manufacturers and their enterprise clients. The latter requires the former. They are indispensable to each other. But enterprise clients often enjoy a better reputation than the contract manufacturers and usually have more bargaining power than the contract manufacturers. Similar to enterprise clients and contract manufacturers, a series of accompanying relationships can be seen in the history of technology chronicled in this book: a client state to a superpower, counterfeits to the authentic, and women assembly workers to multinational enterprises that need to mass-produce electronics. The authentic producers of computer technology consistently drew lines to distinguish themselves from what they called counterfeiters. The existence of the latter justified the "innovation" defined by the former. For the superpower and client state symbiosis, the latter required leveraged power to advance their interest. Although I show in this book how a client state like Taiwan managed to more than survive into the 2020s and how Taiwanese computer manufacturers promoted their compatible computers and became valuable and innovative subcontractors, women assembly workers over time in Taiwan and beyond have had little leverage power, as opposed to their engineer-colleagues and employers, to redefine their accompanying roles. Factory women's tinkering in Taiwan—their skills to manipulate and assemble parts in electronics and increase the yield, did not seem to significantly uplift their status as an accompanying role to the others. Global factory women's low wage and high quality of work fostered the indispensable role of subcontractors to their enterprise clients.

DEFINING SUCCESS

When giving talks on my research for this book, I was often asked, implicitly or explicitly, to offer policy suggestions for the present moment. A few years ago, an enthusiastic electrical engineering PhD student, originally from Côte d'Ivoire and now in the United States, wrote me an email after my talk, inquiring about the first steps his home country could take to develop an electronics manufacturing industry. The overseas 1960s Chiao-Tung engineer alumni discussed in this book were, in their time, puzzled by the same question the student proposed in the twenty-first century. I replied to the student by pointing out that engineering education was important; that was what Chiao-Tung alumni advocated for in the first decades of their exile. What I was unable to explain fully in the email were the other influential, noninstitutional factors that I have delineated in this book, including hands-on experience with technology and the aspiration to build computers that students, engineers, and entrepreneurs developed over six decades. These experiences and aspirations might not be easily replicable. But they have been deemed important in today's classrooms, especially in the United States.

In 2004, five immigrant students from Carl Hayden High School in Phoenix, Arizona—Luis Aranda, Christian Arcego, Lorenzo Santillan, Oscar Vazquez, and Angelica Hernandez—worked on a project that won the Design Elegance Award and was the Overall Winner of the Marine Advanced Technology Remotely Operated Vehicle Competition, in Santa Barbara, California. The underwater robot put together by high school students of limited means outperformed those built by notable teams of college students, including one from MIT. It was unheard of. Luis Aranda was the only student who had legal residency in the United States at that time; the other team members were undocumented. Angelica Hernandez, who also participated in building their robot—Stinky—did not attend the competition due to ROTC obligations.[9]

The news was first covered in depth in Joshua Davis's 2005 *Wired* article, "La Vida Robot." The empowering nature of the story made it the subject of a Hollywood film and two documentaries released from 2014 to 2017.[10] The story was encouraging because of the contrast between public high school students from extremely disadvantaged socioeconomic backgrounds and college students from a private resource-laden university. The appearances of their underwater robots also directly reflected the discrepancy. The MIT team enjoyed a budget of $11,000 and built a robot that was much smaller than others, "densely packed, and [with] a large ExxonMobil sticker emblazoned on the side." Other teams' robots also looked "like pieces of underwater jewelry," built with materials of chrome, stainless steel, or manufactured plastics. The MIT team described their robot as technical "subsystems" interacting with each other. In contrast, Carl Hayden High School focused on "simplicity" due to the limited budget the team had. Stinky, according to Faridodin Lajvardi, one of the team's two advisers, was an $800 "raggedy" robot that "looked like [the team had] just pieced together everything from a junkyard." It was made of cheap PVC pipes from Home Depot, instead of glass syntactic flotation foam, which the team wanted to use but couldn't afford. It was named "Stinky" because of the odor from pipe glue. Hernandez later noted, "if you can't be expensive, be clever." But there were also parts of Stinky that required high precision and tended to be pricey and could not be substituted with cheaper alternatives. For example, the team needed a thermocouple to measure temperature underwater. A few manufacturers donated parts to help the team.[11]

The Carl Hayden team's success shared a couple of similarities with the early history of electronics and computer manufacturing in Taiwan. With limited technical and financial resources, the students studied, designed, improvised, and put together an underwater robot that outperformed their competitors. With limited technical and financial resources, technologists in Taiwan accumulated their hands-on experience with technology to explore their business opportunities in the global market: they probed the first computers on-site, building computers from scratch by using parts available on the island, testing "recipes" on electronics assembly lines, and exploring the mass-manufacturing of calculators and personal computers. Taiwan's computer manufacturers eventually rose in the late 1980s to be a significant player in the industry.

In the documentary *Underwater Dreams*, the MIT team invited the Carl Hayden team to the MIT campus for a visit in 2014, ten years after the competition. Several MIT team members had become engineers in fields ranging from underwater robots to Apple products. In contrast, among the four Carl Hayden students who went to Santa Barbara for the competition, Oscar Vazquez was the only one who as of 2014 had become an employed engineer; he worked at BNSF Railway's locomotive repair and maintenance yard. Both Vazquez and Cristian Arcego lost their scholarship at Arizona State University after Arizona decided to bar undocumented immigrants from receiving state assistance, including the in-state tuition rate. Vazquez managed to find a private scholarship and graduated from college, whereas Arcego did not. Arcego worked at Home Depot and became interested in starting a consumer electronics business. Lorenzo

Santillan was employed at a restaurant and ran his own catering service. Luis Aranda was a janitorial supervisor.[12]

Lajvardi was invited to MIT during the 2014 visit, too. He commented that the conversation about current jobs held by the two teams created "an uncomfortable feeling in the room." In the *Arizona Republic*, Lajvardi also noted,

> "You can't think success is just one sort of result," he said. "Success takes different forms."
> "You can't all be engineers. How boring is that? The point is to be successful at what you're doing."[13]

Lajvardi's open-ended definition of career success is encouraging. His open mind also reflected that, in 2004, he had no presumption about his resourceless students' potentials. But one might also believe that had more legal and educational resources been granted, the team members would have enjoyed more opportunities to shine in an engineering field.

Hernandez was not in Massachusetts for the 2014 MIT visit. But she was featured as an important team member that built Stinky in the 2017 documentary *Dream Big: Engineering Our World*. She was one of the four outstanding engineers featured in *Dream Big*; they came from diverse background and now work on bettering people's lives around the world. Entering the United States with her mother at the age of nine, Hernandez applied for Deferred Action for Childhood Arrival (DACA) and was granted in 2012. She completed her college education at Arizona State University, earned a master's degree from Stanford University, and now works as an engineer in the field of clean energy. The documentary, made for IMAX and giant screen theaters, popular in science museums, aims to inspire young students to engage with STEM education. Hernandez's background and career success made her an example that "dreams big."[14]

Many aspirational individuals in Taiwan tinkered with computers from the 1960s to the 1990s. They taught themselves all about computers, to be wise about distributing their limited financial resource, or to explore manufacturing possibilities. They dreamed big, as did many underwater robot builders on US campuses, including those talented but less well-resourced Carl Hayden High School students in 2004. Despite that aspirational tinkerers' career successes can be heterogeneous, they will not stop in their attempts to open black-boxed technologies or build their own machines.

ACKNOWLEDGMENTS

This book would not have been possible without help from many individuals and institutions. At the University of Minnesota Twin Cities, I thank all my colleagues in the Program in the History of Science, Technology, and Medicine. Jennifer K. Alexander, Mark E. Borrello, Susan D. Jones, Sally Gregory Kohlstedt, Anna Graber, Jennifer Gunn, Jole R. Shackelford, Jeffrey R. Yost, Victor D. Boantza, Michel Janssen, Matthew Reznicek, and Evan Roberts offered their warmest support during my writing of the book. Colleagues at the Department of Electrical and Computer Engineering welcomed me and offered opportunities for me to share my research with the community. I am especially grateful to Randall Victora and Marc Riedel for their guidance and advice. Mostafa Kaveh, Bethanie Stadler, Chris Kim, James Leger, Jian-Ping Wang, Keshab Parhi, David Orser, Sang-Hyun Oh, Murti Salapaka, Emad Ebbini, Kia Bazargan, Ulya Karpuzcu, Vladimir Cherkassky, Sairaj Dhople, Ramesh Harjani, Sachin Sapatnekar, Rhonda Franklin, Tony Low, Sarah Swisher and other faculty of this remarkable department gave me their kindest support.

Several chapters of the book were first drafted at Cornell University. I am very much indebted to Ronald R. Kline, Trevor J. Pinch, and Suman Seth for their insightful comments, practical advice, and unyielding encouragement throughout this project's early stages and the research and writing stages that followed. Members of the Department of Science & Technology Studies at Cornell provided great support. I extend special thanks to S&TS faculty members Stephen Hilgartner, Michael E. Lynch, Christine Leuenberger, Margaret Rossiter, Sara B. Pritchard, and Rachel Prentice, for their distinctive expertise and perspective that I was able to use for this project. I owe much to my colleagues at Cornell for their incredible support when we were in Ithaca and thereafter: Hannah S. Rogers, Kathryn Vignone, Eunjeong Ma, Lisa Onaga, Nicole Nelson,

Darla Thompson, Yulianto Mohsin, Hansen Hsu, Ling-Fei Lin, Tyson Vaughan, Benyah Shaparenko, Jiwon Yun, Seongyeon Ko, Akiko Ishii, Hajimu Masuda, and Samson Lim.

I appreciate very much the opportunities I had at the National University of Singapore's Tembusu College to share my work with Gregory Clancey, Michael M. J. Fischer, Catelijna Coopmans, John P. DiMoia, Michitake Aso, and Haidan Chen. I owe much to colleagues at Department III of the Max Planck Institute for the History of Science. I thank Francesca Bray, Dagmar Schäfer, Nina E. Lerman, Shih-Pei Chen, Martina Siebert, and Martina Schlünder for their comments on my research for the book. The completion of this book also benefited enormously from discussions with colleagues at Earlham College, especially Tom Hamm, Ryan Murphy, Betsy Schlabach, Elana Passman, Joanna Swanger, and Eric Cunningham.

Many colleagues and friends have also read excerpts of early drafts of various parts of this book, or generously offered their comments on my research for this book on different occasions: Suzanne Moon, Thomas J. Misa, Jeffrey R. Yost, William Aspray, Atsushi Akera, Chigusa Kita, Jessica Ratcliff, Corinna Schlombs, Joseph A. November, John Krige, Stuart W. Leslie, Nelly Oudshoorn, Takashi Nishiyama, Fa-ti Fan, Hiromi Mizuno, Zuoyue Wang, Wen-Hua Kuo, Tsui-hua Yang, Sean Hsiang-lin Lei, Sungook Hong, Lars Heide, Wiebe E. Bijker, Martin Campbell-Kelly, April Hughes, Robert E. Murowchick, Ross Bassett, Thomas Haigh and Jason McGrath. My undergraduate adviser Chia-Ling Wu also offered her critically insightful comments and encouragement over the past years.

I gratefully acknowledge my informants for their trust and patience. They kindly shared with me their invaluable memories, personal letters, and photographs and gave me suggestions about how I might proceed with this project. Each of their stories deserves its own chapter. Many librarians and archivists offered their professional assistance during my research trips. Much appreciated are Stephanie H. Crowe at the Charles Babbage Institute, University of Minnesota; Meiyi Hong and Yu-ai Wu at the National Chiao-Tung University Library; Chang-Wai Yang at National Taiwan University's Documentation Division, and archivists at the Institute of Modern History, Academia Sinica, Taiwan; the National Archives at College Park, Maryland; and the United Nations Archives and Records Management Section, United Nations New York City. I also wish to acknowledge the following grants and fellowships that supported the research and writing of this dissertation: Charles Babbage Institute's Adelle and Erwin Tomash Graduate Fellowship, National Science Foundation's Doctoral Dissertation Research Improvement Grant (award no. 0847981), Cornell University's Ta Chung & Ya Chao Liu Memorial Award, Hu Shih Memorial Award, C.V. Starr Fellowships, and various grants and research funds from Earlham College, Max Planck Institute for the

History of Science, Asia Studies Institute at the National University of Singapore, and the University of Minnesota Twin Cities.

I deeply appreciate Katie Helke's guidance in shepherding the publishing process with the MIT Press. I extend my sincere thanks to Justin Kehoe, Suraiya Jetha and their colleagues in the production team. I owe much to the book series editors Thomas J. Misa and William Aspray for their suggestions and encouragement over the years. I am truly grateful to the anonymous reviewers for their engaging, provocative, and thoughtful comments.

My mother Jin-Rui Wu, my late father Tung-Liang Cheng, and my siblings Yi-Chun and Shao-Gu have given me unyielding support during the stages of research, drafting, and writing of this book. Many thanks to Ian-lin, Melody, and Max Liu, too. My late father brought home a PC in my childhood, possibly a knockoff or the so-called self-assembled computer, so my siblings and I could play games. That was the inspiration of this book project. Mom's first job at an asparagus can factory pushed me to write about factory women since my time in college. The Soon family's warm support has played anything but a small role in my completion of this book. My husband Wayne S. Soon has read the drafts of the book probably too many times—I cannot thank him more for his unwavering support, encouragement, and advice. My son Toby C. Soon—I cannot wait to tell you everything about this book!

APPENDIX: ARCHIVES AND OTHER COLLECTIONS

ARCHIVES

Academia Historica, Taipei, Taiwan (AH)
  Jiang Zhongzheng zongtong wenwu (President Chiang Kai-shek files)
  Jiang Jingguo zongtong wenwu (President Chiang Ching-kuo files)
  Yan Jiagan wenwu 1966–1972 (President Chia-kan Yen files, 1966–1972)
  Yan Jiagan wenwu 1978–1993 (President Chia-kan Yen files, 1978–1993)
  Xie Dongmin zongtong wenwu (Vice President Tung-min Hsieh files)
  Xingzhengyuan Files (Executive Yuan files)

Charles Babbage Institute, University of Minnesota, Minneapolis (CBI)
  Computer Product Manuals Collection (CBI 60)
  Control Data Corporation records, news releases (CBI 80)
  James W. Cortada Papers (CBI 185)
  Oral history interviews by Thomas J. Misa
    Robert M. Price (OH 454)
    John Baxter (OH 448)
    Ron G. Bird (OH 442)
    Lyle Bowden and Tony Blackmore (OH 439)
    George Karoly and Marcel Dayan (OH 446)

Institute of Modern History, Academia Sinica, Taipei, Taiwan
  Ministry of Foreign Affairs Files (MFA Files)

National Archives at College Park, College Park, MD, USA (NARA)
  Research Group 59, General Records of the Department of State, Central Foreign Policy Files, 1964–1966

United Nations Archives and Records Management Section, United Nations, NYC, USA
  S-0175-0300 United Nations technical assistance mission in China, folder 4 TE 322/1 CHI
  S-0133-0020 Project Files: Singapore, etc., folder 5 Taiwan—Hydraulic development projects in the Republic of China (Volumes 1 and 2)—technical reports

S-0146-0006 United Nations Secretariat Project Files—Afghanistan etc., folder 10 China
S-0146-0007 United Nations Secretariat Project Files—Afghanistan etc., folder 6 Republic of China

University Records, National Chiao-Tung University Museum, Hsinchu, Taiwan (NCTU University Records)

University Records, National Taiwan University (NTU), Taipei Taiwan (NTU University Records)

ARCHIVES AND ORAL HISTORY INTERVIEWS AVAILABLE ONLINE

Central Intelligence Agency, Freedom of Information Act Electronic Reading Room (CIA FOIA ERR)

Computer History Museum (CHM)
  Chung Laung Liu, interviewed by Doug Fairbairn (X7178.2014)
  George Comstock, interviewed by Gardner Hendrie (X2727.2004)
  Morris Chang, interviewed by Alan Patterson (X4151.2008)
  Taiwanese IT Pioneers Oral History Interviews
    By Craig Addison
      Chang, Chun-Yen (X6262.2012)
      Miau, Matthew (X6264.2012)
      Shih, Jonney (Chong-Tang) (X6263.2012)
    By Ling-Fei Lin
      Lam, Barry (Pak-Lee) (X6260.2012)
      Lee, K. Y. (Kun-Yao) (X6025.2011)
      Shih, Chintay (X6259.2012)
      Shih, Stan (Chen-Jung) (X6261.2012)
      Tsao, Robert H. C. (Shin-Chen) (X6291.2012)
      Yang, Ding-Yuan (X6290.2012)
      Hu, Ding-Hua (X6289.2012)

Oral History Interviews collected by Engineering and Technology History Wiki (https://ethw.org/)
  Erwin Tomash, by William Aspray; and James Early, by David Hochfelder

National Security Archive
  Electronic Briefing Book No. 221
  Electronic Briefing Book No. 20

LIST OF ORAL HISTORY INTERVIEWS CONDUCTED BY AUTHOR

1. Andrew Chew, August 6, 2007
2. Jong-Chuang Tsay, November 20 and December 18, 2007
3. Dean N. Arden, October 4 and November 10 and 11, 2008
4. G. Conrad Dalman, October 24 and 31, 2008; December 4, 2009; February 8 and April 28, 2010
5. Chi-Chang Lee, January 7 and 14, and February 11, 2009
6. Tseng-Yu Lee, January 7, 2009
7. Pin-Yen Lin, Feb 5, 15, and 25, 2009
8. Yun-Tzong Chen, May 24, 2009

9. Ching-Chun Hsieh, May 27 and June 5, 2009

10. Yuan-Kuang Chen, June 2 and 10, 2009

11. Chin-Chi Kao, June 2, June 17, June 20, and July 5, 2009

12. Henry Y. H. Chuang, June 7, 2009

13. Chiong-Yuan Han, June 15 and 22, 2009

14. Chin-Long Chen, July 1 and 8, 2009

15. Chiang-Chang Huang, August 22, 2009

16. Frank Yih, December 29, 2020

SELECTIONS OF SURVEYED PERIODICALS, TRADE JOURNALS, AND NEWSPAPERS

English
Computerworld
Datamation
Forbes
InfoWorld
Los Angeles Times
New York Times
Nikkei Asian Review
Taipei Times
Telecommunication Journal
Japan Times
Washington Post
Wall Street Journal

Chinese
Jiaoda xuekan (Science Bulletin National Chiao-Tung University)
Jin zhoukan (Business Today)
Jingji zibao (Economic Daily News, EDN)
Kexue yuekan (Science Monthly)
Lianhe bao (United Daily News, UDN)
Shangye zhoukan (Business Weekly)
Taidian gongcheng yuekan (Monthly Journal of Taipower's Engineering)
Tianxia zazhi (Commonwealth Magazine)
Lianhe wanbao (United Evening News)
Yi zhoukan (Next Magazine)
Yousheng (Voice of Chiao-Tung Alumni, Voice)

INSTITUTIONAL REPORTS AND YEARBOOKS
NCTU PUBLICATIONS
The Yearbook of Training Course of Electronic Computers, February–June, 1963
The Yearbook of the Institute of Electronics at NCTU, 1964
The 11th Yearbook of Training Courses of Electronic Computers, November–December 1964

THE UNITED NATIONS' PUBLICATIONS AND REPORTS
Yearbook of the United Nations 1948–49 (1950)
Yearbook of the United Nations 1950 (1951)
Yearbook of the United Nations 1958 (1959)
Training and Research Centre for Telecommunications and Electronics, Republic of China: Report (1968)
The Priorities of Progress (1961)

PROQUEST HISTORICAL ANNUAL REPORTS
Control Data Corporation (CDC) Annual Report, 1972
Digital Equipment Corporation Annual Reports, 1971–1974
Ford Annual Reports, 1965–1967
General Instrument (GI) Corporation Annual Reports, 1965–1972
Lockheed Annual Report, 1968
Motorola Annual Reports, 1969–1971
RCA Annual Reports, 1967, 1969
Texas Instruments Annual Reports, 1967, 1969, 1972, 1975, 1981
TRW Annual Report, 1967

ABBREVIATIONS

CBI   Charles Babbage Institute
CDC   Control Data Corporation
CHM   Computer History Museum
CMOS   complementary metal-oxide semiconductor
CPC   Chinese Petroleum Corporation
CSIST   Chung-Shan Institute of Science and Technology
ERSO   Electronics Research and Service Organization
GI   General Instrument
ITRI   Industrial Technology Research Institute
ITU   International Telecommunication Union
MAAG   Military Assistance Advisory Group
NCTU   National Chiao-Tung University
NTU   National Taiwan University
RCA   Radio Corporation of America
TPC   Taiwan Power Company
TSMC   Taiwan Semiconductor Manufacturing Company
UMC   United Microelectronics Corporation

GLOSSARY OF SELECTED CHINESE NAMES

Chang, Hsien-yi (Xianyi Zhang) 張憲義
Chang, Morris C. M. (Zhongmou Zhang) 張忠謀
Chang, Suimei 張水美
Chen, Chin-Long (Jinlong Chen) 陳晉隆
Chen, Jenn-Nan Chen (Zhennan Chen) 陳振楠
Chen, Zhexiong 陳哲雄
Chen, Y. C. (Yazhi Chen) 陳雅枝
Chen, Yuan-Kuang (Yuanguang Chen) 陳垣光
Cheng, Bruce C. H. (Chonghua Zheng) 鄭崇華
Cheng, David K. (Jun Zheng) 鄭鈞
Chew, Andrew Zaixing (Zaixing Qiu) 邱再興
Chieh, Hua-Ting (Huating Xie) 解化亭
Chien, Gisson Chi-Chen (Qichen Qian) 錢其琛
Chu, Lan Jen (Lancheng Zhu) 朱蘭成
Chuang, Henry Y. H. (Yinghuang Zhuang) 莊英煌
Du, Junren 杜俊人
Duh, Chun-Yuan (Junyuan Du) 杜俊元
Fei, Hua 費驊
Feng, Yuanquan 馮源泉
Ga, Congfu 高琮富
Guo, Terry (Taiming Guo) 郭台銘
Han, Chiong-Yuan (Qiongyuan Han) 韓瓊媛
Han, Kuang-wei (Guangwei Han) 韓光渭
Ho, David Li-wei (Liwei He) 賀立維

Hong, Chien-chuan (Jianquan Hong) 洪建全

Hong, Minhong 洪敏弘

Hsieh, Ching-Chun (Qingjun Xie) 謝清俊

Huang, Benyuan 黃本源

Kao, Chin-Chi (Jinji Gao) 高進吉

Kuh, Ernest Shiu-Jen (Shouren Ge) 葛守仁

Lam, Barry Pak-Lee (Baili Lin) 林百里

Lee, Chi-Chang (Qichang Li) 李其昌

Lee, David S. (Xinlin Li) 李信麟

Lee, K. Y. (Kun-Yao Lee; Kunyao Li) 李焜耀

Lee, Regina (Jingxing Li) 李景星

Lee, S. M. (Ximou Li) 李熙謀

Lee, Tseng-Yu (Zengyu Li) 李曾通

Leung, Chee-Chun (Cizhen Liang) 梁次震

Li, K. T. (Kwoh-ting Li, Guoding Li) 李國鼎

Li, Yongzhao 李永炤

Lin, Chaowu 林朝武

Lin, Hung Chang Jimmy (Hongzhang Lin) 凌宏璋

Lin, Jiahe 林家和

Lin, Liqi 林麗琪

Ling, Chih-Bing (Zhiping Lin) 林致平

Ling, Hung-Hsun (Hongxun Ling) 凌鴻勛

Liu, Mark (Deyin Liu) 劉德音

Liu, Yuli 劉玉梨

Lo, Tsung-lo (Zongluo Luo) 羅宗洛

Lo, Y. C. (Yiqiang Lou) 羅益強

Miau, Matthew F. C. (Fengqiang Miau) 苗豐強

Ou, Zhengming 歐正明

Pan, Wen-Yuan 潘文淵

Shih, Jonney Chong-Tang (Chongtang Shi) 施崇棠

Shih, Stan Chen-Jung (Zhenrong Shi) 施振榮

Shen, Chungli Johnny (Zongli Shen) 沈宗李

Soong, Raymond (Gongyuan Song) 宋恭源

Sun, Yun-suan (Yunxuan Sun) 孫運璿

Sze, Simon Min 施敏

Tang, Chun-po (Junbo Tang) 唐君鉑

Tang, Wang 唐望

Tsai, Ing-wen 蔡英文

Tsao, Robert H. C. (Xingcheng Cao) 曹興誠

Tsao, Tsen-Cha (Cengjue Zhao) 趙曾玨

Tsay, Jong-Chuang (Zhongchuan Cai) 蔡中川

Wang, Bo-bo 王渤渤

Wang, Chao Chen (Zhaozheng Wang) 王兆振

Wang, Gibson (Jisong Wang) 王吉松

Wang, Huang San Samuel 王晃三

Wang, Paul (Boyuan Wang) 王伯元

Wang, Stanley T. (Dazhuang Wang) 王大壯

Wei, C. C. (Zhejia Wei) 魏哲家

Wei, Ling-Yun 魏凌雲

Wei, Yingjiao 魏櫻嬌

Wen, Sayling (Shiren Wen) 溫世仁

Xie, Xianghe 謝翔鶴

Yang, Chao-Chih (Chaozhi Yang) 楊超植

Yang, Ching-chu (Qingchu Yang) 楊青矗

Yang, D. Y. (Ding-Yuan Yang) 楊丁元

Ye, Guoyi 葉國一

Ye, Xiang 葉香

Yeh, Carolyn Chi Hua (Zihua Ye) 葉紫華

Yen, Chia-kan (Jiagan Yan) 嚴家淦

Yih, Frank (Shouzhang Ye) 葉守璋

Yin, Zhihou 殷之浩

Yule, David Ta-wei (Dawei Yu) 俞大維

NOTES

INTRODUCTION

1. Peter S. Goodman, "Quake Threatens Tech Biz: Chip Prices Seen Heading Higher as Taiwan Digs Out from Disaster," *CNN Money*, September 20, 1999, https://money.cnn.com/1999/09/20/worldbiz/taiwan/; Peter S. Goodman, "Tech Stocks Swoon in Wake of Taiwan Quake: Triggered a Broad Sell-Off in Technology Stocks Yesterday," *Washington Post*, September 22, 1999, E1; Wayne Arnold, "Taiwan Quake to Be Costly to World Technology Makers," *New York Times*, September 23, 1999, C4.

2. Jefferson Cowie, *Capital Moves: RCA's Seventy-Year Quest for Cheap Labor* (Ithaca, NY: Cornell University Press, 1999), 96; Anthony Spaeth and Robert King, "Computer Counterfeiting Flourishing as Asians Take a Bigger Bite of Apple II," *Wall Street Journal*, June 25, 1982, p. 23. One of the earliest references to four Asian tigers or dragons can be found in Clive Hamilton, "Capitalist Industrialisation in East Asia's Four Little Tigers," *Journal of Contemporary Asia* 13, no. 1 (1983): 35–73.

3. Eden Medina, Ivan da Costa Marques, and Christina Holmes explain that scholars often overlook "processes of reinvention, adaptation, and use" in Latin America if only focusing on the "imported magic" demonstrated by Western technology. Eden Medina, Ivan da Costa Marques, and Christina Holmes, eds., *Beyond Imported Magic: Essays on Science, Technology, and Society in Latin America* (Cambridge, MA: MIT Press, 2014), 2.

4. National Chiao-Tung University merged with National Yang-Ming University in 2021 and is now known as National Yang Ming Chiao Tung University.

5. Stephen G. Craft, "Islands against the Red Tide," *American Justice in Taiwan* (Lexington: University Press of Kentucky, 2016), 13; Yu-Ping Lin, *Taiwan hangkong gongye shi: zhanzheng yuyi xia de 1935 nian–1979 nian* [The history of aviation industry in Taiwan: In the cradle of wars, 1935–1979] (Taipei: Xinrui Wenchuang, 2011), 135; Joyce Kallgren, "Nationalist China's Armed Forces," *China Quarterly* 15 (September 1963): 35–44.

6. Neil Jacoby, *U.S. Aid to Taiwan: A Study of Foreign Aid, Self-Help, and Development* (New York: F. A. Praeger, 1967), 38.

7. Leonard H. D. Gordon, "United States Opposition to Use of Force in the Taiwan Strait, 1954–1961," *Journal of American History* 72, no. 3 (1985): 637–660; Bennett C. Rushkoff, "Eisenhower, Dulles and the Quemoy-Matsu Crisis, 1954–1955," *Political Science Quarterly* 96, no. 3 (1981): 465–480; Henry R. Lieberman, "First Troops Reach Formosa," *New York Times*, February 12, 1955, p. 3; and Michael Szonyi, "The 1954–5 Artillery War," in *Cold War Island: Quemoy on the Front Line* (New York: Cambridge University Press, 2008), 42–49.

8. Michael Szonyi, "The 1958 Artillery War," in *Cold War Island*, 64–78; George Eliades, "Once More unto the Breach: Eisenhower, Dulles, and Public Opinion during the Offshore Island Crisis of 1958," *Journal of American-East Asian Relations* 2, no. 4 (1993): 343–367; and Mitch Meador, "Veterans Who Defended Taiwan Hold Reunion Here," *Lawton Constitution*, October 17, 2015.

9. Shelley Rigger, *Why Taiwan Matters: Small Island, Global Powerhouse* (Lanham, MD: Rowman & Littlefield, 2011).

10. Brian Russell Roberts, *Borderwaters: Amid the Archipelagic States of America* (Durham, NC: Duke University Press, 2021).

11. Bilahari Kausikan, *Singapore Is Not an Island: Views on Singapore Foreign Policy* (Singapore: Straits Times Press, 2019), and *Singapore Is Still Not an Island: More Views on Singapore Foreign Policy* (Singapore: Straits Times Press, 2023).

12. Victor Petrov, *Balkan Cyberia: Cold War Computing, Bulgarian Modernization, and the Information Age behind the Iron Curtain* (Cambridge, MA: MIT Press, 2023).

13. Michael S. Mahoney, "The History of Computing in the History of Technology," *IEEE Annals of the History of Computing* 10, no. 2 (1988): 113–125; William F. Aspray, "The History of Computing within the History of Information Technology," *History and Technology* 11 (1994): 7–19; Paul N. Edwards, *The Closed World: Computers and the Politics of Discourse in Cold War America* (Cambridge, MA: MIT Press, 1996), and "From 'Impact' to Social Process: Computers in Society and Culture," in *Handbook of Science and Technology Studies*, ed. Sheila Jasanoff, Gerald E. Markle, James Petersen, and Trevor Pinch (Thousand Oaks, CA: Sage, 1995), 257–285; Atsushi Akera, *Calculating a Natural World: Scientists, Engineers, and Computers during the Rise of U.S. War Research* (Cambridge, MA: MIT Press, 2007); Jon Agar, *The Government Machine: A Revolutionary History of the Computer* (Cambridge, MA: MIT Press, 2003).

14. Larry Owens, "Where Are We Going, Phil Morse? Changing Agendas and the Rhetoric of Obviousness in the Transformation of Computing at MIT, 1939–1957," *IEEE Annals of the History of Computing* 18, no. 2 (1996): 34–41; David A. Mindell, *Between Human and Machine: Feedback, Control, and Computing before Cybernetics* (Baltimore, MD: Johns Hopkins University Press, 2002); David A. Grier, *When Computers Were Human* (Princeton, NJ: Princeton University Press, 2005); Akera, *Calculating a Natural World*; Andrew Hodges, *Alan Turing: The Enigma* (New York: Touchstone, 1983); Ronald R. Kline, *The Cybernetics Moment: Or Why We Call Our Age the Information Age* (Baltimore, MD: Johns Hopkins University Press, 2015).

15. Paul Forman, "Behind Quantum Electronics: National Security as Basis for Physical Research in the United States, 1940–1960," *Historical Studies in the Physical and Biological Sciences* 18, no. 1

(1987): 149–229; Stuart W. Leslie, *The Cold War and American Science: The Military-Industrial-Academic Complex at MIT and Stanford* (New York: Columbia University Press, 1993); Edwards, *Closed World*; Jennifer Light, *From Warfare to Welfare: Defense Intellectuals and Urban Problems in Cold War America* (Baltimore, MD: Johns Hopkins University Press, 2003); Akera, *Calculating a Natural World*; Janet Abbate, *Inventing the Internet* (Cambridge, MA: MIT Press, 2000); Thomas J. Misa, *Digital State: The Story of Minnesota's Computing Industry* (Minneapolis: University of Minnesota Press, 2013).

16. See Edwards, *Closed World*; Abbate, *Inventing the Internet*; Fred Turner, *From Counterculture to Cyberculture: Stewart Brand, the Whole Earth Network, and the Rise of Digital Utopianism* (Chicago: University of Chicago Press, 2006); Akera, *Calculating a Natural World*; Thomas P. Hughes, *Rescuing Prometheus* (New York: Vintage Books, 2006); Paul Erickson, Judy L. Klein, Lorraine Daston, Rebecca Lemov, Thomas Sturm, and Michael D. Gordin, *How Reason Almost Lost Its Mind* (Chicago: University of Chicago Press, 2013). My definition of technological system draws from Thomas P. Hughes, *Networks of Power: Electrification in Western Society, 1880–1930* (Baltimore, MD: Johns Hopkins University Press, 1983); "Edison and Electric Light," in *The Social Shaping of Technology: How the Refrigerator Got Its Hum*, ed. Donald MacKenzie and Judy Wajcman (Philadelphia, PA: Open University Press, 1999 [1985]), 50–63; and "The Evolution of Large Technological Systems," in *The Social Construction of Technological Systems*, ed. Wiebe Bijker, Thomas P. Hughes, and Trevor J. Pinch (Cambridge, MA: MIT Press, 1987), 50–80.

17. The social and institutional contexts of the Cold War are especially critical in understanding innovations and the business history of computing. See, e.g., Turner, *From Counterculture to Cyberculture*; Ross Knox Bassett, *To the Digital Age: Research Labs, Start-Up Companies, and the Rise of MOS Technology* (Baltimore, MD: Johns Hopkins University Press, 2002); Christophe Lécuyer, *Making Silicon Valley: Innovation and the Growth of High Tech, 1930–1970* (Cambridge, MA: MIT Press, 2006). For examples, beyond the Cold War, that illustrate the co-construction of computing technology and the social changes, see Simon Schaffer, "Babbage's Intelligence: Calculating Engines and the Factory System," *Critical Inquiry* 21, no. 1 (1994): 203–227; and Lorraine Daston, "Enlightenment Calculations," *Critical Inquiry* 21, no. 1 (1994): 182–202.

18. Benjamin Peters, *How Not to Network a Nation: The Uneasy History of the Soviet Internet* (Cambridge, MA: MIT Press, 2016); Slava Gerovitch, "'Mathematical Machines' of the Cold War: Soviet Computing, American Cybernetics and Ideological Disputes in the Early 1950s," *Social Studies of Science* 31, no. 2 (2001): 253–287; *From Newspeak to Cyberspeak: A History of Soviet Cybernetics* (Cambridge, MA: MIT Press, 2002).

19. For computers' central role in unique Cold War–structured, non-US societal contexts, see Eden Medina, *Cybernetic Revolutionaries: Technology and Politics in Allende's Chile* (Cambridge, MA: MIT Press, 2011); Corinna Schlombs, "Toward International Computing History," *IEEE Annals of the History of Computing* 28, no. 1 (2006): 107–108; and *Productivity Machines: German Appropriations of American Technology from Mass Production to Computer Automation* (Cambridge, MA: MIT Press, 2019); Jeffrey R. Yost, ed., "A World of Computers," Special Issue of *IEEE Annals of the History of Computing* 30, no. 4 (2008); Ivan da Costa Marques, ed., "History of Computing in Latin America," *IEEE Annals of the History of Computing* 37, no. 4 (2015); Ross Knox Bassett, "Aligning India in the Cold War Era: Indian Technical Elites, the Indian Institute of Technology at

Kanpur, and Computing in India and the United States," *Technology and Culture* 50, no. 4 (2009): 783–810; Ross Bassett, *The Technological Indian* (Cambridge, MA: Harvard University Press, 2016); Dinesh C. Sharma, *The Outsourcer: The Story of India's IT Revolution* (Cambridge, MA: MIT Press, 2015); Michael Homberg, "Digital India: Swadeshi-Computing in India since 1947," in *Prophets of Computing: Visions of Society Transformed by Computing*, ed. Dick van Lente (New York: ACM Press, 2022), 279–323; Fabian Prieto-Ñañez, "Assembling a Colombian-Cloned Computer: National Development and the Transnational Trade of Electronics Parts in the 1980s," *IEEE Annals of the History of Computing* 44, no. 2 (2022): 55–64; and Colette Perold, "IBM's World Citizens: Valentim Bouças and the Politics of IT Expansion in Authoritarian Brazil," *IEEE Annals of the History of Computing* 42, no. 3 (2020): 38–52. For the history of computers in Japan, see Hidetosi Takahasi, "Some Important Computers of Japanese Design," *IEEE Annals of the History of Computing* 2, no. 4 (1980): 330–337; Sigeru Takahashi, "Early Transistor Computers in Japan," *IEEE Annals of the History of Computing* 8, no. 2 (1986): 144–154; and "A Brief History of the Japanese Computer Industry before 1985," *IEEE Annals of the History of Computing* 18, no. 1 (1996): 76–79; Shigeru Takahashi, "The Rise and Fall of Plug-Compatible Mainframes," *IEEE Annals of the History of Computing* 27, no. 1 (2005): 4–16; Chigusa Kita, "From Technological Mimesis to Creativity: Early Online Rail Reservations in Japan," paper presented at the Annual Meeting of the Society for the History of Technology, Washington, DC, October 2007, pp. 17–21; Hyungsub Choi and Chigusa Kita, "Hiroshi Wada: Pioneering Electronics and Computer Technologies in Postwar Japan," *IEEE Annals of the History of Computing* 30, no. 3 (2008): 84–89; Hyungsub Choi, "Technology Importation, Corporate Strategies, and the Rise of the Japanese Semiconductor Industry in the 1950s," *Comparative Technology Transfer and Society* 6, no. 2 (2008): 103–126; William F. Aspray, "The Intel 4004 Microprocessor: What Constituted Invention?" *IEEE Annals of the History of Computing* 19, no. 3 (1997): 4–15.

20. For computers and the military-industrial-academic complex, see, e.g., Leslie, *Cold War and American Science*; Edwards, *Closed World*; Light, *From Warfare to Welfare*; Akera, *Calculating a Natural World*; and Abbate, *Inventing the Internet*.

21. On US dominance or hegemony that prompted scientific and technological exchanges among the United States and its allies during the Cold War, see Michael Adas, *Dominance by Design: Technological Imperatives and America's Civilizing Mission* (Cambridge, MA: Belknap Press, 2006); John Krige, *American Hegemony and the Postwar Reconstruction of Science in Europe* (Cambridge, MA: MIT Press, 2006). Scholars have argued that international organizations also constructed the portability of experts and expert knowledge; see Donna Mehos and Suzanne Moon, "The Uses of Portability: Circulating Experts and the Technopolitics of Cold War and Decolonization," in *Entangled Geographies: Empire and Technopolitics in the Global Cold War*, ed. Gabrielle Hecht (Cambridge, MA: MIT Press, 2011), 43–74. For transnational knowledge connections with US institutions, scientists, and engineers, see John P. DiMoia, *Reconstructing Bodies: Biomedicine, Health, and Nation-Building in South Korea Since 1945* (Stanford, CA: Stanford University Press, 2013); Bassett, *Technological Indian*.

22. AnnaLee Saxenian's study has revealed the connections since the 1980s between technologists in Taiwan and their Silicon Valley counterparts. The connections greatly contributed to the development of high-tech industry in each of the two places. She argues that Taiwan's electronics

companies are the siblings, parents, and partners of many Silicon Valley companies. My book demonstrates that the network between Taiwanese and US technologists has been in formation since the 1950s and has continued to shape the development of computing activities and computer manufacturing from then to the 2020s. See Saxenian, *The New Argonauts: Regional Advantage in a Global Economy* (Cambridge, MA: Harvard University Press, 2007), chaps. 3–4.

23. See James W. Cortada, *The Digital Flood: The Diffusion of Information Technology across the U.S., Europe, and Asia* (New York: Oxford University Press, 2014); and Aspray, "International Diffusion of Computer Technology, 1945–1955." For the broader scholarship on technology transfer beyond computers, see Bruce Seely, "Historical Patterns in the Scholarship of Technology Transfer," *Comparative Technology Transfer and Society* 1, no. 1 (2003): 7–48.

24. Chalmers Johnson, *MITI and the Japanese Miracle: The Growth of Industrial Policy, 1925–1975* (Stanford, CA: Stanford University Press, 1982); Meredith Jung-En Woo-Cumings, "National Security and the Rise of the Developmental State in South Korea and Taiwan," in *Behind East Asian Growth*, ed. Henry Rowen (New York: Routledge, 1998), 319–337; Bruce Cumings, "Web with No Spider, Spider with No Web: The Genealogy of the Developmental State," in *The Developmental State*, ed. Meredith Woo-Cumings (Ithaca, NY: Cornell University Press, 2000), 61–92; William C. Kirby, "Engineering China: The Origins of the Chinese Developmental State," in *Becoming Chinese*, ed. Wen-hsin Yeh (Berkeley: University of California Press, 2000), 137–160. Another school of scholars considers development as a set of the ideologies, institutions, apparatuses, ideas, academic knowledge production, and identities surrounding "development"; see James Ferguson, *The Anti-Politics Machine: Development, Depoliticization, and Bureaucratic Power in Lesotho* (Minneapolis: University of Minnesota Press, 1994); Akhil Gupta, *Postcolonial Developments: Agriculture in the Making of Modern India* (Durham, NC: Duke University Press, 1998); David C. Engerman, "The Romance of Economic Development and New Histories of the Cold War," *Diplomatic History* 28, no. 1 (2004): 23–54; Nils Gilman, "Modernization Theory: the Highest Stage of American Intellectual History," in *Staging Growth: Modernization, Development, and the Global Cold War*, ed. David Engerman, Nils Nilman, Mark H. Haefele, and Michael E. Latham (Amherst: University of Massachusetts Press, 2003), 251–270; Scott O'Bryan, *The Growth Idea: Purpose and Prosperity in Postwar Japan* (Honolulu: University of Hawai'i Press, 2009); Arturo Escobar, *Encountering Development: The Making and Unmaking of the Third World* (Princeton, NJ: Princeton University Press, 1995); Michael E. Latham, *Modernization as Ideology: American Social Science and "Nation Building" in the Kennedy Era* (Chapel Hill, NC: University of North Carolina Press, 2000), chap. 2.

25. On Taiwanese governmental officials and science and industrial policies, see Thomas B. Gold, *State and Society in the Taiwan Miracle* (Armonk, NY: M. E. Sharpe, 1986); and J. Megan Greene, *The Origins of the Developmental State in Taiwan: Science Policy and the Quest for Modernization* (Cambridge, MA: Harvard University Press, 2008). On the government's role in stimulating the growth of the computer industry after the 1970s, see Chung-Shing Lee and Michael Pecht, *The Taiwan Electronics Industry* (Boca Raton, FL: CRC Press, 1997); Rong-I Wu and Ming-Shen Tseng, "Taiwan's Information Technology Industry," in *Manufacturing Competitiveness in Asia: How Internationally Competitive National Firms and Industries Developed in East Asia*, ed. Jomo K. S. (New York: Routledge, 2003); Willy C. Shih, and Jyun-Cheng Wang, "Upgrading the Economy: Industrial Policy and Taiwan's Semiconductor Industry," Harvard Business School Case 609-089,

February 2009, revised December 2010; National Research Council, *Securing the Future: Regional and National Programs to Support the Semiconductor Industry.* (Washington, DC: The National Academies Press, 2003).

26. Concerned with the future of Taiwan or "Free China," technologists held ambivalent attitudes toward the Nationalist regime led by Chiang Kai-shek and his son Chiang Ching-kuo. In contrast, engineers in Lino Camprubí's study were given the opportunity by the autocratic state to become "mid-level decision makers." Lino Camprubí, *Engineers and the Making of the Francoist Regime* (Cambridge MA: MIT Press, 2014). For recent scholarly work that focuses on engineers, see, e.g., Matthew Wisnioski, *Engineers for Change: Competing Visions of Technology in 1960s America* (Cambridge, MA: MIT Press, 2012); Takashi Nishiyama, *Engineering War and Peace in Modern Japan, 1868–1964* (Baltimore, MD: Johns Hopkins University Press, 2014); Bassett, *Technological Indian.* For technopolitics, see Gabrielle Hecht, *Radiance of France*, 15.

27. James Lin's work especially demonstrates that Taiwanese scientists and technocrats worked together and contributed to agricultural development in the global south. See Lin, "Sowing Seeds and Knowledge: Agricultural Development in Taiwan and the World, 1925–1975," *East Asian Science, Technology and Society: An International Journal* 9, no. 2 (2015): 127–149; *In the Global Vanguard: Agrarian Development and the Making of Modern Taiwan* (Berkeley: University of California Press, forthcoming, 2025).

28. George Basalla, "The Spread of Western Science," *Science* 156, no. 3775 (1967): 611–622; Joseph Needham, "Poverties and Triumphs of the Chinese Scientific Tradition," in *The "Racial" Economy of Science: Toward a Democratic Future*, ed. Sandra Harding (Bloomington: Indiana University Press, 1993 [1969]), 30–46; Charles Coulston Gillispie, *The Edge of Objectivity: An Essay in the History of Scientific Ideas* (Princeton, NJ: Princeton University Press, 1960).

29. Francesca Bray, *Technology and Gender: Fabrics of Power in Late Imperial China* (Berkeley: University of California Press, 1997); Dagmar Schäfer, *The Crafting of the 10,000 Things: Knowledge and Technology in Seventeenth-Century China* (Chicago: University of Chicago Press, 2011); Yulia Frumer, *Making Time: Astronomical Time Measurement in Tokugawa Japan* (Chicago: University of Chicago Press, 2018); Fa-ti Fan, *British Naturalists in Qing China: Science, Empire, and Cultural Encounter* (Cambridge, MA: Harvard University Press, 2004); Gregory Clancey, *Earthquake Nation: The Cultural Politics of Japanese Seismicity, 1868–1930* (Berkeley: University of California Press, 2006); Hiromi Mizuno, *Science for the Empire: Scientific Nationalism in Modern Japan* (Stanford, CA: Stanford University Press, 2009); Daqing Yang, "Colonial Korea in Japan Imperial Telecommunications Network," in *Colonial Modernity in Korea*, ed. Gi-Wook Shin and Michael Edson Robinson (Cambridge, MA: Harvard University Asia Center, 1999), 161–188; Aaron Stephen Moore, *Constructing East Asia: Technology, Ideology, and Empire in Japan's Wartime Era, 1931–1945* (Stanford, CA: Stanford University Press, 2013); Jung Lee, "Invention without Science: 'Korean Edisons' and the Changing Understanding of Technology in Colonial Korea," *Technology and Culture* 54, no. 4 (2013): 782–814; Gregory Clancey, "Hygiene in a Landlord State: Health, Cleanliness and Chewing Gum in Late Twentieth Century Singapore," *Science, Technology and Society* 23, no. 2 (2018): 214–233; Zuoyue Wang, "Saving China through Science: The Science Society of China, Scientific Nationalism, and Civil Society in Republican

China," *Osiris* 17 (2002): 291–322; Thomas Mullaney, *The Chinese Typewriter: A History* (Cambridge, MA: MIT Press, 2017); Jing Tsu, "Chinese Scripts, Codes, and Typewriting Machines," in *Science and Technology in Modern China, 1880s–1940s*, ed. Jing Tsu and Benjamin Elman (Leiden: Brill, 2014), 115–151; Wayne Soon, "Making Blood Banking Work," *Global Medicine in China: A Diasporic History* (Stanford, CA: Stanford University Press, 2020), 95–124; Ying Jia Tan, *Recharging China in War and Revolution, 1882–1955* (Ithaca, NY: Cornell University Press, 2021); Victor Seow, *Carbon Technocracy: Energy Regimes in Modern East Asia* (Chicago: University of Chicago Press, 2021); J. Megan Greene, *Building a Nation at War: Transnational Knowledge Networks and the Development of China during and after World War II* (Cambridge MA: Harvard University Press, 2022)

30. Daiwie Fu, "How Far Can East Asian STS Go? A Position Paper," *East Asian Science, Technology and Society: An International Journal* 1, no. 1 (2007): 1–14; Fa-ti Fan, "East Asian STS: Fox or Hedgehog?" *East Asian Science, Technology and Society: An International Journal* 1 (2007): 243–248; Warwick Anderson, "How Far Can East Asians STS Go? A Commentary," *East Asian Science, Technology and Society: An International Journal* 1, no. 2 (2007): 249–250 (250); Fa-ti Fan, "Modernity, Region, and Technoscience: One Small Cheer for Asia as Method," *Cultural Sociology* 10, no. 3 (2016): 352–368; Michael M. J. Fischer, "Anthropological STS in Asia," *Annual Review of Anthropology* 45 (2016): 181–198 (esp. 181); Michael M. J. Fischer, "Theorizing STS from Asia—Toward an STS Multiscale Bioecology Framework: A Blurred Genre Manifesto/Agenda for an Emergent Field," *East Asian Science, Technology and Society: An International Journal* 12, no. 4 (2018): 519–540; Warwick Anderson, "STS with East Asian Characteristics?" *East Asian Science, Technology and Society: An International Journal* 14, no. 1 (2020): 163–168.

31. Brooke Hindle, *Emulation and Invention* (New York: New York University Press, 1981).

32. Rupert Cox, "Introduction," in *The Culture of Copying in Japan: Critical and Historical Perspectives*, ed. Rupert Cox (London: Routledge, 2007), 1–17; and Coaldrake, "Beyond Mimesis: Japanese Architectural Models at the Vienna Exhibition and 1910 Japan British Exhibition," in *Culture of Copying in Japan*, 199–212.

33. For example, see Warwick Anderson and Vincanne Adams, "Pramoedya's Chickens: Postcolonial Studies of Technoscience," in *The Handbook of Science and Technology Studies*, ed. Edward J. Hackett, Olga Amsterdamska, Michael Lynch, and Judy Wajcman (Cambridge, MA: MIT Press, 2007), 181–203; Ruth Rogaski, "Deficiency and Sovereignty," *Hygienic Modernity: Meanings of Health and Disease in Treaty-Port China* (Berkeley: University of California Press, 2004), 165–192; Moore, *Constructing East Asia*.

34. For a discussion of the rejection of incommensurability, see Suman Seth, "Putting Knowledge in Its Place: Science, Colonialism, and the Postcolonial," *Postcolonial Studies* 12, no. 4 (2009): 373–388; Itty Abraham, *The Making of the Indian Atomic Bomb—Science, Secrecy and the Postcolonial State* (London: Zed Books, 1998). On consumer interpretations of artifacts, Chinese consumers in the Republican Era were not necessarily loyal to either Chinese or Western material cultures; see Frank Dikötter, *Exotic Commodities: Modern Objects and Everyday Life in China* (New York: Columbia University Press, 2006).

35. The school of the social construction of technology (SCOT) has stressed the critical role of the widely neglected users in social and technological change. See Trevor J. Pinch and Wiebe Bijker, "The Social Construction of Facts and Artifacts: Or, How the Sociology of Science and the Sociology of Technology Might Benefit Each Other," in *The Social Construction of Technological Systems*, ed. Wiebe Bijker, Thomas P. Hughes, and Trevor J. Pinch (Cambridge, MA: MIT Press, 1987 [1984]), 17–50; Ronald R. Kline and Trevor J. Pinch, "Users as Agents of Technological Change: The Social Construction of the Automobile in the Rural United States," *Technology and Culture* 37, no. 4 (1996): 763–795; Nelly Oudshoorn and Trevor J. Pinch, "User-Technology Relationships: Some Recent Developments," in *The Handbook of Science and Technology Studies*, 3rd ed., ed. Edward J. Hackett, Olga Amsterdamska, Michael Lynch, and Judy Wajcman (Cambridge, MA: MIT Press, 2007), 541–566; Oudshoorn and Pinch, "How Users and Non-Users Matter," in *How Users Matter: The Co-Construction of Users and Technologies*, ed. Oudshoorn and Pinch (Cambridge, MA: MIT Press, 2003), 1–25. On amateurs and hobbyists, see Susan Douglas, *Inventing American Broadcasting, 1899–1922* (Baltimore, MD: Johns Hopkins University Press, 1989); Kristen Haring, *Ham Radio's Technical Culture* (Cambridge, MA: MIT Press, 2006). See also Zbigniew Stachniak, "Red Clones: The Soviet Computer Hobby Movement of the 1980s," *IEEE Annals of the History of Computing* 37, no. 1 (2015): 12–23. On invisible labor, see Gregory J. Downey "Virtual Webs, Physical Technologies, and Hidden Workers: The Spaces of Labor in Information Internetworks," *Technology and Culture* 42, no. 2 (2001): 209–235; and *Telegraphy Messenger Boys: Labor, Technology, and Geography 1850–1950* (New York: Routledge, 2002). On mediators, see Trevor J. Pinch, "Giving Birth to New Users: How the Minimoog Was Sold to Rock and Roll," in *How Users Matter*, 247–270; Atsushi Akera, "Voluntarism and Occupational Identity," in *Calculating a Natural World*, 249–274; Carolyn M. Goldstein, "From Service to Sales: Home Economics in Light and Power, 1920–1940," *Technology and Culture* 38, no. 1 (1997): 121–115; Ronald R. Kline, *Consumers in the Country: Technology and Social Change in Rural America* (Baltimore, MD: Johns Hopkins University Press, 2000); Ronald R. Kline, "Agents of Modernity: Home Economists and Rural Electrification in the United States, 1925–1950," in *Rethinking Women and Home Economics in the 20th Century*, ed. Sara Stage and Virginia Vincenti (Ithaca, NY: Cornell University Press, 1997), 237–252; Joshua M. Greenberg, *From Betamax to Blockbuster: Video Stores and the Invention of Movies on Video* (Cambridge MA: MIT Press, 2008); Christina Lindsay, "From the Shadows: Users as Designers. Producers, Marketers, Distributors, and Technical Support," in *How Users Matter*, 29–50; Rachel Maines, *Hedonizing Technology: Paths to Pleasure in Hobbies and Leisure* (Baltimore, MD: Johns Hopkins University Press, 2009). On tinkering, see Yuzo Takahashi, "A Network of Tinkerers: The Advent of the Radio and Television Receiver Industry in Japan," *Technology and Culture* 41, no. 3 (2000): 460–484; Kathleen Franz, *Tinkering: Consumers Reinvent the Early Automobile* (Philadelphia: University of Pennsylvania Press, 2005); David Edgerton, *The Shock of the Old: Technology and Global History since 1900* (New York: Oxford University Press, 2007). On maintainers, see Lee Vinsel and Andrew L. Russell, *The Innovation Delusion: How Our Obsession with the New Has Disrupted the Work That Matters Most* (New York: Currency, 2020); Jeffrey R. Yost, "Where Dinosaurs Roam and Programmers Play: Reflections on Infrastructure, Maintenance, and Inequality," *Interfaces: Essays and Reviews in Computing and Culture* 1 (May 2020), https://cse.umn.edu/cbi/past-interfaces; and Tisha Y. Hooks, "Duct Tape and the U.S. Social Imagination" (PhD diss., Yale University, 2015).

36. See, for example, Myles W. Jackson, *Harmonious Triads: Physicists, Musicians, and Instrument Makers in Nineteenth-Century Germany* (Cambridge MA: MIT Press, 2008); Victor D. Boantza, "The Rise and Fall of Nitrous Air Eudiometry: Enlightenment Ideals, Embodied Skills, and the Conflicts of Experimental Philosophy," *History of Science* 51, no. 4 (2013): 377–412; Hyeok Hweon Kang, "Cooking Niter, Prototyping Nature: Saltpeter and Artisanal Experiment in Korea, 1592–1635," *Isis: A Journal of the History of Science Society* 113, no. 1 (2022): 1–21; Simon Werrett, *Thrifty Science: Making the Most of Materials in the History of Experiment* (Chicago: University of Chicago Press, 2019).

37. For a discussion of the limits of garage-based entrepreneurship in explaining the success of founders of contemporary Silicon Valley giants, see Martin Campbell-Kelly et al., "The Shaping of the Personal Computer," in *Computer: A History of the Information Machine*, 4th ed. (New York: Routledge, 2023), 235–255.

38. Turner, *From Counterculture to Cyberculture*; Atsushi Akera, *Calculating a Natural World*; Steven Levy, *Hackers: Heroes of the Computer Revolution* (New York: Penguin, 1994 [1984]). *Hacking Europe: From Computer Cultures to Demoscenes* is also an example in which "playfulness" is the critical lens for understanding how European computer users adopted microcomputers; see Gerard Alberts and Ruth Oldenziel, eds., *Hacking Europe: From Computer Cultures to Demoscenes* (New York: Springer, 2014). For a closer analysis of the Homebrew Computer Club in the United States, see Elizabeth Petrick, "Imagining the Personal Computer: Conceptualizations of the Homebrew Computer Club 1975–1977," *IEEE Annals of the History of Computing* 39, no. 4 (2017): 27–39.

39. Lisa Onaga, "Silkworms, Science, and Nation: A Sericultural History of Genetics in Modern Japan" (PhD diss., Cornell University, 2012); and Clarissa Ai Ling Lee, "Malaysian Physics and the Maker Ethos," *Physics Today* 76, no. 2 (2023): 32–38.

40. David E. Nye, *American Technological Sublime* (Cambridge, MA: MIT Press, 1994).

41. Kline and Pinch, "Users as Agents of Technological Change"; Werrett, *Thrifty Science*.

42. Edgerton, *Shock of the Old*.

43. Joshua Grace, *African Motors: Technology, Gender, and the History of Development* (Durham, NC: Duke University Press, 2022); Chung-hsi Lin, "Jinsheng de jishu—pinzhuang che de meili yu aichou" [The Technology of Silence—The Beauty and Sorrow of Reassembled Car], *Keji bowu* [Technology Museum Studies] 6, no. 4 (2002): 34–58; Atsuro Morita, "Traveling Engineers, Machines, and Comparisons: Intersecting Imaginations and Journeys in the Thai Local Engineering Industry," *East Asian Science, Technology, and Medicine* 7 no. 2 (2013): 221–241.

44. Eugenia Lean, *Vernacular Industrialism in China: Local Innovation and Translated Technologies in the Making of a Cosmetics Empire, 1900–1940* (New York: Columbia University Press, 2020); Paulina Hartono, "Do Radios Have Politics? The Politics of Radio Ownership in China in the 1920s and 1930s" working paper for the Consortium for History of Science and Technology, December 17, 2019, https://www.chstm.org/content/history-technology?page=2; Takahashi, "A Network of Tinkerers."

45. Ivan da Costa Marques, "Cloning Computers: From Rights of Possession to Rights of Creation," *Science as Culture* 14, no. 2 (2005): 139–160; Jenna Burrell, *Invisible Users: Youth in the*

Internet Cafés of Urban Ghana (Cambridge MA: MIT Press, 2012); Anita Say Chan, *Networking Peripheries: Technological Futures and the Myth of Digital Universalism* (Cambridge, MA: MIT Press, 2014); and Lilly U. Nguyen, "Infrastructural Action in Vietnam: Inverting the Techno-Politics of Hacking in the Global South," *New Media & Society* 18, no. 4 (2016): 637–652. Fabian Prieto-Ñañez also researched public expectations of locally made inexpensive microcomputers in Colombia in the 1980s; see "Assembling a Colombian-Cloned Computer." For replicating US computers in the Soviet Union, see Ksenia Tatarchenko, "'Our Past Is Their Future': Rip-Off, Translation, and Trust in the Soviet-American Computing Exchanges during the Cold War," unpublished paper presented at Appreciating Innovation across Countries, Copenhagen Business School, Copenhagen, Denmark, November 2015, pp. 5–6.

46. Silvia M. Lindtner, *Prototype Nation: China and the Contested Promise of Innovation* (Princeton, NJ: Princeton University Press, 2000); Byung-Chul Han, *Shanzhai: Deconstruction in Chinese*, trans. Philippa Hurd (Cambridge, MA: MIT Press, 2017); Ravi Sundaram, *Pirate Modernity: Delhi's Media Urbanism* (New York: Routledge, 2010). See also Homi Bhabha, *The Location of Culture* (New York: Routledge, 1994). For scholars' critiques of the notion of innovation, see Lilly Irani, *Chasing Innovation: Making Entrepreneurial Citizens in Modern India* (Princeton, NJ: Princeton University Press, 2019); Kavita Philip, Lilly Irani, and Paul Dourish, "Postcolonial Computing: A Tactical Survey," *Science, Technology, & Human Values* 37, no. 1 (2012): 3–29; and Fabian Prieto-Ñañez, "Postcolonial Histories of Computing," *IEEE Annals of the History of Computing* 38, no. 2 (2016): 2–4. For discussion on the controversial nature of defining innovations in non-Western contexts, see Medina, da Costa Marques, and Holmes, *Beyond Imported Magic*; Clapperton Chakanetsa Mavhunga, ed., *What Do Science, Technology, and Innovation Mean from Africa?* (Cambridge, MA: MIT Press, 2017).

47. "How Taiwan Missed the Boat to the American Micro Market," *BusinessWeek*, September 24, 1984, pp. 110B–110E; and James McGregor, "IBM Uses Clout to Force Taiwan Cloners to Cough Up Royalties on Its PC Patents," *Wall Street Journal*, September 7, 1988, p. 1.

48. Paul S. Otellini, "Asian Heroes: Stan Shih," *Time*, November 13, 2006, https://content.time .com/time/subscriber/article/0,33009,1554992,00.html.

CHAPTER 1

1. "Fulu er: Tongxuehui benjie zhiyuan lu" [Appendix 2: Council members for Chiao-Tung Alumni Association], *Yousheng* [*The Voice of Chiao-Tung Alumni*, hereafter *Voice*] 1 (April 1952): 10; "Muxiao xiaoqing ceji" [About the celebration of our alma mater's anniversary], *Voice* 2 (May 1952): 8.

2. Morris Chang was born in Zhejiang in 1931. His father, a bank manager, had worked in various locations in China and British colonial Hong Kong; Chang's family thus moved often. Chang ended up attending the Nanyang Model High School in Shanghai for his sophomore and senior years in the 1940s. Morris Chang, *Zhang Zhongmou zichuan, shang, 1931–1964* [Morris Chang autobiography, volume I, 1931–1964] (Taipei: Yuanjian Tianxia [Global Views–Commonwealth], 2020).

3. "Many Functions to Mark 30th Commencement of Chiao-Tung University," *China Press*, June 30, 1930.

4. Jian Wei, "Rujing wensu" [When in Rome, do as the Romans do], *Voice* 51 (February 1957): 10–11.

5. Tingli guanzhu (Wang Guang's pen name), "Benjie xiaoqing nianhui jilu wenxian" [Compilations of documents celebrating alma mater's anniversary], *Voice* 11 (May 1953): 4–11; "Nianhui Nianhui Nianhui" [Anniversary, anniversary, anniversary], *Voice* 8 (February 1953): 16.

6. See J. Megan Greene, *Origins of the Developmental State*, 14–46; Nick Cullather, "'Fuel for the Good Dragon': The United States and Industrial Policy in Taiwan," *Diplomatic History* 20, no. 1 (1996): 8–9; and Tsui-hua Yang, "Hu Shi dui Taiwan kexue fazhan de tuidong" [Promoting Science and Scholarship in Taiwan: Hu-Shih's Dream of "Academic Independence"], *Hanxue yanjiu* [Chinese studies] 20, no. 2 (2002): 327–352.

7. Wang Guang, "Xinzhi Jiaotong daxue de zouyi yu lunkuo" [Proposal for a new Chiao-Tung University], *Voice* 2 (May 1952): 1. For the United States' support for the Free University, see Jeremi Suri, "The Cultural Contradictions of Cold War Education: The Case of West Berlin," *Cold War History* 4, no. 3 (2004): 1–20; and William C. Kirby, *Empires of Ideas: Creating the Modern University from Germany to America to China* (Cambridge, MA: Harvard University Press, 2022), chap. 3.

8. "Diliujie disanci jianlishi fuxiao choubei weiyuanhui disanci huiyi lianxi huiyi jilu" [The combined meeting for the third meeting of the sixth executive committee of Chiao-Tung Alumni Association, and the third meeting on the Preparatory Committee for Re-founding the Alma Mater], *Voice* 8 (February 1953): 7–9.

9. "Sishi'er niandu diyi i lijianshi ganshi lianxi huiyi jilu" [Minutes from the first meeting of the Executive Committee of 1953], *Voice* 11 (June 1953): 11–13. For Taiwan's military expenses in the 1950s, see footnote 6.

10. Fei Hua, "Fahui jiaoda guyou de jingshen" [Revive Chiao-Tung's spirit], *Voice* 6 (September 1952): 1; Gisson Chi-Chen Chien, "Peiyang jiaotong rencai wei dangqian zhi jiwu" [The urgency to train communications personnel], *Voice* 13 (August 1953).

11. "Sishi'er niandu diyi i lijianshi ganshi lianxi huiyi jilu"; and "Disanjie lijianshi ganshi lianyi yuehui huiyi jilu" [Minutes from the monthly gathering of the Executive Committee], *Voice* 15 (September 1953): 13.

12. Qi-xian Yang, "Kanke de fuxiao lucheng" [Arduous journey to reopen the alma mater in Taiwan], a webpage published by Soochow University, accessed November 13, 2020, https://web-ch.scu.edu.tw/huilung/web_page/3310. Educational sociologist Ting-Hong Wong noted in a television interview that the religious backgrounds of Soochow University and Fu Jen Catholic University were crucial for Chiang Kai-shek's agreement on their resuming operations in Taiwan, in 1954 and 1961. Chiang was willing to allow both universities to operate, as they would show his regime was different from the Chinese Communists' control over religion. His interview was part of the episodes "Daxue Shi" [History of universities in Taiwan] and "Taiwan yanyi" [Taiwan's history], aired on Formosa Television, Taiwan, June 28, 2020. For PRC's attitudes toward religious institutions, see, e.g., Frank Dikötter, "Thought Reform," in *The Tragedy of Liberation: A History of the Chinese Revolution, 1945–57* (New York: Bloomsbury Press, 2013). On Soochow University's

prestigious law school in the Republican Era, see Alison W. Conner, "Anglo-American Law at Soochow University," *China's Christian Colleges: Cross-Cultural Connections, 1900–1950*, ed. Daniel H. Bays and Ellen Widmer (Stanford, CA: Stanford University Press, 2009), 147–172.

13. See the official webpages published by Chengchi University about its history at http://archive.nccu.edu.tw/past.htm and http://archive.nccu.edu.tw/past2.htm, accessed June 13, 2018.

14. After Eisenhower gave a speech titled "Atomic Power for Peace," in 1953, Chiang Kai-shek announced that 1954 would be a year for "Science in Taiwan." Li-Yu Fu, "Meiyuan shiqi Taiwan zhongdeng kexue jiaoyu jihua zhi xingcheng yu shishi nianbiao (1951–1965)" [A Chronology of the Initiation and Implement of the Secondary Science Education Project under the US Aid in Taiwan (1951–1965), *Kexue jiaoyu xuekan* [Chinese Journal of Science Education] 14, no. 4 (2006): 447–465, esp. 451 and 452; for Yi-Chi Mei, see Yu-Feng Su, *Kangzhanqian de qinghua daxue: jindai zhongguo gaodeng jiaoyu yanjiu* [Tsing Hua University 1928–1937: A study of modern Chinese higher education] (Taipei: Institute of Modern History, Academic Sinica, 2000), chap. 3.

15. "Youxun" [Alumni news], *Voice* 32 (July 1955): 27.

16. Pi-ling Yeh, "Taibei diguo daxue gongxue bu zhi chuangshe" [The Construction of the Faculty of Engineering at Taihoku Imperial University], *Guoshiguan guankan* [Bulletin of Academia Historica] 52 (June 2017): 73–124. Also see the official webpages of the Electrical Engineering Department at National Taiwan University, at https://web.ee.ntu.edu.tw/about2.php. For the Japanese establishment of educational institutions in Taiwan, see E. Patricia Tsurumi, *Japanese Colonial Education in Taiwan, 1895–1945* (Cambridge MA: Harvard University Press, 1977), 53, 122–124.

17. Yao-te Wang, "Rizhi shiqi tainan gaodeng gongye xuexiao sheli zhi yanjiu" [Founding of Tainan Technical College during Japanese Colonial Era], *Taiwan shi yanjiu* [Taiwanese Historical Research] 18, no. 2 (2011), 53–95; and Li-ling Cheng, "Jindaihua, pingdenghua yu chabiehua zhijian: Taibei gongye xuexiao xuesheng zhi jiuxue yu jiuye (1923–1945)" [Modernization, Equalization, and Differentiation: Education and Employment of Students in Technical Schools of Taipei (1923–1945)], *Taiwan shi yanjiu* [Taiwan Historical Research] 16, no. 4 (2009): 81–114. On National Cheng-Kung University, see its official webpages at http://web.ncku.edu.tw/files/11-1000-48-1.php?Lang=zh-tw. For National Taipei University of Technology, see its official webpages on its history at https://archive.ntut.edu.tw/files/11-1060-634.php?Lang=zh-tw; all accessed June 13, 2018.

18. The unequal educational opportunities provided for Japanese and Taiwanese students were codified in the admissions policy of Taihoku Imperial University. Japanese were given preferential admission to Taihoku Imperial University in Taipei. Ironically, Taiwanese who would have liked to go to college could attend a university in Japan more easily than one in Taiwan. See Tsurumi, *Japanese Colonial Education in Taiwan, 1895–1945*, 124. For the replacement of Japanese faculty members with Chinese faculty members, see Wen-Hsing Wu, Shun-Fen Chen, and Chen-Tsou Wu, "The Development of Higher Education in Taiwan," *Higher Education* 18, no. 1 (1989): 124; and Po-Fen Tai, "Taiwan gaodeng jiaoyu kuozhang lichen zhong de jiaoyu quanli jingying fenxi" [Academic Regime Transformation: An Analysis of Power Elites and Expanding Higher Education], *Taiwan shehuixue xuekan* [Taiwanese Journal of Sociology] 58 (December 2015): 47–93. For

the early post–World War II years of National Taiwan University, see Fan-sen Wang, *Fu Ssu-nien: A Life in Chinese History and Politics* (Cambridge, UK: Cambridge University Press, 2000), 187; and Hsiu-Jung Chang, ed., *Taida Yixue Yuan 1945–1950* [National Taiwan University's medical college, 1945–1950] (Taipei: National Taiwan University Press, 2013), 31–39, 46–47. Even though some faculty members were recruited from China, their expertise might not have perfectly matched the needs of the new department. See Pang-yuan Chi, *The Great Flowing River: A Memoir of China, from Manchuria to Taiwan,* trans. John Balcom (New York: Columbia University Press, 2018), chap. 6.

19. Shanpei Zhu, "kenan fuxiao yundong" [Re-founding the alma mater in a difficult time], *Voice* 25 (October 1954): 7; Shuxi Chen, "Fuxiao yundong zhi wojian" [My opinion on the re-founding of Chiao-Tung University], *Voice* 27 (December 1954): 7–8; "Di ba jie lijianshi linshi huiyi" [The eighth Executive Committee meeting minutes], *Voice* 27 (December 1954): 34–35; Qinbo Ye, "Liangge wei chengshu de jianyi" [Two premature recommendations], *Voice* 31 (June 1955): 22–23; Yuansong Zhou, "Ku Qinbo" [Remembering Qinbo], *Voice* 162 (June 1966): 40.

20. Likun Xiao, "Jiaotong daxue ying zaitai fuxiao" [Chiao-Tung University should resume its operation in Taiwan], *Voice* 32 (July 1955): 1–2.

21. "Zhongyangshe xun " [News reports from the Central News Agency], *Voice* 42 (May 1956): 4.

22. Chi-yun Chang, "Jiaoda jingshen" [Chiao-Tung spirits], *Voice* 42 (May 1956): 2–4; "Jinian dahui jilu" [Notes on the 60th anniversary celebrations], *Voice* 42 (May 1956): 9–12.

23. "Jinian dahui jilu," 12.

24. For the survey, see "Lumei xiaoyou jinxun" [Recent news of American alumni] *Voice* 40 (March 1956): 31. For Tsao's role in the Chiao-Tung Alumni Association, see "Youxun" [Alumni news], *Voice* 13 (August 1953): 11.

25. Tsen-Cha Tsao, *Gongcheng yu kexue* [Engineering and Science] (Taipei: Zhing-hua Books), 80–87, and his "Sanshinianlai huainian muxiao" [Remembering thirty years of the alma mater], *Voice* 41 (April 1956): 33–34; Chiao-Tung Museum, A Memorial Webpage for Tsen-Cha Tsao, http://museum.lib.nctu.edu.tw/chao/#home, accessed November 13, 2020; "Youxun" [Alumni news], *Voice* 32 (July 1955): 29; Wolfgang Saxon, "T. C. Tsao, 99, Educator, Dies; Aided Taiwan Technical College," *New York Times,* June 8, 2001, section C, 13. The US chapter of the Chinese Institute of Engineers was still active in the mid-1940s; see Greene, *Building a Nation at War,* 120–126.

26. Tsen-Cha Tsao, "Huashidai de muxiao yu huashidai de gongcheng kexue" [Alma mater in the context of engineering science in this pivotal era], originally published in 1956, reprinted in *Gongcheng yu kexue,* 54–66; "Jinian dahui jilu," 12; "Jiaotong daxue sheli dianzi yanjiusuo" [National Chiao-Tung University establishes a graduate program in electronics], *Voice* 61 (December 1957): 1; Tsen-Cha Tsao, "Jiaoda fuxiao zhi jingguo ji qi shidai jiazhi zhi zhanwang" [The process of re-founding Chiao-Tung University], in *Jiaotong daxue jiushi nian* [Ninety years of Chiao-Tung], edited by NCTU (Hsinchu, TW: NCTU, 1986), 52–56.

27. Ku graduated from Tsing Hua College (the predecessor of Tsing Hua University). He completed his graduate degree at MIT in 1923 and became a prestigious electrical engineering scholar

in pre–World War II China. He worked for Chiang's post–World War II government in Taiwan briefly but then left Taiwan for a visiting position at MIT in 1950, becoming a professor at the University of Pennsylvania in 1952. "IEEE PES CSEE Yu-Hsiu Ku Electrical Engineering Award," https://www.ieee-pes.org/yu-hsiu-ku-electrical-engineering-award; "Yu Hsiu Ku," https://ethw.org/Yu_Hsiu_Ku; Yu Hsiu Ku; "The Future of Engineering," *Voice* 46 (September 1956): 17; "China Comes to Tech: 1877–1931," http://chinacomestomit.org/student-profiles-2/#/ku-yu-hsiu/, accessed July 6, 2018.

28. "Mei yuanzichang jiang huang diannao" [US nuclear energy plant installs a computer], *Lianhe bao* [United Daily News] (hereafter *UDN*), August 5, 1953, p. 2; "Xinqi de diannao" [The novel computers], *UDN*, September 6, 1954, p. 6; "Meiguo zhicheng kuaisu diannao" [US built fast computers], *UDN*, December 2, 1954, p. 2.

29. "Mei jundaoji jiejidui canguan ji" [A visit to the base of US F-86Fs], *UDN* , April 8, 1955, p. 1; "Fangkong xin liqi: pangda de diannao" [New air defense technology: Colossal computers], *UDN*, March 24, 1956, p. 2.

30. Kashan, "Yingping: diannao fengyun" [Film review: Desk Set], *UDN*, July 28, 1957, p. 6.

31. "Dianzi yanjiusuo choubei gongzuo geng jinyibu" [Progress made in founding a graduate program in electronics], *Voice* 46 (September 1956): 40. Regarding Mei's trip, see National Tsing Hua Library's Digital Archive, "Xinzhu qinghua shiqi" [Hsinchu Tsing Hua era], http://archives.lib.nthu.edu.tw/history/history04.html, accessed May 21, 2018.

32. Yun Tao, "Fuxiao zatan" [Discussions on the re-founding of the alma mater], *Voice* 46 (September 1956): 19–22.

33. "Bianzhe de hua" [From the editor], *Voice* 43 (June 1956): 33–34.

34. Chien, "Yuanzi yu dianzi" [Atoms and electronics], *Voice* 54 (May 1957): 2–4.

35. Bu-Yun Wang, "Dianzi yanjiusuo chouweihui chengli jingguo ji jinzhan qingxing" [The establishment and recent progress of the Graduate Program Preparation Committee], *Voice* 61 (December 1957): 3; "Jiaotong daxue sheli dianzi yanjiusuo," 1; Su-fen Liu, ed., *Li Guoding: Wo de Taiwan Jingyan* [Li Guoding (K. T. Li): My Taiwan experience—oral histories of Li Guoding], (Taipei: Yuan-Liou, 2005), 577–578.

36. "Jiaotong daxue sheli dianzi yanjiusuo," 1. For the United Industrial Research Institute, see Kawaguchi Mitsuo, *Taiwan xigu xungen* [The origin of Taiwan's silicon valley], trans. Liansheng He (Hsinchu, TW: Yanqu Shenghuo zazhishe), 2009.

37. "Yuanzinenghui jusing canhui" [The Atomic Energy Committee meeting], *Voice* 60 (November 1957): 36; "Nobel Prize in Physics 1957—Presentation Speech," the Nobel Prize official website, http://www.nobelprize.org/nobel_prizes/physics/laureates/1957/press.html.

38. "Guoli Jiaotong daxue tongxuehui dianzi yanjiusuo jingfei choumu weiyuanhui diyici huiyi jilu" [The minutes for the first fundraising meeting for Chiao-Tung University's Institute of Electronics], *Voice* 62 (January 1958): 8; and "Liumei Tongxuehui nuli xiezhu choujian dianzi yanjiusuo" [US alumni working hard on planning the graduate institute of electronics], *Voice* 63 (February 1958): 2–8. The exchange rate and the per capita income are from "Guomin suode

tongji changyong ziliao" [National income statistics], published online by Taiwan's Directorate General of Budget, Accounting and Statistics, Executive Yuan, accessed May 31, 2018, https://www.stat.gov.tw/ct.asp?xItem=37407&CtNode=3564&mp=4.

39. "Fengmian shuoming" [cover picture], *Voice* 61 (December 1957): 7.

40. "Fengmian shuoming" [cover picture], *Voice* 62 (January 1958): 4; "Fengmian shuoming" [cover picture], *Voice* 63 (February 1958): 33. The quoted text first appeared in Xuanzhi, "Jinnian shi women de dianzinian" [It is the year of electronics], *Voice* 62 (January 1958): 2.

41. "Bianzhe dehua" [Editor's notes], *Voice* 61 (December 1957): 45.

42. "Liumei tongxuehui nuli xiezhu," 3. Tsao reiterated the same ideas in "Jiaoda yanjiusuo chengli yinian yihou" [One year after the establishment of the graduate program in electronics], *Voice* 82 (September 1959): 28–29.

43. "Liumei tongxuehui nuli xiezhu," 4.

44. Tsao, "Jiaoda yanjiusuo chengli yinian yihou," 29.

45. Lan Jen Chu was a Chiao-Tung alumnus (1934) and obtained an ScD in electrical engineering from MIT in 1938. For Chu, see "Dr. Lan Jen Chu, 59, M.I.T. Engineer Dies," *New York Times*, July 28, 1973, 26; "Biography—Lan Jen Chu," *IRE Transactions on Microwave Theory and Techniques* 6, no. 3 (July 1958): 249; and "Lan Jen Chu," MIT Museum's MIT 150 Exhibition in 2011, accessed August 25, 2020, http://museum.mit.edu/nom150/entries/1297.

46. "Liumei tongxuehui nuli xiezhu"; "Chao Chen Wang (Obituary)," *Asbury Park Press*, September 20, 2012, reprinted at https://www.legacy.com/obituaries/app/obituary.aspx?n=chao-chen -wang&pid=159985084. For Chu and Kuh as consultants, see a letter from the Science Council for Long-Term Development to the Ministry of Education, 1962, November 20, 1962, NCTU University Records no. 511059. I thank Chao Chen Wang's daughter, Mei-Mei Wang, for her help in my research of Chao Chen Wang.

47. Ling-Yun Wei received a master's degree in telecommunications from Shanghai Chiao-Tung University in 1946, and a master's degree in electrical engineering from the University of Illinois Urbana-Champaign, in 1949. He then headed to Taiwan to work as a telecommunications engineer. He left Taiwan in 1956 for the University of Illinois Urbana-Champaign, to complete his PhD and studied with Nobel Prize laureate John Bardeen. His 1958 dissertation was titled, "Diffusion of Silver, Copper, Cobalt and Iron in Germanium." From 1958 to 1960, he worked at the University of Washington in Seattle. He then moved to Waterloo University in 1960 and taught there till 1986. Lillian Hoddeson and Vicki Daitch, *True Genius: The Life and Science of John Bardeen, the Only Winner of Two Nobel Prizes in Physics* (Washington, DC: Joseph Henry Press, 2002), 185; Wei Ling-Yun, "Wenhua kexue yu rensheng zixu" [Culture, science, and life], *Voice* 321 (June 1987): 23–24; Jiaying Shen, "Zhuidao Jianada xiaoyou Wei Lingyun xuezhang" [In memory of Wei Ling-Yun], *Voice* 401 (December 2003): 92–94. For Hsu Yun Fan, see a biography of Purdue University's honorary degree recipient at http://www.physics.purdue.edu/alumni/hondegree/fan .html, accessed June 7, 2018; and "Obituary for Hsu Yun Fan, 1912–2000 (Aged 88)," *Journal and Courier* (Lafayette, Indiana), October 8, 2000, https://www.newspapers.com/clip/42262587 /obituary-for-hsu-yun-fan-1912-2000/.

48. "Liumei tongxuehui nuli xiezhu"; a letter from Kuh to Tsao, January 7, 1958, NCTU University Records no. 470099; and oral history interview with Ernest S. Kuh, conducted by Lisa Rubens 2004–2006, Regional Oral History Office, the Bancroft Library, University of California at Berkeley, see p. 199, https://digitalassets.lib.berkeley.edu/roho/ucb/text/kuh_ernie.pdf. About Kuh, see also, "Ernest S. Kuh, Berkeley Engineering Professor and Dean Emeritus, 1928–2015: Longtime College Leader Was a Pioneer in Electronic Circuit Theory," July 8, 2015, UC Berkeley College of Engineering, https://engineering.berkeley.edu/2015/07/ernest-s-kuh-berkeley-engineering -professor-and-dean-emeritus-1928%E2%80%932015.

49. "Liumei tongxuehui nuli xiezhu."

50. "Liumei tongxuehui nuli xiezhu."

51. "Guoli jiaotong daxue dianzi yanjiusuo choubei weiyuanhui disanci huiyi jilu" [Minutes of the third NCTU preparation committee meeting], *Voice* 72 (November 1958): 3; "Yinianlai de jiaoda dianzi yanjiusuo" [The first year of the Institute of Electronics], *Voice* 82 (September 1959): 31.

52. Hu Shih, "Fazhan Kexue de zhongren he yuanlu" [The responsibility of developing science], *Xin shidai* [New era] 1 (February 1962): 3–4; and Yang, "Hu Shi dui Taiwan."

CHAPTER 2

1. Gisson Chi-Chen Chien, "Jiaoda dianzisuo yu lianheguo jishu xiezhu" [The graduate program in electronics and U.N. technical assistance], *Voice* 98 (Janurary 1961): 4–7.

2. "News Section," *Telecommunication Journal* 28, no. 2 (1961): 74; the information about per capita income is from the survey conducted by Taiwan's Directorate General of Budget, Accounting and Statistics, Executive Yuan (equivalent to the Cabinet or the Council of Ministers).

3. Harry S. Truman, "Inaugural Address of Harry S. Truman, January 20, 1949," in *Foreign Aid and American Foreign Policy: A Document Analysis*, ed. David Baldwin (New York: Frederick A. Praeger, 1966), 60–62, originally from *Public Papers of the Presidents of the United States: Harry S. Truman* (Washington, DC: Government Printing Office, 1964), 112–116.

4. David Webster, "Development Advisors in a Time of Cold War and Decolonization: The United Nations Technical Assistance Administration, 1950–59," *Journal of Global History* 6 (2011): 249–72; see page 9 in *Yearbook of the United Nations 1948–49* (New York: United Nations Publications, 1950), and page 6 in *Yearbook of the United Nations 1950* (New York: United Nations Publications, 1951).

5. "Chapter III: The Economic Development of Under-Developed Countries," in *Yearbook of the United Nations 1958* (New York: United Nations Publications, 1959), 131–142.

6. See Paul Hoffman, "The Critical Role of Special Fund," in *The Priorities of Progress* (New York: United Nations Publications, 1961), 4–5.

7. For UN technical programs in Taiwan, see for example, circular letter no. 340, authored by Sir Alexander MacFarquhar, regional representative for the Far East, UN Technical Assistance Board, May 6, 1960, folder 4 TE 322/1 CHI, box S-0175-0300 United Nations technical

assistance-missions, box number-Q400-R025-SU01, United Nations Archives and Records Management Section. Examples of other boxes about UN technical aid to Taiwan are listed in the appendix. For US-aided dam constructions, see, for example, Neil Jacoby, *U.S. Aid to Taiwan: A Study of Foreign Aid, Self-Help and Development* (New York: F. A. Praeger, 1967), 198.

8. Suzanne M. Moon, "Takeoff or Self-Sufficiency? Ideologies of Development in Indonesia, 1957–1961." *Technology and Culture* 39, no. 2 (1998): 187–212.

9. Edwards, *Closed World*, 81.

10. "Shenqing lianheguo jishu xiezhu tebie jijin jihuashu" [An Application for UN Technical Aid], June 1959, Ministry of Foreign Affairs Files, 635.31 0200, Institute of Modern History, Academia Sinica, Taiwan (hereafter MFA Files).

11. "Jiaotong daxue sheli dianzi yanjiusuo," 1–3. See my discussion in chapter 1.

12. "The Expansion of Electronic Research Institute and Establishment of Telecommunication Advanced Training Centre," June 1959, MFA Files, 635.31 0200.

13. "Expansion of Electronic Research Institute."

14. An earlier example of the vague connection between education and industry can be seen in a 1957 article. In this article, Chien mentioned that Taiwanese engineers should invest in the design of auto-control devices, design of remote-control devices, and digital electronic computing. These fields could improve industrial manufacturing in the long term. But he did not specify the devices on which the university or companies in Taiwan might conduct research. Chien, "Yuanzi yu dianzi."

15. "Expansion of Electronic Research Institute," 5.

16. "Expansion of Electronic Research Institute," 3.

17. Chien, "Dianzi yanjiusuo de qianzhan" [The future of the graduate program in electronics], *Voice* 65 (April 1958): 9–11.

18. S. M. Lee, "Dianzi yanjiusuo jiaoxue fangzhen yu jinhou fazhan" [The directions and development of the graduate program in electronics], *Voice* 82 (September 1959): 11–14.

19. "Dianzi jisuan yi: jiaoda dianzi yanjiusuo jihua yu shiyue zhuangzhi" [NCTU will set up an electronic computer in October], *Voice* 104 (August 1961): 23.

20. Jean Persin to W. A. Lewis, January 12, 1960, MFA Files, 635.31 0200.

21. Jean Persin to W. A. Lewis, January 12, 1960.

22. Sir Alexander MacFarquhar to Tsing-Chang Liu, May 31, 1960, MFA Files, 635.31 0200.

23. Three draft proposals were produced on July 12, circa August, and on November 9, 1960, respectively. See MFA Files, 635.31 0200.

24. Three draft proposals, MFA Files, 635.31 0200.

25. Memorandum written by Sinyu Liu and Guiquan Wang, July 4, 1961, MFA Files, 635.31 0208; original in English.

26. "Chapter III: The Economic Development of Under-Developed Countries," 131–142.

27. The Ministry of Foreign Affairs to NCTU, October 29, 1963, and S. M. Lee to the Ministry of Foreign Affairs, November 20, 1961, MFA File, 635.31 0202.

28. For example, see an announcement of NCTU's first four training courses sent to Ministries of Education and Foreign Affairs, and the Bureau of Telecommunications on September 5, 1962, in MFA Files, 635.31 0202.

29. Greene, *Building a Nation at War.*

30. The Ministry of Foreign Affairs to NCTU, October 29, 1963, MFA File, 635.31 0202.

31. Tsing-Chang Liu to J. N. Corry, November 25, 1963, MFA Files, 635.31 0202.

32. S. M. Lee to the Ministry of Foreign Affairs, November 20, 1961, MFA File, 635.31 0202. The two private companies were Hua-Seng Electronics, which manufactured radio sets, and Taiwan Cyanamid Company, which was a subsidiary of American Cyanamid Company.

33. J. N. Corry to Tsing-Chang Liu, December 6, 1963, MFA Files, 635.31 0202.

34. "Gioli Jiaotong daxue gongxueyuan fushe lianheguo dianzi dianxin xunlian yanjiu zhongxin daixun jishu renyuan tongji biao" [Trainee statistics, NCTU's Training and Research Centre for Telecommunications and Electronics], in *Yinianlai de gongcheng jianshe gaikuang* [The previous year's engineering accomplishments], ed. Zhongguo gongchengshi xiehui (Taipei: Chinese Institute of Engineers, 1970), 92.

35. See page 17 in "Ying haiwai guilai liang guoshi" [A welcome to two overseas alumni], *Voice* 140 (August 1964): 10–20.

36. "Zhide gongkai de jifeng guowai xiaoyou laishu" [Letters from overseas alumni], *Voice* 120 (December 1962): 1–6.

37. "Zhide gongkai de jifeng guowai xiaoyou laishu," 1.

38. For Ta-Chung Liu and his uses of NCTU's mainframe computer for economic planning, see Honghong Tinn, "Modeling Computers and Computer Models: Manufacturing Economic-Planning Projects in Cold War Taiwan, 1959–1968," *Technology and Culture* 59, no. 4 (Supplement 2018): S66–S99; and Hsiang-Ke Chao and Chao-Hsi Huang, "Ta-Chung Liu's Exploratory Econometrics," *History of Political Economy* 43 (Annual Supplement 2011): 140–65.

39. "Zhide gongkai de jifeng guowai xiaoyou laishu" 1–6, esp. 2.

40. "Zhide gongkai de jifeng guowai xiaoyou laishu" 4, 6; Cheng's letter was written in English.

41. "Zhide gongkai de jifeng guowai xiaoyou laishu" 3; R. Norris Shreve, a chemical engineering professor, led a team of sixteen professors from Purdue University to advise scholars at National Cheng-Kung University from 1952 to 1963 on science and engineering education and research.

42. Qinbo Ye, "Gengshang cenglou: zengshe yingyong shuxue yanjiusuo chuyi" [One step further: A recommendation for establishing a program in applied mathematics], *Voice* 100 (April 1961): 30–31.

43. "Expansion of Electronic Research Institute," 5.

44. "Zhide gongkai de jifeng guowai xiaoyou laishu" 6.

45. *Training and Research Centre for Telecommunications and Electronics, Republic of China: Report* (Geneva: International Telecommunication Union, 1968), 6 (hereafter *1968 UN Report*). For the August 1960 draft of the Plan of Operation, see MFA Files, 635.31 0200.

46. *1968 UN Report*, 6.

47. Memorandum from the Second Meeting for Establishing the Training and Research Centre for Telecommunications and Electronics, November 15, 1960, MFA Files, 635.31 0200.

48. Letter from V. R. Sundaram to Gisson Chi-Chen Chien, January 13, 1961, MFA Files, 635.31 0200.

49. Meiling Lin, "Chenggong de nanren he ta shenbian de nüren: Chaowu Lin xuezhang ji xuesao fangwenji" [An interview with Chiao-Tung alumnus Chao-Wu Lin and his wife], *Voice* 356 (June 1996): 58–59; and Tang-Chin Kao, "Danbomingzhi fuwu jiaoda sanshi nian zhi huigu" [Memories of my thirty-year of career at NCTU], July 24, 2010, Jiaoda ren de buluo [NCTU alumni's website], http://blog.alumni.nctu.edu.tw/plate/web/papermsg.jsp?UI=nctu_alumni_voice&PI=13442.

50. "Chengli dianzi jisuan ji zhongxin" [Establishing the computing center], *Voice* 110 (February 1962): 42.

51. Advertisement, *Voice* 111 (March 1962): 37

52. "Muxiao liushiliu zhounian xiaoqing" [Alma mater celebrates 66th anniversary], *Voice* 113 (May 1962): 1–4.

53. For IBM accounting machines used by the Taiwan Sugar Company and the CUSA, see Tinn, "Modeling Computers and Computer Models," S70–S71.

54. "Muxiao liushiliu zhounian xiaoqing."

CHAPTER 3

1. Hughes, *Networks of Power*.

2. See, e.g., Pinch and Bijker, "The Social Construction of Facts and Artifacts"; and Kline and Pinch, "Users as Agents of Technological Change."

3. "Dianzi jisuan yi," 23.

4. *1968 UN Report*.

5. Oral history interview with Henry Y. H. Chuang, June 7, 2009.

6. *1968 UN Report*, 53, 55. The theses included "The Reduction of Minimum Cost Flow Problem to the Solution of Diode-Source Network," "Essential Hazards in Asynchronous Sequential Switching Matchings," and "Bounds on the Error Probability of Digital Modulation Systems."

7. Chi-Chang Lee, oral history interview, January 7, 2009.

8. Dean N. Arden, November 10, 2008, oral history interview; according to the biographical sketch I was able to access with assistance from the MIT Museum, before joining MIT, he was "a

methods analyst at the University of Michigan Tabulating Service and a member of the research staff at the Willow Run Research Center."

9. Dean N. Arden, oral history interview, November 10, 2008.

10. *1968 UN Report*, 53.

11. "A Biographical Sketch of Chih-Bing Ling," in *Collected Papers in Elasticity and Mathematics of Chih-Bing Ling Volume II*, ed. John W. Layman and Harry L. Johnson (Blacksburg: Virginia Polytechnic Institute and State University, 1979), iii.

12. Chih-Bing Ling, "Stresses in a Perforated Circular Ring," *Applied Scientific Research* 29, no. 1 (1974): 99–120; Chih Bing Ling and Chen-Peng Tsai, "Evaluation at Half Periods of Weierstrass' Elliptic Function with Rhombic Primitive Period-Parallelogram," *Mathematics of Computation* 18, no. 87 (1964): 433–440.

13. For his use of the IBM 1620 computer, see Chih-Bing Ling, "Evaluation at Half Periods of Weierstrass' Elliptic Functions with Double Periods 1 and e^{ia}," *Mathematics of Computation* 19, no. 92 (1965): 658–661; *Collected Papers in Elasticity and Mathematics of Chih-Bing Ling Volume II*, 269. In *Collected Papers in Elasticity and Mathematics of Chih-Bing Ling Volume II*, pp. 15, 27, 95, 370, he acknowledged his uses of IBM 7040, 360, and 370.

14. Dean N. Arden, phone interview, October 4, 2008.

15. "Jiaoda tongxuehui si ba niandu di'er qi jiangxuejin dejiang xuesheng zichuan zhuanji" [Biographies of awardees of Chiao-Tung Alumni Association scholarship, 1959], *Voice* 95 (August 1960): 36–38.

16. Oral history interview with Henry Y. H. Chuang, June 7, 2009.

17. *The Yearbook of Training Course of Electronic Computers, February–June, 1963* (National Chiao-Tung University, 1963).

18. "Yanlan rencai jianjie" [Introducing invited overseas talent], *Kexue fazhan yuekan* [Science Development] 15, no. 3 (1987): 421–425.

19. Chi-Chang Lee's unpublished annual report of the Long-Term National Science Development Council (the predecessor of the National Science Council) in Taiwan, written in April 1965; "Ling Hongxun jiangzuo shoujie dejiangren Li Qichang fu jiaoshou" [Chi-Chang Lee won the first Hung-Hsun Ling Award], *Voice* 165 (September 1966): 19; oral history interview with Chi-Chang Lee, January 7, 2009.

20. *The Yearbook of the 13th Training Course of Electronic Computers, 1965* (Hsinchu, TW: National Chiao-Tung University, 1965).

21. Oral history interview with Chi-Chang Lee, January 7, 2009; "Jiaoda tongxuehui si ba niandu"; Chi-Hsiang Wong, Tung-Mou Yen, and Tseng-Yu Lee, "The Crystal Structure of Uranium Chloride π-tricyclopentadienyl," *Acta Crystallographica* 18, part 3 (1965): 340–345 (the manuscript was received on January 13, 1964).

22. Oral history interview with Tseng-Yu Lee, January 7, 2009; Wong, Yen, and Lee, "The Crystal Structure of Uranium Chloride π -tricyclopentadienyl"; Chi-Hsiang Wong, Tseng-Yu Lee, and

Yuen-Tseh Lee, "The Crystal Structure of Tris (Cyclopentadieny) Samarium (III)," *Acta Crystallographica* B25, part 12 (1969): 2580–2584.

23. The prize was jointly awarded to Dudley R. Herschbach, Yuan-Tseh Lee, and John C. Polanyi. According to the Council for the Lindau Nobel Laureate meetings, Lee was awarded for his work on using "information derived from angular and velocity measurements of elementary reactions in vacuum to understand the dynamics of chemical reactions." See the website of the Lindau Nobel Laureate meetings, http://www.lindau-nobel.org, and the official website of the Nobel Prize, accessed November 30, 2011, http://nobelprize.org/nobel_prizes/chemistry/laureates/1986/press .html.

24. See page 341 in E. G. Cox and G. A. Jeffrey, "The Use of 'Hollerith' Computing Equipment in Crystal-Structure Analysis," *Acta Crystallographica* 2, part 6 (1949): 341–343. In this paper, Cox and Jeffrey cited Wallace J. Eckert's 1940 book, *Punched Card Methods in Scientific Computations* (New York: Thomas J. Watson Astronomical Computing Bureau, Columbia University, 1940). For the computing of crystallography and related fields, see, for example, Jon Agar, "What Difference Did Computers Make?" *Social Studies of Science* 36, no. 6 (2006): 869–907 and Evan Hepler-Smith, "'A Way of Thinking Backwards': Computing and Method in Synthetic Organic Biology," *Historical Studies in Natural Sciences* 48, no. 3. (2018): 300-337.

25. Oral history interview with Tseng-Yu Lee, January 7, 2009.

26. Oral history interview with Tseng-Yu Lee and Chi-Chang Lee, January 7, 2009; "Meizhou zong xiaoyouhui zhixing weiyuanhui huiyi jilu" [Memorandum of the recent Chiao-Tung alumni's US chapter meeting], *Voice* 181 (January 1968): 19.

27. Oral history interview with Yuan-Kuang Chen, August 22, 2009; Yuan-Kuang Chen, "Jiandan bizhao jisuanji zhi zucheng ji qi yingyong" [The applications of analog computers], *Taidian gongcheng Yuekan* (hereafter *Monthly Journal of Taipower's Engineering*) 144 (1960): 4–30. Chang-Yuan, Zhang, "Liangzhi jisuan chi qiu chidu fa" [Using the slide rule method for the sag of transmission lines], *Monthly Journal of Taipower's Engineering* 145 (1960): 35–38.

28. Yuan-Kuang Chen, "Dianli xitong jingji yunyong gailun" [An introduction to the coordination of power systems], *Monthly Journal of Taipower's Engineering* 166 (1962): 18–25; Jia-Hui Zhou, "Dianyuan jingji kaifa fangshi yu dianzi jisuanji zhi yingyong" [Electricity developing and the application of electronic computers], *Monthly Journal of Taipower's Engineering* 172 (1963): 5–8.

29. Oral history interview, February 5, 15, and 25, 2009, with Pin-Yen Lin, who helped me find out which departments these trainees worked in the Taiwan Power Company.

30. Yuan-Kuang Chen, "Jiandan bizhao jisuanji," and "Dianli xitong jingji."

31. Yuan-Kuang Chen, "Shuzhi dianzi jisuanji zhi chengxu jihua" [Programming on electronic digital computers], *Monthly Journal of Taipower's Engineering* 184 (1963): 21–27.

32. Hong-Ren Li and Yuan-Kuang Chen, "Dianzi jisuanji dianli chaoliu jisuan chengxu ji qi yingyong (1)" [Electricity programming and its applications, part I], *Monthly Journal of Taipower's Engineering* 199 (1965): 4–10; Chiang-Chang Huang, "IBM 360 dianzi jisuanji yunyong zhi jiben gainian (1)" [Using the IBM 360, part I], *Monthly Journal of Taipower's Engineering* 213 (1966):

34–37; "IBM 360 dianzi jisuanji yunyong zhi jiben gainian (2)" [Using the IBM 360, part II], *Monthly Journal of Taipower's Engineering* 214 (1966): 33–38.

33. NTU's decision to rent an IBM was made by February 1963. See correspondence between Chi-fan Chen (Zhifan Chen) and NTU president Shih-liang Chien on February 2 and March 27, 1963, University Records no. 03010000010015, 052001239, NTU. Chi-fan Chen is a famous essay writer in Taiwan, who held a master's degree in science and taught at Christian Brothers College in Tennessee at that time. Chen attempted to persuade Chien to rent IBM 1620 model II instead of model I. NTU decided to go for model I, considering its rental fee was only half of model II.

34. "Shiyou gongsi tankanchu paiyuan lai suo canguan jisuanji" [Chinese Petroleum Corporation personnel visited NCTU for computers], *Voice* 111 (March 1962): 42.

35. *1968 UN Report*, 54.

36. *1968 UN Report*, 57.

37. Lee, "Correlation of Sediments by Using an Electronic Digital Computer," *Petroleum Geology of Taiwan*, August 1963, 138; originally in English.

38. The IBM 650 system at NCTU included an IBM 650 console, IBM 655 power unit, and IBM 533 card reader-punch unit. These were the most basic and minimal required components for the IBM 650 system. In September 1956, IBM provided a new system called "the IBM 650 RAMAC," which "combines the inherent data processing capacity of the IBM 650 with the facility for large-capacity random access storage." But what NCTU obtained was the original model. See "Computer Comparison Chart," *Updating Supplement* no. 27, March 1961, published by Automation Consultants Inc., page D24), James W. Cortada Papers (Charles Babbage Institute, hereafter CBI, 185), box 28; "IBM 650 Chronology," IBM Archives, accessed November 30, 2011, http://www-03.ibm.com/ibm/history/exhibits/650/650_ch1.html; "IBM 650 Dianzi ziliao chu-liji yingyong jianjie" [An introduction to the IBM 650 computer], *Voice* 117 (September 1962): 40–44; "IBM 650 RAMAC," published by IBM in 1956, Computer Product Manuals Collection (CBI 60), box 101, folder IBM 650 RAMAC, CBI.

39. The IBM 650 system at Cornell's computing center was replaced by the Burroughs 220 computer in 1959, and a Control Data 1604-160A computer system was installed in 1962. See Richard C. Lesser, "Richard C. Lesser's Recollections: The Cornell Computing Center, the Early Years, 1953 to 1964," 1996, on the website of oral and personal histories of computing at Cornell University, accessed November 30, 2011, http://www2.cit.cornell.edu/computer/history/.

40. William F. Aspray and Bernard O. Williams, "Arming American Scientists: NSF and the Provision of Scientific Computing Facilities for Universities, 1950–1973," *IEEE Annals of the History of Computing* 16, no. 4 (1994): 60–74.

41. Atsushi Akera, "Research and Education: The Academic Computing Centers at MIT and the University of Michigan," *Calculating a Natural World*, 277–311.

42. John W. Rudan, *The History of Computing at Cornell University* (Ithaca, NY: Cornell Internet-First University Press, 2005), accessed November 30, 2011, http://hdl.handle.net/1813/82, see page 272; Lyle Wadell, "History of the Northeast Dairy Records Processing Laboratory, 1948–1985,

with Additional Comments by J. D. Burke and H. Wilmot Carter," oral and personal histories of computing at Cornell, accessed November 30, 2011, http:// www2.cit.cornell.edu/computer/ history.

43. Aspray and Williams, "Arming American Scientists," 70; Akera, "Research and Education."

44. Aspray and Williams, "Arming American Scientists," 70; Akera, "Research and Education."

45. Aspray and Williams, "Arming American Scientists," 65.

46. *The Yearbook of Training Course of Electronic Computers, February–June, 1963* (Hsinchu, TW: National Chiao-Tung University, 1963).

47. *1968 UN Report*, 28.

48. Letter from NCTU to the Ministry of Foreign Affairs and Hsinchu Police Office, September 3, 1963, 635.31 0202, MFA Files; and "Technical Co-operation, Back from China," *Telecommunication Journal* 31, no. 10 (1964): 271.

49. *1968 UN Report*, 28.

50. Letter from NCTU to the Ministry of Foreign Affairs and Hsinchu Police Office, September 3, 1963, 635.31 0202, MFA Files; oral history interview with Dalman, October 24, 2009.

51. Leslie, *Cold War and American Science*.

52. "Jiaoda dianzi yanjiusuo yu gongyejie yanjiu hezuo zuotanhui jilu" [Memorandum, a workshop for NCTU administrators and industrial representatives], *Voice* 139 (July 1964): 1–7.

53. See, for example, page 372 in Martin H. Weik, *A Third Survey of Domestic Electronic Digital Computing Systems* (hereafter *BRL Report 1961*), Report No. 1115, March 1961, published by the Ballistic Research Laboratories, Aberdeen Proving Ground, Maryland, available through the Computer History Museum (hereafter CHM). An online full text version was accessed November 30, 2011, http://ed-thelen.org/comp-hist/BRL61-ibm065.html#IBM-650-RAMAC.

54. "Dianzisuo jinkuang" [Recent news from NCTU's Institute of Electronics], *Voice* 135 (March 1964): 36.

55. *1968 UN Report*, 29.

56. *1968 UN Report*, 7.

57. *1968 UN Report*, 28.

58. Oral history interview with Chi-Chang Lee, January 7, 2009.

59. Kao, email communication (original in English), June 2, 2009.

60. Kao, email communication, June 2, 2009. For the economic-planning project, see Tinn, "Modeling Computers and Computer Models."

61. William Wesley Peterson, Frank Dow Vickers, J. Robert Meachem, and Leonard Stein (the Statistical Laboratory at the University of Florida), "FLATRAN: Florida Translator for IBM 650," Computer Product Manuals Collection (CBI 60), box 101, CBI.

62. William Wesley Peterson, *Error-Correcting Codes* (Cambridge, MA: MIT Press, 1961).

63. Martin H. Weik pointed out that transistorized computers were easier to maintain than vacuum-tube computers, because "tube count and a knowledge of tube operating characteristics may yield an approximate estimation of some of the problems that may be encountered in the operation of the system." The numbers of vacuum tubes in early electronic digital computers could be up to twenty thousand, and the IBM 650 had approximately two thousand vacuum tubes. It was difficult to identify a burned-out tube among such a large number of tubes or, even worse, a failing but not yet burned-out tube. To avoid the hassles associated with efforts to identify the tube needing replacement, computer manufacturers sometimes provided preventive-maintenance routines. Because transistors consumed less power than vacuum tubes and because their printed circuits and packaging techniques continued to be improved, Weik pointed out that "the question of reliability is rapidly being resolved." See Martin H. Weik, *A Survey of Domestic Electronic Digital Computing Systems*, Report No. 971, (Aberdeen: Ballistic Research Laboratories, Aberdeen Proving Ground, Maryland, 1955), 64; and Weik, *BRL Report 1961*, 1032–1033 and 1072–1073.

64. It was a vocational school, established by the Japanese colonial government as Prefectural Taipei Industrial Institute in 1912; see chapter 1. In general, at that time, students from the Provincial Taipei Institute of Technology were more likely to stay and work in Taiwan, as opposed to graduates of Taiwan's three undergraduate programs in electrical engineering, at Chiao-Tung, National Taiwan University, and National Cheng-Kung University, who tended to go abroad to study after graduation.

65. The company was Zhongguo dianqi qicai kongsi.

66. Oral history interview with Kao, June 20, 2009.

67. Oral history interview with Kao, June 20, 2009.

68. Oral history interview with Kao, June 2, 2009.

69. Oral history interview with Kao, June 2, 2009. Kao pointed out that NCTU also bought an 870 Document Writing System, which arrived at the same time with the IBM 1620 in 1964.

70. Oral history interview with Kao, June 2, 2009.

71. "Returning the signed contract of NTU's leading of IBM peripheral machines," a letter from IBM to NTU, March 9, 1965, University Records no. 25003000060002, 054000501, NTU.

72. Oral history interview with Kao, June 23, 2009.

73. *1968 UN Report*, 53 and 55. Peterson's student Chin-Long Chen pointed out that the course title should be "Information Theory."

74. William Wesley Peterson, "Preface," in *Computing with the IBM 1620* (Taipei: Central Book Company, 1966).

75. Private letter from Peterson to Lee, April 20, 1965, collected by Chi-Chang Lee.

76. Jennifer Light, "Programming," in *Gender and Technology: A Reader*, ed. Nina Lerman, Ruth Oldenziel, and Arwen P. Mohun (Baltimore, MD: Johns Hopkins University Press, 2003), 295–328; originally published as "When Computers Were Women," *Technology and Culture* 40, no. 3 (1999):

455–483. For discussion of women computer operators, see Mar Hicks, *Programmed Inequality: How Britain Discarded Women Technologists and Lost Its Edge in Computing* (Cambridge, MA: MIT Press, 2017); Corinna Schlombs, "A Gendered Job Carousel: Employment Effects of Computer Automation," in *Gender Codes: Why Women Are Leaving Computing*, ed. Thomas J. Misa (Wiley and IEEE Computer Society, 2010), 75–93. For women's participation in computing, see Thomas J. Misa, ed., *Gender Codes*; and Janet Abbate, *Recoding Gender: Women's Changing Participation in Computing* (Cambridge, MA: MIT Press, 2012).

77. David A. Grier, *When Computers Were Human*, 276; see also Light, "Programming."

78. Private letter from Lee to Peterson collected by Chi-Chang Lee, November 4, 1965.

79. Private letter from Lee to Peterson collected by Chi-Chang Lee, January 18, 1966.

80. Peterson, "Preface."

81. Oral history interview with Kao, June 2, 2009.

82. Email communication with Chin-Long Chen, July 8, 2009.

83. Email communication with Chin-Long Chen, July 1, 2009.

84. *The Yearbook of the Institute of Electronics at NCTU, 1964* (Hsinchu, TW: National Chiao-Tung University, 1964), 12.

85. Email communication with Chin-Long Chen, July 1, 2009.

CHAPTER 4

1. Paul Edwards, "SAGE," *Closed World*, 75–111; Thomas P. Hughes, "MIT as System Builder," *Rescuing Prometheus*, 15–68; Martin Campbell-Kelly et al., *Computer*, 73–76, 152–155; Thomas J. Misa, "Military Needs, Commercial Realities, and the Development of the Transistor, 1948–1958," in *Military Enterprise and Technological Change*, ed. Merritt Roe Smith (Cambridge, MA: MIT Press, 1985), 253–287; Forman, "Behind Quantum Electronics"; Charles Phipps, "The Early History of ICs at Texas Instruments: A Personal View," *IEEE Annals of the History of Computing* 34, no. 1 (2012): 37–47; and Leslie, *Cold War and American Science*.

2. The Taiwanese military at that time often referred to themselves as the ROC military. In this chapter, I use both terms interchangeably.

3. Hughes, *Networks of Power*.

4. Hanting Chen and Shunde Luo, *Guofang buzhang Yu Dawei* [Biography of the Minister of Defense David Ta-wei Yule] (Taipei: Zhuanji Wenxue, 2015), 121–126, and 142–144.

5. For the Military Assistance Program's objectives and a review of the projected military equipment to transfer in 1957 and 1958 from the US to ROC forces, see "Country Statement: Government of the Republic of China (Taiwan)," October 22, 1956, folder 2.500 010 Military (MAAG) Program incl. DFS 1956, box 13: Bureau of Far Eastern Affairs/ office of Chinese Affairs], record group 59: Department of State, NARA. For navy ship procurements, see Li Chang, *Wu Shiwen xiansheng fangwen jilu* [The Reminiscences of Adm. Wu Shih-wen] (Taipei: Academia Sinica,

2017), 151. For Chiang and ministers of defense's negotiation with Americans, see Chen and Luo, *Guofang buzhang Yu Dawei*; and Jay Taylor, *The Generalissimo: Chiang Kai-shek and the Struggle for Modern China* (Cambridge, MA: Harvard University Press, 2009).

6. Yongzhou Li, *Hangkong, hangkong wushinian: qier yiwang* [Fifty years in the aeronautics industry: A memoir at age of seventy-two] (Taipei: Daosheng, 1987), 76–82, 86–87; and Lin, *Taiwan hangkong gongye shi*, 151.

7. Chen and Luo, *Guofang buzhang Yu Dawei*, 119, 137–138, 141–200.

8. Li, *Hangkong, hangkong wushinian*; and Lin, *Taiwan hangkong gongye shi*, 68, 76–78. For Qian Xuesen, see John Krige, "Representing the Life of an Outstanding Chinese Aeronautical Engineer: A Transnational Perspective," *Technology's Stories*, March 12, 2018, https://www.technologystories .org/chinese-engineer/.

9. Lin, *Taiwan hangkong gongye shi*, 68–69, 155–159, 165–167; and Li, *Hangkong, hangkong wushinian*, 81 and 99.

10. Lin, *Taiwan hangkong gongye shi*, 158; and Li, *Hangkong, hangkong wushinian*, 81.

11. Li, *Hangkong, hangkong wushinian*. From January to September 1946, the ROC government sent Yongzhao Li to study mechanical engineering at the University of Michigan, Ann Arbor. For Li's research in Michigan, see "Determining Cutting Tool Temperatures," *Tool Engineer* 23 (October 1949): 32. For UH-1H helicopters, see Lin, *Taiwan hangkong gongye shi*, 158; Li, *Hangkong, hangkong wushinian*, chap. 6; "GRC Helicopter Co-production," a memorandum of conversation of the Department of State, August 22, 1968, folder DEF 12–4 CHINAT 1/1/67, box no. 1531, Central Foreign Policy Files, 1967–1969, RG 59, Central Records of the Department of State, NARA; and Zhezheng Hong, "Yunbu, jiuzai, zhanxun ta douhang" [The helicopter for supply, rescue, military exercise], *Lianhe Wanbao* [United Evening News], February 6, 2019, A3. Chiang Ching-kuo also called the MAAG to urge consideration of helicopter manufacture before he left his position as the minister of defense in July 1968 to become the vice premier. See the telegram from US Embassy Taipei to Secretary of State, June 29, 1968, folder DEF 12–5 CHINAT 1/1/67, box no. 1531, Central Foreign Policy Files, 1967–1969, RG 59, Central Records of the Department of State, NARA.

12. Nancy Bernkopf Tucker, "Back to the Strait," *The China Threat: Memories, Myths, and Realities in the 1950s* (New York: Columbia University Press, 2012), 141; Appu K. Soman, "'Who's Daddy' in the Taiwan Strait," *Journal of American-East Asian Relations* 3, no. 4 (1994): 387–388; Melvin Gurtov, "The Taiwan Strait Crisis Revisited: Politics and Foreign Policy in Chinese Motives," *Modern China* 2, no. 1 (1976): 49–103, see 68–69; Stephen G. Craft, "Islands against the Red Tide," *American Justice in Taiwan* (Lexington: University Press of Kentucky, 2016), 18; John Finney, "Taiwan to Get U.S. Missile Unit, First Atomic Arm in the Far East," *New York Times*, May 7, 1957, p. 1; Hans M. Kristensen, "Nukes in the Taiwan Crisis," May 13, 2008, Federation of American Scientists blog, https://fas.org/blogs/security/2008/05/nukes-in-the-taiwan-crisis/.

13. The Second Taiwan Strait Crisis prompted politicians, military leaders, and the US public to debate whether the United States should use nuclear weapons to resolve the crisis. General Nathan Twining, chairman of the Joint Chiefs of Staff, considered strikes in cities as distant as

Shanghai. See Tucker, "Back to the Strait," 145; Soman, "'Who's Daddy' in the Taiwan Strait," 380; Kristensen, "Nukes in the Taiwan Crisis"; McGeorge Bundy, *Danger and Survival: Choices about the Bomb in the First Fifty Years* (New York: Vintage Books, 1990), 280; and Morton H. Halperin, *The 1958 Taiwan Straits Crisis: A Documented History* (Santa Monica, CA: RAND Corporation, 1966), 113–114.

14. Tang Tsou, "The Quemoy Imbroglio: Chiang Kai-Shek and the United States," *Western Political Quarterly* 12, no. 4 (1959): 1084–1085; Tucker, "Back to the Strait," 145; Ron Westrum, "In Combat," *Sidewinder: Creative Missile Development at China Lake* (Annapolis, MD: Naval Institute Press, 1999), 207–219; Halperin, *1958 Taiwan Straits Crisis*, 64; Soman, "'Who's Daddy' in the Taiwan Strait," 376–378; Gurtov, "The Taiwan Strait Crisis Revisited," 49–103, see p. 75; Stephen I. Schwartz, ed., *Atomic Audit: The Costs and Consequences of U.S. Nuclear Weapons since 1940* (Washington, DC: Brookings Institution Press, 1998), 282; Tomasz Smura, "In the Shadow of Communistic Missiles—Air and Missile Defence in Taiwan," Casimir Pulaski Foundation, https://pulaski.pl/en/in-the-shadow-of-communistic-missiles-air-and-missile-defence-in-taiwan/, accessed May 15, 2019; Meador, "Veterans Who Defended Taiwan"; and Nike Historical Society, "Selected Nike Missile-Related Links," http://nikemissile.org/links.shtml, accessed May 15, 2019.

15. Nye, *American Technological Sublime*.

16. On the prominence of analog computers in the 1950s, see James S. Small, "Engineering, Technology and Design: The Post–Second World War Development of Electronic Analogue Computers," *History and Technology* 11, no. 1 (1993): 33–48; "General-Purpose Electronic Analog Computing: 1945–1965," *IEEE Annals of the History of Computing* 15, no. 2 (1993): 8–18; and *The Analogue Alternative: The Electronic Analogue Computer in Britain and the USA, 1930–1975* (New York: Routledge, 2013), 109. For analog computing applications, see also David C. Brock, "From Automation to Silicon Valley: The Automation Movement of the 1950s, Arnold Beckman, and William Shockley," *History and Technology* 28, no. 4, 375–340. See also Edwards, *Closed World*, 66–68, 104; W. H. C. Higgins, B. D. Holbrook, and J. W. Emling, "Electrical Computers for Fire Control," *IEEE Annals of the History of Computing* 3, no. 3 (1982), 233. Ronald R. Kline has noted a "digital progress narrative" from 1945 the early 1970s, even though "the viability of analog computing" in research and industry did not end until 1970. Ronald R. Kline, "Inventing an Analog Past and a Digital Future in Computing," in *Exploring the Early Digital*, ed. Thomas Haigh (Cham: Springer Nature Switzerland AG, 2019), 33. For the analog option for the Apollo Guidance Computer, see David Mindell, *Digital Apollo: Human and Machine in Spaceflight* (Cambridge, MA: MIT Press, 2009), 139.

17. Donald MacKenzie, *Inventing Accuracy: A Historical Sociology of Nuclear Missile Guidance* (Cambridge, MA: MIT Press, 1993), 17; David K. Allison, "U.S. Navy Research and Development since World War II," in *Military Enterprise and Technological Change*, ed. Merritt Roe Smith (Cambridge, MA: MIT Press, 1985), 314–320; Ron Westrum, "Struggles with Infrared," *Sidewinder*, 47–61; Edwards, *Closed World*, 104; Higgins et al., "Electrical Computers for Fire Control," 233; David A. Laws, "A Company of Legend: The Legacy of Fairchild Semiconductor," *IEEE Annals of the History of Computing* 32, no. 1 (2010): 60–74; Phipps, "Early History of ICs at Texas Instruments"; Christopher McDonald, "From Art Form to Engineering Discipline?: A History of US Military Software Development Standards, 1974–1998," *IEEE Annals of the History of Computing* 32, no. 4 (2010):

32–45; Nike Historical Society, "Hercules Missile Function Description," accessed May 15, 2019, http://nikemissile.org/IFC/HERCFUNCTION.shtml; Thomas B. Cochran, William M. Arkin, and Milton M. Hoenig, *Nuclear Weapons Databook, Volume 1: U.S. Nuclear Forces and Capacities* (Cambridge, MA: Ballinger, 1984), 287; and CHM, "Minuteman Missile Guidance Computer," https://www.computerhistory.org/revolution/real-time-computing/6/128/531, accessed May 15, 2019.

18. For McGuire Air Force Base, see Edwards, *Closed World*, chap. 3.

19. For Dongfeng missiles, see Krige, "Representing the Life of an Outstanding Chinese Aeronautical Engineer;" Hsi-hua Cheng, "Airpower in the Taiwan Strait," in *The Chinese Air Force: Evolving Concepts, Roles, and Capabilities*, ed. Richard P. Hallion, Roger Cliff, and Phillip C. Saunders (Washington, DC: National Defense University, 2012), 328; Federation of Atomic Scientists, "DF-1," June 10, 1998, https://fas.org/nuke/guide/china/theater/df-1.htm; and Mike Gruntman, *Blazing the Trail: The Early History of Spacecraft and Rocketry* (Reston, VA: American Institute of Aeronautics & Astronautics, 2004), 440.

20. For the components of a Nike Hercules missile system, see the Nike Historical Society, "Lesson 1, Introduction to the Improved Nike Hercules Missile System," http://www.nikemissile.org/MMS-150-Ch01.pdf, accessed June 8, 2020; and "1950's Nike Hercules Flight and Intercept Computers. Battery site SF-88 #15," posted on March 9, 2010, by YouTube user TilTuli on a tour of a retired Nike Hercules system, https://www.youtube.com/watch?v=y0o_oUKTQBk. For the public demonstration of the Nike Hercules system in Taiwan, see "Feidan yanxi sanxiang mude" [The three goals of the Nike Hercules missile exercise], *UDN*, May 6, 1961, p. 1.

21. "Country Statement: Government of the Republic of China (Taiwan)," NARA. On the difficulties of procuring naval ships, see Chen and Luo, *Guofang buzhang Yu Dawei*, 119.

22. Lin, *Taiwan hangkong gongye shi*, 124, 151–154, 164–166; and Kallgren, "Nationalist China's Armed Forces," 39.

23. Lin, *Taiwan hangkong gongye shi*, 155.

24. For Black Cat Squadron, see pages 20–32, and 60, in Wei-Bin Chang, *Kuaidao jihua jiemi: Heimao zhongdui yu taimei gaokong zhencha hezuo neimu* [Project RAZOR: The Black Cat Squadron and collaboration in high altitude reconnaissance tasks between Taiwan and the United States] (Taipei: Xinrui wenchuang, 2012); Wai Yip, "RB-57A and RB-57D in Republic of China Air Force Service," *American Aviation Historical Society Journal* 44, no. 4 (1999): 259–274; and Paul H. Tai, "A Slice of Cold War History: The Soaring Cat," *Journal of Chinese Studies* 13, no. 1 (2006): 89–95; Bob Bergin, "The Growth of China's Air Defenses: Responding to Covert Overflights, 1949–1974," *Studies in Intelligence* 57, no. 2 (June 2013): 19–28; Hsichun Mike Hua, "The Black Cat Squadron," *Air Power History* 49, no. 1 (2002): 4–19; Dino A. Brugioni, "Tactical Use of the U-2 and Related Technical Developments," in *Eyes in the Sky: Eisenhower, The CIA and Cold War Aerial Espionage* (Annapolis, MD: Naval Institute Press), 260–326. For Black Bat Squadron, see Chin-Shou Wang, *Heibianfu zhi lian* [Chain of black bats] (Taipei: Linking Books), 2011; Bergin, "Growth of China's Air Defenses"; Chris Pocock, *The Black Bats: CIA Spy Flights over China from Taiwan 1951–1969* (Atglen, PA: Schiffer, 2010); Wen-lu Huang and Ziching Li, *Feiyue dihou 3000 li: Heibianfu zhongdui yu dashidai de women* [Three hundred miles into the territory of the enemy: The Black Bat

Squadron and our time] (Taipei: Xinrui wenchuang, 2018); "Heibianfu zhongdui jinianguan daolan duanpian" [A tour at Black Bat Squadron Museum], YouTube, toured by Chongshan Li, a former black bat squadron member, https://www.youtube.com/watch?v=Bz8oSwz4Ywo&t=758s, accessed April 23, 2019; and Deputy Director of Plans Richard M. Bissell Jr.'s Memorandum for Chief, Research and Development Branch, DP Division, "Program Approval—Repeater Jammer for P2V-7U," May 28, 1959, CIA FOIA ERR, https://www.cia.gov/library/readingroom/document /cia-rdp33-02415a000100060030-0.

25. Lin, *Taiwan hangkong gongye shi*, 166; and Wen-Hiao Liu, *Shiluo de wudu feixing yuan* [The Lost Voodoo Pilot] (Taipei: Bingqi zhanshu tushu, 2014), 69.

26. "TVBS Kanban renwu: heimao zhongdui" [Billboard celebrities: Black Cat Squadron], an interview with Zongli Shen and two other Black Cat Squadron members, aired on the TVBS channel, November 14, 2010, https://www.youtube.com/watch?v=9ffP3tOmq18; Brugioni, *Eyes in the Sky*, 307; and Lin, *Taiwan hangkong gongye shi*, 165–166.

27. Wang, *Heibianfu zhi lian*, 17, 19, 37, 38, 48, 51, 54, 94–95,103.

28. "Minguo wushier nian liuyue Jiang jingguo yu naerxun huitan" [Memorandum of Chiang Ching-kuo's meeting with Nelson, June 1963], doc. no. 005-010301-00004-017, June 19, 1963, Jiang Jingguo zongtong wenwu wenjian jiedai binke huitanjiyao [President Chiang Ching-kuo documents, receptions, and memos], Jiang Jingguo zongtong wenwu [President Chiang Ching-kuo files], Academia Historica, Taiwan (hereafter AH); and "Minguo wushier nian qiyue Jiang jingguo yu naerxun huitan" [Chiang Ching-kuo's meeting with Nelson, July 1963], doc. no. 005-010301-00005-003, July 30, 1963, Jiang Jingguo zongtong wenwu wenjian jiedai binke huitanjiyao" [President Chiang Ching-kuo documents, receptions, and memos], Jiang Jingguo zongtong wenwu [President Chiang Ching-kuo files], AH; Chung-ting Huang, "The Military Assistance of the Republic of China to the Republic of Vietnam during the Vietnam War," *Bulletin of the Institute of Modern History Academia Sinica* 79 (March 2013): 137–172; "Nelson Takes Over NACC; Ray Cline Departs," *China News*, May 31, 1962, CIA FOIA ERR, https://www.cia.gov/library /readingroom/docs/CIA-RDP75-00001R000100330028-1.pdf. For Ray S. Cline and Chiang Ching-kuo, see Jay Taylor, *The Generalissimo's Son: Chiang Ching-kuo and the Revolutions in China and Taiwan* (Cambridge, MA: Harvard University Press, 2000), 239, 253–265, and 273. For Lanzhou's Gaseous Diffusion Plant and the railway survey of the surrounding area, see "Cheng-chou to Lan-chou Railroad Study," a memorandum from Chief, CIA/PID (NIPC) to Chief, Manufacturing and Services Division, ORR, CIA FOIA ERR, https://www.cia.gov/library/readingroom/docs/CIA -RDP78T05439A000300210011-7.pdf, accessed June 1, 2020; and Federation of American Scientists, "Lanzhou: Lanzhou Nuclear Fuel Complex, 504 Plant," https://fas.org/nuke/guide/china /facility/lanzhou.htm, accessed June 1, 2020.

29. "Luzong zhaokao jiqi zuoye renyuan" [Army headquarters recruit machine operators], *UDN*, June 4, 1964, p. 2; and "Diannao yongtu kuoda zhuanjia fanguo yanjiang" [An expert returned home to offer a talk on the expanding applications of computers], *UDN*, July 1, 1965, p. 2.

30. "Diannao yongtu kuoda"; and "Lu gongbu qiyong diannao, sudu kuai zuoyeliang da" [Army's Logistic Command installs a computer, high speed and large volume of processing], *UDN*, August 3, 1967, p. 2.

31. "Lu gongbu qiyong diannao"; "Han mei jiu junguan, ditai xue diannao" [Nine Korean and American military officers arrived to learn computerization], *Jingji ribao* [Economic Daily News , hereafter *EDN*], August 27, 1967, p. 5; and "Wuqi zhuangbei jingliang, lu un zhanli jiaqiang" [Army is well-equipped with sophisticated weapon and its combat power is fortified], *UDN*, September 3, 1967, p. 6. For US Air Force Logistics Command's attempt to build a "real-time computer network for highly complex logistics," see Jeffrey R. Yost, "Materiel Command and the Materiality of Commands: An Historical Examination of the US Air Force, Control Data Corporation, and the Advanced Logistics System," in *History of Computing: Lessons from the Past*, ed. Arthur Tatnall (London: Springer, 2010), 89–100.

32. "Mingnian qiyue zhuang diannao" [Computer installation scheduled in July next year], August 27, 1968, *EDN*, p. 4; "Diannao shensou Le Meixing" [Mr. Computer Meixing Le], December 23, 1982, *EDN*, p. 12. For Executive Yuan's computerization, see "Xingzhengyuan di yiyisiwu huiyu" [Executive Yuan Meeting no. 1145], November 16, 1969, doc. no. 014-000205-00351-001, "Xingzhengyuan huiyi yishilu tai di sanersi ce yiyisiwu zhi yiyisiliu" [The Executive Yuan's Meeting Minutes, volume 324, 1145–1146], Xingzhengyuan Files [Executive Yuan files], AH, Taiwan; C. D. (Dan) Mote Jr., "Ruth M. Davis," *Memorial Tributes: National Academy of Engineering of the United States of America, Volume 17* (Washington, DC: National Academies Press, 2013), 74–79.

33. For CDC computer installations, see "Taiwan Railroad Orders Control Data Computer System, January 19, 1971," box 4, folder 1, Control Data Corporation records. News releases, CBI 80, series 13, CBI Archives.

34. "Jiasu wo gongshang qiye diannaohua Fan Quangling ti kexing fangxiang" [Fan Quangling proposes suggestions on speeding up computerization in industries in Taiwan], *EDN*, May 6, 1970, p. 6; "Huahang kaishi zuyong zhongdian gongsi diannao" [China Airlines begins to lease a computer from the Chunghua Computer Company], *EDN*, April 24, 1971, p. 5; "Meishang diannao gongsi, jiaoyu zhongxin luocheng, juxing qingzhu jiuhui" [An American computer corporation celebrates the opening of the educational center], *EDN*, June 12, 1971, p. 8. In the United States, Control Data Corporation also sold hardware and services to military clients starting in 1958; see Jeffrey R. Yost, *Making IT Work: A History of the Computer Services Industry* (Cambridge, MA: MIT Press), 205. For the numbers of Chile's computer installations, see Eden Medina, "Designing Freedom, Regulating a Nation: Socialist Cybernetics in Allende's Chile," *Journal of Latin America Studies* 38, no. 3 (2006): 586.

35. "Jiaxin guanxi qiye guanli zuoye naru diannaoxitong" [Chia Hsin computerizes its management], *EDN*, October 26, 1971, p. 6; "Wenhua xueyuan jiaiang dianhua jiaoxue, gou CDC diannao xitong" [Chinese Culture College strengthens its computer education with a new CDC computer], *EDN*, September 7, 1977, p. 9; Department of Computer Science, Soochow University, Taiwan, http://www.csim.scu.edu.tw/department/history.htm; and Information and Networking Center, National Taipei University of Business, https://inc.ntub.edu.tw/p/405-1011-76515,c3993 .php?Lang=zh-tw, accessed June 1, 2020. For the linear programming workshop, see "Xianxing guihua yantao mingtian juhang" [Linear programming workshop will take place tomorrow], *EDN*, November 5, 1971, p. 9. For the thirty-five IBM installations in Taiwan, see Cortada, *Digital Flood*, 419 and 421. For the survey of computers in late-1970s Taiwan, see Jiulong Wang, "Dianzi

jisuanji wang: dianzi jisuanji xitong de xinfazhan" [Computer network: New development of computer systems], *Xinxin jikan* [Hsin Hsin Quarterly] 7, no. 3 (1979): 80–86; the journal is one of the academic journals published by CSIST.

36. "Sanyiwuling xing diannao shiyu gongshang jie shiyong" [Model 3150 computer suitable for various industries], *EDN*, August 7, 1969. For the TR-48 computer, see Kuang-wei Han, in *Xuexi de rensheng: Han Guangwei huiyilu* [A life of learning: The memoirs of Kuang-wei Han], ed. by Li Chang (Taipei: Academia Sinica, 2010), 533. For the second department of CSIST, see Shu-En Hsu, *Cailiao ye shenqi: Keji xuezhe Xu Shuen de yisheng* [The marvels of material science: The memoirs of Shu-En Hsu] (Taipei: Xiuwei, 2004).

37. David Li-wei Ho, *Hedan MIT: yige shangwei jieshu de gushi* ["A" bomb made in Taiwan: An unfinished story] (Taipei: Women chuban, 2015), 95; Feng Wang, "Yiselie hedan zhi fu mizhu Jiang Jieshi fazhan hewu neiqing" [Inside information about the secret assistance offered by the father of Israel's nuclear bomb to Chiang Kai-shek's development of nuclear weapons], *Yazhou zhoukan* [Asia Newsweek], no. 15 (April 2010): 12–18; Yitzhak Shichor, "The Importance of Being Ernst: Ernst David Bergmann and Israel's Role in Taiwan's Defense," *CIRS Asia Papers* no. 2 (Doha, QA: Center for International and Regional Studies, 2016); and CIA, "Taipei's Capabilities and Intentions Regarding Nuclear Weapons Development," Special National Intelligence Estimate 43-1-72, November 16, 1972, *U.S. Opposed Taiwanese Bomb during 1970s*, ed. William Burr, National Security Archive's Electronic Briefing Book No. 221, June 15, 2007, https://nsarchive2 .gwu.edu/nukevault/ebb221/#1.

38. For the US embassy report, see US Embassy Taipei, Airgram 1037, June 20, 1966, "Indications GRC Continues to Pursue Atomic Weaponry," Subject-Numeric 1964–66, DEF 12-1 Chinat, *New Archival Evidence on Taiwanese "Nuclear Intentions," 1966–1976*, ed. William Burr, National Security Archive's Electronic Briefing Book No. 20, October 13, 1999, https://nsarchive2.gwu.edu /NSAEBB/NSAEBB20/docs/doc18.pdf.

39. Ho, *Hedan MIT*; Shichor, "Importance of Being Ernst"; Yi-shen Chen, Mengtao Peng, and Jiahui Jian, *Hedan! Jiandie? CIA: Zhang xianyi fangwen jilu* [Nuke! Spy? CIA: An interview with Hsien-yi Chang] (Taipei: Yuanzu, 2016). Yi-shen Chen is a historian at the Institute of Modern History at Academia Sinica, Taiwan. For the conflicts between the ROC Ministry of Foreign Affairs and the CSIST, see Fredrick Foo Chien, *Qianfu huiyilu* [The memoirs of Fredrick Foo Chien], vol. 1 (Taipei: Tianxia [Commonwealth], 2005).

40. For Control Data's 3000 series, see "Sanyiwuling xing diannao shiyu gongshang jie shiyong"; pp. 80, 92, 100–102, 414, in oral history interview with Robert M. Price, OH 454, by Thomas J. Misa in 2009, CBI; CBI, "CDC Product Timeline," http://www.cbi.umn.edu/collections/cdc/prod-timeline.html; and "Control Data 3000 Series: Total Technical Excellence Now! Hardware/Software/Personnel," a 3000 series brochure, https://www.computerhistory.org/collections/catalog /102646247, accessed June 1, 2020. For Control Data in Australia, Thailand, and Japan, see oral history interview with John Baxter, OH 448; and oral history interview with Ron G. Bird, OH 442, by Thomas J. Misa, 2013, CBI. For Control Data's sales in Germany and Mexico, see Simon Donig, "Appropriating American Technology in the 1960s: Cold War Politics and the GDR Computer Industry," *IEEE Annals of the History of Computing* 32, no. 2 (2010): 32–45; and Bernardo Bátiz-Lazo

and Thomas Haigh, "Engineering Change: The Appropriation of Computer Technology at Grupo ICA in Mexico (1965–1971)," *IEEE Annals of the History of Computing* 34, no. 2 (2012): 20–33.

41. For the Ministry of Finance's computerization of taxes, see Tinn, "Modeling Computers and Computer Models." For CDC installations in Taiwan, see "Control Data Installs Largest University Computer System in Taiwan," August 11, 1969, box 3, folder 5, Control Data Corporation records, news releases, CBI 80, series 13, CBI; "Sanyiwuling xing diannao shiyu gongshang jie shiyong"; "Jiasu wo gongshang qiye diannaohua"; "Meishang diannao gongsi, jiaoyu zhongxin luocheng"; and "Yuandong fangzhi gongsi shezhi diannao zhongxin" [Far Eastern Textile sets up a computer center], *EDN*, June 18, 1970, p. 6.

42. "Taiwan Research Institute Orders $1.2 Million Control Data Computer System, May 12, 1972," box 4, folder 10, "Control Data Corporation records, news releases," CBI 80, Series 13, CBI. "Fuzongtong jian xingzheng yuanzhang Yan Jiagan jiejian meiguo ziliao guanzhi diannao gongsi fuzongcai Miller" [Vice President and Premier Chia-kan Yen receives Control Data's Vice President Miller], May 20, 1971, doc. no. 006-030203-00049-023, "Minguo liushi nian Yan Jiagan fuzongtong huodong ji (san)" [Vice President Chia-kan Yen events in 1971, Part 3]; and "Fuzongtong jian xingzheng yuanzhang Yan Jiagan jiejian meiguo Control Data diannao gongsi fuzongcai M. R. Swoman [*sic*]" [Vice President and Premier Chia-kan Yen receives Control Data's Vice President M. R. Swenson], May 12, 1972, doc. no. 006-030203-00057-027, "Minguo liushiyi nian Yan Jiagan fuzongtong huodong ji (san)" [Vice President Chia-kan Yen activities in 1972, Part 3]; both in Yan Jiagan wenwu 1966–1972 [President Chia-kan Yen files, 1966-1972], AH. See also "Shiwenxun laitai zhuwo yanjiu diannao ziliao de guanli jishu" [Swenson arrived in Taiwan to assist in management of computer data], *EDN*, May 11, 1972, p. 6; and "Control Data Corporation Annual Report," 1972, ProQuest Historical Annual Reports. The corporation annual reports cited in this book are from the electronic database ProQuest Historical Annual Reports.

43. For US Air Force Logistics Command's plan to use CDC Cyber 70 computers, see Jeffrey R. Yost, "Materiel Command and the Materiality of Commands." For the CDC Cyber 72/14 computer in Israel and Poland, see Y. L. Varol, "Some Remarks on Computer Acquisition," *Computer Journal* 19, no. 2 (1976): 127–131; and "COINS Files," August 8, 1973, CIA FOIA ERR, https://www.cia.gov/library/readingroom/document/cia-rdp82m00531r000400190025-8. For the CDC Cyber series, see the following CBI oral history interviews, oral history interview with Lyle Bowden and Tony Blackmore, OH 439, and oral history interview with George Karoly and Marcel Dayan, OH 446, esp. p. 36, by Thomas J. Misa, November 2013. According to Karoly and Dayan, who worked for Control Data Australia, they considered the Cyber series the "development machine" and the 6600 series the commercial machine, and the 6600 series was the successor of the Cyber series. For CDC installations for nuclear research at Los Alamos, see Nicholas Lewis, "Purchasing Power: Rivalry, Dissent, and Computing Strategy in Supercomputer Selection at Los Alamos," *IEEE Annals of the History of Computing* 39, no. 3 (2017): 25–40.

44. For Control Data's Cybernet, see oral history interview with Robert M. Price, 163 and 314; Misa, *Digital State*, 99–134; Yost, *Making IT Work*, 196–197; and "Control Data Corporation," Minnesota Computing History Project, https://mncomputinghistory.com/control-data-corporation/, accessed June 1, 2020. For the computer network between the CSIST and National

Tsing Hua University, see "About Us," the website of the National Tsing Hua University's computer center, http://www.cc.nthu.edu.tw/p/412-1285–1208.php?Lang=en, accessed February 5, 2016.

45. CIA, "National Intelligence Survey 39B; Nationalist China; Science," April 1974, CIA FOIA ERR, https://www.cia.gov/library/readingroom/document/cia-rdp01-00707r000200080023-2.

46. Ho, *Hedan MIT*. Li-Wei Ho, "Perturbation Theory in Nuclear Fuel Management Optimization," (PhD diss., Iowa State University, 1981); and CIA, "Taipei's Capabilities and Intentions."

47. Jenn-Nan Chen published a short biography on the official website of the dean of the College of Computer Science at the China University of Technology [Zhongguo keji daxue], whose campuses are located in Taipei and Hsinchu, https://www.cute.edu.tw/ccs, accessed May 1, 2020. Jenn-Nan Chen, "Verification and Translation of Distributed Computing System Software Design," (PhD diss., Northwestern University, 1987).

48. For Sieling's visit, see "Xiningwen ren Taiwan keji dayou fazhan" [Sieling believes that Taiwan's technological development is promising], *EDN*, March 16, 1978, p. 9; "CDC zhixing fuzongcai Xiningwen zuo laihua kaocha" [Control Data's vice president Sieling arrived in Taiwan], *EDN*, March 15, 1978, p. 9; and "Mei CDC gongsi jiangshe shiba wanglu diannao zhongxin" [Control Data plans to set up cybernet center], *EDN*, November 26, 1979, p. 8. For Kaohsiung Harbor's CDC computer, see "Gaogang zhanchu yewu mairu diannao zuoye" [Kaohsiung Harbor computerizes its storage facilities operation], *EDN*, February 8, 1980, p. 3; and "Computerized Harbor Management System Unclogs Taiwan Port, January 1979," box 7, folder 1, Control Data Corporation records, news releases, CBI 80, series 13, CBI. For Tsing Hua's new CDC computer, see "About Us."

49. Ho, *Hedan MIT*, 64–71; and Yi-shen Chen et al., *Hedan! Jiandie? CIA*, 169 and 171. For Fujitsu's expansion in the United States, see Yost, *Making IT Work*, 248; and Scientific Supercomputer Subcommittee, Committee on Communications and Information Policy, US Activities Board, IEEE, "U.S. Supercomputer Vulnerability," August 8, 1988. A copy of this report can be accessed through CIA's FOIA site, https://www.cia.gov/library/readingroom/docs/CIA-RDP90G01353R001100170002 -4.pdf. For Control Data's supercomputers, see Misa, *Digital State*, chap. 4.

50. Han, *Xuexi de rensheng*, 527. For the US embassy report, see "Indications GRC Continues to Pursue Atomic Weaponry"; for Bergmann's connections with Taiwan, see Shichor, "Importance of Being Ernst"; and Han, *Xuexi de rensheng*, chap. 15. For rocket research at the Aviation Research Institute, see Lin, *Taiwan hangkong gongye shi*, 158.

51. Han, *Xuexi de rensheng*, chaps. 9, 11, and 15; Han and George J. Thaler, "Phase-Space Analysis and Design of Linear Discontinuously Damped Feedback Control Systems," *Transactions of the American Institute of Electrical Engineers, Part II: Applications and Industry* 80, no. 4 (1961): 196–203.

52. Han, *Xuexi de rensheng*, chap. 15.

53. Han, *Xuexi de rensheng*, chap. 15. Navy Admiral Shi-wen Wu briefly noted an analog gun fire control system in ROC navy ships in an oral history interview; see Li Chang, *Wu Shiwen xiansheng fangwen jilu*, 154–155.

54. Han, *Xuexi de rensheng*, chap. 15, esp. 556–561, and 564–565; Lu Cheng, "Laodangyizhuang de wujin sanxing quzhujian" [ROC Navy's Wu-Chin III conversion destroyers], *Jianduan keji* [Defense Technology Monthly] 148 (December 1996): 54–61.

55. Han, *Xuexi de rensheng*, 588–590, a short text authored by Zheng to celebrate Han's achievements but included in Han's biography; and Jialong Ye, "Tongzidian keji yu shuweihua zhanchang xianqu: zhuanfang benyuan qian zixun tongxin yanjiusuo suozhang Zheng Mingjie boshi" [Forerunner on the battlefield of telecommunications, informatics, electronics, and digitalization: An interview with Dr. Mingjia Zheng, former director of the Information and Communications Research Division, CSIST], *Xinxin jikan* [Hsin Hsin Quarterly] 44, no. 2 (2016): 4–7.

56. Han, *Xuexi de rensheng*, chap. 15.

57. Han, *Xuexi de rensheng*, chap. 15. The Gabriel II missiles was also discussed in a CIA report, see page 4 of "Taipei's Capabilities and Intentions." For the similarity between Gabriel and Hsiung Feng, see Shichor, "Importance of Being Ernst."

CHAPTER 5

1. "Dianzi jisuan yi," 23. For the epigraph, please see footnote 10.

2. For the ACS, see Campbell-Kelly et al., *Computer*, 224.

3. Takahasi, "Some Important Computers of Japanese Design"; Takahashi, "Early Transistor Computers in Japan."

4. For example, their articles discuss the 1950s university transfer of computer-related technology to companies without charging fees and university-built computers with corporate assistance.

5. Kita, "From Technological Mimesis to Creativity." Ksenia Tatarchenko also has discussed Russian efforts in making compatible computers of IBM mainframes, especially the IBM 360 computer after 1968. Ksenia Tatarchenko, "'Our Past Is Their Future': Rip-off, Translation, and Trust in the Soviet-American Computing Exchanges during the Cold War," workshop presentation at Appreciating Innovation across Countries, Copenhagen Business School, Denmark, November 5–6, 2015.

6. "Application to the United Nation Special Fund for the Establishment of a Training Center for Technicians in Telecommunications and Electronics, Directorate General of Telecommunications, Ministry of Communications, Republic of China," May 1964. MFA Files, 635.31 0201, IMH archives.

7. Letter from Knut H. Winter, Resident Representative and Director of Special Fund Program in China, to Chih-ming Kao, Director, International Organizations Department, Ministry of Foreign Affairs, Taipei, Republic of China, August 18, 1964. MFA Files, 635.31 0201, IMH archives.

8. Minutes, Special Meeting held by Ministry of Communications for Applying to the United Nations Special Fund, July 7, 1964, MFA Files, 635.31 0201, IMH archives.

9. Knut H. Winter to Chih-ming Kao, August 18, 1964.

10. Jong-Chuang Tsay "Jiaotong daxue shouzhi chenggong, woguo diyibu xiaoxing dianzi jisuanji jianjie" [National Chiao-Tung University's first domestically manufactured computer], 1971, accessed November 30, 2011, http://www.csie.nctu.edu.tw/~jctsay/; Tsay, "Woguo diyibu xiaoxing dianzi jisuanji jianjie, Jiaotong daxue shouzhi chenggong" [National Chiao-Tung University's

success in building the first domestic electronic digital computer], *Kexue yuekan* [Science Monthly] 21 (September 1971): 48–53. Tsay's *Kexue yuekan* article, with the same title but some light revisions, was also published in *Voice* 222 (April 1972): 38–44. "Diyibu guoren zizhi diannao shunli wancheng" [The first domestically manufactured computer], *EDN*, July 17, 1971, 1. It is unclear who applied for research funds from Taiwan's National Science Council to support the project. But it is likely that one or more NCTU faculty members used their research funds to support various parts of the minicomputer building project.

11. Oral history interview with Jong-Chuang Tsay, December 18, 2007, and oral history interview with Ching-Chun Hsieh, May 27, 2009.

12. Oral history interview with Hsieh, May 27, 2009. Ju Ching Tu received his PhD in electrical engineering from the University of Michigan at Ann Arbor in 1956; his dissertation is titled, "An Algebraic Approach to the Synthesis of Equalizers for a Prescribed Frequency Response." According to his obituary, his specialty was logic design. He had set up Multi-Fineline Electronix, Inc., in Anaheim, California, in 1984, which later established assembly factories in China after 1994 and became a publicly traded company on the NASDAQ in 2004. See his obituary at http://forestlawn.tributes.com/show/Ju-Ching-Tu-93964480, and Multi-Fineline Electronix's official website, https://www.mflex.com/about-mflex/, accessed August 23, 2020.

13. Oral history interview with Hsieh, May 27, 2009; Xiaofan (pen name), "Cong yizhi shoushang de yanjing shuoqi" [Speaking of the eye injury incident], *UDN*, November 20, 1972, p. 9.

14. The types of transistors he used were 2n402 and 2n403. Ching-Chun Hsieh, "Dianzi jisuanji zhi sheji" [The design of a general-purpose computer], *Jiaoda xuekan* [Science Bulletin National Chiao-Tung University] 2, no. 2 (1967): 105–118.

15. Ching-Chun Hsieh, "Chuxi jianada yijiuliuqi nian guoji dianzi nianhui zuandu lunwen baogao" [My presentation at the 1967 International Electronics Conference in Canada], *Voice* 179 (November 1967): 14–18.

16. For example, see Hildi Kang, ed., *Under the Black Umbrella: Voices from Colonial Korea 1910–1945* (Ithaca, NY: Cornell University Press, 2005), 28. Ena Chao developed an analysis of Taiwanese elites' journals on their travels in the United States during the Cold War; see Ena Chao, "Guancha meiguo Taiwan jingying bixiao de meiguo xingxiang yu jiaoyu jiaohuan jihua 1950–1970" [Observing America: American Images in Taiwan elites' writing and American education exchange programs], *Taida lishixue xuebao* [Historical Inquiry] 48 (December 2011): 97–163.

17. Hsieh, "Chuxi jianada yijiuliuqi nian guoji dianzi nianhui."

18. Oral history interview with Hsieh, May 27, 2009; oral history interview with Chew, August 6, 2007; Andrew Chew, *Shede: Dianziye xianqu Qiu Zai-xing de shiye yu zhiye* [Willing: The calling of the pioneer of electronics industry, Zaixing Chew] (Taipei: Yuanshen, 2015), 74; oral history interview with Hsieh, May 27, 2009; Xiaofan, "Cong yizhi shoushang de yanjing shuoqi"; Lin, "Chenggong de nanren he ta shenbian de nüren"; "Shexia IC ban gongye longtou zhuiqiu meide tansuo—Qiu Zaixing xuechang" [Giving up ICs, shifting to his pursuit for arts—Zaixing Chew], *Jiaoda jiazu julebu* [NCTU Family] 48 (December 2006), http://140.113.39.126/13/2010-07-06/nctufamily.nctu.edu.tw/nctu_club/2006/12/p01.html.

19. Tsay, "Woguo diyibu xiaoxing dianzi jisuanji jianjie."

20. The term "technologist" refers to *keji rencai* in the original text.

21. In 1972, *Business Week* reported that Digital Computer Controls made "equivalents" of DEC's PDP-8 minicomputers in 1970 and Digital General Corporation's Nora in 1972. PDP-8 was the most popular mini, and Nora was the second most popular mini at that time. See "Building Models of Hot Computers," *Business Week*, February 5, 1972, pp. 73–74; "The Newer Systems Set the Pace," *Business Week*, July 7, 1973, p. 72; "Bell Labs Swing to Minis," *Business Week*, June 5, 1971, pp. 118–119; "Mini diannao: Shiyong jianbian jiage dilian" [Minicomputers: Easy uses and sweet deals], *EDN*, January 8, 1968, p. 4. For minicomputers and DEC, see Campbell-Kelly et al., *Computer*, 222–225; and Paul E. Ceruzzi, *A History of Modern Computing*, 2nd ed. (Cambridge, MA: MIT Press, 2003), 124–136, 191.

22. James Brinton, "Computer Has Calculator Price Tag," *Electronics*, February 2, 1970, http://wang3300.org/docs/WangProgrammerV5N5.5-71.pdf; "Wang Delivers First Mini-Computer," *Wang Laboratories Programmer*, May 1971, pp. 13–15, http://wang3300.org/docs/3300press.1970.pdf.

23. Digital Equipment Corporation Annual Reports, 1971, 1972, 1973, and 1974.

24. "NCR xiaoxing mini diannao" [NCR minicomputers], *EDN*, December 25, 1974, p. 7; "Huipu yu wojiaoda hezuo zhanchu zhongwen diannaohua yanjiu chengguo" [Chinese-language computers displayed and developed by Hewlett-Packard and with Chiao-Tung], *EDN*, August 16, 1973, p. 6; "Wang An mini diannao shiyong gaoji chengshi" [Wang Laboratories' new minicomputers, commanded with high-level programming languages], *EDN*, September 26, 1972, p. 6; "Jisuanji xinxingshi shengbao mini diannao" [Sampo announces its new minicomputers], *EDN*, September 23, 1972, p. 6; Lili Wang, "Dianzi jisuanji xiaoxing de jinnia xiaolu kanhao" [Smaller electronic calculators finds a good market during recent years], *EDN*, May 13, 1972, p. 9; "Meishang huipu gongsi zai tai she fengongsi" [Hewlett-Packard sets up a subsidiary in Taipei], *EDN*, October 21, 1970, p. 6; "Huipu gongsi zhuanjia Bobuxin shiyong xiaoxing diannao shifan" [Mr. Ernie Poblacion demonstrates Hewlett-Packard's minicomputers], *EDN*, August 4, 1971, p. 6. The donation of Hewlett-Packard computers to Chiao-Tung was approved by the company's Taiwan office manager, Leji Lin. But another HP employer, Pierre Loisel, whose Chinese name is Lixue Liu, might have contributed to the technical aspect of the Chinese-language computer designed at Chiao-Tung in 1973. See Pei-huang Chen, "Rang Jiang Jing-guo qizha de laowai—Liu Li-xue" [The foreigner that irritated Chiang Ching-Kuo: Lixue Liu], *Pingguo Ribao* [Apple Daily], September 8, 2018, https://tw.appledaily.com/headline/20180908/AE6DATW3NBNCYIHNOIEO5MGCWU/; and Cai-hong Li, "Qian shentong diannao fuzong bian shou-p-u-en-lang jianada laowai Liu Li-xue ba chuyu bian heijin" [The former VP of MiTAC now works on composting, Lixue Liu a Canadian to turn compost to gold], *Shangye zhoukan* [Business Weekly] 696 (March 22, 2001): 78–83.

25. As early as 1957, K. T. Li, as an official of the Executive Yuan, conceptualized the idea to set up an export processing zone. He and Chia-kan Yen had visited Trieste in Italy, which had a special trade zone. He also researched a small foreign trade zone in Shannon Airport in North Ireland. In September 1963, he proposed the idea to several high-level officials, and they visited

the site near the Kaohsiung Harbor that would soon become the designated zone. See Liu, ed., *Li Guoding: Wo de Taiwan Jingyan*, 331–340; and "Zhengfu yanni zai Gaoxiong shezhi jiagong chukou qu, youguan renyuan shicha xiangguan zuoye qingxing" [The government considers setting up an export processing zone in Kaohsiung, staff visit the site], photos taken by photographers of *Taiwan xinwen Bao*, 156-030101-0007-004, September 9, 1963, AH.

26. Philco-Ford mentioned its investment in Taiwan in 1965 in its annual report. But its Tamsui factory was completed in the summer of 1966 and its Kaohsiung factory was completed in December in 1966; see "Gaoxiong jiagong chukuoqu jin juxeng jiancheng dianli" [Kaohsiung export processing zone is inaugurated today], *UDN*, December 3, 1966, p. 2; and "Feige shoupi chanpin jintian kongyun xiao mei" [First batch of Philco-Ford's products are shipped to the US today], *UDN*, August 16, 1966, p. 5. See also Ford Annual Report, 1965, p. 14; Ford Annual Report, 1966, 31–32; Ford Annual Report, 1967, p. 12. Philco-Ford's Tamsui plant was also known as Feige Electronics, which was a direct transliteration of Philco in Mandarin.

27. For General Microelectronics, see Bassett, *To the Digital Age*, 155 and esp. chapter 5; Lécuyer, *Making Silicon Valley*, 240; Christophe Lécuyer and David C. Brock, *Makers of the Microchip: A Documentary History of Fairchild Semiconductor* (Cambridge, MA: MIT Press, 2010), 44. For General Instrument Corporation (GI), see Bassett, *To the Digital Age*, e.g., 222 and 239. For the importance of MOS technology, see Bassett, *To the Digital Age*, 1–11. More discussion on GI's operation in Taiwan is in chapter 6.

28. An engineer—one of the five engineers with master's degrees working at Kaohsiung Electronics—shared that in 1968 the company manufactured three types of products. The first was transistors made from silicon, instead of germanium, for radios. The second was linear integrated circuits for wireless radio communications devices and color TV sets. The third was diode-transistor logic integrated circuits, for products such as computers. "Gaoxiong dianzi gongsi meishang touzi guoren zhuchi" [Kaohsiung Electronics invested by Americans and managed by Chinese], *EDN*, November 11, 1968, p. 4; and Hongpu Zhou, "Zhizao bandaoti he dianjingti de gaoxiong dianzi gongsi" [Kaohsiung Electronics makes semiconductors and transistors], *EDN*, February 23, 1968, p. 4.

29. Oral history interview with Ching-Chun Hsieh, May 27, 2009; Tsay, "Woguo diyibu xiaoxing dianzi jisuanji jianjie."

30. Eric A. Weiss, "Elogue: An Wang, 1920–1990," *IEEE Annals of the History of Computing* 15, no. 1 (1993): 60–69.

31. "Taiwan Wangan diannao jiji kuoda chanxiao" [Wang Laboratories' Taiwanese factory expanded its production and export], *EDN*, May 23, 1968, p. 1.

32. Weiss, "Elogue: An Wang," 60–69; "Meiguo wangshi shiyansuo laitai shechang chanzhi dianzi lingjian: Feng Yuanquan jin lai zuguo qiashang" [American company Wang Laboratories set up an electronics plant in Taiwan: Yuanquan Feng will arrive his homeland for arrangements], *EDN*, June 1, 1967, p. 2; Zai-xing Zheng, "Dianzizhan zhong Wang An gongsi zhanchu guoren zizhi de diannao" [Wang Laboratories showcased computers made domestically], *EDN*, November 26, 1969, p. 6; "Anyuan maoyi gongsi zaitai kaiye jingxiao wangshi jiduanji ji diannao" [Anyuan

Trade Company begins to market and sell Wang Laboratories computers and calculators], *EDN*, August 6, 1971, p. 6; and An Wang and Eugene Linden, *Lessons: An Autobiography* (Reading, MA: Addison-Wesley, 1986), 137.

33. For Wang's contribution to the development of magnetic core memory, see Weiss, "Elogue: An Wang"; and Wang and Linden, *Lessons*.

34. Oral history interviews with Hsieh and Tsay; "Meiguo wangshi shiyansuo laitai shechang chan-zhi dianzi lingjian," 2; "Wang An diannao gongsi shenqing laitai shechang" [Wang Laboratories applies for permission to set up an plant in Taiwan], *EDN*, October 5, 1967, p. 2; "Meizhou xiaoyou zonghui tongxun" [US chapter news], *Voice* 185 (June 1968): 30; Tinglan Li, "Feng Yuanquan yu Wang An diannao" [Feng Yuanquan and Wang Laboratories], *EDN*, October 17, 1974, p. 12.

35. Yuanquan Feng, "Fazhan Taiwan diannao shiye de zhanwang" [A vision of the computer-industry development in Taiwan], *EDN*, January 1, 1970, p. 14.

36. "Diyibu guoren zizhi diannao shunli wancheng."

37. Chao-Chih Yang's doctoral dissertation at Northwestern University awarded in 1966 was to develop associative memory, as opposed to random access memory, in circuits made of cryotrons and cutpoint cells. The associative memory system was compatible with the batch fabrication of integrated circuits. "Zongtong zhaojian dianyansuo biyesheng Yang Chaozhi boshi" [President receives an alumnus, Chao-Chih Yang], *Voice* 175 (July-August 1970): 54; Chao-Chih Yang, "Asso-ciative Memory Systems and Their Applications," (PhD diss., electrical engineering, Northwestern University, 1966); Yang, *Woguo diannao de yingyong yu xiandaihua* [The application and mod-ernization of computers in Taiwan] (Taipei: Ministry of Education, 1971); and Donald G. Fink, *Diannao yu rennao* [Computers and the human mind] (Taipei: Shangwu, 1971), trans. Chao-Chih Yang; originally published as *Computers and the Human Mind: An Introduction to Artificial Intelli-gence* (Garden City, NY: Doubleday Anchor, 1966). For Donald G. Fink, see Donald Christiansen, "Donald Glen Fink, 1911–1996," *Memorial Tributes: National Academy of Engineering of the United States of America*, vol. 9 (Washington, DC: National Academies Press, 2001), 82–87. Yang's time at the Naval College of Engineering was briefly mentioned in Shu-En Hsu's autobiography; Hsu, *Cailiao ye shenqi*, 125.

38. Oral history interview with Ching-Chun Hsieh, June 5, 2009; oral history interview with Jong-Chuang Tsay, December 18, 2007; Jong Chuang Tsay, "Design and Experimentation of a 4X4X2 Memory," a reprinted paper from a 1969 research report, submitted to the Engineering Science Research Center, Taiwan, National Science Council, 595–608; and "Jiaotong daxue shouzhi chenggong." For Philips' plant in Taiwan, see "Qiao-wai touzian zhun shiyijian" [Eleven foreign or overseas Chinese investment applications are granted], *UDN*, April 22, 1966, p. 5; "Gaoxiong jiagong chukouqu fazhan dianzi gongye lixiang yuandi" [Kaohsiung Export Processing Zone is the cradle for the electronics industry], *EDN*, October 25, 1967, p. 5.

39. Hsieh, "Dianzi jisuanji zhi sheji."

40. Chung Laung Liu won several awards from various prestigious societies, including the Asso-ciation of Computer Machinery (ACM) and the IEEE Computer Society; see his biography at the Engineering and Technology History Wiki, maintained by professional societies including the IEEE,

https://ethw.org/Chung_Laung_Liu; and an oral history interview with Liu conducted by Doug Fairbairn, on June 2, 2014, CHM, https://www.youtube.com/watch?v=e7sb8gKFyhM, or https:// archive.computerhistory.org/resources/access/text/2015/06/102739932-05-01-acc.pdf. Francis Fan Lee's research focus at that time was on "grapheme-to-phoneme translation of English," aiming to develop a mechanism to transform text to speech. For Francis F. Lee, see *Massachusetts Institute of Technology Bulletin: Report of The President* 105, no. 3 (1969): 25 and 481. For Lee and Liu's research for the Project MAC, see "Project MAC Progress Report, July 1965 to July 1966," 94 and 208, https:// ban.ai/multics/doc/MAC-PR-03-648346.pdf.

41. Jong-Chuang Tsay, "Design and Experimentation of a 4X4X2 Memory"; "Woguo diyibu xiaoxing dianzi jisuanji jianjie"; oral history interview with Jong-Chuang Tsay, December 18, 2007; oral history interview with Ching-Chun Hsieh, May 27, 2009.

42. For the UN report, see Frank Roy Gustavson, "Survey of the Status of the Electronic Industry of the Republic of China" (hereafter "Gustavson Survey"), October 23, 1967, p. 4, MFA Files 635.31 0192, IMH archives; Daniela K. Rosner, Samantha Shorey, Brock Craft, and Helen Remick, "Making Core Memory: Design Inquiry into Gendered Legacies of Engineering and Craftwork," *Proceedings of the 2018 CHI Conference on Human Factors in Computing Systems (CHI'18)*, Association for Computer Machinery (ACM), April 2018, 1–13; and Samantha Shorey and Daniela Rosner, "Making Core Memory—An Experiment in Troubling Computing Histories," *Technology's Stories* 7, no. 2 (2019), https://www.technologystories.org/making-core-memory/. For An Wang's choice of Taiwan to set up its overseas plant, see Wang and Linden, *Lessons*, 137.

43. Oral history interview with Hsieh, May 27, 2009.

44. Tsay, "Woguo diyibu xiaoxing dianzi jisuanji jianjie," 40. Barry Gilbert also considered wiring among the most difficult tasks in making the computing units for an X-ray computed tomography machine for Mayo Clinic in the 1970s. Gilbert, "The Origin and Evolution of X-Ray Computed Tomography at Mayo Clinic, 1960–1980," Department of Electrical and Computer Engineering Colloquium, University of Minnesota Twin Cities, April 20, 2023.

45. Tsay, "Woguo diyibu xiaoxing dianzi jisuanji jianjie," 39. Ernest Shiu-Jen Kuh is discussed in chap. 1.

46. Harry M. Collins, "The TEA Laser," *Changing Order: Replication and Induction in Scientific Practice* (Chicago: University of Chicago Press, 1992 [London: Sage, 1985]).

47. Tsay, "Jiaotong daxue shouzhi chenggong," 1.

48. Tsay's personal webpage at NCTU, https://people.cs.nctu.edu.tw/~jctsay/jctsay.html, accessed July 8, 2009. Shu-Xun Zhao, "Qieer bushe jianku zhuojue" [An exemplar of perseverance and strenuousness: An interview with Jong-Chuang Tsay], *Voice* 395 (December 2002): 46–50. "Dou Zulie yuanshi jianli [Academician Julius T. Tou], https://academicians.sinica.edu.tw/index.php?r=academician -n%2Fshow&id=405; Julius T. Tou, "Learning Control via Associative Retrieval and Inference," in *Pattern Recognition and Machine Learning*, ed. K. S. Fu (New York: Plenum Press, 1971), 243–251.

49. Hsieh, "A Study on Periodic Sequences," *Jiaoda xuekan* 5, no. 2 (1972): 103–115; "Zhongwen zigen de zhucun he zhongwenzi de hecheng" [The memory storage of radicals of Chinese characters and the retrieval of the Chinese characters], *Jiaoda xuekan* 6, no. 1 (1973): 122–131;

"Jisuan yu kongzhi xi de huigu yu zhanwang" [News from the Department of Computer and Control Engineering], *Voice* 222 (April 1972): 34; Hsieh's webpage at Academia Sinica, Taiwan, https://www.iis.sinica.edu.tw/pages/hsieh/vita_zh.html, accessed August 28, 2020; "Xie tongxue Qingjun huode boshi xuewei" [Ching-Chun Hsieh awarded a doctoral degree], *Voice* 226 (October 1972); "Guonei xuezhe tongli hezuo zhongwen diannao zhengzhong wenshi" [Researchers work together, and their Chinese-language computer debuts], *UDN*, November 29, 1972, p. 3; "Xie Qingjun xiansheng zhi jianjie yu fangtan jianyao" [A synopsis of an interview with Ching-Chun Hsieh], in *Zhongyang yanjiuyuan zixun kexue yanjiusuo ershi zhounianqing tekan* [The 20th anniversary of the Institute of Information Science at Academia Sinica], interviewed on July 26, 2001, published in October 2002 at the Institute of Information Science's official website, https://www.iis.sinica.edu.tw/page/aboutus/20thAnniversary/i05.html.

50. Oral history interview with Hsieh, May 27, 2009.

51. The computer was probably the 2114b minicomputer, mistakenly noted as 214B in Yang's report. Chao-Chih Yang, "Yinianlai de jisuan yu kongzhixue xi (The first year of the Department of Computer and Control Engineering)," *Voice* 211 (April 1971): 17.

52. "Diyibu guoren zizhi diannao."

53. "Diyibu guoren zizhi diannao."

54. Jong-Chuang Tsay, email communication, November 20, 2007.

55. "Diyibu guoren zizhi diannao."

56. Tsay, "Jiaotong daxue shouzhi chenggong."

57. Tsay, "Jiaotong daxue shouzhi chenggong," 1.

58. Hsieh, "Jisuan yu kongzhi xi de huigu yu zhanwang."

59. Oral history interview with Tsay, December 18, 2007.

60. Hsieh, "Jisuan yu kongzhi xi de huigu yu zhanwang."

61. "Disidai dianzi jisuanji chengda shisheng yanzhi Chenggong" [The fourth-generation calculator, an achievement by Cheng-Kung faculty and students], *UDN*, January 25, 1972, p. 3; "Dianzi jisuan jiqi zhizuo jishu tigao" [Enhanced techniques in manufacturing calculators], *UDN*, January 30, 1973, p. 3.

62. Wang Tang, "Wode sanpian xuewei lunwen: jin yi benwen xiangei enshi Dr. Anthony N. Michel zaitianzhiling" [My three degree theses: In memory of my advisor Dr. Anthony N. Michel], *Xiangxun* [Newsletters of San Diego Taiwanese Cultural Association], March 11–17.

63. Yageo Corporation, founded in 1977, was one of the largest manufacturers in chip-resistor, multilayer ceramic capacitors and other passive devices, such as inductors, used in electronics. See Yageo's official website at https://www.yageo.com/en/Html/Index/orgnization.

64. CTT Inc.'s official website, http://www.cttinc.com/; College of Electrical Engineering and Computer Science; "Dianjixi cengren chan-guan-xue-yan gaojie zhiwei zhi xiyou" [Prestigious alumni in industry, government, university and institute], College of Electrical Engineering and

Computer Science, http://www.eecs.ncku.edu.tw/var/file/20/1020/img/2199/113738714.pdf, accessed August 20, 2020.

65. Wang Tang, "Wode sanpian xuewei lunwen"; "Jiaoyu buzhang mian chengda xuesheng" [Minister of Education sends words of encouragement to Cheng-Kung students], *UDN*, January 26, 1972 p. 3; "Chengda shengchan jisuanji Cai Ming-yuan tigong zijin" [Cheng-Kung makes a calculator, Mingyuan Cai offers funds for a scale-up], *UDN*, January 26, 1972, p. 3; "Disidai dianzi jisuanji chengda shisheng yanzhi chenggong" [Cheng-Kung's success of developing fourth generation electronic calculator], *UDN*, January 25, 1972, p. 3.

66. "Guokehui jueding yu chengda hezuo yanzhi lianghzong teshu yongtu diansuanji" [National Science Council works with Cheng-Kung on developing two specific-purpose computers], *EDN* February 26, 1973, p. 2; "Guokehui gongcheng kexue yanjiu zhongxin jihua yu Gaoxiong dianzi hezuo yanzhi xiaoxing dianzi jisuanji" [Engineering Science Center at the National Science Council develops small computers with Kaohsiung Electronics], *EDN*, March 12, 1970, p. 2.

67. "Taida shisheng yanzhi dianzi jisuanji" [National Taiwan University faculty and students develops a computer], *UDN*, March 28, 1972, p. 3; Barry Pak-Lee Lam, "Design of a Digital Minicomputer: The NTUEC-1000," (master's thesis, National Taiwan University, June 1972); see the foreword. Lam also thanked Mr. Tong and Miss Grace Wong for their assistance, though their Chinese full names were not included in his thesis.

68. Lam, "Design of a Digital Minicomputer," 3, 32; oral history interview with Barry (Pak-Lee) Lam, by Ling-Fei Lin, Taiwanese information technology (IT) pioneers, March 2, 2011, ref.: X6260.2012, CHM; Robert L. Morris and John R. Miller, eds., *Designing with TTL Integrated Circuits: Prepared by the IC Applications Staff of Texas Instruments Incorporated* (New York: McGraw-Hill, 1971).

69. "Guokehui weituo taida yanjiu zizhi diansuanji" [National Science Council collaborates with the National Taiwan University to develop a computer domestically], *EDN*, April 13, 1972, p. 2; "Taida shisheng yanzhi dianzi jisuanji"; Xiu-zhen Xu, "Lin Baili kaijiang: sanliu xuesheng biancheng jiechu xiaoyou" [Perspectives from Pak-Lee Lam: A struggling student becomes an outstanding alumnus], *Jin zhoukan* [Business Today], April 15, 1999, p. 102; Meidong (pen name), "Qiuxuezhong chuangye xueyizhiyong de kaimo" [Founding a start-up company while pursuing a degree; an exemplar of putting learning into practice], *EDN*, March 29, 1973, p. 7; oral history interview with Barry (Pak-Lee) Lam, by Lin, CHM.

70. Oral history interview with Barry (Pak-Lee) Lam, CHM, 4. "Huipu gongsi zhuanjia Bobuxin," 6. For the HP 2114 and 2116 computers, see "Hewlett-Packard's First Computer," Hewlett-Packard Company Archives Virtual Vault, https://www.hewlettpackardhistory.com/item/hewlett-packards-first-computer/, accessed September 22, 2020; and "2114B," in the "Early 2000" section of the website of the HP Computer Museum (not affiliated with HP Inc. or with Hewlett Packard Enterprise), Melbourne, Australia, https://www.hpmuseum.net/display_item.php?hw=97, accessed September 22, 2020. A Hewlett-Packard 2114B was demonstrated in the workshop, though the computer in use had been sold to Lunghwa College of Technology, a private college that offered a five-year course of study, with senior high school and college classes aimed at vocational training.

The workshop attracted eighteen attendees from National Taiwan University, National Tsing Hua University, Chung Cheng Institute of Technology (a military academy), ROC Air Force School of Telecommunications and Electronics, Tatung Institute of Technology, National Taipei College of Technology, and Lunghwa College. National Taipei College of Technology is known as the National Taipei University of Technology now; see chap. 1.

71. Letter from Knut H. Winter to Shih-ming Kao, May 3, 1966, MFA Files, 635.31 0199, IMH archives; letter from Knut H. Winter to Meng-Hsien Wang, International Organizations Department, Ministry of Foreign Affairs, November 21, 1966, MFA Files, 635.31.0192, IMH archives.

72. Frank Roy (LeRoy) Gustavson was born in 1921. He received a certificate from the US Navy's Radio Technician School right before World War II ended. He then pursued a BS in electronics at the University of California, Berkeley, from 1946 to 1948, and did graduate work in business administration there from 1950 to 1951. He founded Royce Instruments in Palo Alto, California, in 1955. Though it was not clear what the company specialized in, Gustavson's company's revenue reached $1.5 million US in 1959, as he noted in his résumé. He then worked for Lockheed Aircraft in Sunnyvale, California, on satellite construction, and for IAEA in Vienna on various tasks including training Iranian students and instructors on nuclear electronics and instruments. Personal history statement of Frank LeRoy Gustavson, June 1967, MFA Files 635.31 0192, IMH archives.

73. Skoumal was born in 1920. He attended the University of Prague from 1938 to 1947, earning an undergraduate degree in telecommunications electronics. He worked for Elektro-Signal in Prague from 1947 to 1949, and moved to Coventry, England, to work for the General Electric Company from 1950 to 1966. He began as a development engineer, but from 1960 his work shifted to research and analysis of potential computer customers and data transmission systems. He moved to Essex in 1966 and worked for the Marconi Company as a deputy manager of long-term planning. Personal history statement of S. Skoumal, attached with a letter from W. Roy Lucas, Resident Representative and Director of Special Fund Program in China, to Y. S. Che, Director, International Organizations Department, Ministry of Foreign Affairs, Taipei, Republic of China, March 25 1970, MFA Files 635.31 0133, IMH archives.

74. See "Gustavson Survey." For Ta Tung's manufacturing contract for IBM, see "Dianzi zhanlan, gaosu dajia, zhe shidai, shi 'zhengti dianlu' de xinshidai: yi cui xiao shuazi, naiwei dianrongqi, jikuai feitiepi, queshi jisuanji!" [Electronics expo shows it is now the age of integrated circuits: The small brush is a capacitor, and the scrap metal is a calculator!], *EDN*, October 30, 1967, p. 2.

75. "Gustavson Survey," 2, 6–8, 15. While the idea of a research park was not realized in the late 1960s, it was a precursor of the Hsinchu Science Park established in 1979 and thereafter known as Taiwan's Silicon Valley.

76. "Gustavson Survey," 3–4, 12, 14.

77. S. Skoumal, "Electronic Industry of the Republic of China: Final Report of the UNIDO Expert S. Skoumal, Dipl.-Ing" (hereafter "Skoumal Report"), October 1970, p. 3. MFA Files 635.31 0133, IMH archives.

78. "Gustavson Survey," 13,19, 28–31.

79. "Skoumal Report," e.g., 8, 10–11.

80. "Gustavson Survey," 3–4.

81. "Gustavson Survey," 16–18. Pacific Wire and Cable was the company that set up Lunghwa College, which purchased a Hewlett-Packard minicomputer in 1971.

82. "Skoumal Report," 1, 17–19.

83. "Skoumal Report," 14–15, 23.

84. I thank Jeffrey R. Yost for his suggestions on Melvin Conway's work. Melvin E. Conway, "How Do Committees Invent?" *Datamation* 14, no. 4 (1968): 28–31; and Frederick P. Brooks Jr., *The Mythical Man-Month: Essays on Software Engineering*, anniv. ed. (Reading, MA: Addison-Wesley, 1995 [1975]).

85. Melvin E. Conway, "Conway's Law," https://www.melconway.com/Home/Conways_Law .html, accessed August 12, 2020.

86. For SCOT, see, e.g., Pinch and Bijker, "The Social Construction of Facts and Artifacts," and Kline and Pinch, "Users as Agents of Technological Change."

CHAPTER 6

1. "Fu zongtong jian xingzheng yuanzhang Yan jiagan zhuchi Gaoxiong jiagong chukou qu jiemu" [Vice President and Premier Chia-kan Yen unveils a statue at the inauguration of the Kaohsiung Export Processing Zone], December 3, 1966, doc. no. 006-030203-00001-115, "Minguo wushiwu nian Yan Jiagan fuzongtong huodong ji (yi)" [Vice President Chia-kan Yen events in 1966, part 1], Yan Jiagan wenwu 1966–1972 [President Chia-kan Yen files, 1966–1972], AH; and "Gaoxiong jiagong chukuoqu jin juxeng jiancheng dianli"; see also fn 26, chap. 5.

2. Hicks, *Programmed Inequality*.

3. For the electronics highway, see "Cong dianzi zhongzhen dao dianzi gonglu" [From electronics town to electronics highway], *EDN*, April 20, 1971, p. 2. For Philco-Ford, see fn 26, chap. 5.

4. GI Annual Reports, 1965, 1967, and 1971. RCA set up a factory in Taiwan in 1967; see "Guonei dianzi zhipin guanmo zhanlan jintian zai taibei shi jiemu" [Expo of domestic electronics products inaugurates in Taipei today], *EDN*, October 26, 1967, p. 1; RCA Annual Report 1967, 1969; Cowie, *Capital Moves*, see 94, 96, and 127. For Philips' TV component production, see "Feilipu gongsi zai Zhubei jianchang houtian potu" [Philips' Zhubei plant schedules its groundbreaking the day after tomorrow], *EDN*, April 26, 1970, p. 5.

5. GI Annual Reports, 1967 and 1971; "Gaoxiong jiagong chukouqu fazhan dianzi gongye lixiang yuandi"; Zhou, "Zhizao bandaoti he dianjingti de Gaoxiong dianzi gongsi"; and "Gaoxiong dianzi gongsi meishang touzi guoren zhuchi."

6. "Gustavson Survey," 4.

7. Lisa Nakamura, "Indigenous Circuits: Navajo Women and the Racialization of Early Electronic Manufacture," *American Quarterly* 66, no. 4 (2014): 919–941, esp. 920 and 926; and Lisa Nakamura, "Economies of Digital Production in East Asia iPhone Girls and the Transnational Circuits of Cool," *Media Fields Journal: Critical Explorations in Media and Space* 2 (2011), http://mediafieldsjournal.org/economies-of-digital/. For a discussion of the construction of the image of women's manual dexterity in East Asia and Malaysia, see also Esther Ngan-ling Chow, *Transforming Gender and Development in East Asia* (New York: Routledge, 2002), 16; and Shruti Rana, "Fulfilling Technology's Promise: Enforcing the Rights of Women Caught in the Global High-Tech Underclass," *Berkeley Women's Law Journal* 15 (2000): 272–311, see esp. 284 and 291. Rana has emphasized how Asian immigrants were considered the most suitable workers for some assembly work that was subcontracted to individual households in Silicon Valley in the late 1990s. For immigrant women's employment in Silicon Valley in the 1980s, see Karen J. Hossfeld, "Their Logic against Them: Contradictions in Sex, Race, and Class in Silicon Valley," in *Women Workers and Global Restructuring*, ed. Kathryn Ward (New York: Cornell University Press, 1990), 149–178.

8. RCA 1967 Annual Report, p. 31, and 1969 Annual Report, p. 20.

9. "Gaoxiong jiagong chukouqu fazhan dianzi gongye lixiang yuandi," 5; Mincheng Liu, "Mantan huaqiao yu wairen touzi" [A discussion of overseas Chinese and foreigners' investment], *EDN*, January 1, 1968, p. 10.

10. For Tatung, see "Dianzi zhanlan, gaosu dajia, zhe shidai," 2; "Gustavson Survey," 21; and "Li Guoding yu gongshangye jie ying duo liyong diannao cujing jingji fanrong" [K. T. Li urges the industrialists and businesspeople to utilize computers for improving economic performance], *UDN*, November 5, 1968 p. 8. The contractor's name was not revealed in part because K. T. Li probably wanted during the inauguration ceremony to emphasize IBM's investment in its new office building instead of the company that won the deal. The term "electronic memory circuit" used in the news report alludes to semiconductor memory. But it was more likely a magnetic core memory. According to historian Ross Knox Bassett, IBM began to explore ways to manufacture MOS memory from 1967 to 1969, but large-scale production was not realized at that time. For IBM's MOS technology, see Bassett, *To the Digital Age*, chap. 7; and Jeffrey R. Yost, "Manufacturing Mainframes: Component Fabrication and Component Procurement at IBM and Sperry Univac, 1960–1975," *History and Technology* 25, no. 3 (2009): 219–235.

11. "Gustavson Survey," 4.

12. See "Qiao wai touzi anzhun shiyi jian"; "Zhongou maoyi jianru jiajing" [Trade with Europe is coming up roses], *EDN*, February 20, 1971 p. 4; "Feilinpu gongsi zai Zhubei jianchang houtian potu." According to Y. C. Lo, the board members of Philips thought Taiwan was poor and lacked the resources to support the production, but Frederik J. Philips held his own opinion against that of the majority. K. van Driel testified that most of his colleagues and friends in the Netherlands did not expect to see the subsidiary achieve much. Van Driel noted that he felt much relieved two years after Philips' operation in Taiwan, as the subsidiary proved itself in many aspects. One indicator of the company's success in Taiwan is that Philips' Kaohsiung factory did so well that the company began to build another factory in Hsinchu in 1970. See Yazi Chen, "Ji Fan Deli, Fang Peihan zhi Taiwan jixing" [A travel journal of K. van Driel and Ludo van Bergen's Taiwan

trip], Caituan faren Taiwan feilipu pinzhi wenjiao jijinhui [Philips Taiwan Quality Foundation website], a webpage published in 2007, http://www.ptqf.org.tw/declaratory.asp?lang=1&id=9, accessed August 16, 2020. For Y. C. Lo, see Manpeng Diao, *Jingli rensheng: Luo Yiqiang wan quanqiu qiye de lequ* [Life of a manager: Y. C. Lo's fun with Philips' global enterprise] (Tianxia Wenhua [Commonwealth Publishing], 2001); and Liang-rong Chen, "The Late Y.C. Lo: A Catalyst of Taiwan's Electronics Industry," *Tianxia zazhi* [Commonwealth Magazine] website, May 29, 2015, https://english.cw.com.tw/article/article.action?id=234.

13. Yap held an academic position at Erasmus University in Rotterdam in the 1960s and maintained a good relationship with Taiwan's diplomats as early as 1961 and met Chiang Kai-shek in person in 1965. In 1963, he also sent Taiwanese officials a one-page information sheet on what he knew about the PRC's trade officials' visit to the Netherlands and their interest in the field, including machinery, petroleum, and so on. See "The Month in Free China," *Taiwan Today*, April 1, 1969, https://taiwantoday.tw/news.php?unit=4&post=5908; "Ye Jihan jilai zai helan suo souji youguan zhonggong maoyi fangwentuan zhi ziliao yifen" [A letter from Kie-han Yap, forwarded from China-Europe Industrial Cooperation Association to the Executive Yuan], 014-070900-0085, June 4, 1963, Xingzhengyuan Files [Executive Yuan files], AH; and "Zongtong Jiang Zhongzheng jiejian helan qingnian qiaoling Ye Jihan hou heying" [Photograph of President Chiang Kai-shek receiving young overseas Chinese leader Kie-han Yap from the Netherlands], 002-050101-00059-015, June 10, 1965, lingxiu zhaopian ziliao jiji (wushiqi) [Leader's Photos and Documents, Part 57], Jiang Zhongzheng zongtong wenwu [President Chiang Kai-shek files], AH.

14. Hongpu Zhou, "Zhuanmen zhi jiyipan de jianyuan dianzi gongsi" [Jianyuan specifies memory plane manufacturing], *EDN*, February 24, 1968, p. 4; "Kaoxiong jiagong chukouqu fazhan dianzi gongye lixiang yuandi" [Kaohsiung Export Processing Zone is ideal for electronics industry], *EDN*, October 25, 1967, p. 5; "Feilipu gongsi zai Zhubei jianchang houtian potu," 5; Deren Liu, "Dianzi lingzujian woguo chanping you duoshao?" [How many components of electronics products are made domestically?], *EDN*, April 28, 1972, p. 2. For Chiao-Tung alumni hired by Philips, see Hsieh, "Chuxi jianada yijiuliuqi nian guoji dianzi nianhui"; Lin, "Chenggong de nanren he ta shenbian de nüren"; Xiaofan, "Cong yizhi shoushang de yanjing shuoqi."

15. Zhou, "Zhuanmen zhi jiyipan de jianyuan dianzi gongsi."

16. Zheng, "Dianzizhan zhong Wang An gongsi zhanchu guoren zizhi de diannao" 6.

17. Digital Equipment Corporation Annual Report, 1972.

18. National Research Council, Computer Science and Telecommunications Board, Committee on Innovations in Computing and Communications: Lessons from History, *Funding a Revolution: Government Support for Computing Research* (Washington, DC: The National Academies Press, 1999), 95; and Jube Shiver Jr, "Electronic Memories and Titan Systems Plan Merger," *Los Angeles Times*, March 5, 1985. Though the specific date is unclear, EMM was founded by a group of engineers and marketing employers of Telemeter Magnetics (where Erwin Tomash was between 1956 and 1962). Oral history interview of Erwin Tomash, by William Aspray in 1993, the Engineers as Executives Oral History Project, sponsored by the Center for the History of Electrical Engineering, The Institute of Electrical and Electronics Engineers, Inc., available at https://ethw.org/Oral-History:Erwin_Tomash.

19. EMM had decided to set up factories in Hong Kong and Singapore in 1966. For their third factory in East and Southeast Asia, they considered Korea, Taiwan, and Indonesia, settling in Taiwan in 1971. See "Taiwan dianzi diannao gongsi kaigong shengchan" [Taiwan Electronics and Computers Corporation begins to operate], *EDN*, February 27, 1972, p. 5; and "Meishang touzi Taiwan zai dianzi gongye shang de chengjiu (xia)" [American investment in Taiwan's electronics industry, part II], *EDN*, January 11, 1969, p. 3; Yuyi Liu, "Xianggang dianzi gongye de fazhan" [Electronics Industry in Hong Kong], *EDN*, June 20, 1968, p. 3. For EMM's Taiwanese subsidiary, see "Meiguo dianzi diannao gongsi jue zai Taizhong shechang shengchan diannao jiyipan waixiao" [EMM decides to set up a plant in Taiching to make computer memory planes for export], *EDN*, August, 3, 1971, p. 5; "Meiguo dianzi diannao gongsi huozhun zai Taizhong jiagong chukouqu shechang" [EMM is permitted to set up a plant in Taichung Export Processing Zone], *EDN*, August 7, 1971, p. 5; "Taiwan dianzi diannao gongsi zhouliu kaimu zhi dianrongqi waixiao" [EMM inaugurates on Saturday, making capacitors for export], *EDN*, February 24, 1972, p. 5.

20. For Ampex, see "Anpei dianzi gongsi jiang zai guishan shechang" [Ampex sets up its factory in Guishan], *EDN*, March 12, 1968, p. 1; Zhang, "Cong dianzi zhongzhen dao dianzi gonglu," 2; "Guoneiwai canzhan changshang mingdan jiqi chanpin" [A list of exhibition participants and products], *EDN*, November 17, 1973, p. 9.

21. For the protests and riots in Hong Kong in 1967, see Gary Ka-wai Cheung, "Impact of the 1967 Riots," in *Hong Kong's Watershed: The 1967 Riots* (Hong Kong: Hong Kong University Press, 2009), 131–142; and Christian Loh, *Underground Front: The Chinese Communist Party in Hong Kong* (Hong Kong: Hong Kong University Press, 2010), 99–124.

22. "Taiwan Creates a Place for Foreign Money," *Business Week*, July 6, 1968, pp. 88–90.

23. For "core house," see Robert Price's oral history interview, 101–102. Lam's oral history interview also mentioned that he witnessed electronics factories in Hong Kong in the 1970s, set up by companies such as Ampex, National Semiconductor, and Texas Instruments; see Oral history interview with Barry (Pak-Lee) Lam, by Lin, CHM. For Lockheed in Hong Kong, see Lockheed Annual Report 1968, p. 28.

24. Lécuyer, *Making Silicon Valley*, 204–207; and Brock and Lécuyer, "Company Profile: Solid State Journal, Volume 1, Number 2, September-October 1960," *Makers of the Microchip*, 216–222.

25. Miaofen Lü, Hongyuan Liao, Su-fen Liu, Xiaorong Tian, Jiaen Liao, Guanjie Ceng, and Jingwei Yuan, *Hongji jingyan yu Taiwan dianzi ye—Shi Zhengrong xiansheng fangwen jilu* [Acer and Taiwan's electronics industry: The reminiscences of Stan Shih] (Taipei: Institute of Modern History, Academia Sinica, 2018), 6.

26. Lili Yu, "Jiangqiu pinzhi de Zheng Xingdao" [Xingdao Zheng, striving for quality], *EDN*, December 13, 1976, p. 12.

27. "Zai dianzi chang gongzuo yao xiang bu you jiyi de diannao" [You got to think like a computer with memory units to work in an electronics plant], *EDN*, September 29, 1972, p. 9.

28. For computers and programming languages serving as metaphors to interpret individual or societal activities, see, e.g., Edwards, *Closed World*; and Héctor Beltrán, "Code Work: Thinking with the System in México," *American Anthropologist* 122, no. 3 (2020): 487–500. For material objects'

metaphorical roles in knowledge making, see, e.g., Emily Martin, "Toward an Anthropology of Immunology: The Body as Nation State," *Medical Anthropology Quarterly* (New Series) 4, no. 4 (1990): 410–426; Christine Leuenberger, "Constructions of the Berlin Wall: How Material Culture Is Used in Psychological Theory," *Social Problems* 53, no. 1 (2006): 18–37; Stefan Sperling, "Managing Potential Selves: Stem Cells, Immigrants, and German Identity," *Science and Public Policy* 31, no. 2 (2004): 139–149; and Honghong Tinn, "Between 'Magnificent Machine' and 'Elusive Device': Wassily Leontief's Interindustry Input-output Analysis and its International Applicability," *Osiris* 38 (2023): 129–146.

29. Guisheng Xu, "Zuo fangzhigong, xiang suanpan xiang jisuanji" [Textile factory workers act like an abacus or a computer], *EDN*, September 22, 1972, p. 9.

30. Van Bergen held a master's degree in mechanical engineering from Delft University of Technology.

31. Chen, "Ji Fan Deli, Fang Peihan zhi Taiwan." Van Driel stayed in Taiwan for only two years. The head of the human resources department, Maling Zhuang, gave van Driel and Van Bergen their Chinese names. Zhuang married Chaowu Lin. For Philips' operations, see "Qiao wai touzi shenyi weihui hezhun heshang touzi zhi dianzi qicai" [Foreign investment committee grants Philips' IC production], *EDN*, September 12, 1969, p.1; and "Feilipu gongsi zai Zhubei jianchang houtian potu"; and Diao, "Jingli rensheng."

32. Lécuyer, *Making Silicon Valley*, 204; and Brock and Lécuyer, "Company Profile: Solid State Journal, Volume 1, Number 2, September-October 1960," *Makers of the Microchip*, 216–222. Yield in the semiconductor production refers to the ratio of working dies to the total dies. Morris Chang compared the hourly rate of operators in 1958 to what he could make in 1952 as a research assistant at MIT in his last year as a college student there. Doing his best to support himself in college, Chang was so excited to learn that his hourly salary in 1952 made affording scrambled eggs with shrimp in Chinese restaurants possible. Chang, *Zhang Zhongmou zichuan*, 71, 117–121.

33. "Feige yuangong de xiuxian huodong" [Leisure time for Feige employees], *EDN*, July 20, 1969, p. 5. "Gaoxiong dianzi gongsi xibie Ye Shouzhang jinwan you shenghui" [Kaohsiung Electronics hosts a banquet to send off Frank Yih], *EDN*, September 13, 1970, p. 6. For Kaohsiung Electronics Corporation's semiconductor production, see fn 28, chap. 5.

34. Yiling Xiao, *Jinchaiji: Qianzhen jiagongqu nüxing laogong de koushu jiyi* [Gold hairpins: Oral histories of women laborers employed at Cianjhen Export Processing Zone] (Kaohsiung: Bureau of Cultural Affairs of Kaohsiung Municipal Government and Liwen wenhua Publishing, 2014), esp. 52. For the purchase of Philco-Ford's facilities in Kaohsiung, see, e.g., "Gaoxiong dianzi gongsi gaicheng" [Koahsiung Electronics renamed], *EDN*, June 18, 1971, p. 5.

35. Bassett, *To the Digital Age*; GI Annual Reports, 1965–1971; "Taiwan Creates a Place for Foreign Money," 88–90.

36. GI Annual Report, 1972, pp. 4, 12, 13.

37. GI Annual Reports, 1965, 1966, 1970–1974; Hsin-Hsing Chen, "Dazao diyige quanqiu zhuang-peixian: Taiwan tongyong qicai gongsi yu chengxiang yimin, 1964–1990" [To make the first global assembly line: General Instrument, Taiwan, and the institutions of rural-urban migration, 1964–1990], *Zhengda laodong xuebao* [Bulletin of Labour Research] 20 (2006): 1–48, esp. 20 and 27.

38. Readers might wonder where Intel was in the report. Sperry Rand had considered a joint venture with Intel. Yost, "Manufacturing Mainframes," 229.

39. For Motorola, see Motorola Annual Reports 1969, 1970, and 1971; "Han dianzi chanpin shuchu shangbannian jin qibaiwan meiyuan" [Korean electronics export reaches seven billion US dollars for the first half of the year], *EDN*, July 4, 1968, p. 3; "Motuoluola dianzi gongsi tuchengchang kaigong zhizao dianshiji deng" [Motorola's Tucheng plant begins to make TV sets and other products], *EDN*, November 5, 1970, p. 5; "Motuoluola zhongguo fenchang paiyuan chuxi mangu huiyi" [Motorola's ROC plant sends staff to attend the company's Bangkok meeting], *EDN*, November 4, 1973, p. 6; "Canzhan changshang mingdan ji zhanchu de chanpin" [List of companies and their products in electronics expo], *EDN*, April 28, 1972, p. 10. For Texas Instruments, see Texas Instruments Annual Report 1969; "Dezhou yiqi gongsi juban yuangong zhiqian Jiangxi" [Texas Instruments holds an orientation for employees], *EDN*, July 11, 1969, p. 6; "Shenyihui hezhun zhonghua maoyi kaifa gongsi liangxiang duiwai touzi" [Investment review committee grants two investment applications from China Trade and Development Corporation], *EDN*, March 14, 1969, p. 1; "Dezhou yiqi gongsi canzhan jingmi chanpin" [Texas Instruments participated in this year's electronics expo], *EDN*, December 17, 1970, p. 6; "Qiao wai touzi ji jishu hezuo an zuo xu tongguo shisi jian" [Fourteen applications granted, bringing more foreign and overseas Chinese investment and technical cooperation], *EDN*, May 19, 1972, p. 2. Morris Chang was a vice president at Texas Instruments in the early 1970s, where he was involved in setting up factories in Taiwan and Hong Kong; see Ruxin Zhang, *Xishuo Taiwan: Taiwan bandaoti chanye chuanqi* [Speaking of silicon in Taiwan: Legends about Taiwan's semiconductor industry] (Taipei: Pan Wenyuan Foundation and Tianxia wenhua [Commonwealth Publishing], 2006),186–187.

40. Frank Yih was sent from California to work in Hong Kong at the end of 1962 to set up a plant there. In 1965, he moved to another company that was subsequently acquired by Philco-Ford; he was then assigned to manage the plant in Taiwan in 1966. See later sections of this chapter and fn 62.

41. "Taiwan Creates a Place for Foreign Money," 88–90. Minister of Economic Affairs K. T. Li also arranged a ceremony in 1968 to publicly thank Moses Shapiro, vice chairman of the board, when he visited Taiwan, for GI's success in Taiwan that had attracted increasing foreign investment. Shapiro had visited Taiwan in May 1964 to evaluate whether the company should set up a plant in Taiwan. Shapiro was formally a lawyer involved in labor arbitration but joined Automatic Mfg. Co. in 1952. "Meiguo tongyong gongsi fu dongshizhang Xia Biluo zuo huo Li Guoding ban jiangzhuang" [K. T. Li awards GI's Vice Chairman Shapiro], *UDN*, September 6, 1968, p. 8; and "General Instrument Bides Its Time," *Business Week*, November 29, 1969, p. 118-120.

42. Regarding the years that "dapin de gongren" [Assiduous Workers] was released and re-released in Chen's albums, I consulted the copyright department of Asia Records (Yazhou changpian), a non-academic phonograph record researcher; Changle Lin, *Lai tingge ba! dang taiyu laoge yushang heijiao changpian zhanlan zhuankan* [Let's listen to songs! Classic Taiwanese songs on phonograph records], an exhibition catalog published by the Kaohsiung Museum of History in 2013; and "Zhang Ying choupai xinpian gunü de yuanwang" [Ying Zhang prepares new film, An Orphan's Wish], *UDN*, February 26, 1961, p. 6. The storyline of the movie *An Orphan's Wish* was primarily

about a Japanese woman's romance with a Taiwanese man, from the colonial era to the post–World War II era. The "orphan" referred to their daughter, who had to support herself since her mother left for Japan after World War II and her father became sick.

43. Huey-Fang Wu, "Tianzhujiao shengyanhui de she fuwu shiye: yi xindian dapinglin dehua nüzi gongyu weili (1968–1988)" [The Social Services Delivered by the Catholic Divine Word Missionaries: A Case Study of Dehua Girls' Hostel at Dapinglin, Xindian, 1968–1988," *Guoli zhengzhi daxue lishi xuebao* [The Journal of History, NCCU] 44 (November 2015): 223–280, see 235.

44. Hsin-Hsing Chen, "Dazao diyige quanqiu zhuangpeixian," 35; "Jian nügong sushe, tongyong gongsi juankuan baiwan" [GI donates a million dollor to build a dormitory for women workers], *UDN*, December 23, 1966, p. 2; "Fu Jike rexin gongyi, zeng nüzi sushe dianshan" [Fu Jike donates fans to women workers' dormitory], *EDN*, March 28, 1970, p. 6. For the dormitory in Kaohsiung, see Liu, ed., *Li Guoding: Wo de Taiwan Jingyan*, 351; and Taiwan dianying wenhua gongsi, "Gaoxiong jiagong qu nuzi xiushe" [Kaohsiung export processing zone women's dormitory], Shuwei diancang yu shuwei xuexi lianhe mulu [Taiwan e-learning and digital archives program], accessed December 1, 2020, http://catalog.digitalarchives.tw/item/00/31/a8/fa.html. In the second export processing zone set up in Kaohsiung in 1969, known as Nanzih Export Processing Zone, a dormitory for women was built as well. See Meizi Shen, "EPZs: Ideal Industrial Climate," *Taiwan Panorama*, December 1980, https://www.taiwan-panorama.com/Articles/Details?Guid=d974d0da-492d-4db4-9c3e-66cf8df8d67d&langId=3&CatId=11.

45. "Feige yuangong de xiuxian huodong"; "Wande tongkuai! Gande qijing!" [Have fun on the excursion! Have fun at work!], *EDN*, June 24, 1971, p. 11; "Jianyuan dianzi gongsi gongying yuangong xiannai" [Jianyuan supplies employees with free milk], *EDN*, August 4, 1973, p. 6; and "Yuangong jiankang yu yingyang: Ke Weilian tan jingyingzhe he laomuji" [Employee health and nutrition: Weilian Ke talks on running a company like a hen], *EDN*, August 12, 1973, p. 6.

46. "Feige gongsi nügong huan guaibing, gongkuang jianchahui jinxing diaocha" [Philco women workers contract mysterious illness, industrial and mining committee begins investigations], *UDN*, October 21, 1972, p. 3; "Gaoxiong yiming nügong cuishi, yisi zhiye 'guaibing' suozhi" [A factory woman passed away, possibily caused by a mysterious occupational disease], *UDN*, November 2, 1972, p. 3; "You yi dianzi nügong binggu jiazhang zuo xiang fayuan shenga" [Another electronics factory women dies, parents filed a suit], *UDN*, November 7, 1972, p. 3; "Nügong lihuan guaibing siwang, xiyin sanlüyixi zhongdu" [Mysterious deaths of women workers were caused by trichloroethylene], *UDN*, November 30, 1972, p. 3; "Zhuanjia zaikan sanmei gongsi, caiqu nügong niaoye huayan" [Experts order urine tests at Mitsumi], *UDN*, November 5, 1972, p. 3; "Gaoxiong jiagongqu tongzhi changshang gaishan quwuji" [Kaohsiung Export Processing Zone notified companies to stop using trichloroethylene], *EDN*, November 3, 1972, p. 6; "Jiagong chukou qu de dianzi gongye sanmei gongsi zuiju guimo" [Mitsumi is the most developed electronics company in the export processing zone], *EDN*, February 21, 1968, p. 4.

47. Yi-Ping Lin, "Sile jiwei dianzichang nügong zhihou: youji rongji de jiankang fengxian zhengyi" [After the Death of Some Electronic Workers: The Health Risk Controversies of Organic Solvents], *Keji yiliao yu shehui* [Taiwanese Journal for Studies of Science Technology and Medicine] 12 (April 2011): 61–112; and Yu-ling Ku, "Human Lives Valued Less Than Dirt: Former RCA Workers

Contaminated by Pollution Fighting Worldwide for Justice (Taiwan)," in *Challenging the Chip: Labor Rights and Environmental Justice in the Global Electronics Industry*, ed. Ted Smith, David A. Sonnenfeld, and David Naguib Pellow (Philadelphia, PA: Temple University Press, 2006), 181–190. For the most recent verdict on the RCA case, see Jason Pan, "High Court Orders RCA to Pay 24 of 246 Plaintiffs," *Taipei Times*, March 7, 2020, p. 2. Scholars have also reviewed pollution cases caused by local electronics manufacturers set up after the late 1970s, see, e.g., Shenglin Chang, Hua-mei Chiu, and Wen-ling Tu, "Breaking the Silicon Silence: Voicing Health and Environmental Impacts within Taiwan's Hsinchu Science Park," in *Challenging the Chip*, 170–180. For the pollution cases associated with electronics manufacturing in the United States, see Phaedra C. Pezzullo, "What Gets Buried in a Small Town: Toxic E-Waste and Democratic Frictions in the Crossroads of the United States," in *Histories of the Dustheap: Waste, Material Cultures, Social Justice*, ed. Stephanie Foote and Elizabeth Mazzolini (Cambridge, MA: MIT Press, 2012), 119–146; and Peter C. Little, *The Toxic Town: IBM, Pollution, and Industrial Risks* (New York: New York University Press, 2014). For the health risks faced by immigrant workers in Silicon Valley's assembly factories, see David N. Pellow and Lisa Sun-Hee Park, *The Silicon Valley of Dreams: Environmental Injustice, Immigrant Workers, and the High-Tech Global Economy* (New York: New York University Press, 2002).

48. For Yang's family background, see "Jiuban zixu" [Preface to the first edition], in *Gongchang ren* [Factory folks] (Taiwan: Shuiling wenchuang, 2019 [1975]); Huichhen Fang et al., "Wo zai Gaoxiong lianyouchang de rizi: sanwei gongren de mingyun jiaochadian" [My time at the Kaohsiung oil refinery: How three workers met], *Yancong zhi dao: women yu shihua gongcun de liangwan ge rizi* [A Smoking Island: Petrochemical Industry, Our Dangerous Companion more than Fifty Years] (Taiwan: Chunshan chuban, 2019), 63–77; and Thomas B. Gold, ed., *Selected Stories of Yang Ch'ing-Ch'u*, trans. Thomas B. Gold (Kaohsiung, TW: First Publishing, 1983). For the literary works about women factory workers, see Jiaxin Lai, "*Gongchang nü'erquan*—lun 1970–80 niandai Taiwan wenxue zhong de nügong yangmao" [Womanland of factory women: The literary images of women factory workers in Taiwan, 1970–1980], (master's thesis, History department, National Taiwan Normal University, 2007). For the themes in Yang's work, see Jiaxin Lai, "Nü'er guodu de meili yu aichou—lun Yang Qingchu *Gongchang nü'erquan* de nügong qunxiang" [The beauty and the sorrow in the womenland: An analysis of the images of women factory workers in Yang's *Womanland of Factory Women*], *Lishi jiaoyu* [History Education] 12 (June 2008): 131–186.

49. "Qiuxia's Sick Leave" [Qiuxia de bingjia], and "Our Own Manager" [Ziji de jingli] are both in *Gongchang nü'erquan* [Womanland of Factory Women] (Taipei: Shuiling wenchuang, 2019 [1978]), 23–40, 41–51. *Nü'erquan* refers to historical myths about matrilineal societies with no men in classic Chinese novels, and stories created for movies and TV dramas in Hong Kong, Taiwan, and China. I translate the term *nü'erquan* into womanland.

50. Gold, *State and Society in the Taiwan Miracle*, 89; and Lydia Kung, "Worker Consciousness," *Factory Women in Taiwan* (Ann Arbor, MI: UMI Research Press, 1987 [1978]), 171–180. Gold and Kung noticed that Taiwanese women factory workers displayed few characteristics of working-class consciousness and, hence, few demands to raise wages. Kung conducted an ethnography at three Taiwan plants owned by a US electronics manufacturer. She found that while many women workers understood that "all factories just use our labor to make money," that understanding did not develop into collective action for higher wages.

51. Lai, "Nü'er guodu de meili yu aichou," 178–180. For personal strategies against racial, class-based, and gendered management used by laborers in Silicon Valley to achieve short-term individual goals, see Hossfeld, "Their Logic against Them." For women workers' resistance to "scientific management" in export processing zones in Colombia, see Cynthia Truelove, "Disguised Industrial Proletarians in Rural Latin America: Women's Informal Sector Factory Work and the Social Reproduction of Coffee Farm Labor in Colombia," in *Women Workers and Global Restructuring*, 48–63, esp. 55–56.

52. "Seguo de banheng" [Specks on the astringent fruit], "Da dushi" [Metropolis], and "Jiandiao banbianxiang" [A portrait cut in half] are three short stories in *Waixiang nü* [Women from Other Towns] (Taipei: Shuiling wenchuang, 2019), 4–91. The stories collected in the anthology were originally published between 1983 and 1990.

53. The second export processing zone was set up in Nanzih, Kaohsiung, in 1969, known as Nanzih Export Processing Zone. Zhijie Cai, "Gongzuo de shijian jilü: butong niandai de laodong shizuo" [Time and discipline at work: poems across decades], Kulao Wang (a website that pioneered publishing news and commentaries about workers' rights), July 22, 2019, https://www.coolloud.org.tw/node/93237.

54. The poem is translated by the author and reprinted with permission from Xiang Ye. The plum snacks are *suanmei*, which can be dried or pickled, salted or sweetened. They are founded in various food cultures in the Middle East and Asia. The taste and texture is similar to dried cranberries or apricots.

55. "Shengchanxian shang," in *Gongchang nü'erquan*, 3. Xiang Ye is a pen name of Yuexiang Hu.

56. *Fly Up with Love*, directed by Shu-Sheng Chang (Shusheng Zhang), was issued by the Central Motion Picture Corp. in 1978.

57. The overlooking of women's contribution to science and technology shared some similarities with the overlooking of African Americans' contribution to the development of space science, as the NAACP's executive secretary Roy Wilkins stated in his letter to President Kennedy in 1963. In the letter he noted that African Americans who were employed at Cape Canaveral and Patrick Air Force Base were proud of being part of the NASA's space projects, but the pride was "tinged with growing bitterness," because the US government failed to ensure them "equality of access to homes, job advancement, and schooling for their children" in the area. Richard Paul and Steven Moss, "'There Was a Lot of History There': Theodis Ray," in *We Could Not Fail: The First African Americans in the Space Program* (Austin: University of Texas Press, 2015), 32–49, see esp. 45.

58. Guisheng Xu, "Dianzichang nügong buru zhuangpeiyuan haoting" [Assemblers sounds betters than electronics factory women workers], *EDN*, July 20, 1972, p. 9.

59. Ching-chu Yang, "Guipabi, shuibengshan" [Turtles climbing up the cliff, water washing away soil on the slope], in *Gongchang nü'erquan*, 73–100. The short story was first published in a newspaper in 1976. Yang meant to point out that education was not the major factor in deciding if one could start one's own business in, at least, the textile industry. The supervisor reveals to Qinglan that, while most of the women factory workers in the factory graduated from a junior high school, the owner of the factory, who is a volatile newly rich boss, only completed elementary school but was able to sell his family's land to raise funds for his textile business.

60. Kung, *Factory Women in Taiwan,* 88–111.

61. For the lack of acknowledgment of women in science and technology fields, see, e.g., Margaret Rossiter, *Women Scientists in America: Struggles and Strategies to 1940* (Baltimore, MD: Johns Hopkins University Press, 1982); and Light, "Programming." For research about women and the national economy, scholars have noted that national economic development can be complemented by a higher participation of women; see Valentine Moghadam, "Gender Dynamics of Restructuring in the Semi-Periphery," in *Engendering Wealth & Well-Being: Empowerment for Global Change* ed. R. L. Blumberg, C. A. Rakowski, I. Tinker, and M. Monteón (San Francisco, CA: Westview Press, 1995).

62. For Yih, see Zhaoyue Wang, "Gaoxiong dianzi zouguo banshiji: baiming lao yuangong 'hui niangjia' huanju" [Hundreds of Kaohsiung Electrinics employees joined a reunion], *UDN,* November 15, 2019. Yih's BS and MS in chemical engineering were from University of London and MIT (1960), respectively. He then worked at Fairchild as a process engineer and participated in introducing "plastic materials to reduce IC production manufacturing costs by a wide margin." He was assigned to set up a subsidiary for Fairchild in Hong Kong in 1962, which was "the very first semiconductor company in Southeast Asia," according to his biographical statement in 2014. In 1964, he helped the Zau family establish a company called Microelectronics in Hong Kong to make transistors, beginning from silicon wafer to finished products. Yih emphasized that the production was the earliest in "this part of world." See his biographical statement, pp. 6 and 7 in "Conference Program of the 5th World Business Ethics Forum," December 9–11, 2014, University of Macau, https://www.um.edu.mo/fba/wbef2014/doc/WBEF_Program03.pdf. I also interviewed Frank Yih on December 29, 2020. For Andrew Chew and Jisong Wang, see Chew, *Shede,* 68, 81–98; and Hsieh, "Chuxi jianada yijiuliuqi nian guoji dianzi nianhui."

63. Chew, *Shede,* 81–98.

64. "TRW [*sic*] laitai shechang zhi dianzi gongye lingjian" [TRW sets up a factory in Taiwan to make components of electronics products], *UDN,* April 2, 1966, p. 5. By September 1967, TRW had hired 377 employees in Taiwan; among them, five held a BS in EE; see "Gustavson Survey," 19. Furthermore, Lite-On also trained new entrepreneurs too. Two Provincial Taipei Institute of Technology alumni, Yinfu Ye and Maogui Lin, worked as engineers at Lite-On on LEC packaging for years, but they left to found and join electronics manufacturers Everlight Electronics (Yiguang) and Chicony Electronics (Qunguang), respectively. The two companies have been as profitable as Lite-On over time and have made computer peripherals since the 1980s.

65. Yeu-wen Chang, *Shizai de liliang: Zheng Chonghua yu taidadian de jingying zhihui* [Solid power: The business philosophy of Bruce Cheng and Delta Electronics] (Taipei: Tianxia wenhua [Commonwealth Publishing], 2010), chap. 2; and TRW Annual Report 1967.

66. Chang, *Shizai de liliang,* chap. 2; and TRW Annual Report 1967.

67. Chang, *Shizai de liliang,* chaps. 3–4; TRW Annual Report 1967; and Kathrin Hille, "Taiwanese Apple and Tesla Contractor Cuts China Headcount by Almost Half," *Financial Times,* March 18, 2021, http://www.ft.com/content/194de653-608f-480b-9871-3ebdfb6bcbbb.

68. For Y. C. Lo's employment at Philips, see Diao, *Jingli rensheng;* Chen, "The Late Y.C. Lo."

69. Diao, *Jingli rensheng*, chap. 2.

70. Chujing Peng, "Diangong 57 ji, heshang ASML Taiwan zongcai Chen Zhexiong xuezhang zhuanfang" [Interview with Zhexiong Chen, CEO of ASML's Taiwan subsidiary, class of 1968], *Voice* 424 (October 2007): 39–46; and Diao, *Jingli rensheng*, chap. 4. Quaiquai is also the brand name of a popular snack for children in Taiwan. Because quaiquai refers to being "disciplined" and "well-behaved," it is common to see maintenance engineers in the semiconductor industry in Taiwan now use Quaiquai snacks to "worship" their machineries or bring good luck to ensure the proper running of machinery. Based on a supernatural belief, the lucky charm can tame the machinery and ward off glitches. Hope Ngo, "The 'Good Luck' Snack That Makes Taiwan's Technology Behave," BBC, April 15, 2021, https://www.bbc.com/worklife/article/20210414-the-good-luck-snack-that-makes-taiwans-technology-behave.

71. Diao, *Jingli rensheng*, chaps. 3 and 4.

72. Diao, *Jingli rensheng*, 52–53.

73. Chen, "The Late Y.C. Lo"; Diao, *Jingli rensheng*, esp. chap. 6; Zhang, *Xishuo Taiwan*, 199.

74. Orient Semiconductor Electronics was founded in 1971 by two electrical engineering PhDs who received their degrees from US universities. As of 1973, Orient's products included ceramic dual in-line package chips, LED display panels, metal can ICs and transistor, hermetic ceramic packaged MOS, and bipolar LSI and MSI, and it expected to expand to plastic dual in-line package chips and LCD modules for electronic watches. Orient was the eighth largest IC packaging company in Taiwan in 2010 and 2011. Cofounder Chun-Yuan Duh (Junyuan Du) received his PhD in 1967 from Stanford University, where he studied under William B. Shockley, one of the first inventors of the transistor, and John L. Moll, a pioneer in solid-state physics. Duh was then employed by the IBM Watson Research Center at Yorktown Heights, New York, to work on semiconductor research. After realizing in 1968 that his father's health was deteriorating, he returned to Taiwan and began to work for IBM's Taiwan office. Cofounder Minhong Hong held a PhD from Michigan State University and returned to Taiwan in 1969. Hong's father, Chien-chuan Hong (Jianquan Hong), began his business by fixing radios in Taipei after World War II. He purchased parts from Japan, specifically from Kōnosuke Matsushita, the founder of Panasonic. Hong's father then invested with Matsushita in 1962, to set up Matsushita Taiwan, some product lines of which were branded under the name "National." The company is now known as the Panasonic Taiwan Group. Chew's Unitron began operations in 1969. Because Hong had known Chew when they were undergraduates at National Taiwan University, Hong visited Unitron, inspiring Hong to start his own company on IC packaging and testing. Zhengming Ou and a few of his colleagues at Kaohsiung Electronics were then recruited to Orient. See "Huatai zhuanzhi diannao diansuanji lingzujian" [Orient Semiconductor makes computer componets], *EDN*, August 5, 1973, p. 5; and Joseph Ya-Ming Lee, *Cong bandaoti kan shijie* [A World Perspective Based on the Development of the Semiconductor Industry] (Taipei: Yuanjian Tianxia [Global Views-Commonwealth Publishing], 2012), 167–171. Chun-Yuan Duh's PhD dissertation was titled "Hot Electron Transport in Silicon at High Electric Fields," Stanford University; he received his degree in 1967. For Duh, see Jianfu Jiang, "Du Chun-Yuan boshi" [Dr. Chun-Yuan Duh], *Taida dianji zhi you* [Friends of NTU's Electrical Engineering Department], published online in 2013, see https://alumni.ee.ntu.edu.tw

/?p=200, accessed October 28, 2020; and Zongwen Li, "Yizhansuoxue de Du Junyuan" [Chun-Yuan Duh plays to his strengths], *EDN*, September 3, 1976, p. 12. For Minhong Hong, see Xiumei Lin, "Jianhong jituan zongcai Minhong Hong de chuangye zhexue" [Entrepreneurial philosophy of Jianhong Group's CEO Minhong Hong], *Taida xiaoyou shuangyuekan* [NTU Alumni Bimonthly] 89 (September 2013): 59–63. For Minhong Hong's father Chien-chuan Hong and his business, see also Gold, *State and Society in the Taiwan Miracle*, 84.

75. Yunzhen Cai, "Ou Zhengming diandian chi sanwan gongban" [Zhengming Ou lays low], *Jin zhoukan* [Business Today], March 3, 2000, p. 26.

76. Wang moved to Taiwan with his family from China in 1949 when he was six years old. "Wang Dazhuang juan taida yi yi xintaibi [Stanley Wang donates 1 million dollars to NTU]," *Shijie ribao* [World Journal], March 16, 2015, https://www.pressreader.com/usa/world-journal-san-francisco /20150316/282518656980916; Maggie Li (Qiaoju Li), "Chuangzao zuida huaren siren qiye Pantronix de Wang Dazhuang (Stanley Wang)" [Entrepreneur of the largest Chinese American enterprise: Stanley Wang], in *Yu Tian Changlin jiangzuo dashi men tanxin* [The hidden laws of success: Meet masters of Tien Forum] (Taipei: Da hangjia, 2010), 77–90.

77. "Gaoxiong dianzi gongsi xibie Ye Shouzhang jinwan you shenghui"; also see fn 62. For Intersil, see Lécuyer, *Making Silicon Valley*, 264, 276–279; and Brock and Lécuyer, *Makers of the Microchip*, 236.

78. Junming Li et al., "Gongyuan Song," "Yinfu Ye," and "Maogui Lin," in *Kuashiji de chanye tuishou: ershi ge yu Taiwan gongtong chengzhang de gushi* [Remarkable industry leaders in Taiwan: Twenty stories interwoven with Taiwan's history] (Taipei: Yuanjian Tianxia [Global Views-Commonwealth Publishing], 2016), 28–43, 88–103, 150–163. "Tai Xing liangdi dezhou yiqi gongchang jiang fenbie caijian yuangong yiqian ming" [Texas Instruments is planning layoffs in Taiwan and Singapore], *EDN*, October 18, 1974, p. 6; Lingwen Yang, "Guangbaoke zhuanxing jidi luojiao Gaoxiong Song Gongyuan zheme shuo" [Lite-On's new business is based in Kaohsiung: Why does Soong choose Kaohsiung], Juheng Wang (financial news website), August 15, 2016, https://news.cnyes.com/news/id/2156610; "Lite-On Technology Corp.," *Nikkei Asia*, https://asia .nikkei.com/Companies/Lite-On-Technology-Corp, accessed November 6, 2020.

79. Song was born in Taiwan in 1949 after his father moved from China to Taiwan. He worked for RCA for a few years and was promoted to be a senior engineer, and for Ampex as a manager for eleven years. At Ampex, he moved across five of six factories in Taiwan. He later held managerial positions at AST Research and Kingston Technology in Taiwan. AST Research, Inc., was founded in California in 1980. Kingston Technology Corporation primarily makes flash memory products and other computer-related memory products and was founded in California in 1987. Both founders of Kingston, John Tu and David Sun, spent a significant time in Taiwan before they immigrated to the United States in the 1970s. Junming Li et al., "Song Fuxiang," "Cai Dugong," and "Chang Suimei,"in *Kuashiji de chanye tuishou*, 61–73, 104–118, 286–301; and "BYD Tops Bloomberg Businessweek's 12th Annual Tech 100 List," *Bloomberg Businessweek*, May 20, 2010, http://www.businessweek.com/technology/special_reports/20100520tech_100.htm.

80. Cai was born to a Taiwanese family in Taichung in 1950. Junming Li et al., "Cai Dugong," in *Kuashiji de chanye tuishou*, 104–118.

81. Junming Li et al., "Chang Suimei," in *Kuashiji de chanye tuishou*, 286–301.

82. Thomas J. Misa, *Leonardo to the Internet: Technology and Culture from the Renaissance to the Present*, 3rd ed. (Baltimore, MD: Johns Hopkins University Press, 2022), esp. chap. 10.

CHAPTER 7

1. Chung C. Tung, "The 'Personal Computer': A Fully Programmable Pocket Calculator," *Hewlett-Packard Journal* 25, no. 9 (May 1974): 7–14. Unlike HP-35, HP-45, and HP-65, HP 9100 calculator is a programmable desk calculator available in 1968.

2. "Wo lümei kexuejia Dong Zhong fazhan cheng zuixiaoxing jiasuanji" [Expatriate scientist Tung Chung developed a small computer], *EDN*, August 20, 1974, p. 2; "Ready for Its Closeup," Hewlett-Packard Company Archives Virtual Vault, accessed January 14, 2021, https://www.hewlettpackardhistory.com/item/ready-for-its-closeup/. Thomas Haigh and Paul E. Ceruzzi, *A New History of Modern Computing* (Cambridge, MA: MIT Press, 2021), 169. The concept of personal computers, different from the Apple and IBM computers that later dominated the market, may refer to the computers that a few companies and electronics enthusiasts developed in the 1970s. For example, Xerox PARC (Palo Alto Research Center) developed the Alto computer in 1973, Ed Roberts began to sell the Altair 8000 in 1975, and members of the Homebrew Computer Club, founded in 1975, created diverse products. In this context, Chung C. Tung's use of the term personal computer in 1974 reflects an early enthusiasm for ownership of a "personal" computer.

3. Paul E. Ceruzzi, "Moore's Law and Technological Determinism: Reflections on the History of Technology," *Technology and Culture* 46, no. 3 (2005): 584–593, see 590.

4. Dejan Ristanović and Jelica Protić, "Once Upon a Pocket: Programmable Calculators from the Late 1970s and Early 1980s and the Social Networks around Them," *IEEE Annals of the History of Computing* 34, no. 3 (2012): 55–66.

5. Stewart Brand, "Spacewar: Fanatic Life and Symbolic Death among the Computer Bums," December 7, 1972, *Rolling Stone*, 55–56; and Turner, *From Counterculture to Cyberculture*, 116.

6. Akihiko Yamada, "IPSJ Third Information Processing Technology Heritage Certification," *IEEE Annals of the History of Computing* 33, no. 2 (2011): 109. In another article, Yamada reviewed an exhibition in 2013 at the Museum of Science at the Tokyo University of Science, which included many electronic calculators made by Sharp, Canon, Sony, and Busicom. See Akihiko Yamada, "UNIVAC 120 in Japan," *IEEE Annals of the History of Computing* 35, no. 3 (2013): 73–74.

7. Aspray, "The Intel 4004 Microprocessor: What Constituted Invention?"; Yongdo Kim, "Interfirm Cooperation in Japan's Integrated Circuit Industry, 1960s–1970s," *Business History Review* 86, no. 4 (2012): 773–792; and Takahashi Dean, "Canon to Introduce Portable Computer with Built-In Printer Technology: An Announcement Is Expected Monday from the Costa Mesa-Based Company" *Los Angeles Times*, April 11, 1993, p. 7. Historian Yongdo Kim noted a negative effect of the Japanese calculator industry on the IC industry. While calculator companies worked closely with IC manufacturers in Japan in the 1970s, the collaboration weakened the design capacity of IC companies, because calculator companies preferred not to share critical technical information with

364

NOTES TO CHAPTER 7

IC companies. Additionally, IC companies focused merely on supplying products for the rather short-term and competitive calculator market, ignoring their research and development capacity.

8. Ksenia Tatarchenko, "'The Man with a Micro-Calculator': Digital Modernity and Late Soviet Computing Practices," in *Exploring the Early Digital*, ed. Thomas Haigh (Cham: Springer Nature Switzerland AG, 2019), 179–199.

9. Richard N. Langlois, "External Economies and Economic Progress: The Case of the Microcomputer Industry," *Business History Review* 66, no. 1 (1992): 1–50, on 17. Campbell-Kelly et al., *Computer*, 241, and Paul Freiberger and Michael Swaine, *Fire in the Valley: The Making of the Personal Computer* (Berkeley, CA: Osborne/McGraw-Hill, 1984), 40.

10. Lijuan (Janet) Wang, *Shi Min yu shuwei shidai de gushi* [Min Sze and stories of the digital age] (Keelung, TW: Hongjin Shuwei Keji, 2013), 6; Chew, *Shede*, 3–4, 90. The founding of Unitron preceded that of Wanbang Electronics (in 1970) and Orient Semiconductor Electronics (in 1971; see chap. 6).

11. Mostafa Kaveh, professor of electrical and computer engineering at the University of Minnesota, Twin Cities, recalled that he had Sze's textbook in a graduate course at University of California, Berkeley, when he received his master's degree in 1970.

12. Wang, *Shi Min yu shuwei shidai de gushi*, 16, 29–30; and Zhang, *Xishuo Taiwan*, 41–43. At the University of Washington, Sze worked on his thesis on solid-state electronics with Ling-Yun Wei, who was a Chiao-Tung alumnus. For his Bell Labs employment, Sze was interviewed and recruited by James Early at Bell Labs. For James Early, see https://ethw.org/Oral-History:James _Early. The award citation was from the IEEE Electron Devices society's website, https://eds.ieee .org/awards/j-j-ebers-award/past-j-j-ebers-award-winners.

13. Wang, *Shi Min yu shuwei shidai de gushi*, 6; Chew, *Shede*, 3–4, 90; oral history interview with Stan (Chen-Jung) Shih, by Ling-Fei Lin, Taiwanese information technology (IT) pioneers, March 21, 2011, ref.: X6261.2012, CHM, see p. 10.

14. Min-Chiuan Wang and Hsiao-Wen Zan, *Chuanqi rensheng: Zhang Junyan zhuan* [Legendary life: Chang Chun-yen] (Hsinchu, TW: National Chiao Tung University Press, 2019), 79. Tung Chaoyung founded the Orient Overseas Line shipping company. He was the father of Tung Chee-hwa, the first chief executive of Hong Kong after the sovereignty of Hong Kong was transferred from the UK to the PRC in 1997.

15. Wang, *Shi Min yu shuwei shidai de gushi*, 74–75; Chew, *Shede*, 90–98.

16. Chew, *Shede*, 90–98.

17. Wang, *Shi Min yu shuwei shidai de gushi*, 74–75. "Fuzongtong Yan Jiagan canguan Huanyu dianzi gongye gufen youxian gongsi" [Vice president Chia-kan Yen visits Huanyu Electronics], Minguo liushier nian Yan Jiagan fuzongtong huodong ji (shiyi) [Vice President Chia-kan Yen events in 1973, part 11], Yan Jiagan wenwu [President Chia-kan Yen Files, 1966–1972], archival document no. 006-030204-00021-004, July 10, 1973, AH.

18. Wang, *Shi Min yu shuwei shidai de gushi*, 74; Lü et al., *Hongji jingyan yu Taiwan dianzi ye*, 4, 6; oral history interview with Chew, August 6, 2007; oral history interview with Stan (Chen-Jung) Shih, by Lin, CHM.

19. Zhengxian Zhou, *Shi Zhenrong de diannao chuanqi* [Stan Shih: The computer legend] (Taipei: Linking Books, 1996), 51; Lü et al., *Hongji jingyan yu Taiwan dianzi ye*, 4, 6; Wang, *Shi Min yu shuwei shidai de gushi*, 74–75; "Huanyu gongsi choushe yiguan zuoye xitong chengli dianzi yanjiu zhongxin" [Unitron plans to fabricate transistors and sets up a research center], *EDN*, February 7, 1971, p. 5; Deren Liu, "Qiu Zaixing tan fazhan guoren dianzi gongye" [Zaixing Chew comments on domestic electronics industry], *EDN*, December 2, 1972, p. 6; oral history interview with Stan (Chen-Jung) Shih, by Lin, CHM.

20. In Chew's autobiography, when speaking of the walkie-talkie manufacture that led him to try calculator assembly, he alluded that Lam shared a similar career trajectory to his: Lam used to work on calculator manufacturing and assembly in the early 1970s, and later founded Quanta to work on computer assembly. Chew, *Shede*, 96; Lü et al., *Hongji jingyan yu Taiwan dianzi ye*, 6; "Huanyu gongsi jiang kuojian changfang zhizao dianzi yiqi" [Unitron expands its factories to make electronics instruments], *EDN*, December 25, 1970, p. 5.

21. "Guochan dianzi jisuanji huanyu gongsi yanzhi chenggong" [Unitron successfully makes domestic calculators], *EDN*, March 4, 1972, p. 6.

22. "Guochan dianzi jisuanji huanyu gongsi yanzhi chenggeng"; Liu, "Qiu Zaixing tan fazhan guoren dianzi gongye"; and Wang, *Shi Min yu shuwei shidai de gushi*, 74–75.

23. Chew, *Shede*, 96.

24. Chew, *Shede*, 92–95, 101–102.

25. For a history of the fax machine, see Jonathan Coopersmith, *Faxed: The Rise and Fall of the Fax Machine* (Baltimore, MD: Johns Hopkins University Press, 2015).

26. Liu, "Qiu Zaixing tan fazhan guoren dianzi gongye"; and Wang, *Shi Min yu shuwei shidai de gushi*, 74–75.

27. William Aspray has pointed out that, soon after Mostek was founded in 1969, the company supplied LSI integrated circuits for the Japanese manufacturer Busicom's calculators. Busicom meant to compete with Sharp, whose US partner was Rockwell. In addition to Rockwell, Sharp also worked with NEC and Hitachi, co-developing ICs for its calculators in the late 1960s. According to Yongdo Kim, Japanese companies tended to use US ICs at the turn of the 1970s. However, they began to purchase ICs domestically in the early 1970s, because US IC makers also got into the market of calculators and immediately became their competitors. Aspray, "The Intel 4004 Microprocessor: What Constituted Invention?"; and Kim, "Interfirm Cooperation in Japan's Integrated Circuit Industry, 1960s–1970s." For more on Mostek, see Bassett, *To the Digital Age*, 164, 192. For Mostek chips' application in calculators, see Nigel Tout, "The Arrival of the 'Calculator-on-a-Chip,'" Vintage Calculators Web Museum, 2015, http://www.vintagecalculators.com/html/the_calculator-on-a-chip.html.

28. Chew, *Shede*, 96–97; "Guochan dianzi jisuanji huanyu gongsi yanzhi chenggong"; "Huanyu yanzhi dianzi jisuanji zhengqiu mingming" [Unitron solicits names for its calculator], *EDN*, March 25, 1972, p. 8; "Diyi bu guochan dianzi jisuanji" [First locally made calculator], *EDN*, April 30, 1972, p. 5; "Woguo chan de diansuanqi qiantu guangming meili" [Locally made calculator has a strong market], *EDN*, January 7, 1974, p. 2; oral history interview with Stan (Chen-Jung) Shih, by Lin, CHM; Lü et al., *Hongji jingyan yu Taiwan dianzi ye*, 6.

29. Lü et al., *Hongji jingyan yu Taiwan dianzi ye*, 6–8; and Zhou, *Shi Zhenrong de diannao chuanqi*, 52–53; "Guochan dianzi jisuanji huanyu gongsi yanzhi chenggong."

30. Chew, *Shede*, 90–106; Wang, *Shi Min yu shuwei shidai de gushi*, 60, 74–75; Zhou, *Shi Zhenrong de diannao chuanqi*, 53–54; and "Not My Style—Acer's Shih Is Latest CEO to Turn to Print," March 26, 1998, Tech Monitor, https://techmonitor.ai/technology/not_my_style_acers_shih_is _latest_ceo_to_turn_to_print. Chew's autobiography did not specify Sen Lin, but it was revealed in Shih's oral history interview at CHM. See oral history interview with Stan (Chen-Jung) Shih, by Lin, CHM, 7. A news article noted that Peiyuan Lin was the general manager of the company as of March 4, 1972. See "Guochan dianzi jisuanji huanyu gongsi yanzhi chenggeng." Sze noted that the ITT that acquired Huanyu was the Canadian subsidiary of the ITT. But the news articles I surveyed indicated that it was the US ITT.

31. Chew, *Shede*, 99–101.

32. Chew, *Shede*, 109, 121–124. For the absence of discussion of ITT's acquisition of Huanyu, see, e.g., "Qiao wai touzi zuo xuzhun 22 jian" [Twenty-two foreign or overseas Chinese investment applications are granted], *EDN*, February 9, 1974, p. 2; "Mei guoji dianbao dianhua gongsi zong-cai Kang Liwei yixing fanghua" [International Telephone and Telegraph Company CEO visits ROC], *EDN*, April 1, 1976, p. 2. For Chew's time in East Germany, he noted that it was the era when the Intel 80286 and Intel 80386 (released in 1982 and 1985) were widely used for personal computers.

33. Chew, *Shede*, 99–101, Wang, *Shi Min yu shuwei shidai de gushi*, 74–75.

34. It was likely the TMS1802NC chip released in 1971. See Ken Shirriff, "The Surprising Story of the First Microprocessors," *IEEE Spectrum*, August 30, 2016, https://spectrum.ieee.org/tech-history /silicon-revolution/the-surprising-story-of-the-first-microprocessors; Tout, "The Arrival of the 'Calculator-on-a-Chip.'"

35. "Taiwan sheng zhengfu zhuxi Xie Dongmin fangwen Taibei shi Rongtai dianzi gufen youxian gongsi" [Provincial Chairperson Tung-min Hsieh visits Qualitron (Rongtai electronics)], archival document no. 009-030204-00058-085, December 24, 1975, "Xie Dongmin xiansheng sheng-zhengfu zhuxi shiqi zhao (wushiba)" [Provincial chairperson Tung-min Hsieh photographs], Xie Dongmin zongtong wenwu [Vice President Tung-min Hsieh Files], AH.

36. Lü et al., *Hongji jingyan yu Taiwan dianzi ye*, 7–12; Zhou, *Shi Zhenrong de diannao chuanqi*, 54–64, 69.

37. Zhou, *Shi Zhenrong de diannao chuanqi*, 58–64, 69.

38. Zhou, *Shi Zhenrong de diannao chuanqi*, 59–69. For Mao-pang Chen's tenure as the president of the association, see "Yazhou dianzi huiyi" [Asia Meeting of Electronics], *EDN*, November 24, 1969, p. 12; and Chen Kan, "Dianzi gongye mairu xin lichen, chanxiao jiegou zhengzai tuibian Zhong" [Electronics industry's milestone: Structural changes of production and consumption], *EDN*, October 28, 1980, p. 3. Mao-pang Chen vocally urged the government to aid the local elec-tronics industry in the 1970s; see Gold, *State and Society in the Taiwan Miracle*, 84; "Dianzi lingzujian gongchang fangwen tuan canguan shengbao disan chang" [Electronics industry representatives visit Sampo's third factory], *EDN*, December 11, 1971, p. 8; "Jiti dianlu ji dian jingti, wo ying tiqian

wancheng zizhi" [We should consider manufacturing ICs and transistors], *EDN*, July 18, 1973, p. 2; Mao-pang Chen, "Dui diwu qi jingjian jihua dianzi gongye fazhan de qiwang" [My thoughts on the electronics industry in the fifth economic planning project], *EDN*, January 1, 1969, p. 11; Mao-pang Chen, "Riben ren xinmu zhong de Taiwan dianzi gongye xianzhuang" [Japan's viewpoints on the status of Taiwanese electronics industry], *EDN*, December 1, 1969, p. 6.

39. Wang, "Dianzi jisuanji xiaoxing de jinnia xiaolu kanhao."

40. For TECO, see Qiukun Chen and Chengtian Guo, interviews, "Wo yu dongyuan dianji gongsi—Lin Changcheng xiansheng fangwen jilu" [TECO and I: Oral history interview with Zhangcheng Lin], *Koushu lishi* [Oral History] 2 (Feburary 1991): 200; and TECO Technology Foundation and Stephanie Shih, *Dongli dongyuan: Mada zhuanchu wuxian shengji* [The Empire of Motor: TECO] (Taipei: Tianxia wenhua [Commonwealth Publishing], 1996), 69–74. For an earlier attempt to make a computer to process Chinese characters, see fn 24, chap. 5. For Chinese language input system for computers, see Thomas Mullaney's *The Chinese Computer: A Global History of the Information Age* (Cambridge, MA: MIT Press, 2024).

41. "Woguo chan de diansuanqi qiantu guangming meili"; Lili Yu, "Dianzi zhanlan hui toushi" [An overview of electronics expo], *EDN*, November 12, 1974, p. 7; Zaixing Zheng, "Guochan dian suanqi chanxiao mairu xin jieduan" [A new era for domestic calculator production and consumption], *EDN*, June 19, 1974, p. 3; "Jisuanqi ye pan gongji yuan zaori kaifa jiti dianlu" [Calculator industry urges ITRI to develop ICs], *EDN*, July 24, 1980, p. 2; Zhengsheng Yao, "Dianzi jisuanji kongye kuaisu chengzhang" [Strong growth of electronic calculator industry], *EDN*, January 26, 1980; "Taiwan Eyes Jump in Calculator Sales," *Japan Times*, September 2, 1975, p. 9; "Systech Failure Adds to Calculator Fears," *Japan Times*, October 25, 1976, p. 5.

42. Guozhu Jiang, "Chen Jinzhang yu Haozhou dian suanji de dansheng" [The story of Chen Jinzhang, inventing Haozhou calculator], *EDN*, October 22, 1973, p. 6.

43. "Zhangcheng dianzi jishu fuwu" [Zhangcheng Electronics offers technical service], *EDN*, November 2, 1968, p. 4.

44. "Zhangcheng dianzi gongsi zhizao chawuyi, ke tixuan dian jingti" [Zhangcheng Electronics sells transistor testing equipment], *EDN*, May 7, 1972, p. 5; "Huali dian suanji Zhangcheng chanzhi jieshou dinggou" [Zhangcheng built premium desk calculators for customers], *EDN*, May 18, 1973, p. 6; "Gezhong shoushang xing dian suanji" [Review of handheld calculators], *EDN*, November 14, 1974, p. 7.

45. Shouying Yuan, "Mini diansuanqi zai jiating zhong pai yongchang" [Mini calculators come in handy at home], *EDN*, September 7, 1975, p. 5.

46. "Shengbao xiaoxing diansuanpan Zhendanhang xin jinkou" [Sampo's electronic abacuses are now available at Aurora shops], *EDN*, November 8, 1973, p. 6; "Zhendanhang zuotian juban dianzi suanpan caozuosai" [Aurora organizes speed contests for electronic abacuses], *EDN*, December 6, 1973, p. 6; Yuanmei Ouyang, "Suanpan huibuhui moluo?" [Will abacuses diminish?], *UDN*, August 24, 1973, p. 8; Chengrong Chang, "Chengyuan zhongxue juban biechuxincai bisai, suanpan zhansheng le dianzi jisuanqi" [Abacuses defeat calculators in a speed contest at Chengyuan Middle School], *UDN*, December 13, 1973, p. 6.

47. "Bianyi de xiao diansuanji daochu maidedao" [Inexpensive calculators are available everywhere], *EDN*, October 21, 1975, p. 6.

48. "Changcheng dianzi qing chuangye san zhounian juban suanpan huan dian suanji huodong" [Changcheng celebrates its third anniversary: Trade in your abacus for a calculator], *EDN*, August 20, 1975, p. 7; and "Shengbao pai diansuanji banli taijiuhuanxin" [Sampo sales: Trade in your old calculators for credit], *EDN*, July 5, 1976, p. 8.

49. Yu, "Jiangqiu pinzhi de Zheng Xingdao"; "Diwen shezhang wu zhounian, jin juxing yuangong lianhuan" [Dee Van employees gather for the company's fifth anniversary], *EDN*, January 18, 1976, p. 8; and Yuzhen Cai, "Diwen qiye bushi putong juese" [Dee Van is not a mediocre company], *Jin zhoukan* [Business Today], July 20, 2000, p. 102.

50. Xiuchen Wu, "Chengjiu Taiwan biji xing diannao wangguo de 'san'ai bang'—Lin Baili: meiyou san'ai, wo zhishi xiao gongchengshi" [The Santron team build Taiwan's kingdom of laptops—Barry Lim: without Santron, I would have been only a minor engineer], *Shangye zhoukan* [Business Weekly], April 25, 2002, pp. 90–94; "Pinhu bian fuhao: Ye guoyi, Wen Shiren chuangye chuanqi" [Millionaires from humble backgrounds: Legends of entrepreneurs Guoyi Ye and Sayling Wen], *Yi zhoukan* [Next Magazine], November 6, 2003, https://tw.nextmgz.com/realtimenews/news/13646921; Meidong, "Qiuxuezhong chuangye"; oral history interview with Barry (Pak-Lee) Lam, by Lin, CHM.

51. Oral history interview with Barry (Pak-Lee) Lam, by Lin, CHM, 6; Meidong, "Qiuxuezhong chuangye"; and Wu, "Chengjiu Taiwan biji xing diannao wangguo de 'san'ai bang.'"

52. "Shi'er weishu dianzi jisuanji, Lin Baili, Wen Shiren yanjiu Chenggong" [Twelve-digit calculator invented by Barry Lam and Saylin Wen], *EDN*, March 12, 1973, p. 5; "Pinhu bian fuhao: Ye Guoyi, Wen Shiren chuangye chuangqi."

53. Wu, "Chengjiu Taiwan biji xing diannao wangguo de 'san'ai bang'"; Zhou, *Shi Zhenrong de diannao chuanqi*, 57.

54. Oral history interview with Barry (Pak-Lee) Lam, by Lin, CHM; Wu, "Chengjiu Taiwan biji xing diannao wangguo de 'san'ai bang.'"

55. Oral history interview with Barry (Pak-Lee) Lam, by Lin, CHM; Wu, "Chengjiu Taiwan biji xing diannao wangguo de 'san'ai bang'"; Liu, "Qiu Zaixing tan fazhan guoren dianzi gongye"; Hongzhi, "Qiye xinsheng: dianzi gongye dai jiejue de nanti" [Corporate voice: Problems to be solved in the electronics industry], *EDN*, July 30, 1970, p. 6.

56. "Jiang yuanzhang mian youxiu qingnian" [Premier Chiang receives young talents], *EDN*, March 30, 1973, p. 6; "Bawei jiechu qingnian huode qingnian jiangzhang" [China Youth Corps awards eight young talented adults], *EDN*, March 21, 1972, p. 6; Meidong, "Qiuxuezhong chuangye."

57. Oral history interview with Barry (Pak-Lee) Lam, by Lin, CHM; Nigel Tout, "List of Vintage Hand-Held Calculators & Addendum to *The Collector's Guide to Pocket Calculators 'S*,'" Vintage Calculators Web Museum, http://www.vintagecalculators.com/html/calculator_book_addendum_s.html.

58. Oral history interview with Barry (Pak-Lee) Lam, by Lin, CHM. Wu, "Chengjiu Taiwan biji xing diannao wangguo de 'san'ai bang,'" 90.

59. Wu, "Chengjiu Taiwan biji xing diannao wangguo de 'san'ai bang'"; Hongwen Lin, "Xiu Chaoying shi zixun ye de Meng Changjun" [Xiu Chaoying is Lord Mengchang of the information industry], *Jin zhoukan* [Business Today], August 16, 2001, p. 46.

60. Lili Yu, "Guiguo qiaosheng Lin Baili, Liang Cizhen hezuo kaifa xin chanpin" [Overseas Chinese students Barry Lam and Chee-Chun Leung developed a brand new product], *EDN*, April 3, 1974, p. 6; "Jinbao pai diansuan ji tuichu liang xing jizhong" [Two new models of Kinpo calculators available], *EDN*, September 18, 1973, p. 6; Guozhu Jiang, "Guochan diansuan ji jiji zhengqu waixiao, jinbao zhipin jinnian chukou shiji hao" [Domestic calculator makers expand their export market, Kinpo's overseas sales excel], *EDN*, September 16, 1975, p. 7.

61. "Dianzi xianshi yiqi faguang erjiti zizhi chenggong shengchan gongying" [Light-emitting diode are now made in Taiwan], *EDN*, November 2, 1975, p. 3; "Dianzi gongye fazhan de xin fangxiang er, diannao shi dianzi shouyin ji qianchengsijin" [New directions of electronics industry, no. 2: The market of electronic cash registers is booming], *EDN*, March 10, 1977, p. 3. Morris Chang left Texas Instruments in 1983 and moved to Taiwan in 1985, so the contract must have been prior to 1983, and more likely before 1980, after which Kinpo began to manufacture terminal monitors.

62. "Faming Taiwan diyi tai diannao: pinmin Lin Baili bian baiyi fuweng" [Inventor of Taiwan's first computer: Millionaire Barry Lam came from a humble background], *Yi zhoukan* [Next Magazine], March 18, 2019, https://tw.nextmgz.com/realtimenews/news/464326. According to *EDN*'s reports on the periodical review of foreign and overseas Chinese investment, a company named "Guobao Computer" was set up in 1978 with investment from overseas Chinese; its major business was to make computer terminals; see, e.g., "Shengchan waixiao diannao zhongduan ji" [Manufacturing computer terminals for the export market], *EDN*, March 9, 1978, p. 2.

63. Oral history interview with Barry (Pak-Lee) Lam, by Lin, CHM. "Pinhu bian fuhao." A brief history of Santron and the companies funded by Lam and Ye to manufacture laptops, after they left Santron, can be found in Howard H. Yu and Willy C. Shih, "Taiwan's PC Industry, 1976–2010: The Evolution of Organizational Capabilities," *Business History Review* 88, no. 2 (2014): 341–343. For Quanta's manufacturing innovations, see Ling-Fei Lin, "Design Engineering or Factory Capability? Building Laptop Contract Manufacturing in Taiwan," *IEEE Annals of the History of Computing* 38, no. 2 (2016): 22–39.

64. Tony Beaumont, "Price War in the Calculator Business," *New Scientist*, June 29, 1972, pp. 751–748.

65. Lewis H. Young, "Why U.S. Companies Can Compete," *Journal of International Affairs* 28, no. 1 (1974): 81–90, see p. 87.

66. Nigel Tout, "The Story of the Race to Develop the Pocket Electronic Calculator," Vintage Calculators Web Museum, accessed April 9, 2021, http://www.vintagecalculators.com/html/the_pocket_calculator_race.html#Ref10.

67. "Dianzi xianshi yiqi faguang erjiti."

68. Oral history interview with Matthew F. C. Miau, by Craig Addison, Taiwanese information technology (IT) pioneers, February 10, 2011, ref.: X6264.2012, CHM. See also MiTAC's official website, https://www.mitac.com/zh-TW/company_leadership/index. "Tianzhong Chenxiong renwei dianzi chanpin yong weisuanji daiti luoji xianlu [Tanaka Tatsuo suggests the replacement of logic circuits with microprocessors in electronics products], *EDN*, August 17, 1975, p. 7. Matthew Miau switched to a marketing department of Intel before he went to Taiwan to found MiTAC.

69. "Fazhan dianzi gongye ying kaituo bandaoti shengchan" [Semiconductor manufacturing is key to electronics industry], *EDN*, October 22, 1974, p. 7; "Tianzhong Chenxiong renwei dianzi chanpin yong weisuanji daiti luoji xianlu"; Lü et al., *Hongji jingyan yu Taiwan dianzi ye*, 11, 22–23.

CHAPTER 8

1. The news video can be accessed in Harry McCracken, "Watch a TV News Report on the Wacky Apple II Knockoffs of the 1980s," *Fast Company*, March 12, 2021, https://www.fastcompany.com/90614203/watch-a-tv-news-report-on-the-wacky-apple-ii-knockoffs-of-the-1980s. The news video's earliest release date is unknown. According to *Fast Company*'s editor McCracken, the news program was aired in 1985. But a closer look at the content of the news video indicates that it was more likely to have been produced in 1983. The attorneys and US Customs Service officials interviewed in the video show an overlap with witnesses who testified in hearings before a subcommittee of the House of Representatives in summer 1983. The materials and narratives in the news program resembled topics brought up in those hearings. See *Unfair Foreign Trade Practices, Part 1*, Hearings before the Subcommittee on Oversight and Investigations of the Committee on Energy and Commerce, House of Representatives, Ninety-Eighth Congress, First Session, June 27 and July 27, 1983, serial No. 98–74.

2. Zhou, *Shi Zhenrong de diannao chuanqi*, 117–119, 133–134; oral history interview with Stan (Chen-Jung) Shih, by Lin, CHM, 17–18; Lü et al., *Hongji jingyan yu Taiwan dianzi ye*, 28–47; "Apple Counterattacks the Counterfeiter," *BusinessWeek*, August 16, 1982, p. 82; "Apple Computer Gets ITC to Investigate Infringement Charge," *Wall Street Journal*, March 1, 1983, p. 7. See pages 161 and 350 in *Unfair Foreign Trade Practices, Part 1*.

3. I use "incompatibility" to refer to "incommensurability," by which historians of science and technology describe the difference of two scientific knowledge systems, as Thomas Kuhn proposed in his *The Structure of Scientific Revolutions* (Chicago: University of Chicago Press, 2012 [1962]), e.g., 156. See also Eunjeong Ma and Michael Lynch, "Constructing the East-West Boundary: The Contested Place of a Modern Imaging Technology in South Korea's Dual Medical System," *Science, Technology, & Human Values* 39, no. 5 (2014): 639–665. Incommensurability also refers to the discrepancy of Western and Eastern worldviews on science and instruments; see Simon Schaffer, "Instruments as Cargo in the China Trade," *History of Science* 44, no. 2 (2006): 217–246, on 218; or, on architectural models, see Coaldrake, "Beyond Mimesis," in *Culture of Copying in Japan*, 199–212.

4. Zhou, *Shi Zhenrong de diannao chuanqi*, 117–119.

5. See "Jia pingguo yijia luanzhen" [Fake apples look genuine], *EDN*, December 16, 1982, p. 10; Shuwen Zheng, "Xiaojiaoshou de zhenhan" [The awe of micro-professor II], *EDN*, August 20, 1983, p. 11; and an advertisement of Micro-Professor II, *Guanli Zazhi* [Management Magazine] 100 (1982).

6. Takashi Oka, "Fitting China's Alphabet into 24 Computer Keys," *Christian Science Monitor*, August 20, 1982. Zhou, *Shi Zhenrong de diannao chuanqi*, 100–103, 135; letter from Stan Shih to John D. Dingell, chairman of the subcommittee on oversight and investigation, September 12, 1983, *Unfair Foreign Trade Practices, Part 2*, Hearings before the Subcommittee on Oversight and Investigations of the Committee on Energy and Commerce, House of Representatives, Ninety-Eighth Congress, First Session, August 2, September 21 and 23, 1983, serial No. 98–77, pp. 318–323; oral history interview with Jonney (Chong-Tang) Shih, by Craig Addison, Taiwanese information technology (IT) pioneers, February 15, 2011, ref.: X6263.2012, CHM, see pp. 6–7; and Nona Yates, "History of Microsoft," *Los Angeles Times*, April 4, 2000, https://www.latimes.com/archives/la-xpm-2000-apr-04-fi-15769-story.html.

7. Zhou, *Shi Zhenrong de diannao chuanqi*, 79–82. Lü et al., *Hongji jingyan yu Taiwan dianzi ye*, 19–20.

8. "Wei diandong wanju gongye qingming" [Speak for the arcade-console Industry: Please save our information industry], *0 & 1 Technology* 13 (May 1982): 50–52; and "Lin laoshifu xinxiang" [Letter to teacher Lin], *0 & 1 Technology* 13 (May 1982): 217. Apple's attorney, C. V. Chen, noted that arcade game console makers had shifted to manufacturing Apple II counterfeits; see Michael Park, "High-Tech Pirates Sell Look-Alikes," *Los Angeles Times*, October 13, 1982, pp. 1, 10–11. The ban on arcade game consoles and its impact on the growing computer industry was also identified by later critics and industry observers; see for example, Rone Tempest, "Taiwan, the Republic of Computers," *Los Angeles Times*, December 21, 1995, p. 1. For the arcade games ban, see Honghong Tinn, "From DIY Computers to Illegal Copies: The Controversy over Tinkering with Microcomputers in Taiwan, 1980–1984," *IEEE Annals of the History of Computing* 33, no. 2 (2011): 75–88. For Shih's note on the ban, see "Dianwan Ye hequhecong" [Where should arcade game console industry head to?], *EDN*, April 9, 1982, p. 2.

9. Peter H. Lewis, "Peripherals: Chess Program Beckons," *New York Times*, August 12, 1986, C5; Zhou, *Shi Zhenrong de diannao chuanqi*, 73–74, 94; Lü et al., *Hongji jingyan yu Taiwan dianzi ye*, 15–25. Multitech also held competitions to invite programmers to develop computer programs to play Go with and defeat human players. Stan Shih even offered a million-dollar prize to Go programmers who would meet the challenge in 1986. He explained that, because computer power in the 1980s was unlikely to beat human players, he decided to be generous about the prize, confident it was not going to be won. It is worth noting that Shih was a member of the bridge club at Chiao-Tung, and he plays Go.

10. "New peripherals," *InfoWorld*, August 9, 1982, p. 54; Zhou, *Shi Zhenrong de diannao chuanqi*, 79–95, 109, 111–117, 115; Zheng, "Xiaojiaoshou de zhenhan," 11; oral history interview with K. Y. (Kun-Yao) Lee, by Ling-Fei Lin, Taiwanese information technology (IT) pioneers, March 8, 2011, ref.: X6025.2011, CHM, 8.

11. "Diangong qicai gonghui diannao xiaozu" [News from the computer division of the Association for Electrical and Mechanical Manufacturers], *UDN*, December 28, 1981, p. 2.

12. "Non-English Basics Available on Cartridge," *InfoWorld*, August 8, 1982, p. 6; and oral history interview with Jonney (Chong-Tang) Shih, by Addison, CHM, see 6–7; oral history interview with K. Y. (Kun-Yao) Lee, by Lin, CHM, 19; oral history interview with Stan (Chen-Jung) Shih, by Lin, CHM, 26-27.

13. Zhou, *Shi Zhenrong de diannao chuanqi*, 117–118, 133; Lü et al., *Hongji jingyan yu Taiwan dianzi ye*, 28–29, 43–47.

14. Zhou, *Shi Zhenrong de diannao chuanqi*, 117–118, 133, 169–170; Lü et al., *Hongji jingyan yu Taiwan dianzi ye*, 28–29, 43–47. For Jonney Chong-Tang Shih's use of the term "reverse engineering," see pp. 6–7 in oral history interview with Jonney (Chong-Tang) Shih, by Addison, CHM. For cleanroom design, see, e.g., D. R. Siegel, "Functional Compatibility and the Law: From the Necessity of Cleanroom Development of Computer Software to the Copyrightability of Computer Hardware," *Digest of Paper: COMPCON Spring 88 Thirty-Third IEEE Computer Society International Conference*, San Francisco, CA, 1988, pp. 361–367; and Rachel Parker, "Lower Cost, More Power Are Key to Making Mac Compatibles," *InfoWorld*, February 4, 1991, p. 46.

15. Syntek semiconductor was formed in 1982; see Zhang, *Xishuo Taiwan*, 150; Lü et al., *Hongji jingyan yu Taiwan dianzi ye*, 30.

16. Zhou, *Shi Zhenrong de diannao chuanqi*, 61.

17. "New Peripherals," *InfoWorld*, August 9, 1982, p. 54; David Needle, "Multitech Announces 1st Chinese Home Computer," *InfoWorld*, August 30, 1982, pp. 1, 6; Scott Mace, "Speech Tech, Mice Draw Crowds at Mini/Micro 82," *InfoWorld*, October 11, 1982, pp. 1, 6; "Circle November for Applefest/SF," *InfoWorld*, October 11, 1982, 5; Kathy Chin and Scott Mace, "Applefest/San Francisco," *InfoWorld*, December 20, 1982, pp. 5–6; "Japanese Micros Wow NCC Crowd in Houston," *InfoWorld*, June 28, 1982, p. 5; "New peripherals," *InfoWorld*, November 22, 1982, p. 65; Yueqing Zhuang, "Guochan gerenyong diannao zaimei zhachu shou huanying" [Domestic computers welcomed in an expo in the United States], *EDN*, November 21, 1982, p. 2; oral history interview with Matthew F. C. Miau, by Addison, CHM; and MiTAC's official corporation history webpage, "Lianhua shentong jituan yu IT chanye zhi ershi nian lishi huigu" [Lianhua and the MiTAC group's history of first two decades in the IT industry], Lianhua Industrial Corporation, http://www.msgroup.com.tw/history_of_IT_biz_zh.asp; Teng-li Lin, "Taiwan diannao gongyeshi wangzhan jianzhi jihua yanjiu baogao" [A report of the construction of a website of the history of the Taiwanese computer industry], National Science and Technology Museum, November 14, 2007, p. 26.

18. A Micro-Professor II computer advertisement, *InfoWorld*, March 21, 1983, p. 47; "Reviews," *Your Computer Magazine* 2 no. 10 (cover, October 1982): 22, 28–29; and a Micro-Professor II computer advertisement, *Electronics Australia*, February 1983, p. 109; "Multitech: MPF-II," https://www.old-computers.com/museum/computer.asp?c=276&st=1.

19. A Micro-Professor II computer advertisement, *Electronics Australia*, 109.

20. A Micro-Professor II computer advertisement, *Electronics Australia*, 109

21. Zhou, *Shi Zhenrong de diannao chuanqi*, 117–119, 133–134; oral history interview with Stan (Chen-Jung) Shih, by Lin, CHM, 17–18; Lü et al., *Hongji jingyan yu Taiwan dianzi ye*, 28–47; "Apple Computer Gets ITC to Investigate Infringement Charge," *Wall Street Journal*, March 1,

1983, p. 7; letter from Stan Shih to John D. Dingell, chairman of the Subcommittee on Oversight and Investigation, September 12, 1983, in *Unfair Foreign Trade Practices, Part 2*, p. 320.

22. Zhou, *Shi Zhenrong de diannao chuanqi*, 117–119, 133–134.

23. Spaeth and King, "Computer Counterfeiting Flourishing as Asians Take a Bigger Bite of Apple II," 23.

24. Advertisements for Focus Corporation and Rong-Kuan Corporation, *0 & 1 Technology* 9 (1982): 25, 31; see Tinn, "From DIY Computers to Illegal Copies."

25. Yong-Hu Zheng, "Pinguo pai diannao fangmaopin chongchi" [The illegal copies are everywhere], *EDN*, December 16, 1982, p. 10; Meiping Xu, "Diandong wanju jinzhi yicheng dingju" [The ban on arcade games is confirmed], *UDN*, March 28, 1982, p. 3.

26. McCracken, "Watch a TV News Report on the Wacky Apple II Knockoffs of the 1980s"; see fn 1 of this chapter. *Unfair Foreign Trade Practices, Part 1*, pp. 219–222.

27. Spaeth and King, "Computer Counterfeiting Flourishing as Asians Take a Bigger Bite of Apple II," 23; Yueqing Zhuang, "Qinfan zhuanli jiangshi eming yuanbo" [Copyright infringement makes the country notorious], *EDN*, October 19, 1982, p. 3.

28. McCracken, "Watch a TV News Report on the Wacky Apple II Knockoffs of the 1980s"; see also *Unfair Foreign Trade Part 1*, pp. 292–297.

29. Tinn, "From DIY Computers to Illegal Copies."

30. Marilyn Chase, "Apple Computer Says Small Asian Firms Are Producing Counterfeit Apple II Models," *Wall Street Journal*, March 17, 1982, p. 8.

31. Gerardo Con Diaz, *Software Rights: How Patent Law Transformed Software Development in America* (New Haven, CT: Yale University Press, 2019), 197–203; Dana E. Miles, "Copyrighting Computer Software after Apple v. Franklin," *IEEE Software* 1, no. 2 (April 1984): 84–87; and "Franklin Loses Suit, Apple Has Ace in the Hole with ROM," *InfoWorld*, September 26, 1983, pp. 1, 8.

32. Chen, "The Rise and Fall of Illegal Copies of Apple II Computers," 12; "Meishang kong wofang fangmao diannao chanpin" [US companies accused local companies of copyright infringement], *EDN*, August 14, 1983, p. 2; Zhuang, "Qinfan zhuanli jiangshi eming yuanbo." I have discussed Apple's lawsuits in Taiwan in Tinn, "From DIY Computers to Illegal Copies."

33. *Unfair Foreign Trade Practices, Part 1*, p. 266; and "Apple Hits the Microprofessor II," *Popular Computing Weekly* 2, no. 32 (August 1983): 5.

34. *Unfair Foreign Trade Practices, Part 1*, p. 5.

35. *Unfair Foreign Trade Practices, Part 1*, pp. 222, 224.

36. *Unfair Foreign Trade Practices, Part 1*, pp. 189, 225. "Two Who Smuggled Counterfeit Computers Get Prison and Fines," *Wall Street Journal*, May 1, 1984, p. 16. Virginia Inman and Carrie Dolan, "Fake Apple Computers Were Smuggled from Taiwan, U.S. Grand Jury Charges," *Wall Street Journal*, February 10, 1984, p. 1.

37. "Shexian zousi fangmao diannao, wu gongsi zaimei bei qisu" [Five companies prosecuted for smuggling counterfeit computers], *EDN*, February 11, 1984, p. 3.

38. *Unfair Foreign Trade Practices, Part 1*, p. 350; Robert Batt, "Wins Inquiry by ITC: Apple Continues Fight against Counterfeits," *Computerworld*, March 7, 1983, p. 77; *Silicon Cowboys*, 2016, film directed by Jason Cohen; "Hongqi xiaojiaoshou weidiannao neixiao waixiao gongbuyingqiu" [Domestic and export markets express a high demand for Multitech's microcomputers], *EDN*, September 22, 1982, p. 5.

39. Park, "High-Tech Pirates Sell Look-Alikes," 11.

40. "High-Tech Entrepreneurs Create a Silicon Valley in Taiwan," *BusinessWeek*, May 1, 1983, pp. 34–35.

41. *Unfair Foreign Trade Practices, Part 1*, p. 272.

42. *Unfair Foreign Trade Practices, Part 1*, p. 272.

43. Zhou, *Shi Zhenrong de diannao chuanqi*, 116, 188. For Microtek, see "Taiwan Developing High Technology," *New York Times*, September 7, 1982; and Damon Darlin, "Unlikely Leader: Taiwan, Long Noted for Cheap Imitations, Becomes an Inventor," *Wall Street Journal*, June 1, 1990, A1; Zheng, "Xiaojiaoshou de zhenhan," 11.

44. Robert King, "Taiwan Set to Invade United States and Europe with Cut-rate Computers," *Asian Wall Street Journal Weekly*, June 6, 1983; cited from *Unfair Foreign Trade Practices, Part 1*, pp. 348–349.

45. David Needle, "Multitech Announces 1st Chinese Home Computer," *InfoWorld*, August 30, 1982, pp. 1, 6.

46. *Unfair Foreign Trade Practices, Part 2*, pp. 182–371.

47. *Unfair Foreign Trade Practices, Part 2*, pp. 188, 208. For Matthew Miau, see oral history interview with Matthew F. C. Miau, by Addison, CHM; and Shih-Chang Hung and Richard Whittington, "Strategies and Institutions: A Pluralistic Account of Strategies in the Taiwanese Computer Industry," *Organization Studies* 18, no. 4 (1997): 551–575, esp. 566. Matthew Miau's father, Yuxiu Miau, was the founder of Lian Hwa Foods, since the 1950s the largest flour supplier in Taiwan. He later invested in the petroleum and other chemical industries.

48. Letter from Stan Shih to John D. Dingell, *Unfair Foreign Trade Practices, Part 2*, p. 322.

49. Letter from Stan Shih to John D. Dingell, *Unfair Foreign Trade Practices, Part 2*, p. 319.

50. Letter from Stan Shih to John D. Dingell, *Unfair Foreign Trade Practices, Part 2*, p. 318.

51. Letter from Stan Shih to John D. Dingell, *Unfair Foreign Trade Practices, Part 2*, p. 319.

52. Letter from Stan Shih to John D. Dingell, *Unfair Foreign Trade Practices, Part 2*, p. 319.

53. For example, a US Customs officer noted that Sunyu Corporation's staff stated that the company's computers, seized by the US Customs, should be released and re-ported to Canada, and brought up the unsettled Franklin case to substantiate the statement. *Unfair Foreign Trade Practices, Part 1*, pp. 229–265, esp. 237.

54. Needle, "Multitech Announces 1st Chinese Home Computer."

55. Zhou, *Shi Zhenrong de diannao chuanqi*, 133; oral history interview with Stan (Chen-Jung) Shih, by Lin, CHM, 17–18; Lü et al., *Hongji jingyan yu Taiwan dianzi ye*, 28–30, 43–47; a Micro-Professor

II computer advertisement, *Electronics Australia*, February 1983, p. 109. "Pingguo gongsi fangfeng-sheng, diannao gongshi disanbo" [Apple computer states it will file the third wave of lawsuits], *UDN*, April 22, 1984, p. 3.

56. For the MCZ-1 and ZDS-1 systems, Tunnell was possibly referring to Zilog's microcomputers, the Z80 Micro Computer System and Z80 Development System. Multitech might have sold Zilog's microcomputers in Taiwan since Multitech was the official agent of Zilog in Taiwan.

57. *Unfair Foreign Trade Practices, Part 2*, p. 208.

58. Letter from Stan Shih to John D. Dingell, *Unfair Foreign Trade Practices, Part 2*, pp. 319–320. Maria Shao and Adi Ignatius, "Sold to China: Despite Political Differences, China and Taiwan Find Their Markets Are Well Suited to One Another," *Wall Street Journal*, August 8, 1985, p. 1. Historian Bo An shared with me his knowledge of microcomputers and bogus Apples brought in the early 1980s from Taiwan to Fujian, the closest Chinese province to Taiwan. Dongwon Jo's research also shows that Taiwanese fake Apples were brought or sold to South Korea and encouraged South Korea companies to manufacture computers. Dongwon Jo, "Vernacular Technical Practices beyond the Imitative/Innovative Boundary: Apple II Cloning in Early-1980s South Korea," *East Asian Science, Technology and Society: An International Journal* 16, no. 2 (2022): 157–180. These Taiwanese counterfeit Apples were believed to be sold in South Africa, South America, and Southeast Asia in October 1982. See Park, "High-Tech Pirates Sell Look-Alikes," *Los Angeles Times*.

59. *Unfair Foreign Trade Practices, Part 2*, p. 321.

60. *Unfair Foreign Trade Practices, Part 2*, pp. 188, 208.

61. "Mei yezhe niezao shishi eyi zhongshang" [American companies fabricate facts and damage Taiwanese companies], *EDN*, September 24, 1983, p. 2; for MiTAC, see "Lianhua shentong jituan yu IT chanye zhi ershi nian lishi huigu."

62. *Unfair Foreign Trade Practices, Part 2*, pp. 205, 207, 211; "Mei yezhe niezao shishi eyi zhongshang"; Teng-li Lin, "Taiwan diannao gongyeshi wangzhan jianzhi jihua yanjiu baogao," 26; and oral history interview with D. Y. (Ding-Yuan) Yang, by Ling-Fei Lin, Taiwanese information technology (IT) pioneers, Feburary 23, 2011, ref.: X6290.2012, CHM, 18.

63. *Einstein User: An Independent Quarterly Magazine* 1, no. 1 (November 1984): 2, 12. The magazines can be found on a website built by a computer user in 2006; see http://www.tatungeinstein.co.uk/.

64. *Unfair Foreign Trade Practices, Part 2*, p. 196; "Mei yezhe niezao shishi eyi zhongshang"; Anthony Spaeth, "Hong Kong's computer rip-off market draws throngs of buyers and gawkers," *Wall Street Journal*, October 26, 1982, p. 37.

65. Christopher Wipple, "Fakes! *Life* visits Taiwan's Booming Business in Counterfeit Goods," *Life*, September 1984, pp. 44–48; Shaolong Guo, "Wo yao chengqing, wo bushi fangmaozhe" [Let me clarify, I am not a counterfeiter], *EDN*, October 16, 1984, p. 2; "Chen Zongyuan zhun-bei konggao Shenghuo zazhi" [Chen Zongyuan plans to sue *Life* magazine], *EDN*, October 17, 1984, p. 1; "Meiguo shenghuo zazhi biaoshi si jiehuo wofang xinhan hou tamen jiangyao kaihui

shangliang" [Life magazine will discuss Chen's case once Chen's letter arrives], *EDN*, October 18, 1984, p. 2; "Konggao shenghuo zazhi feibangan meiguo lushi you xingqu chengban" [American law firms express interest in assisting Chen to take actions against *Life* magazine], *EDN*, February 1, 1985, p. 2.

66. Wipple, "Fakes! *Life* visits Taiwan's booming business in counterfeit goods."

67. Edward Said, *Orientalism* (New York: Vintage Books, 1994 [1978]); and Timothy Mitchell, *Colonising Egypt* (Cambridge, UK: Cambridge University Press, 1988).

68. Gyan Prakash, "Staging Science," "Body and Governmentality," and "Technologies of Government," in *Another Reason: Science and the Imagination of Modern India* (Princeton, NJ: Princeton University Press, 1999), 17–49, 123–158, 159–200; Warwick Anderson, "'Where Every Prospect Pleases and Only Man Is Vile': Laboratory Medicine as Colonial Discourse," *Critical Inquiry* 18 no. 3 (1992): 506–529; and "Excremental Colonialism: Public Health and the Poetics of Pollution," *Critical Inquiry* 21, no. 3 (1995): 640–669.

69. Rogaski, "Deficiency and Sovereignty," in *Hygienic Modernity*, 192. For Anderson's research, see the preceding note. See also da Costa Marques, "Cloning Computers: From Rights of Possession to Rights of Creation"; and my discussion of some Taiwanese social groups' recognition of imitation as an acceptable strategy for coping with the technological gap between Taiwan and other countries in the early 1980s in Tinn, "From DIY Computers to Illegal Copies."

70. "Tuohuang qingnian shijie shijie" [Ten pioneering youth], *EDN*, October 12, 1983, p. 3; Yanqun Zhou, "Lao'er zhuyi xia de RD" [R&D for number-two-ism], *EDN*, March 17, 1986, p. 6; "Shi zhenrong: yanjiu fazhan shi jushouzhilao" [Stan Shih: R&D is a piece of cake], *Commonwealth Magazine* 75, August 1, 1987, https://www.cw.com.tw/article/5038845.

71. "High-Tech Entrepreneurs Create a Silicon Valley in Taiwan"; "A Well-Worn Road to Dominance," *Economist*, September 17, 1988, p. 76; David E. Sanger, "PC Powerhouse (Made in Taiwan)," *New York Times*, September 28, 1988, D1; and Tempest, "Taiwan, the Republic of Computers."

CHAPTER 9

1. *Silicon Cowboys*, 2016, directed by Jason Cohen.

2. Lü et al., *Hongji jingyan yu Taiwan dianzi ye*, 45–47; oral history interview with Stan (Chen-Jung) Shih, by Lin, CHM, 17–18; oral history interview with D. Y. (Ding-Yuan) Yang, by Lin, CHM, 18–26; "Shiliu weiyuan diannao denglu meiguo chenggong" [16-bit computers successfully landed in the United States], *EDN*, July 22, 1984, p. 3; "How Taiwan Missed the Boat"; Alexander Besher, "Hong Kong's Microcomputer Industry: Pirates Eye Market for IBM PC Clones," *InfoWorld*, March 12, 1984, pp. 79–80.

3. Lü et al., *Hongji jingyan yu Taiwan dianzi ye*, 45–47; Zhou, *Shi Zhenrong de diannao chuanqi*, 131–132; oral history interview with Stan (Chen-Jung) Shih, by Lin, CHM, 17-18; "Shuishuo taiwan shiqu liangji" [Who asserts that Taiwan has missed the opportunity], *EDN*, September 19, 1984, p. 3; "Shiliu weiyuan diannao denglu meiguo chenggong"; Denis Caruso, "IBM Wins Disputes

over PC Copyrights," *InfoWorld*, February 27, 1984. For ADDS, see also Alan Alper, "Once Leader of Low-End Mart, NCR's ADDS Reverses Course," *Computerworld*, August 11, 1986, p. 93.

4. ERSO, in its capacity as an agency to assist the industry, had negotiated with Microsoft to get a better deal to license its products to Taiwanese companies. However, Digital Research was then a strong competitor of Microsoft, and owned products that performed better than MS-DOS, which ran the IBM XT PCs. For the company Digital Research, see Martin Campbell-Kelly, "Not Only Microsoft: The Maturing of the Personal Computer Software Industry, 1983–1995," in *From Airline Reservations to Sonic the Hedgehog: A History of the Software Industry* (Cambridge, MA: MIT Press, 2003), 165–177.

5. Lü et al., *Hongji jingyan yu Taiwan dianzi ye*, 46; oral history interview with Stan (Chen-Jung) Shih, by Lin, CHM, 18; Zhou, *Shi Zhenrong de diannao chuanqi*, 131–132; "Shiliu weiyuan diannao denglu meiguo chenggong"; Huichuan You, "Tupo meiguo haiguan zhongwei, hongqi xiaojiao-shou zuoxianfeng" [Multitech computers become pioneers to enter the US market], *EDN*, July 22, 1984, p. 3; Peter Ping Li, "The Evolution Of Multinational Firms from Asia: A Longitudinal Study of Taiwan's Acer Group," *Journal of Organizational Change Management* 11, no. 4 (1998): 321–337.

6. Oral history interview with D. Y. (Ding-Yuan) Yang, by Lin, CHM, 18–16.

7. "How Taiwan Missed the Boat," 110B–110E.

8. Oral history interview with D. Y. (Ding-Yuan) Yang, by Lin, CHM, 18–26; oral history interview with Stan (Chen-Jung) Shih, by Lin, CHM, 18.

9. Oral history interview with D. Y. (Ding-Yuan) Yang, by Lin, CHM, 18–26, esp. 20; see "Gongyeju pan qita changshang xiang IBM zhuce" [ERSO urges companies to obtain IBM's approval], *Information Biweekly* 4 (June 1984): 5; Meiping Xu and Xun Hu, "Buguo shibi shengyi" [It is just business], *EDN*, April 7, 1984, p. 3; "Qianzhan keji kaifa gongsi" [Qianzhan research and development corporation], *EDN*, May 23, 1984, p. 2; Denis Fred Simon, "Taiwan's Emerging Technological Trajectory: Creating New Forms of Competitive Advantage," in *Taiwan: Beyond the Economic Miracle*, ed. Denis Fred Simon and Michael Ying-Mao Kau (London: M. E. Sharpe, 1992), 123–152, esp. 141–142.

10. "How Taiwan Missed the Boat"; "Shuishuo taiwan shiqu liangji."

11. Oral history interview with Stan (Chen-Jung) Shih, by Lin, CHM, 20–22; oral history interview with Matthew F. C. Miau, by Addison, CHM, 23–25; "How Taiwan Missed the Boat." For David S. Lee's career in Friden, Diablo, and Qume, see oral history interview with George Comstock, CHM, ref: X2727.2004, by Gardner Hendrie, August 13, 2003, CHM; Robert Batt, "Analyst Urged U.S. Firms to Produce Offshore," August 22, 1983, *Computerworld*, 74; Kim Bergheim, "ITT enters Micro Market," *InfoWorld* June 11, 1984, p. 15; Yunpeng Yin and Yingchun Wu, "Xigu chuangtianxia de zhongguoren—Li Xinlin" [A Chinese entrepreneur in Silicon Valley—David S. Lee], *Commonwealth Magazine* 17, October 1, 1982, https://www.cw.com.tw/article/5103341. According to AnnaLee Saxenian, David S. Lee also sent a team to "teach Taiwanese engineers how to manufacture and test IBM PCs," see Saxenian, *New Argonauts*, 158.

12. Zhihao Yin was the founder of the Continental Engineering Corporation (Dalu Gongcheng), one of the largest civil engineering corporations in Taiwan up to the present. Taiwan's high speed

train system was built by the company in the 1990s. Oral history interview with Stan (Chen-Jung) Shih, by Lin, CHM, 20–22.

13. Oral history interview with K. Y. (Kun-Yao) Lee, by Lin, CHM, 10–16.

14. Zhou, *Shi Zhenrong de diannao chuanqi*, 47, 139–130, 181–182; oral history interview with Stan (Chen-Jung) Shih, by Lin, CHM, 20–22; oral history interview with K. Y. (Kun-Yao) Lee, by Lin, CHM, 12; and "How Taiwan Missed the Boat." For DEC's factory in Taiwan, see chap. 5, fn 23.

15. Oral history interview with K. Y. (Kun-Yao) Lee, by Lin, CHM, 10–16.

16. Oral history interview with K. Y. (Kun-Yao) Lee, by Lin, CHM, 10–16.

17. Oral history interview with K. Y. (Kun-Yao) Lee, by Lin, CHM, 14; oral history interview with Matthew F. C. Miau, by Addison, CHM, 18; "How Taiwan Missed the Boat"; Tempest, "Taiwan, the Republic of Computers"; Chunrui Wang, "Kangbo diannao jue zaitai dacaigou" [Compaq orders a large quantity of computers from Taiwanese companies], *EDN*, September 13, 1995, p. 2.

18. "How Taiwan Missed the Boat," 110B–110E; Batt, "Analyst Urged U.S. Firms to Produce Off-shore," 74.

19. Batt, "Analyst Urged U.S. Firms to Produce Offshore," 74; Robert Batt, "Second-Tier Countries Seen Challenging U.S. Vendors," *Computerworld*, June 11, 1984, pp. 151–152.

20. "High-Technology Jobs Going Overseas as U.S. Costs Rise," *New York Times*, March 19, 1983, p.1.

21. Kathy Chin, "Computer Firms Step Up Shift to Overseas Production," *InfoWorld*, April 11, 1983, pp. 14–15; Patricia Bellew Gray, "Calculated Move: Asian Computer Firms Invade the U.S. Market for Personal Machines," *Wall Street Journal*, January 10, 1986, pp. 1, 18; Brenton R. Schlender, "Higher-Quality IBM 'Clones' Put New Pressure on Computer Prices," *Wall Street Journal*, May 1, 1986, p. 1; Yingchun Wu, "IBM zai Taiwan—Shikedigu" [IBM in Taiwan—As Large as a Country], *Commonwealth Magazine*, November 1, 1986, https://www.cw.com.tw/article/5104081. For Leading Edge Hardware Products Inc., see Michael W. Miller, "IBM PC Clones Multiply Amid Price Battles: A Free-Lance Designer Sees Commodity-Type Market," *Wall Street Journal*, June 17, 1986, p. 6. Being more resourceful than start-ups, Korean conglomerates, of which Daewoo was one, invested in making personal computers and peripherals in the 1980s. In 1984, IBM also considered contracting Hyundai to make IBM's Multistation 5550 computer, which used the Intel 8086 chips, for its East Asia market. Hyundai had established its electronics subsidiary only the year before. IBM wanted to use imported parts to assemble the computer in Korea, but Hyundai ambitiously sought to make its monitors and printers in Korea. See James R. Schiffman, "IBM Mulls a Computer-Making Venture in South Korea, Government Sources Say," *Wall Street Journal*, June 26, 1984, p. 1.

22. Peter Ping Li, "Evolution of Multinational Firms," 328; Zhou, *Shi Zhenrong de diannao chuanqi*, 158–161; Jeffrey Hoffman, "Taiwan: Industry Aims to Turn Technology-Intensive," *Wall Street Journal*, October 12, 1988, B15; and Eduardo Lachica, "U.S. Mulls Easing 16-Bit Computer Export Controls," *Wall Street Journal*, August 13, 1987, p. 48.

23. Chung-Shing Lee and Michael Pecht, *The Taiwan Electronics Industry* (Boca Raton, FL: CRC Press, 1997), 33; Peter Ping Li, "Evolution of Multinational Firms," 328; Zhou, *Shi Zhenrong de*

diannao chuanqi, 158–161; Sanger, "PC Powerhouse"; Douglas R. Sease, "Trade Tangle: Taiwan's Export Boom to U.S. Owes Much to American Firms," *Wall Street Journal*, May 27, 1987, p. 1.

24. Zhou, *Shi Zhenrong de diannao chuanqi*, 158–161; oral history interview with Jonney (Chong-Tang) Shih, by Addison, CHM, 7–9; oral history interview with Stan (Chen-Jung) Shih, by Lin, CHM, 24–25; Lü et al., *Hongji jingyan yu Taiwan dianzi ye*, 78; Saxenian, *New Argonauts*, 156, 158; Curt Suplee, "Vegas' High-Tech Bargaining Chips: Wheeling and Dealing at the Computer Convention," *Washington Post*, November 15, 1986, C1; Tengli Lin, "The Social and Economic Origins of Technological Capacity: A Case Study of the Taiwanese Computer Industry," PhD dissertation, Temple University, 2000, 172–175; Patrick Waurzyniak and Hank Bannister, "Multitech to Introduce $4000 386 at Comdex," *InfoWorld*, November 3, 1986, pp. 1, 3; Tom Moran and Scott Mace, "Convergence Plans 386 Workstations; Zenith Readies 386-Based Z-348 PC," *InfoWorld*, November 3, 1986, pp. 1, 3, 6, 8; "Computer Expo Looks to Future Features," *Sun Sentinel*, November 15, 1986, p. 10C; "Vendors Supplying 80386-Based Products," *InfoWorld*, November 10, 1986, pp. 77–79; Calvin Sims, "Compaq Introduces More Powerful PC," *New York Times*, September 10, 1986, D5; "Acer sanshier weiyuan chaoji geren diannao ACER" [Acer's 32-bit premier personal computer], *EDN*, May 29, 1987, p. 9; Qiming Chen, "Hongqi kainao xingdong zaidu gaojie" [Acer's recent R&D project wins another victory], *EDN*, March 9, 1987, p. 12.

25. Oral history interview with Jonney (Chong-Tang) Shih, by Addison, CHM, 7–8.

26. Suplee, "Vegas' High-Tech Bargaining Chips," C8; Oral history interview with Jonney (Chong-Tang) Shih, by Addison, CHM, 6–7, 11, 15; oral history interview with Stan (Chen-Jung) Shih, by Lin, CHM, 24–25; David Lieberman "Motherboard Arena Widens," *Electronic Engineering Times*, December 19, 1994, p. 25; Leslie Chang, "Intel Challenges Taiwan on Its Own Turf," *Wall Street Journal*, October 31, 1995, A14.

27. Lü et al., *Hongji jingyan yu Taiwan dianzi ye*, 76–81; oral history interview with Jonney (Chong-Tang) Shih, by Addison, CHM, 14–16; oral history interview with Stan (Chen-Jung) Shih, by Lin, CHM, 25–27; oral history interview with D. Y. (Ding-Yuan) Yang," by Lin, CHM, 27; oral history interview with K. Y. (Kun-Yao) Lee, by Lin, CHM, 18–19; Jim Forbes, "Vendors Cook Up Clones," *InfoWorld*, April 29, 1985, 69–70; Priscilla M. Chabal and Edward Warner, "Boards Bring 80386 Power to PCs, XTs," *InfoWorld*, June 1, 1987, p. 1; Zhou, *Shi Zhenrong de diannao chuanqi*, 169.

28. Wu and Tseng, "Taiwan's Information Technology Industry," 85; Chabal and Warner, "Boards Bring 80386 Power to PCs, XTs"; Forbes, "Vendors Cook Up Clones," 69–70.

29. Ching-lou Liaw, "A Young Man's Dream Comes True," *Jingji zazhi* [Economic Magazine] 32 (December 1986), https://classictech.files.wordpress.com/2010/03/1987-acer-presskit-introducing -thenew-386-based-multitech-1100-5-8-87.pdf.

30. Zhou, *Shi Zhenrong de diannao chuanqi*, 172–173; John A. Matthews and Charles C. Snow, "A Conversation with Acer Group's Stan Shih on Global Strategy and Management," *Organization Dynamics* 27, no. 1 (1998): 65–74; Don Shapiro, "Ronald McDonald, Meet Stan Shih," *Sales & Marketing Management* 147, no. 11 (1995): 85.

31. Zhou, *Shi Zhenrong de diannao chuanqi*, 165–186; Paul B. Carroll, "IBM Demands Clone Firms Pay Retroactive Fee," *Wall Street Journal*, May 17, 1988, p. 1; McGregor, "IBM Uses Clout to

Force," 1; Carla Lazzareschi, "IBM Will Pursue Back Fees for 'Clones' of PCs," *Los Angeles Times*, May 18, 1988, p. 2; "A Well-Worn Road to Dominance," 76; *Silicon Cowboys*, 2016, directed by Jason Cohen.

32. McGregor, "IBM Uses Clout."

33. McGregor, "IBM Uses Clout"; Carroll, "IBM Demands Clone Firms Pay Retroactive Fee"; G. Pascal Zachary, "Dell to Pay IBM Royalty on Clones," *Journal of Commerce Online*, June 15, 1988, https://www.joc.com/article/dell-pay-ibm-royalty-clones_19880615.html.

34. Zhou, *Shi Zhenrong de diannao chuanqi*, 165–171; McGregor, "IBM Uses Clout"; Ron Copeland, "Compaq Gets MCA License but Denies Plans to Use It," *InfoWorld*, July 17, 1989, p. 1, 93; Patricia Keefe and Michael Alexander, "Compaq to Pay IBM Patent Bill," July 17, 1989, pp. 1, 101; "IBM quanlijin jiama zougao" [IBM raises royalties], *EDN*, August 12, 1988, p. 21; Zhengxian Zhou, "Yinying yulai yuduo de quanlijin xusuo" [Responses to the rising royalties], *EDN*, July 30, 1990, p. 3; "Hongqi jiaota liangchuan wenzhawenda" [Acer ensures its participation in two alliances], January 24, 1989, p. 16.

35. Jiayu Huang and Qiqing Wang, "Chongti tiaogao quanlijin IBM xianbo" [IBM raises royalties again, generating controversy], *EDN*, October 2, 1993, p. 12.

36. Zhou, *Shi Zhenrong de diannao chuanqi*, 165–171, 325–326; Sanger, "PC Powerhouse"; Peter Ping Li, "The Evolution of Multinational Firms," 328; Jingwen Zhang and Jiehan Chen, "Hongqi IBM jueding hejie" [Acer and IBM reach a settlement], *EDN*, January 4, 1990, p. 2; Zhengxian Zhou, "Hongqi jiang fen wunian zhifu peichang feiyong" [Acer will pay settlement over five years], *EDN*, May 28, 1990, p. 10; "Yang Shijian kan chanye waiyi" [Shijian Yang comments on offshoring], *EDN*, September 25, 1990, p. 10; oral history interview with Jonney (Chong-Tang) Shih, by Addison, CHM, 7–8. Award Software's business in developing BIOS systems for IBM compatibles was mentioned in two columns in *InfoWorld*. See John Dvorak, "Incredible Rumors Abound at Tandy," *InfoWorld*, March 4, 1985, p. 63; and Scott Mace, "Competitor's Problem: To Clone or Not to Clone," *InfoWorld*, March 11, 1985, p. 27.

37. Peter Ping Li, "Evolution of Multinational Firms," 329; Shapiro, "Ronald McDonald, Meet Stan Shih"; Matthews and Snow, "A Conversation with Acer Group's Stan Shih on Global Strategy and Management," 66; Zhou, *Shi Zhenrong de diannao chuanqi*, 392–398; and Tempest, "Taiwan, the Republic of Computers."

38. Shapiro, "Ronald McDonald, Meet Stan Shih"; Tempest, "Taiwan, the Republic of Computers."

39. Lin, "Design Engineering or Factory Capability?"; Zhou, *Shi Zhenrong de diannao chuanqi*, 121; Tempest, "Taiwan, the Republic of Computers"; Ashlee Vance, "Acer's Everywhere: How Did That Happen?" *New York Times*, June 27, 2009, ProQuest; and Paul Mozur, "Trouble in Taiwan's Tech Sector: Old Guard of a Hardware Boom Is Now Seen as a Hindrance," *New York Times*, January 18, 2016, B1; "Gartner Says Strong Mobile Sales Lift Worldwide PC Shipments to 12 Percent Growth in 2004," January 18, 2005, http://www.gartner.com/newsroom/id/492098.

40. Jiayu Huang, "Pingguo diannao daigong dingdan yongyue dengtai" [OEM contract from Apple comes to Taiwan], *EDN*, July 29, 1993, p. 12; Tempest, "Taiwan, the Republic of Computers."

41. Wu and Tseng, "Taiwan's Information Technology Industry," 85; Tempest, "Taiwan, the Republic of Computers"; Mark Landler, "The Silicon Godfather: The Man Behind Taiwan's Rise in the Chip Industry," *New York Times*, February 1, 2000, C1.

42. Steve Jobs and Bill Gates talked about the differences between their attitudes toward compatible systems in 2011, before Jobs passed away. See Walter Isaacson, *Steve Jobs* (New York: Simon & Schuster, 2011), 554 and Campbell-Kelly et al., *Computer*, 303.

43. Letter from Stan Shih to John D. Dingell, *Unfair Foreign Trade Practices, Part 2*, p. 319.

44. Liaw, "A Young Man's Dream Comes True."

45. Otellini, "Asian Heroes: Stan Shih"; Rod Canion, *Open: How Compaq Ended IBM's PC Dominance and Helped Invent Modern Computing* (Dallas, TX: BenBella Books, 2013); Sanger, "PC Powerhouse"; Zhou, *Shi Zhenrong de diannao chuanqi*, 524.

CHAPTER 10

1. "Sanxing TSMC qiatan hezuo keneng cong weizhi jiyiti kaishi" [Samsung and TSMC consider allying on memory chip production], *EDN*, January 26, 1989, p. 16; "Zhang Zhongmou zhengshi 1989 nian shi Sanxing Li Jianxi ceng yanlan ta" [Lee Kun-hee invited Morris Chang to work for Samsung in 1989], *Jin zhoukan* [Business Today], November 4, 2017, https://www.businesstoday.com.tw/article/category/80394/post/201711040002/; Stan Shih, *Hongqi de shiji biange* [Acer's millennium transformation] (Taipei: Tianxia wenhua, Commonwealth Publishing, 2004), 61–63; Bailu Wang, *Taijidian weishenme shen* [Why does TSMC triumph] (Taipei: Shibao chuban, 2021), 230.

2. Shih, *Hongqi de shiji biange*, 61–63; Clair Brown and Greg Linden, *Chips and Change: How Crisis Reshapes the Semiconductor Industry* (Cambridge, MA: MIT Press, 2011); Wang, *Taijidian weishenme shen*, 49; TSMC Press Release, "TSMC'S Aggressive Expansion," May 2, 2000, https://pr.tsmc.com/english/news/2222.

3. Shih, *Hongqi de shiji biange*, 61–63; Zhongxian Wu, *Toushi Taijidian* [A thorough analysis of TSMC] (Taipei: Wunan, 2006), 5.

4. Wang and Zan, *Chuanqi rensheng*, 68–75.

5. Wang and Zan, *Chuanqi rensheng*, 89, 102–108, 222; oral history interview with Chun-yen Chang, by Craig Addison, Taiwanese information technology (IT) pioneers, February 16, 2011, ref.: X6262.2012, CHM; and oral history interview with Ding-Hua Hu, by Ling-Fei Lin, Taiwanese information technology (IT) pioneers, March 9, 2011, ref.: X6289.2012, CHM; and Zhang, *Xishuo Taiwan*, 76–77.

6. Hung Chang Jimmy Lin had worked for RCA and Westinghouse by 1966. He obtained his MA and PhD in the United States in 1956. For a US Air Force contract at Westinghouse, he invented the lateral transistor that incorporated both NPN and PNP transistors in an integrated circuit, which was crucial for avoiding extraterrestrial radiation. Lin won the J. J. Ebers prize in 1978, awarded by IEEE Electron Devices Society; see "Past J.J. Ebers Award Winners," https://eds.ieee.org/awards/j-j-ebers-award/past-j-j-ebers-award-winners. Lin retired in 1990 as professor

of electrical engineering at the University of Maryland; see https://eng.umd.edu/ihof/hung-lin. For H. C. Lin, see Zhang, *Xishuo Taiwan*, 34, 66, 70, 92; H. C. Lin et al., "Lateral Complementary Transistor Structure for the Simultaneous Fabrication of Functional Blocks," *Proceedings of the IEEE* 52, no. 12 (1964): 1491–1495; "Ling Hongzhang xuezhang ronghuo Aiboshi dianzi jiangzhang" [Alumnus Hung Chang Jimmy Lin won IEEE's J.J. Ebers Award], *Voice* 273 (January 1979); Hung Chang Lin, "Huoxuan zhonghua minguo diershisan jie zhongyanyuan yuanshi ganyan" [Acceptance speech from an elected academician of Academia Sinica], *Voice* 382 (October 2000): 85–86.

7. METS was initiated by Hua Fei, a Chiao-Tung alumnus and official of the Ministry of Communications, and Wen Yuan Pan, an RCA engineer in Princeton, New Jersey, and an old friend of Fei's from college. Based in New York City, the Chinese Institute of Engineers, whose active members were predominantly Chiao-Tung alumni, also greatly contributed to the METS. Fei and Pan first met when they studied at pre–World War II Chiao-Tung University. Pan, an electrical engineering major, was class of 1935, and Fei, a civil engineering major, was class of 1934. Fei was impressed with Pan's organization of basketball and football games on campus. They had reconnected during the mobilization for reopening Chao-Tung University in Taiwan in the 1950s. Hua Fei, "Shenme shi Jiaoda de chuantong jingshen" [What is the ethos of Chiao-Tung], *Voice* 299 (May 1983): 13–16; Tsen-Cha Tsao, "Tongdao Fei Hua xuezhang" [In memorial of Hua Fei], *Voice* 305 (April 1984): 37–38; Zhang, *Xishuo Taiwan*, 68.

8. For METS, see oral history interview with Ding-Hua Hu, by Lin, CHM; and "Lijie jindai gongcheng jishu taolunhui jianbao" [Briefing documents of METS], December 1984, 006-010907-00047-004, folder "Xieren zongtong hou: jindai gongcheng taolunhui ziliao" [Post presidency: METS], series Yan Jiagan zongtong wenwu [President Chia-kan Yen Files, 1978–1993], AH. James D. Meindl joined the department of electrical engineering at Stanford University in 1967 and became provost of Rensselaer Polytechnic Institute in 1986; see https://news.stanford.edu/2020/06/15/james-d-meindl-master-integrated-circuits-dies-87/.

9. For Mao-pang Chen's suggestions on the government's role in assisting the industry, see "Jiti dianlu ji dian jingti, wo ying tiqian wancheng zizhi," 2.

10. Geoffrey Cain, *Samsung Rising: The Inside Story of the South Korean Giant That Set Out to Beat Apple and Conquer Tech* (New York: Currency, 2020), 52–54; Anthony Michell, *Samsung Electronics and the Struggle for Leadership of the Electronics Industry* (Hoboken, NJ: Wiley, 2011), 19–21.

11. Zhang, *Xishuo Taiwan*, 82–85; and Chen Huiling and Lin Qiyue, *Fang Xianqi zhuan: dianxin zhifu, keji tuishou* [Fang Xianqi: Father of telecommunications, trailblazer of technology] (Taipei: Yuanjian Tianxia [Global Views-Commonwealth Publishing], 2016), chap. 9; Texas Instruments Annual Report, 1975.

12. Zhang, *Xishuo Taiwan*, 84–94; Yang, *Sun Yunxuan zhuan*, 100–103; Ruiqun Li, "Woguo dianzi gongye youyao xiangqian mai yidabu" [A big step of Taiwan's electronics industry], *EDN*, December 29, 1975, p. 2. Hughes Aircraft Company became a prominent company in "avionics, radar, guided missiles, and missile and weapons guidance systems" in mid-1950s; see Peter Michel, Brooks Whittaker, and Chris Bruce, "Guide to the Hughes Electronics Corporation Records," University Libraries, Special Collections and Archives, University of Nevada, Las Vegas, December 27, 2018, p. 4, https://www.library.unlv.edu/speccol/finding-aids/MS-00485.pdf.

13. Zhang, *Xishuo Taiwan*, 89–94, Xianqi Fang, "Huai Pan Wenyuan xiong ersanshi" [Remembering Wen Yuan Pan], *Voice* 351 (August 1995): 48–49; "Gongye jishu yanjiuyuan dianzi gongye yanjiusuo jianbao ziliao" [Briefing documents of ERSO, ITRI], May 17, 1979, 006-010902-00006-001, Xieren zongtong hou: Gongye jishu yanjiuyuan dianzi gongye yanjiusuo jianbao [Post Presidency: ERSO, ITRI], Yan Jiagan zongtong wenwu 1978–1993 [President Chia-kan Yen Files, 1978-1993], AH; Texas Instruments Annual Report 1975.

14. But it turned out that the NMOS factory could not offer the training at some point. RCA then compensated these engineers with bipolar manufacturing techniques. Zhang, *Xishuo Taiwan*, 96–105; and Aizhu Chen, "Jiti dianlu huandang shanglu" [Integrated circuits shift gears], *Voice* 350 (June 1995): 29–34.

15. Zhang, *Xishuo Taiwan*, 106–111, 200; oral history interview with Ding-Hua Hu, by Lin, CHM, 15, 23–24, 33; oral history interview with D. Y. (Ding-Yuan) Yang, by Lin, CHM, 13–14.

16. Zhang, *Xishuo Taiwan*, 153–165; oral history interview with Robert H. C. (Shin-Chen) Tsao, by Ling-Fei Lin, Taiwanese information technology (IT) pioneers, February 17, 2011, ref.: X6291.2012, CHM.

17. Commonwealth Magazine, *Cao Xingcheng: Liandian de baye chuanqi* [Robert H. C. Tsao: UMC and its empire] (Taipei: Commonwealth, 1999), 106.

18. Xiuxian Chen, "Jiyi jingpian chuzou rihan" [Memory chip designers make chips in Japan and Korea], *Commonwealth Magazine* 76, September 1, 1987, https://www.cw.com.tw/article/5038802; oral history interview with Robert H. C. (Shin-Chen) Tsao, by Lin, CHM, 6–8.

19. Zhang, *Xishuo Taiwan*, 174–177; oral history interview with Robert H. C. (Shin-Chen) Tsao, by Lin, CHM, 6–8.

20. Zhang, *Xishuo Taiwan*, 176.

21. "Semiconductor Giants Shake Hands in Relationship Ice-Breaker," Overseas Community Affair Council English News, December 16, 2020, https://ocacnews.net/article/268175.

22. Chang, *Zhang Zhongmou Zichuan*.

23. Chang, *Zhang Zhongmou Zichuan*; and Tekla S. Perry, "Foundry Father," *IEEE Spectrum* 48, no. 5 (2011): 46–50.

24. Chang, *Zhang Zhongmou Zichuan*.

25. Texas Instruments Annual Reports, 1967 and 1972.

26. See chap. 6; Zhang, *Xishuo Taiwan*, 186–187. Morris Chang noted his visits to Taiwan in the 1970s in a talk on April 21, 2021, accessed through YouTube video, "Zhang Zhongmou: Lu fei duishou shenfang hanguo [Morris Chang: Instead of China, Korea is the rival], TVBS television channel, https://www.youtube.com/watch?v=tCErzGqPO5k.

27. Morris Chang, oral history interview of Morris Chang, by Alan Patterson, August 24, 2007, ref.: X4151.2008, CHM; Mark Liu, "Taiwan and the Foundry Model," *Nature Electronics* 4, no. 5 (2021): 318–320, see 318; Morris Chang, Jude Blanchette, and Ryan Hass, "Can Semiconductor Manufacturing Return to the US?" Brookings, April 14, 2022, https://www.brookings.edu/podcast-episode/can-semiconductor-manufacturing-return-to-the-us/.

28. For Chang's proposal at Texas Instruments, see Chris Miller, *Chip War: The Fight for the World's Most Critical Technology* (New York: Scribner, 2022), 165.

29. For Chang's proposal at General Instrument, see Zhang, *Xishuo Taiwan*, 176.

30. Chang, oral history of Morris Chang, 2007, by Patterson, CHM; oral history interview with Chintay Shih, by Ling-Fei Lin, Taiwanese information technology (IT) pioneers, February 21, 2011, ref.: X6259.2012, CHM. Carver Mead and Lynn Conway, *Introduction to VLSI System* (Boston, MA: Addison-Wesley Longman, 1979); Marco Casale-Rossi et al., "Panel: The Heritage of Mead & Conway: What Has Remained the Same, What Was Missed, What Has Changed, What Lies Ahead," in Proceedings of the Conference on Design, Automation and Test in Europe (San Jose, CA: EDA Consortium, 2013), pp. 171–175. Mead visited Taiwan after the publication of the textbook. In March 1981, the then ITRI director, Xianqi Fang, had a friend whose son studied with Mead at Caltech. ITRI thus invited Mead to Taiwan. His visit further inspired ITRI to plan on IC design training projects in the years to follow; see Zhang, *Xishuo Taiwan*, 123.

31. Chang, oral history of Morris Chang, 2007, by Patterson, CHM; and Zhang, *Xishuo Taiwan*, 195.

32. Chang, oral history of Morris Chang, 2007, by Patterson, CHM; Landler, "The Silicon God-father," C1. Shelley Rigger also argued that "TSMC's origins lie in a strategic debate, . . . should Taiwan stretch toward high-tech innovation or build on its existing strengths?" in *The Tiger Leading the Dragon: How Taiwan Propelled China's Economic Rise* (Lanham, MD: Rowman & Little-field, 2021), 76.

33. Chen, "The Late Y.C. Lo"; Diao, *Jingli rensheng*, esp. chap. 6; Zhang, *Xishuo Taiwan*, 199.

34. Chang, oral history of Morris Chang, 2007, by Patterson, CHM, 14; Zhang, *Xishuo Taiwan*, 200–203, 207; Andrew Tanzer, "Silicon In, Cash Out," *Forbes*, March 13, 1995, pp. 54–56; "Seizing an Opportunity," *Forbes*, June 1, 1998, p. 126; and "Made in Taiwan," *Forbes*, May 19, 2000, p. 186. The quote of Jensen Huang is from a comic artwork he created as a gift for Morris Chang on June 2, 2018; see Nvidia's company website at https://blogs.nvidia.com.tw/2018/06/28/nvidia-ceo-talk-about-friendship-with-tsmc-ceo-morris-chang/.

35. Tanzer, "Seizing an Opportunity"; Brown and Linden, *Chips and Change*; "How TSMC Has Mastered the Geopolitics of Chipmaking," *Economist*, May 1, 2021; "TSMC to Get 84% Share of Global Tech Market," *China Post*, September 28, 2014; Yen Nee Lee, "2 Charts Show How Much the World Depends on Taiwan for Semiconductors," CNBC, March 15, 2021, https://www.cnbc.com/2021/03/16/2-charts-show-how-much-the-world-depends-on-taiwan-for-semiconductors.html.

36. Tanzer, "Seizing an Opportunity," 126.

37. Don Clark, "Pentagon, with an Eye on China, Pushes for Help from American Tech," *New York Times*, October 25, 2019, ProQuest; Ting-Fang Cheng, "TSMC Tells US Clients It Can Meet Pentagon Security Standards," *Nikkei Asian Review*, November 2, 2019, https://asia.nikkei.com/Business/Companies/TSMC-tells-US-clients-it-can-meet-Pentagon-security-standards; Lauly Li and Ting-Fang Cheng, "Washington Pressures TSMC to Make Chips in US," *Nikkei Asian Review*, January 23, 2020, Westlaw; Don Clark and Ana Swanson, "T.S.M.C. Is Set to Build a U.S. Chip Facility, a Win for Trump," *New York Times*, May 14, 2020, https://www.nytimes.com/2020/05

/14/technology/trump-tsmc-us-chip-facility.html; and Katie Tarasov, "Inside TSMC, the Taiwanese Chipmaking Giant That's Building a New Plant in Phoenix," October 16, 2021, *CNBC News*, https://www.cnbc.com/2021/10/16/tsmc-taiwanese-chipmaker-ramping-production-to-end-chip-shortage.html.

38. Alan Crawford, Jarrell Dillard, Helene Fouquet, and Isabel Reynolds, "The World Is Dangerously Dependent on Taiwan for Semiconductors," *Bloomberg*, January 25, 2021, https://www.bloomberg.com/news/features/2021-01-25/the-world-is-dangerously-dependent-on-taiwan-for-semiconductors; Jim Tankersley and Ana Swanson, "Biden Orders a Review of Supply Chains in Manufacturing," *New York Times*, February 25, 2021, A16.

39. National Security Commission on Artificial Intelligence, Final Report, March 1, 2021, p. 217; https://reports.nscai.gov/final-report

40. National Security Commission on Artificial Intelligence, Final Report, March 1, 2021, p. 214.

41. White House Report, "Building Resilient Supply Chains, Revitalizing American Manufacturing, and Fostering Broad-Based Growth, 100-Day Reviews under Executive Order 14017," June 2021, see p. 39. https://www.whitehouse.gov/wp-content/uploads/2021/06/100-day-supply-chain-review-report.pdf.

42. Pat Gelsinger, "More Than Manufacturing: Investments in Chip Production Must Support U.S. Priorities," *Politico*, June 24, 2021, https://www.politico.com/sponsor-content/2021/06/24/more-than-manufacturing-investments-in-chip-production-must-support-us-priorities.

43. Dan Catchpole, "Intel Aims to Vault Ahead of Competition during Chip Shortage," Fortune.com, December 2, 2021, https://fortune.com/2021/12/02/intel-aims-to-vault-ahead-of-competition-during-chip-shortage/; Eric Chang, "Intel Says US Chipmakers Should Be Priority over TSMC, Samsung," *Taiwan News*, December 3, 2021, https://www.taiwannews.com.tw/en/news/4362993; Debby Wu, "TSMC and Intel Get Into a Rare Public Spat over U.S. Chipmaking," December 7, 2021, *Bloomberg*, https://www.bloomberg.com/news/newsletters/2021-12-07/tsmc-intel-trade-barbs-over-u-s-chipmaking-fully-charged.

44. Chun-mei Hwang and Luisetta Mudie, "Taiwan Should 'Keep a Tight Hold,' Defend Its Chip Industry: Morris Chang," *Radio Free Asia*, April 21, 2021, https://www.rfa.org/english/news/china/chip-04212021113729.html. The original talk could be accessed through a YouTube video by TVBS, April 21, 2021, https://www.youtube.com/watch?v=tCErzGqPO5k.

45. Ting-Fang Cheng and Lauly Li, "The Resilience Myth: Fatal Flaws in the Push to Secure Chip Supply Chains," *Nikkei Asian Review*, July 28, 2022, Westlaw; Yu Nakamura, "Taijidian chuang-shiren Zhang Zhongmou pi meiguo zhengfu he yingteer" [TSMC founder Morris Chang questioned Intel and US government], October 28, 2021, https://zh.cn.nikkei.com/industry/itelectric-appliance/46491-2021-10-28-08-41-06.html; Matthew Strong, "TSMC Says Few People Will Fall for Intel Claim That Taiwan Is Unstable," *Taiwan News*, December 3, 2021, https://www.taiwannews.com.tw/en/news/4364198; Landler, "The Silicon Godfather."

46. Guo sought the KMT ticket to run against incumbent President Tsai Ing-wen of the DPP in 2019. However, Guo did not win the primary, and Guo's opponent eventually lost the election to

Tsai. In summer 2023, Guo announced his entrance into the presidential race as an independent candidate, but withdrew in November the same year. Gou founded Foxconn in Taiwan in 1974, manufacturing turner knobs for televisions. In the 1980s, Foxconn's subcontracting products included connectors for game consoles. Foxconn made desktop computer chassis for Dell Computer and Compaq in the 1990s, and has been a contractor of Apple as early as 2002. Campbell-Kelly et al., *Computer*, 307-308.

47. "Taiwan tech giants Foxconn and TSMC to buy 10m Covid jabs," July 12, 2021, *BBC News*, https://www.bbc.com/news/business-57801031; "TSMC, Hon Hai/YongLin Foundation Donate BNT Vaccine to Taiwan CDC for COVID-19 Epidemic Prevention," TSMC Press release, July 12, 2021, https://pr.tsmc.com/english/news/2847; Natalie Tso, "Taiwan Vaccine Deal Terry Gou Delivers Where Tsai Could Not," *Time*, July 26, 2021, https://time.com/6083748/taiwan-vaccines -terry-gou-foxconn/; Samson Ellis, "Taiwan Recruits TSMC, Foxconn to Help Secure BioNTech Vaccines," *Bloomberg*, June 18, 2021, https://www.bloomberg.com/news/articles/2021-06-18 /taiwan-recruits-tsmc-foxconn-to-help-secure-biontech-vaccines; Kathrin Hille, "TSMC and Foxconn Join Forces to Secure Vaccines for Taiwan," *Financial Times*, July 11, 2021, https://www.ft .com/content/a5b8cf73-0d46-4ed8-a75f-69902b8a951c; Honghong Tinn, "Technological Exceptionalism during the Coronavirus Pandemic in Taiwan," annual meeting of the Association for Asian Studies (AAS), virtual panels, March 27, 2022.

48. Demetri Sevastopulo, "US House Speaker Nancy Pelosi to Visit Taiwan Next Month amid China Tensions," *Financial Times*, July 18, 2022, https://www.ft.com/content/20c8b5e7-e98a -494e-bcad-fc5dd2b35899#post-9ccddf82-7b47-4388-b046-25f0c928896f; Tom Mitchell, Edward White, and Felicia Schwartz, "Xi Warns Biden Not to 'Play with Fire' ahead of Taiwan Trip by Pelosi," *Financial Times*, July 28, 2022, https://www.ft.com/content/10542e5d-f27e-4212-938d -3b6956c6eb09#post-c768a63b-3420-4a48-b807-a0ad760cdb9d; Marianna Sotomayor, "Pelosi's Taiwan Trip a Culmination of Decades of Challenging China," *Washington Post*, August 2, 2022, https://www.washingtonpost.com/politics/2022/08/02/pelosis-taiwan-trip-culmination-decades -challenging-china/.

49. *Fareed Zakaria GPS*, CNN, July 31, 2022, https://transcripts.cnn.com/show/fzgps/date/2022-07 -31/segment/01.

50. Sun Yu, Gloria Li, and Tom Mitchell, "'Take Down Pelosi's Plane': Chinese Social Media Users React to Taiwan Visit," *Financial Times*, August 3, 2022, https://www.ft.com/content/78c12485 -3dcc-45f8-b779-7dd581cd66b9; Cate Cadell and Jeanne Whalen, "Pelosi Dined with Taiwan Computer Chip Executives during Her Brief Visit," *Washington Post*, August 3, 2022, Westlaw; Paul Mozur and Raymond Zhong, "In the Tug of War over Taiwan, Chips Play a Decisive Role," *New York Times*, August 3, 2022, Proquest; Liyan Qi and Rachel Liang, "Nancy Pelosi's Plane Took Three-Hour Detour to Get to Taiwan," *Wall Street Journal*, August 2, 2022, https://www.wsj .com/articles/nancy-pelosis-plane-took-a-three-hour-detour-to-get-to-taiwan-11659482515; Paul Mozur, John Liu, and Raymond Zhong, "'The Eye of the Storm': Taiwan Is Caught in a Great Game over Microchips," *New York Times*, August 29, 2022, B1.

51. Jinfang Xie, "Taiwan zao fengsuo you duo yanzhong" [How disastrous a blockade of Taiwan would be?], *Xinxinwen* [The Journalist] 1850, September 1, 2022, pp. 68–73 (originally published

online on August 19, 2022); Lesley Stahl, "As Tensions with China Rise, Life Goes on in Taiwan," *60 Minutes*, CBS News, October 9, 2022, https://www.cbsnews.com/news/taiwan-china-lesley-stahl-60-minutes-2022-10-09/.

52. Wenyi Li, "Zhiji! Taijidian 300 ren baoji jiekai meiguo taojing zhanzheng" [TSMC reserved a plane for 300 employees and their families to go to Arizona—Chip rush], *Shangye zhoukan* [Business Weekly], Nov. 16, 2022, pp. 68–80.

53. "Economy Minister Urges Calm on TSMC," *Taipei Times*, October 14, 2022, p. 12; Fengqi Lin, "Taijidian fumei kong taokong Taiwan bandaoti? [TSMC's US investment will undermine Taiwan's semiconductor industry?], *Yuanjian* [Global Views], December 6, 2022, https://www.gvm.com.tw/article/97277.

54. Live streaming of the TSMC's "first tool-in" ceremony in Arizona through CNBC Television, December 6, 2022, https://www.youtube.com/watch?v=k3a42WWcPaw. Huang and Su are Taiwanese Americans. They were born in Taiwan but moved to the United States with their parents during childhood. TSMC manufactures chips for their companies.

55. Tim Culpan, "Sorry, USA, $40 Billion Won't Buy Chip Independence," *Bloomberg*, December 6, 2022, https://www.bloomberg.com/opinion/articles/2022-12-07/-40-billion-us-semiconductor-plant-won-t-buy-independence?leadSource=uverify%20wall; Tae Kim, "The Chip War Can't Be Solved by Taiwan Semi's $40 Billion," *Barron's*, December 6, 2022, https://www.barrons.com/articles/taiwan-semi-chips-biden-apple-arizona-51670349300; "Sheping: taijidian bian "meijidian," jingzhong yijing qiaoxiang" [Alarm is sounded after TSMC becomes "USSMC": Global Times editorial], *Huanqiu Shibao* [Global Times], December 8, 2022, https://m.huanqiu.com/article/4AnFGHFue28, or the English version at https://www.globaltimes.cn/page/202212/1281494.shtml.

EPILOGUE

1. Landler, "The Silicon Godfather," C1.

2. See Tinn, "From DIY Computers to Illegal Copies"; and Campbell-Kelly et al., *Computer*, 316–317.

3. "RUN! Don't Walk! Xianka jiaofu Huang Renxun taida bidian zhici" [RUN! Don't Walk! The godfather of graphics processing card industry Jensen Huang's Commencement Address at National Taiwan University], recorded by TTV News (Taiwan Dianshi Gongsi [Taiwan Television Enterprise]), https://www.youtube.com/watch?v=_sftvrqIfIU&ab_channel=%E5%8F%B0%E8%A6%96%E6%96%B0%E8%81%9ETTVNEWS; and Ben Cohen, "The $1 Trillion Company That Started at Denny's," June 1, 2023, *Wall Street Journal*, https://www.wsj.com/articles/nvidia-ai-chips-jensen-huang-dennys d3226926?st=7kdruaqttt33hqz&reflink=desktopwebshare_permalink.

4. See, for example, William C. Kirby, Michael Shih-ta Chen, and Keith Wong, "Taiwan Semiconductor Manufacturing Company Limited: A Global Company's China Strategy," Harvard Business Review Case Study, 2015; and William C. Kirby, Billy Chang, and Daw H. Lau, "Taiwan Semiconductor Manufacturing Company Limited: A Global Company's China Strategy (B)," Harvard Business Review Case Study, 2020.

5. Ting-Fang Cheng and Lauly Li, "Taiwan's Acer in a Sticky Situation," *Nikkei Asian Review*, February 26, 2014, Westlaw; Kazunari Yamashita, "Taiwan's Business Model Ready for a Revamp," *Nikkei Asian Review*, April 16, 2014, Westlaw; Ting-Fang Cheng and Lauly Li, "Apple's Nightmare before Christmas: Supply Chain Crisis Delays Gift Deliveries," *Nikkei Asian Review*, December 9, 2021, Westlaw; "Acer Remains Focused on Gaming, Chromebook Products," February 13, 2019, ETMAG.com (*Eurotrade Magazine*'s online version), Factiva. The revenues of Acer, Quanta, and Asus in 2021 are from companiesmarketcap.com.

6. "Guangda diannao kao butiepai sifuqi tuwei" [Quanta rises by their made-to-order servers], July 10, 2015, *Nikkei*'s Chinese-language news service, https://zh.cn.nikkei.com/industry/itelectric-appliance/15109-20150710.html; Quincy Liang, "Quanta Aims for Top Spot in Notebook PC Industry in 4Q," *Taiwan Economic News*, October 11, 2000, Factiva; Drew Hasselback, "iPhone Stakeout," *National Post*, June 9, 2008, FP2; "Silicon Valley Plant Named as Apple Manufacturer," *Agence France Presse*, January 25, 2013, Factiva; "Quanta Expects Significant Growth in Server Business in 2014," ETMAG.com, March 6, 2014, Factiva; Shira Ovide, "Web Weapon: No-Frills Servers," *Dow Jones Institutional News*, June 9, 2014, Factiva; Yu and Shih, "Taiwan's PC Industry, 1976–2010."

7. See fn 46, chap. 10 for a brief history of Foxconn's subcontacting business.

8. Yijie Chen et al., "Yimiao zuo pinpai shi duzhu" [Local company's risky bid on making vaccines], *Huiliu xinwen wang* [CNEWS], May 27, 2021, https://www.youtube.com/watch?v=gLE8s-8rlFTo; "Yimiao bumai bu daigong que zichan" [Why they chose to make vaccines locally instead of procuring or subcontracting], *Yi dianshi* [Next TV], June 18, 2021, https://www.youtube.com/watch?v=9X6d-6icrR8.

9. Richard Ruelas, "10 Years Ago They Beat MIT. Today, It's Complicated," azcentral.com, *Arizona Republic*'s website, July 17, 2014, https://www.azcentral.com/story/life/az-narratives/2014/07/17/phoenix-high-school-win-mit-resonates-decade-later/12777467/; "Phoenix High School's 'Stinky' Robot Displayed at Smithsonian for Documentary Premiere," azcentral.com, February 16, 2017, https://www.azcentral.com/story/news/local/arizona-science/2017/02/16/phoenix-high-schools-stinky-robot-displayed-smithsonian-documentary-premiere/97953442/.

10. Joshua Davis, "La Vida Robot," *Wired*, April 1, 2005, https://www.wired.com/2005/04/la-vida-robot/. The Hollywood film is *Spare Parts* (directed by Sean McNamara, 2015). The two documentaries are Mary Mazzio's *Underwater Dreams* in 2014 and Greg MacGillivray's *Dream Big: Engineering Our World* in 2017. Joshua Davis later published *Spare Parts: Four Undocumented Teenagers, One Ugly Robot, and the Battle for the American Dream* (New York: Farrar, Straus and Giroux, 2014).

11. Mazzio, *Underwater Dreams*; MacGillivray, *Dream Big*; and Davis, "La Vida Robot."

12. Ruelas, "10 Years Ago They Beat MIT."

13. Ruelas, "10 Years Ago They Beat MIT."

14. For Hernandez's career, see the page created by the production team of *Dream Big*, at https://dreambigfilm.com/team/angelica-hernandez/.

BIBLIOGRAPHY

Abbate, Janet. *Inventing the Internet*. Cambridge, MA: MIT Press, 2000.

Abbate, Janet. *Recoding Gender: Women's Changing Participation in Computing*. Cambridge, MA: MIT Press, 2012.

Abraham, Itty. *The Making of the Indian Atomic Bomb: Science, Secrecy and the Postcolonial State*. London: Zed Books, 1998.

Adas, Michael. *Dominance by Design: Technological Imperatives and America's Civilizing Mission*. Cambridge, MA: Belknap Press, 2006.

Adas, Michael. *Machines as the Measure of Men: Science, Technology, and Ideologies of Western Dominance*. Ithaca, NY: Cornell University Press, 1989.

Agar, Jon. *The Government Machine: A Revolutionary History of the Computer*. Cambridge, MA: MIT Press, 2003.

Agar, Jon. "What Difference Did Computers Make?" *Social Studies of Science* 36, no. 6 (2006): 869–907.

Akera, Atsushi. *Calculating a Natural World: Scientists, Engineers, and Computers during the Rise of U.S. War Research*. Cambridge, MA: MIT Press, 2006.

Alberts, Gerard, and Ruth Oldenziel, ed. *Hacking Europe: From Computer Cultures to Demoscenes*. New York: Springer, 2014.

Allen, Joseph R. *Taipei: City of Displacements*. Seattle: University of Washington Press, 2012.

Allison, David K. "US Navy Research and Development since World War II." In *Military Enterprise and Technological Change*, edited by Merritt Roe Smith, 314–320. Cambridge, MA: MIT Press, 1985.

Anderson, Warwick. "Excremental Colonialism: Public Health and the Poetics of Pollution." *Critical Inquiry* 21, no. 3 (1995): 640–669.

Anderson, Warwick. "From Subjugated Knowledge to Conjugated Subjects: Science and Globalisation, or Postcolonial Studies of Science?" *Postcolonial Studies* 12, no. 4 (2009): 389–400.

Anderson, Warwick. "How Far Can East Asians STS Go? A Commentary." *East Asian Science, Technology and Society: An International Journal* 1, no. 2 (2007): 249–250.

Anderson, Warwick. "STS with East Asian Characteristics?" *East Asian Science, Technology and Society: An International Journal* 14, no. 1 (2020): 163–168.

Anderson, Warwick. "'Where Every Prospect Pleases and Only Man Is Vile': Laboratory Medicine as Colonial Discourse." *Critical Inquiry* 18, no. 3 (1992): 506–529.

Anderson, Warwick, and Adams Vincanne. "Pramoedya's Chickens: Postcolonial Studies of Technoscience." In *The Handbook of Science and Technology Studies*, 3rd ed., edited by Edward J. Hackett, Olga Amsterdamska, Michael Lynch, and Judy Wajcman, 181–204. Cambridge, MA: MIT Press, 2007.

Aspray, William F. "The History of Computing within the History of Information Technology." *History and Technology* 11, no. 1 (1994): 7–19.

Aspray, William F. "The Intel 4004 Microprocessor: What Constituted Invention?" *IEEE Annals of the History of Computing* 19, no. 3 (1997): 4–15.

Aspray, William F. "International Diffusion of Computer Technology, 1945–1955." *IEEE Annals of the History of Computing* 8, no. 4 (1986): 351–360.

Aspray, William F, and Bernard O. Williams. "Arming American Scientists: NSF and the Provision of Scientific Computing Facilities for Universities, 1950–1973." *IEEE Annals of the History of Computing* 16, no. 4 (1994): 60–74.

Baldwin, David. *Foreign Aid and American Foreign Policy: A Document Analysis*. New York: Frederick A. Praeger, 1966.

Bardini, Thierry, and August Horvath. "The Social Construction of the Personal Computer User." *Journal of Communication* 45, no. 3 (1995): 40–65.

Barry, Pak-Lee Lam. "Design of a Digital Minicomputer: The NTUEC-1000." Master's thesis, National Taiwan University, 1972.

Basalla, George. "The Spread of Western Science." *Science* 156, no. 3775 (1967): 611–622.

Bashe, Charles J., Lyle R. Johnson, John H. Palmer, and Emerson W. Pugh. *IBM's Early Computers*. Cambridge, MA: MIT Press, 1985.

Bassett, Ross Knox. "Aligning India in the Cold War Era: Indian Technical Elites, the Indian Institute of Technology at Kanpur, and Computing in India and the United States." *Technology and Culture* 50, no. 4 (2009): 783–810.

Bassett, Ross Knox. *The Technological Indian*. Cambridge, MA: Harvard University Press, 2016.

Bassett, Ross Knox. *To the Digital Age: Research Labs, Start-Up Companies, and the Rise of MOS Technology*. Baltimore, MD: Johns Hopkins University Press, 2002.

Bátiz-Lazo, Bernardo, and Haigh Thomas. "Engineering Change: The Appropriation of Computer Technology at Grupo ICA in Mexico (1965–1971)." *IEEE Annals of the History of Computing* 34, no. 2 (2012): 20–33.

Beltrán, Héctor. "Code Work: Thinking with the System in México." *American Anthropologist* 122, no. 3 (2020): 487–500.

Bergin, Bob. "The Growth of China's Air Defenses: Responding to Covert Overflights, 1949–1974." *Studies in Intelligence* 57, no. 2 (June 2013): 19–28.

Bhabha, Homi. *The Location of Culture*. New York: Routledge, 1994.

Boantza, Victor D. "The Rise and Fall of Nitrous Air Eudiometry: Enlightenment Ideals, Embodied Skills, and the Conflicts of Experimental Philosophy." *History of Science* 51, no. 4 (2013): 377–412.

Brand, Stewart. "Spacewar: Fanatic Life and Symbolic Death Among the Computer Bums." *Rolling Stone*, December 7, 1972, pp. 50–57.

Bray, Francesca. "Fabrics of Power: The Canonical Meanings of Women's Work." In *Technology and Gender: Fabrics of Power in Late Imperial China*, 183–205. Berkeley: University of California Press, 1997.

Brock, David C. "From Automation to Silicon Valley: The Automation Movement of the 1950s, Arnold Beckman, and William Shockley." *History and Technology* 28, no. 4 (2012): 375–340.

Brock, David C., and Christophe Lécuyer. *Makers of the Microchip: A Documentary History of Fairchild Semiconductor*. Cambridge, MA: MIT Press, 2010.

Brugioni, Dino A. "Tactical Use of the U-2 and Related Technical Developments." In *Eyes in the Sky: Eisenhower, The CIA and Cold War Aerial Espionage*, 260–326. Annapolis, MD: Naval Institute Press.

Bundy, McGeorge. *Danger and Survival: Choices about the Bomb in the First Fifty Years*. New York: Vintage Books, 1990.

Burr, William, ed. *New Archival Evidence on Taiwanese "Nuclear Intentions," 1966–1976*. National Security Archive's Electronic Briefing Book No. 20, October 13, 1999.

Burr, William, ed. *U.S. Opposed Taiwanese Bomb during 1970s*. National Security Archive's Electronic Briefing Book No. 221, June 15, 2007.

Burrell, Jenna. *Invisible Users: Youth in the Internet Cafés of Urban Ghana*. Cambridge, MA: MIT Press, 2012.

Cain, Geoffrey. *Samsung Rising: The Inside Story of the South Korean Giant That Set Out to Beat Apple and Conquer Tech*. New York: Currency, 2020.

Campbell-Kelly, Martin. *From Airline Reservations to Sonic the Hedgehog: A History of the Software Industry*. Cambridge, MA: MIT Press, 2003.

Campbell-Kelly, Martin, William F. Aspray, Jeffrey Yost, Honghong Tinn, Gerardo Con Díaz, and Nathan Ensmenger. *Computer: An Information Machine*. 4th ed. New York: Routledge, 2023.

Camprubí, Lino. *Engineers and the Making of the Francoist Regime*. Cambridge, MA: MIT Press, 2014.

Canion, Rod. *Open: How Compaq Ended IBM's PC Dominance and Helped Invent Modern Computing*. Dallas, TX: BenBella Books, 2013.

Casale-Rossi, Marco, Alberto Sangiovanni-Vincentelli, Luca Carloni, Bernard Courtois, Hugo de Man, Antun Domic, and Jan M. Rabaey. "Panel: The Heritage of Mead & Conway: What Has Remained the Same, What Was Missed, What Has Changed, What Lies Ahead." In Proceedings of the Conference on Design, Automation and Test in Europe. 171–175. San Jose, CA: EDA Consortium, 2013.

Ceruzzi, Paul E. "Are Historians Failing to Tell the Real Story about the History of Computing?" *IEEE Annals of the History of Computing* 36, no. 3 (2014): 94–95.

Ceruzzi, Paul E. *A History of Modern Computing.* 2nd ed. Cambridge, MA: MIT Press, 2003.

Ceruzzi, Paul E. "Moore's Law and Technological Determinism: Reflections on the History of Technology." *Technology and Culture* 46, no. 3 (2005): 584–593.

Chambers, T. H. E., and M. J. Whitmarsh-Everiss. *Advances in Nuclear Science and Technology.* Volume 17 in the series Simulators for Nuclear Power, edited by Jeffrey Lewins and Martin Becker. New York: Plenum Press New, 1978.

Chan, Anita Say. *Networking Peripheries: Technological Futures and the Myth of Digital Universalism.* Cambridge, MA: MIT Press, 2014.

Chang, Hsiu-Jung, ed. *Taida Yixue Yuan 1945–1950* [National Taiwan University's medical college, 1945–1950]. Taipei: National Taiwan University Press, 2013.

Chang, Li. *Wu Shiwen xiansheng fangwen jilu* [The Reminiscences of Adm. Wu Shih-wen]. Taipei: Academia Sinica, 2017.

Chang, Morris. *Zhang Zhongmou zichuan, shang, 1931—1964* [Morris Chang autobiography, vol. I, 1931–1964]. Taipei: Yuanjian Tianxia [Global Views-Commonwealth Publishing Group], 2020.

Chang, Wei-Bin. *Kuaidao jihua jiemi: Heimao zhongdui yu taimei gaokong zhencha hezuo neimu* [Project RAZOR: The Black Cat squadron and collaboration in high altitude reconnaissance tasks between Taiwan and the United States]. Taipei: Xinrui Wenchuang, 2012.

Chang, Yeu-wen. *Shizai de liliang: Zheng Chonghua yu taidadian de jingying zhihui* [Solid power: The business philosophy of Bruce Cheng and Delta Electronics]. Taipei: Tianxia Wenhua [Commonwealth Publishing], 2010.

Chao, Ena. "Guancha meiguo Taiwan jingying bixia de meiguo xingxiang yu jiaoyu jiaohuan jihua 1950–1970" [Observing America: American Images in Taiwan Elites' Writing and American Education Exchange Programs]. *Taida lishixue xuebao* [Historical Inquiry] 48 (December, 2011): 97–163.

Chao, Hsiang-Ke, and Chao-Hsi Huang. "Ta-Chung Liu's Exploratory Econometrics." *History of Political Economy* 43 (Annual supplement 2011): 140–165.

Chen, Hanting, and Luo Shunde. *Guofang buzhang Yu Dawei* [Biography of the Minister of Defense David Ta-wei Yule]. Taipei: Zhuanji Wenxue, 2015.

Chen, Hsin-Hsing. "Dazao diyige quanqiu zhuangpeixian: Taiwan tongyong qicai gongsi yu chengxiang yimin, 1964–1990" [To Make the First Global Assembly Line: General Instrument, Taiwan, and the Institutions of Rural-Urban Migration, 1964–1990]. *Zhengda laodong xuebao* [Bulletin of Labour Research] 20 (July 2006): 1–48.

Chen, Huiling, and Lin Qiyue. *Fang Xianqi zhuan: Dianxin zhifu, keji tuishou* [Fang Xianqi: Father of telecommunications, trailblazer of technology]. Taipei: Yuanjian Tianxia [Global Views-Commonwealth Publishing], 2016.

Chen, Qiukun, and Chengtian Guo. "Wo yu dongyuan dianji gongsi—Lin Changcheng xiansheng fangwen jilu" [TECO and I: Oral history interview with Zhangcheng Lin]. *Koushu lishi* [Oral history] 2 (Feburary 1991): 185-206.

Chen, Yi-shen, Mengtao Peng, and Jiahui Jian. *Hedan! Jiandie? CIA: Zhang xianyi fangwen jilu* [Nuke! Spy? CIA: An interview with Hsien-yi Chang]. Taipei: Yuanzu, 2016.

Cheng, Hsi-hua. "Airpower in the Taiwan Strait." In *The Chinese Air Force: Evolving Concepts, Roles, and Capabilities*, edited by Richard P. Hallion, Roger Cliff, and Phillip C. Saunders. Washington, DC: National Defense University, 2012.

Cheng, Li-ling. "Jindaihua pingdenghua yu chabiehua zhijian: Taibei gongye xuexiao xuesheng zhi jiuxue yu jiuye (1923–1945)" [Modernization, Equalization and Differentiation: Education and Employment of Students in Technical Schools of Taipei (1923–1945)].*Taiwan shi yanjiu* [Taiwan historical research] 16, no. 4 (2009): 81–114.

Cheng, Lu. "Laodangyizhuang de wujin sanxing quzhujian" [R.O.C. Navy's Wu-Chin III conversion destroyers]. *Jianduan keji* [Defense Technology Monthly] 148 (December 1996): 54–61.

Cheung, Gary Ka-wai. "Impact of the 1967 Riots." In *Hong Kong's Watershed: The 1967 Riots*. 131–142. Hong Kong: Hong Kong University Press, 2009.

Chew, Andrew. *Shede: Dianziye xianqu Qiu Zai-xing de shiye yu zhiye* [Willing: The calling of the pioneer of electronics industry, Zaixing Chew]. Taipei: Yuanshen, 2015.

Chi, Pang-yuan. *The Great Flowing River: A Memoir of China, from Manchuria to Taiwan*. Translated by John Balcom. New York: Columbia University Press, 2018.

Chien, Fredrick Foo. *Qianfu huiyilu* [The memoirs of Fredrick Foo Chien]. Vol. 1. Taipei: Tianxia [Commonwealth Publishing], 2005.

Choi, Hyungsub. "Technology Importation, Corporate Strategies, and the Rise of the Japanese Semiconductor Industry in the 1950s." *Comparative Technology Transfer and Society* 6, no. 2 (2008): 103–126.

Choi, Hyungsub, and Chigusa Kita, "Hiroshi Wada: Pioneering Electronics and Computer Technologies in Postwar Japan." *IEEE Annals of the History of Computing* 30, no. 3 (2008): 84–89.

Chow, Esther Ngan-ling. *Transforming Gender and Development in East Asia*. New York: Routledge, 2002.

Christiansen, Donald. "Donald Glen Fink, 1911–1996." *Memorial Tributes: National Academy of Engineering of the United States of America*. Vol. 9, pp. 82–87. Washington, DC: National Academies Press, 2001.

Clancey, Gregory. *Earthquake Nation: The Cultural Politics of Japanese Seismicity, 1868–1930*. Berkeley: University of California Press, 2006.

Clancey, Gregory. "Hygiene in a Landlord State: Health, Cleanliness and Chewing Gum in Late Twentieth Century Singapore." *Science, Technology and Society* 23, no. 2 (2018): 214–233.

Coaldrake, William H. "Beyond Mimesis: Japanese Architectural Models at the Vienna Exhibition and 1910 Japan British Exhibition." In *The Culture of Copying in Japan: Critical and Historical Perspectives*, edited by Rupert Cox, 199–212. New York: Routledge, 2007.

Cochran, Thomas B., William M. Arkin, and Milton M. Hoenig. *Nuclear Weapons Databook: U.S. Nuclear Forces and Capacities*. Vol. 1. Cambridge, MA: Ballinger, 1984.

Cohen, I. Bernard. *Howard Aiken: Portrait of a Computer Pioneer*. Cambridge, MA: MIT Press, 1999.

Collins, Harry M. "The TEA Laser." In *Changing Order: Replication and Induction in Scientific Practice*, 51–78. Chicago: University of Chicago Press, 1992 [1985].

Commonwealth Magazine. *Cao Xingcheng: Liandian de baye chuanqi* [Robert H. C. Tsao: UMC and its empire]. Taipei: Commonwealth, 1999.

Conner, Alison W. "Anglo-American Law at Soochow University." In *China's Christian Colleges: Cross-Cultural Connections, 1900–1950*, edited by Daniel H. Bays and Ellen Widmer, 147–172. Stanford, CA: Stanford University Press, 2009.

Coopersmith, Jonathan. *Faxed: The Rise and Fall of the Fax Machine.* Baltimore, MD: Johns Hopkins University Press. 2015.

Conway, Melvin E. "How Do Committees Invent?" *Datamation* 14, no. 4 (1968): 28–31.

Cortada, James W. *Digital Flood: The Diffusion of Information Technology across the U.S., Europe, and Asia.* New York: Oxford University Press, 2012.

Cortada, James W. "Patterns and Practices in How Information Technology Spread around the World." *IEEE Annals of the History of Computing* 30, no. 4 (2008): 4–25.

Cowan, Ruth Schwartz. "The Consumption Junction: A Proposal for Research Strategies on the Sociology of Technology." In *The Social Construction of Technological Systems*, edited by Wiebe Bijker, Thomas P. Hughes, and Trevor J. Pinch, 261–80. Cambridge, MA: MIT Press, 1987.

Cowie, Jefferson. *Capital Moves: RCA's Seventy-Year Quest for Cheap Labor.* Ithaca, NY: Cornell University Press, 1999.

Cox, Rupert. "Introduction." *The Culture of Copying in Japan: Critical and Historical Perspectives*, edited by Rupert Cox, 1–17. New York: Routledge, 2007.

Craft, Stephen G. "Islands against the Red Tide." In *American Justice in Taiwan*, 11–18. Lexington: University Press of Kentucky, 2016.

Cullather, Nick. "'Fuel for the Good Dragon': The United States and Industrial Policy in Taiwan, 1950–1965." *Diplomatic History* 20, no. 1 (1996): 1–25.

Cumings, Bruce. "Web with No Spider, Spider with No Web: The Genealogy of the Developmental State." In *The Developmental State*, edited by Meredith Woo-Cumings, 61–92. Ithaca, NY: Cornell University Press, 2000.

Da Costa Marques, Ivan. "Cloning Computers: From Rights of Possession to Rights of Creation." *Science as Culture* 14, no. 2 (2005): 139–60.

Da Costa Marques, Ivan, ed. "History of Computing in Latin America." *IEEE Annals of the History of Computing* 37, no. 4 (2015): 10–12.

Daston, Lorraine. "Enlightenment Calculations." *Critical Inquiry* 21, no. 1 (1994): 182–202.

Davis, Joshua. *Spare Parts: Four Undocumented Teenagers, One Ugly Robot, and the Battle for the American Dream.* New York: Farrar, Straus, and Giroux, 2014.

Dennis, Michael Aaron. "Our First Line of Defense: Two University Laboratories in the Postwar American State." *Isis* 85, no. 3 (1994): 427–455.

Diao, Manpeng. *Jingli rensheng: Luo Yiqiang wan quanqiu qiye de lequ* [Life of a manager: Y. C. Lo's fun with Philips' global enterprise]. Taipei: Tianxia wenhua [Commonwealth Publishing], 2001.

Diaz, Gerardo Con. *Software Rights: How Patent Law Transformed Software Development in America.* New Haven, CT: Yale University Press, 2019.

Dikötter Frank. *Exotic Commodities: Modern Objects and Everyday Life in China.* New York: Columbia University Press, 2006.

Dikötter Frank. *The Tragedy of Liberation: A History of the Chinese Revolution, 1945–57*. New York: Bloomsbury Press, 2013.

DiMoia, John. "Atoms for Sale? Cold War Institution-Building and the South Korean Atomic Energy Project, 1945–1965." *Technology and Culture* 51, no. 3 (2010): 589–618.

DiMoia, John. *Reconstructing Bodies: Biomedicine, Health, and Nation-Building in South Korea Since 1945*. Stanford, CA: Stanford University Press, 2013.

Donig, Simon. "Appropriating American Technology in the 1960s: Cold War Politics and the GDR Computer Industry." *IEEE Annals of the History of Computing* 32, no. 2 (2010): 32–45.

Douglas, Susan. *Inventing American Broadcasting, 1899–1922*. Baltimore, MD: Johns Hopkins University Press, 1989.

Downey, Gregory J. "Jumping Contexts of Space and Time." *IEEE Annals of the History of Computing* 26, no. 2 (2004): 94–96.

Downey, Gregory J. *Telegraphy Messenger Boys: Labor, Technology, and Geography 1850–1950*. New York: Routledge, 2002.

Downey, Gregory J. "Virtual Webs, Physical Technologies, and Hidden Workers: The Spaces of Labor in Information Internetworks." *Technology and Culture* 42, no. 2 (2001): 209–235.

Duara, Prasenjit. *Sovereignty and Authenticity*. New York: Rowman & Littlefield, 2003.

Edgerton, David. *The Shock of the Old: Technology and Global History since 1900*. New York: Oxford University Press, 2007.

Edwards, Paul N. *The Closed World: Computers and the Politics of Discourse in Cold War America*. Cambridge, MA: MIT Press, 1996.

Edwards, Paul N. "From 'Impact' to Social Process: Computers in Society and Culture." In *Handbook of Science and Technology Studies*, edited by Sheila Jasanoff, Gerald E. Markle, James Petersen, and Trevor Pinch, 257–285. Thousand Oaks, CA: Sage, 1995.

Eliades, George. "Once More unto the Breach: Eisenhower, Dulles, and Public Opinion during the Offshore Island Crisis of 1958." *Journal of American-East Asian Relations* 2, no. 4 (1993): 343–367.

Engerman, David C. "The Romance of Economic Development and New Histories of the Cold War." *Diplomatic History* 28, no. 1 (2004): 23–54.

Engerman, David, Nils Nilman, Mark H. Haefele, and Michael E. Latham, eds. *Staging Growth: Modernization, Development, and the Global Cold War*. Boston: University of Massachusetts Press, 2003.

Erickson, Paul, Judy L. Klein, Lorraine Daston, Rebecca Lemov, Thomas Sturm, and Michael D. Gordin. *How Reason Almost Lost Its Mind: The Strange Career of Cold War Rationality*. Chicago, IL: Chicago University Press, 2013.

Escobar, Arturo. *Encountering Development: The Making and Unmaking of the Third World*. Princeton, NJ: Princeton University Press, 1995.

Evans, Robert. "Soothsaying or Science? Falsification, Uncertainty and Social Change in Macroeconomic Modelling." *Social Studies of Science* 27, no. 3 (1997): 395–438.

Evans, W. Duane, and Marvin Hoffenberg. "The Interindustry Relations Study for 1947." *Review of Economics and Statistics* 34, no. 2 (1952): 97–142.

Fan, Fa-ti. *British Naturalists in Qing China: Science, Empire, and Cultural Encounter.* Cambridge, MA: Harvard University Press, 2004.

Fan, Fa-ti. "East Asian STS: Fox or Hedgehog?" *East Asian Science, Technology and Society: An International Journal* 1, no. 2 (2007): 243–248.

Fan, Fa-ti. "Modernity, Region, and Technoscience: One Small Cheer for Asia as Method." *Cultural Sociology* 10, no. 3 (2016): 352–368

Fang, Huichen, Rongzing He, Yuyou Lin, and Yiting Jiang. "Wo zai Gaoxiong lianyouchang de rizi: sanwei gongren de mingyun jiaochadian" [My time at the Kaohsiung oil refinery: How three workers met]. In *Yancong zhi dao: women yu shihua gongcun de liangwan ge rizi* [A Smoking Island: Petrochemical Industry, Our Dangerous Companion more than Fifty Years], 63–77. Taiwan: Chunshan chuban, 2019.

Ferguson, James. *The Anti-Politics Machine: Development, Depoliticization, and Bureaucratic Power in Lesotho.* Minneapolis: University of Minnesota Press, 1994.

Fink, Donald G. *Diannao yu rennao* [Computers and the Human Mind]. Translated by Chao-Chih Yang. Taipei: Shangwu, 1971. Originally published as *Computers and the Human Mind: An Introduction to Artificial Intelligence.* Garden City, NY: Doubleday Anchor, 1966.

Fischer, Michael M. J. "Anthropological STS in Asia." *Annual Review of Anthropology* 45 (2016): 181–198.

Fischer, Michael M. J. "Theorizing STS from Asia—Toward an STS Multiscale Bioecology Framework: A Blurred Genre Manifesto/Agenda for an Emergent Field." *East Asian Science, Technology and Society: An International Journal* 12, no. 4 (2018): 519–540.

Foote, Stephanie, and Mazzolini Elizabeth, eds. *Histories of the Dustheap: Waste, Material Cultures, Social Justice.* Cambridge, MA: MIT Press, 2012.

Forman, Paul. "Behind Quantum Electronics: National Security as Basis for Physical Research in the United States, 1940–1960." *Historical Studies in the Physical and Biological Sciences* 18, no. 1 (1987): 149–229.

Franz, Kathleen. *Tinkering: Consumers Reinvent the Early Automobile.* Philadelphia: University of Pennsylvania Press, 2005.

Frederick, P. Brooks, Jr. *The Mythical Man-Month: Essays on Software Engineering.* Anniversary edition (first edition 1975). Reading, MA: Addison-Wesley, 1995.

Freiberger, Paul, and Michael Swaine. *Fire in the Valley: The Making of the Personal Computer.* Berkeley, CA: Osborne/McGraw-Hill, 1984.

Frumer, Yulia. *Making Time: Astronomical Time Measurement in Tokugawa Japan.* Chicago: University of Chicago Press, 2018.

Fu, Daiwie. "How Far Can East Asian STS Go? A Position Paper." *East Asian Science, Technology and Society: An International Journal* 1, no. 1 (2007): 1–14.

Fu, Li-Yu. "Meiyuan shiqi Taiwan zhongdeng kexue jiaoyu jihua zhi xingcheng yu shishi nianbiao (1951–1965)" [A Chronology of the Initiation and Implement of the Secondary Science

Education Project under the U.S. Aid in Taiwan (1951–1965)]. *Kexue jiaoyu xuekan* [Chinese Journal of Science Education] 14, no. 4 (2006): 447–465.

Gerovitch, Slava. *From Newspeak to Cyberspeak: A History of Soviet Cybernetics.* Cambridge, MA: MIT Press, 2002.

Gerovitch, Slava. "'Mathematical Machines' of the Cold War: Soviet Computing, American Cybernetics and Ideological Disputes in the Early 1950s." *Social Studies of Science* 31, no. 2 (2001): 253–287.

Gillispie, Charles Coulston. *The Edge of Objectivity: An Essay in the History of Scientific Ideas.* Princeton, NJ: Princeton University Press, 1960.

Gilman, Nils. "Modernization Theory: The Highest Stage of American Intellectual History." In *Staging Growth: Modernization, Development, and the Global Cold War,* edited by David Engerman, Nils Nilman, Mark H. Haefele, and Michael E. Latham, 251–270. Boston: University of Massachusetts Press, 2003.

Gold, Thomas B., ed. *Selected Stories of Yang Ch'ing-Ch'u.* Translated by Thomas B Gold. Kaohsiung, TW: First Publishing, 1983.

Gold, Thomas B. *State and Society in the Taiwan Miracle.* Armonk, NY: M. E. Sharpe, 1986.

Goldstein, Carolyn M. "From Service to Sales: Home Economics in Light and Power, 1920–1940." *Technology and Culture* 38, no. 1 (1997): 121–152.

Gordon, Leonard H. D. "United States Opposition to Use of Force in the Taiwan Strait, 1954–1961." *Journal of American History* 72, no. 3 (1985): 637–660.

Grace, Joshua. *African Motors: Technology, Gender, and the History of Development.* Durham, NC: Duke University Press, 2022.

Greenberg, Joshua M. *From Betamax to Blockbuster: Video Stores and the Invention of Movies on Video.* Cambridge, MA: MIT Press, 2008.

Greene, J. Megan. *Building a Nation at War: Transnational Knowledge Networks and the Development of China during and after World War II.* Cambridge, MA: Harvard University Press, 2022.

Greene, J. Megan. *The Origins of the Developmental State in Taiwan.* Cambridge, MA: Harvard University Press, 2008.

Grier, David A. *When Computers Were Human.* Princeton, NJ: Princeton University Press, 2005.

Grier, David A., and Yamada Akihiko. "Events and Sightings." *IEEE Annals of the History of Computing* 35, no. 3 (2013): 72–76.

Gruntman, Mike. *Blazing the Trail: The Early History of Spacecraft and Rocketry.* Reston, VA: American Institute of Aeronautics & Astronautics, 2004.

Gupta, Akhil. *Postcolonial Developments: Agriculture in the Making of Modern India.* Durham, NC: Duke University Press, 1998.

Gurtov, Melvin. "The Taiwan Strait Crisis Revisited: Politics and Foreign Policy in Chinese Motives." *Modern China* 2, no. 1 (1976): 49–103.

Haigh, Thomas, ed. *Exploring the Early Digital.* Cham: Springer Nature Switzerland AG, 2019.

Haigh, Thomas and Paul E. Ceruzzi. *A New History of Modern Computing.* Cambridge, MA: MIT Press, 2021.

Halperin, Morton H. *The 1958 Taiwan Straits Crisis: A Documented History.* Santa Monica, CA: RAND Corporation, 1966.

Hamilton, Clive. "Capitalist Industrialisation in East Asia's Four Little Tigers." *Journal of Contemporary Asia* 13, no. 1 (1983): 35–73.

Han, Byung-Chul. *Shanzhai: Deconstruction in Chinese.* Translated by Philippa Hurd. Cambridge, MA: MIT Press, 2017.

Han, Kuang-wei. *Xuexi de rensheng: Han Guangwei huiyilu* [A Life of Learning: The Memoirs of Kuang-wei Han]. Edited by Li Chang. Taipei: Academia Sinica, 2010.

Han, Kuang-wei, and George J. Thaler. "Phase-Space Analysis and Design of Linear Discontinuously Damped Feedback Control Systems." *Transactions of the American Institute of Electrical Engineers, Part II: Applications and Industry* 80, no. 4 (1961): 196–203.

Haring, Kristen. *Ham Radio's Technical Culture.* Cambridge, MA: MIT Press, 2006.

Hartono, Paulina. "Do Radios Have Politics? The Politics of Radio Ownership in China in the 1920s and 1930s." Working paper for the Consortium for History of Science and Technology, December 17, 2019. https://www.chstm.org/content/history-technology?page=2.

Hecht, Gabrielle, ed. *Entangled Geographies: Empire and Technopolitics in the Global Cold War.* Cambridge, MA: MIT Press, 2011.

Hecht, Gabrielle. *The Radiance of France: Nuclear Power and National Identity after World War II.* Cambridge, MA: MIT Press, 2009.

Hepler-Smith, Evan. "'A Way of Thinking Backwards': Computing and Method in Synthetic Organic Biology." *Historical Studies in Natural Sciences* 48, no. 3. (2018): 300-337.

Hicks, Mar. *Programmed Inequality: How Britain Discarded Women Technologists and Lost Its Edge in Computing.* Cambridge, MA: MIT Press, 2017.

Higgins, W. H. C., B. D. Holbrook, and J. W. Emling. "Electrical Computers for Fire Control." *IEEE Annals of the History of Computing* 3, no. 3 (1982): 233.

Hindle, Brooke. *Emulation and Invention.* New York: New York University Press, 1981.

Ho, David Li-wei. *Hedan MIT: Yige shangwei jieshu de gushi* ["A" bomb made in Taiwan: An unfinished story]. Taipei: Women Chuban, 2015.

Hoddeson, Lillian, and Vicki Daitch. *True Genius: The Life and Science of John Bardeen, the Only Winner of Two Nobel Prizes in Physics.* Washington, DC: Joseph Henry Press, 2002.

Hodges, Andrew. *Alan Turing: The Enigma.* New York: Touchstone, 1983.

Hoffman, Paul. "The Critical Role of Special Fund." In *The Priorities of Progress*, edited by the United Nations Special Fund, 4–5. New York: The United Nations, 1961.

Homberg, Michael. "Digital India. Swadeshi-Computing in India since 1947." In *Prophets of Computing: Visions of Society Transformed by Computing*, edited by Dick van Lente, 279–323. New York: ACM Press, 2022.

Hooks, Tisha Y. Duct Tape and the U.S. Social Imagination. PhD diss., Yale University, 2015.

Hossfeld, Karen J. "Their Logic against Them: Contradictions in Sex, Race, and Class in Silicon Valley." In *Women Workers and Global Restructuring*, edited by Kathryn Ward, 149–178. New York: Cornell University Press, 1990.

Hsieh, Ching-Chun. "Dianzi jisuanji zhi sheji" [A design of a general-purpose computer]. *Jiaoda xuekan* [Science Bulletin National Chiao-Tung University] 2, no. 2 (1967): 105–118.

Hsu, Shu-En. *Cailiao ye shenqi: Keji xuezhe Xu Shuen de yisheng* [The marvels of material science: The memoirs of Shu-En Hsu]. Taipei: Xiuwei, 2004.

Hu, Shih. "Fazhan Kexue de zhongren he yuanlu" [The responsibility of developing science]. *Xin shidai* [New Era] 1 (February 1962): 3–4.

Hua, Hsichun Mike. "The Black Cat Squadron." *Air Power History* 49, no. 1 (2002): 4–19.

Huang, Chung-ting. "The Military Assistance of the Republic of China to the Republic of Vietnam during the Vietnam War." *Bulletin of the Institute of Modern History Academia Sinica* 79 (March 2013): 137–172.

Huang, Wen-lu, and Li Ziching. *Feiyue dihou 3000 li: Heibianfu zhongdui yu dashidai de women* [Three hundred miles into the territory of the enemy: The Black Bat Squadron and our time]. Taipei: Xinrui wenchuang, 2018.

Hughes, Thomas P. "Edison and Electric Light." In *The Social Shaping of Technology: How the Refrigerator Got Its Hum*, edited by Donald MacKenzie and Judy Wajcman, 50–63. Philadelphia, PA: Open University Press, 1999 [1985].

Hughes, Thomas P. "The Evolution of Large Technological Systems." In *The Social Construction of Technological Systems*, edited by Wiebe Bijker, Thomas P. Hughes, and Trevor J. Pinch, 50–80. Cambridge, MA: MIT Press, 1987.

Hughes, Thomas P. *Networks of Power: Electrification in Western Society, 1880–1930*. Baltimore, MD: Johns Hopkins University Press, 1983.

Hughes, Thomas P. *Rescuing Prometheus*. New York: Vintage Books, 2006.

Hung, Shih-Chang, and Richard Whittington. "Strategies and Institutions: A Pluralistic Account of Strategies in the Taiwanese Computer Industry." *Organization Studies* 18, no. 4 (1997): 551–575.

Irani, Lilly. *Chasing Innovation: Making Entrepreneurial Citizens in Modern India*. Princeton, NJ: Princeton University Press, 2019.

Isaacson, Walter. *Steve Jobs*. New York: Simon and Schuster, 2011.

Jackson, Myles W. *Harmonious Triads: Physicists, Musicians, and Instrument Makers in Nineteenth-Century Germany*. Cambridge, MA: MIT Press, 2008.

Jacoby, Neil. *U.S. Aid to Taiwan: A Study of Foreign Aid, Self-Help, and Development*. New York: F. A. Praeger, 1967.

Jeon, Chihyung. "A Road to Modernization and Unification: The Construction of the Gyeongbu Highway in South Korea." *Technology and Culture* 51, no. 1 (2010): 55–79.

Jo, Dongwon. "Vernacular Technical Practices beyond the Imitative/Innovative Boundary: Apple II Cloning in Early-1980s South Korea." *East Asian Science, Technology and Society: An International Journal* 16, no. 2 (2022): 157–180.

Johnson, Chalmers. *MITI and the Japanese Miracle: The Growth of Industrial Policy, 1925–1975*. Stanford, CA: Stanford University Press, 1982.

Kallgren, Joyce. "Nationalist China's Armed Forces." *China Quarterly* 15 (September 1963): 35–44.

Kang, Hildi, ed. *Under the Black Umbrella: Voices from Colonial Korea 1910–1945.* Ithaca, NY: Cornell University Press, 2005.

Kang, Hyeok Hweon. "Cooking Niter, Prototyping Nature: Saltpeter and Artisanal Experiment in Korea, 1592–1635." *Isis: A Journal of the History of Science Society* 113, no. 1 (2022): 1–21.

Kausikan, Bilahari. *Singapore Is Not an Island: Views on Singapore Foreign Policy.* Singapore: Straits Times Press, 2019.

Kausikan, Bilahari. *Singapore Is Still Not an Island: More Views on Singapore Foreign Policy.* Singapore: Straits Times Press, 2023.

Kim, Seong-Jun. "Technology Transfer behind a Diplomatic Struggle: Reappraisal of South Korea's Nuclear Fuel Project in the 1970s." *Historia Scientiarum* 19, no. 2 (2009): 184–93.

Kim, Tae-Ho. "New Rice for Unification and Independence: Tongil Rice and South Korean Agronomy in the 1970s." Paper presented at the Annual Meeting of the Society for the History of Technology, Lisbon, Portugal, October 11–14, 2008.

Kim, Yongdo. "Interfirm Cooperation in Japan's Integrated Circuit Industry, 1960s–1970s." *Business History Review* 86, no. 4 (2012): 773–792.

Kirby, William C. "Engineering China: The Origins of the Chinese Developmental State." In *Becoming Chinese*, edited by Wen-hsin Yeh, 137–160. Berkeley: University of California Press, 2000.

Kirby, William C. *Empires of Ideas: Creating the Modern University from Germany to America to China.* Cambridge, MA: Harvard University Press, 2022.

Kirby, William C., Billy Chang, and Daw H. Lau. "Taiwan Semiconductor Manufacturing Company Limited: A Global Company's China Strategy." Harvard Business Review Case Study, 2020.

Kirby, William C., Michael Shih-ta Chen, and Keith Wong. "Taiwan Semiconductor Manufacturing Company Limited: A Global Company's China Strategy." Harvard Business Review Case Study, 2015.

Kita, Chigusa. "Character Codes and Local Writing Cultures." A presentation in the symposium Just Code: Power, Inequality, and the Global Political Economy of IT. CBI, University of Minnesota, October 23–24, 2020.

Kita, Chigusa. "Events and Sightings." *IEEE Annals of the History of Computing* 33, no. 2 (2011): 106–110.

Kita, Chigusa. "From Technological Mimesis to Creativity: Early Online Rail Reservations in Japan." Paper presented in the Annual Meeting of the Society for the History of Technology, Washington, DC, October 17–21, 2007.

Kline, Ronald R. "Agents of Modernity: Home Economists and Rural Electrification in the United States, 1925–1950." In *Rethinking Women and Home Economics in the 20th Century*, edited by Sara Stage and Virginia Vincenti, 237–252. Ithaca, NY: Cornell University Press, 1997.

Kline, Ronald R. *Consumers in the Country: Technology and Social Change in Rural America.* Baltimore, MD: Johns Hopkins University Press, 2000.

Kline, Ronald R. "Inventing an Analog Past and a Digital Future in Computing," In *Exploring the Early Digital*, edited by Thomas Haigh, 19–39. Cham: Springer Nature Switzerland AG, 2019.

Kline, Ronald R. *The Cybernetics Moment: Or Why We Call Our Age the Information Age*. Baltimore, MD: Johns Hopkins University Press, 2015.

Kline, Ronald R., and Trevor Pinch. "Users as Agents of Technological Change: The Social Construction of the Automobile in the Rural United States." *Technology and Culture* 37, no. 4 (1996): 763–795.

Krige, John. *American Hegemony and the Postwar Reconstruction of Science in Europe*. Cambridge, MA: MIT Press, 2006.

Krige, John. "Representing the Life of an Outstanding Chinese Aeronautical Engineer: A Transnational Perspective." *Technology Stories*, March 12, 2018. https://www.technologystories.org/chinese-engineer/.

Kristensen, Hans M. "Nukes in the Taiwan Crisis." Federation of American Scientists blog, May 13, 2008. https://fas.org/blogs/security/2008/05/nukes-in-the-taiwan-crisis/.

Kuhn, Thomas. *The Structure of Scientific Revolutions*. Chicago: University of Chicago Press, 2012 [1962].

Kung, Lydia. "Worker Consciousness." In *Factory Women in Taiwan*, 171–180. Ann Arbor, MI: UMI Research Press, 1987.

Lai, Jiaxin. "Gongchang nü'erquan—lun 1970–80 niandai Taiwan wenxue zhong de nügong yangmao" [Womanland of factory women: The literary images of women factory workers in Taiwan, 1970–1980]. Master's thesis, National Taiwan Normal University, 2007.

Lai, Jiaxin. "Nü'er guodu de meili yu aichou—lun Yang Qingchu *Gongchang nü'er quan* de nügong qunxiang [The beauty and the sorrow in the nations of girls: An analysis of the images of female factory workers in Yang's *Womanland of Factory Women*]. *Lishi jiaoyu* [The History Education] 12 (June 2008): 131–186.

Langlois, Richard N. "External Economies and Economic Progress: The Case of the Microcomputer Industry." *Business History Review* 66, no. 1 (1992): 1–50.

Latham, Michael E. "American Social Science, Modernization Theory, and the Cold War." *Modernization as Ideology: American Social Science and "Nation Building" in the Kennedy Era*. Chapel Hill: University of North Carolina Press, 2000.

Latour, Bruno. *Science in Action*. Cambridge, MA: Harvard University Press, 1987.

Laws, David A. "A Company of Legend: The Legacy of Fairchild Semiconductor." *IEEE Annals of the History of Computing* 32, no. 1 (2010): 60–74.

Lay, David C. *Linear Algebra and Its Applications*. 3rd ed. New York: Addison Wesley, 2003.

Layman, John W., and Harry L. Johnson, eds. *Collected Papers in Elasticity and Mathematics of Chih-Bing Ling Volume II*. Blacksburg: Virginia Polytechnic Institute and State University, 1979.

Lean, Eugenia. *Vernacular Industrialism in China: Local Innovation and Translated Technologies in the Making of a Cosmetics Empire, 1900–1940*. New York: Columbia University Press, 2020.

Lécuyer, Christophe. *Making Silicon Valley: Innovation and the Growth of High Tech, 1930–1970*. Cambridge, MA: MIT Press, 2006.

Lécuyer, Christophe, and David C. Brock, *Makers of the Microchip: A Documentary History of Fairchild Semiconductor*. Cambridge, MA: MIT Press, 2010.

Lee, Chung-Shing, and Michael Pecht. *The Taiwan Electronics Industry*. Boca Raton, FL: CRC Press, 1997.

Lee, Clarissa Ai Ling. "Malaysian Physics and the Maker Ethos." *Physics Today* 76, no. 2 (2023): 32–38.

Lee, Joseph Ya-Ming. *Cong bandaoti kan shijie* [A World Perspective Based on the Development of the Semiconductor Industry]. Taipei: Yuanjian Tianxia [Global Views-Commonwealth Publishing], 2012.

Lee, Jung. "Invention without Science: 'Korean Edisons' and the Changing Understanding of Technology in Colonial Korea." *Technology and Culture* 54, no. 4 (2013): 782–814.

Leslie, Stuart W. *The Cold War and American Science: The Military-Industrial-Academic Complex at MIT and Stanford*. New York: Columbia University Press, 1993.

Leslie, Stuart W., and Robert Kargon. "Exporting MIT: Science, Technology, and Nation-Building in India and Iran." *Osiris* 21, no. 1 (2006): 110–130.

Lesser, Richard C. "Richard C. Lesser's Recollections: The Cornell Computing Center, the Early Years, 1953 to 1964." Oral and Personal Histories of Computing at Cornell. Written in 1996. Accessed November 30, 2011. http:// www2.cit.cornell.edu/computer/history/.

Leuenberger, Christine. "Constructions of the Berlin Wall: How Material Culture Is Used in Psychological Theory." *Social Problems* 53, no. 1 (2006), 18–37.

Levy, Steven. *Hackers: Heroes of the Computer Revolution*. New York: Penguin Books, 2010.

Lewis, Nicholas. "Purchasing Power: Rivalry, Dissent, and Computing Strategy in Supercomputer Selection at Los Alamos." *IEEE Annals of the History of Computing* 39, no. 3 (2017): 25–40.

Li, Junming, Cuiqing Li, Qixun Xie, Weiqiong Fu, Qinyu Shen, and Yinjun Wang. *Kuashiji de chanye tuishou: Ershi ge yu Taiwan gongtong chengzhang de gushi* [Remarkable industry leaders in Taiwan: Twenty stories interwoven with Taiwan's history]. Taipei: Yuanjian Tianxia [Global Views-Commonwealth Publishing], 2016.

Li, Maggie. "Chuangzao zuida huaren siren qiye Pantronix de Wang Dazhuang (Stanley Wang)" [Entrepreneur of the largest Chinese American enterprise: Stanley Wang]. In *Yu Tian Changlin jiangzuo dashi men tanxin* [The hidden laws of success: Meet masters of Tien Forum], 77–90. Taipei: Da hangjia, 2010.

Li, Peter Ping. "The Evolution of Multinational Firms from Asia: A Longitudinal Study of Taiwan's Acer Group." *Journal of Organizational Change Management* 11, no. 4 (1998): 321–337.

Li, Yongzhou. "Determining Cutting Tool Temperatures." *Tool Engineer* 23 (October 1949): 32.

Li, Yongzhou. *Hangkong, hangkong wushinian: qier yiwang* [Fifty years in the aeronautics industry: A memoir at age of seventy-two]. Taipei: Daosheng, 1987.

Light, Jennifer. *From Warfare to Welfare: Defense Intellectuals and Urban Problems in Cold War America*. Baltimore, MD: Johns Hopkins University Press, 2003.

Light, Jennifer. "Programming." In *Gender and Technology: A Reader*, edited by Nina Lerman, Ruth Oldenziel, and Arwen P. Mohun, 295–328. Baltimore, MD: Johns Hopkins University Press, 2003. Originally published as "When Computers Were Women." *Technology and Culture* 40, no. 3. (1999): 455–483.

Lin, Chung-hsi. "Jinsheng de jishu—pinzhuang che de meili yu aichou" [The Technology of Silence—the Beauty and Sorrow of Reassembled Car]." *Keji bowu* [Technology Museum Studies] 6, no. 4 (2002): 34–58.

Lin, Hung Chang Jimmy, T. B. Tan, G. Y. Chang, B. Van der Leest, and N. Formigoni. "Lateral Complementary Transistor Structure for the Simultaneous Fabrication of Functional Blocks." *Proceedings of the IEEE* 52, no. 12 (1964): 1491–1495.

Lin, James. *In the Global Vanguard: Agrarian Development and the Making of Modern Taiwan.* Berkeley: University of California Press, forthcoming (2025).

Lin, James. "Sowing Seeds and Knowledge: Agricultural Development in Taiwan and the World, 1925–1975." *East Asian Science, Technology and Society: An International Journal* 9, no. 2 (2015): 127–149.

Lin, Ling-Fei. "Design Engineering or Factory Capability? Building Laptop Contract Manufacturing in Taiwan." *IEEE Annals of the History of Computing* 38, no. 2 (2016): 22–39.

Lin, Tengli. *The Social and Economic Origins of Technological Capacity: A Case Study of the Taiwanese Computer Industry.* PhD diss., Temple University, 2000.

Lin, Xiumei, "Jianhong jituan zongcai Minhong Hong de chuangye zhexue" [Entrepreneurial philosophy of Jianhong Group's CEO Minhong Hong]. *Taida xiaoyou shuangyuekan* [NTU Alumni Bimonthly], no. 89 (September 2013): 59–63.

Lin, Yi-Ping. "Sile jiwei dianzichang nügong zhihou: Youji rongji de jiankang fengxian zhengyi" [After the Death of Some Electronic Workers: The Health Risk Controversies of Organic Solvents]. *Keji yiliao yu shehui* [Taiwanese Journal for Studies of Science Technology and Medicine] 12 (April 2011): 61–112.

Lin, Yu-Ping. *Taiwan hangkong gongye shi: zhanzheng yuyi xia de 1935 nian—1979 nian* [The history of aviation industry in Taiwan: In the cradle of wars, 1935–1979]. Taipei: Xinrui Wenchuang, 2011.

Lindsay, Christina. "From the Shadows: Users as Designers. Producers, Marketers, Distributors, and Technical Support." In *How Users Matter: The Co-Construction of Users and Technology,* edited by Nelly Oudshoorn and Trevor J. Pinch, 29–50. Cambridge, MA: MIT Press, 2003.

Lindtner, Silvia M. *Prototype Nation: China and the Contested Promise of Innovation.* Princeton, NJ: Princeton University Press, 2000.

Ling, Chih-Bing. *Collected Papers in Elasticity and Mathematics of Chih-Bing Ling Volume I.* Taipei: Institute of Mathematics, Academia Sinica, 1963.

Ling, Chih-Bing. "Evaluation at Half Periods of Weierstrass' Elliptic Functions with Double Periods 1 and e^{ia}." *Mathematics of Computation* 19, no. 92 (1965): 658–661.

Ling, Chih-Bing, and Chen-Peng Tsai. "Evaluation at Half Periods of Weierstrass' Elliptic Function with Rhombic Primitive Period-Parallelogram." *Mathematics of Computation* 18, no. 87 (1964): 433–440.

Little, Peter C. *Toxic Town: IBM, Pollution, and Industrial Risks.* New York: New York University Press, 2014.

Liu, Mark. "Taiwan and the Foundry Model." *Nature Electronics* 4, no. 5 (2021): 318–320.

Liu, Su-fen, ed. *Li Guoding: Wo de Taiwan Jingyan* [Li Guoding (K. T. Li): My Taiwan experience—oral histories of Li Guoding]. Taipei: Yuan-Liou, 2005.

Liu, Wen-Hiao. *Shiluo de wudu feixing yuan* [The Lost Voodoo Pilot]. Taipei: Bingqi zhanshu tushu, 2014.

Loh, Christine. *Underground Front: The Chinese Communist Party in Hong Kong.* Hong Kong: Hong Kong University Press, 2010.

Lü, Miaofen, Hongyuan Liao, Su-fen Liu, Xiaorong Tian, Jiaen Liao, Guanjie Ceng, and Jingwei Yuan. *Hongji jingyan yu Taiwan dianzi ye—Shi Zhengrong xiansheng fangwen jilu* [Acer and Taiwan's electronics industry: The reminiscences of Stan Shih]. Taipei: Institute of Modern History, Academia Sinica, 2018.

Ma, Eunjeong, and Michael Lynch. "Constructing the East–West Boundary: The Contested Place of a Modern Imaging Technology in South Korea's Dual Medical System." *Science, Technology, & Human Values* 39, no. 5 (2014): 639–665.

MacKenzie, Donald. *Inventing Accuracy: A Historical Sociology of Nuclear Missile Guidance.* Cambridge, MA: MIT Press, 1993.

Mahoney, Michael S. "The History of Computing in the History of Technology." *IEEE Annals of the History of Computing* 10, no. 2 (1988): 113–125.

Maines, Rachel. *Hedonizing Technology: Paths to Pleasure in Hobbies and Leisure.* Baltimore, MD: Johns Hopkins University Press, 2009.

Martin, Emily. "Toward an Anthropology of Immunology: The Body as Nation State." *Medical Anthropology Quarterly*, New Series 4, no. 4 (1990): 410–426.

Matthews, John A., and Charles C. Snow. "A Conversation with Acer Group's Stan Shih on Global Strategy and Management." *Organization Dynamics* 27, no. 1 (1998): 65–74.

Mavhunga, Clapperton Chakanetsa, ed. *What Do Science, Technology, and Innovation Mean from Africa?* Cambridge, MA: MIT Press, 2017.

McDonald, Christopher. "From Art Form to Engineering Discipline? A History of US Military Software Development Standards, 1974–1998." *IEEE Annals of the History of Computing* 32, no. 4 (2010): 32–45.

Mead, Carver, and Lynn Conway. *Introduction to VLSI System.* Boston, MA: Addison-Wesley Longman, 1979.

Medina, Eden. *Cybernetic Revolutionaries: Technology and Politics in Allende's Chile.* Cambridge, MA: MIT Press, 2011.

Medina, Eden. "Designing Freedom, Regulating a Nation: Socialist Cybernetics in Allende's Chile." *Journal of Latin America Studies* 38, no. 3 (2006): 571–606.

Medina, Eden, Ivan da Costa Marques, and Christina Holmes, eds. *Beyond Imported Magic: Essays on Science, Technology, and Society in Latin America.* Cambridge, MA: MIT Press, 2014.

Mehos, Donna, and Suzanne Moon. "The Uses of Portability: Circulating Experts and the Technopolitics of Cold War and Decolonization." In *Entangled Geographies: Empire and Technopolitics in the Global Cold War*, edited by Gabrielle Hecht, 43–74. Cambridge, MA: MIT Press, 2011.

Michell, Anthony. *Samsung Electronics and the Struggle for Leadership of the Electronics Industry.* Hoboken, NJ: Wiley, 2011.

Miller, Chris. *Chip War: The Fight for the World's Most Critical Technology.* New York: Scribner, 2022.

Mindell, David A. *Between Human and Machine: Feedback, Control, and Computing before Cybernetics.* Baltimore, MD: Johns Hopkins University Press, 2002.

Mindell, David A. *Digital Apollo: Human and Machine in Spaceflight.* Cambridge, MA: The MIT Press, 2009.

Mirowski, Philip. *Machine Dreams: Economics Becomes a Cyborg Science.* Cambridge, UK: Cambridge University Press, 2002.

Misa, Thomas J. *Digital State: The Story of Minnesota's Computing Industry.* Minneapolis: University of Minnesota Press, 2013.

Misa, Thomas J., ed. *Gender Codes: Why Women Are Leaving Computing.* Hoboken, NJ: Wiley and IEEE Computer Society, 2010.

Misa, Thomas J. *Leonardo to the Internet: Technology and Culture from the Renaissance to the Present.* 3rd ed. Baltimore, MD: Johns Hopkins University Press, 2022.

Mitchell, Timothy. *Colonising Egypt.* Cambridge, UK: Cambridge University Press, 1988.

Mitsuo, Kawaguchi. *Taiwan xigu xungen: Rizhi shiqi Taiwan gaokeji chanye shihua* [The Origin of Taiwan's Silicon Valley: High-Tech Industry in Taiwan Under Japanese Rule]. Translated by Liansheng He. Hsinchu, TW: Yanqu Shenghuo zazhishe, 2009.

Mizuno, Hiromi. *Science for the Empire: Scientific Nationalism in Modern Japan.* Stanford, CA: Stanford University Press, 2009.

Mody, Cyrus C. M. *Instrument Community: Probe Microscopy and the Path to Nanotechnology.* Cambridge, MA: MIT Press, 2011.

Moghadam, Valentine. "Gender Dynamics of Restructuring in the Semi-Periphery." In *Engendering Wealth & Well-Being. Empowerment for Global Change*, edited by R. L. Blumberg, C. A. Rakowski, I. Tinker, and M. Monteón, 17-37. San Francisco, CA: Westview Press, 1995.

Moon, Suzanne M. "Justice, Geography, and Steel: Technology and National Identity in Indonesian Industrialization." *Osiris* 24 (2009): 253–277.

Moon, Suzanne M. "Takeoff or Self-Sufficiency? Ideologies of Development in Indonesia, 1957–1961." *Technology and Culture* 39, no. 2 (1998): 187–212.

Moon, Suzanne M. *Technology and Ethical Idealism: A History of Development in the Netherlands East Indies.* Leiden: CNWS Publications, 2007.

Moore, Aaron Stephen. *Constructing East Asia: Technology, Ideology, and Empire in Japan's Wartime Era, 1931–1945.* Stanford, CA: Stanford University Press, 2013.

Moore, Gordon. "Cramming More Components onto Integrated Circuits." *Electronics Magazine* 38, no. 8 (April 1965): 114–117.

Morita, Atsuro. "Traveling Engineers, Machines, and Comparisons: Intersecting Imaginations and Journeys in the Thai Local Engineering Industry." *East Asian Science, Technology, and Medicine* 7 no. 2 (2013): 221–241.

Morris, Robert L., and John R. Miller, eds. *Designing with TTL Integrated Circuits: Prepared by the IC Applications Staff of Texas Instruments Incorporated.* New York: McGraw-Hill, 1971.

Mote, C. D. (Dan) Jr. "Ruth M. Davis," *Memorial Tributes: National Academy of Engineering of the United States of America*. Vol. 17, pp. 74–79. Washington, DC: National Academies Press, 2013.

Mullaney, Thomas S. *The Chinese Typewriter: A History*. Cambridge, MA: MIT Press, 2017.

Mullaney, Thomas S. *The Chinese Computer: A Global History of the Information Age*. Cambridge, MA: MIT Press, 2024.

Nakamura, Lisa. "Economies of Digital Production in East Asia iPhone Girls and the Transnational Circuits of Cool." *Media Fields Journal: Critical Explorations in Media and Space* 2 (2011). http://mediafieldsjournal.org/economies-of-digital/.

Nakamura, Lisa. "Indigenous Circuits: Navajo Women and the Racialization of Early Electronic Manufacture." *American Quarterly* 66, no. 4 (2014): 919–941.

National Research Council. *Funding a Revolution: Government Support for Computing Research*. Washington, DC: National Academies Press, 1999.

National Research Council. *Securing the Future: Regional and National Programs to Support the Semiconductor Industry*. Washington, DC: National Academies Press, 2003.

Needham, Joseph. "Poverties and Triumphs of the Chinese Scientific Tradition." In *The "Racial" Economy of Science: Toward a Democratic Future*, edited by Sandra Harding, 30–46. Bloomington: Indiana University Press, 1993 [1969].

Nelson, Alondra, Thuy Linh Nguyen Tu, and Alicia Headlam Hines, eds. *Technicolor: Race, Technology, and Everyday Life*. New York: New York University Press, 2001.

Nguyen, Lilly U. "Infrastructural Action in Vietnam: Inverting the Techno-Politics of Hacking in the Global South." *New Media & Society* 18, no. 4 (2016): 637–652.

Nishiyama, Takashi. *Engineering War and Peace in Modern Japan, 1868–1964*. Baltimore, MD: Johns Hopkins University Press, 2014.

Norberg, Arthur L. *Computers and Commerce: A Study of Technology and Management at Eckert-Mauchly Computer Company, Engineering Research Associates, and Remington Rand, 1946–1957*. Cambridge, MA: MIT Press, 2005.

November, Joseph. *Biomedical Computing: Digitizing Life in the United States*. Baltimore, MD: Johns Hopkins University Press, 2012.

Nye, David E. *American Technological Sublime*. Cambridge, MA: MIT Press, 1994.

O'Bryan, Scott. *The Growth Idea: Purpose and Prosperity in Postwar Japan*. Honolulu: University of Hawai'i Press, 2009.

Onaga, Lisa. "Silkworms, Science, and Nation: A Sericultural History of Genetics in Modern Japan." PhD diss., Cornell University, 2012.

Oudshoorn, Nelly, and Trevor J. Pinch. "How Users and Non-Users Matter." In *How Users Matter: The Co-Construction of Users and Technologies*, edited by Nelly Oudshoorn and Trevor J. Pinch, 1–25. Cambridge, MA: MIT Press, 2003.

Oudshoorn, Nelly, and Trevor J. Pinch. "User-Technology Relationships: Some Recent Developments." In *The Handbook of Science and Technology Studies*, 3rd ed, edited by Edward J. Hackett, Olga Amsterdamska, Michael Lynch, and Judy Wajcman, 541–566. Cambridge, MA: MIT Press, 2007.

Owens, Larry. "Where Are We Going, Phil Morse? Changing Agendas and the Rhetoric of Obviousness in the Transformation of Computing at MIT, 1939–1957." *IEEE Annals of the History of Computing* 18 no. 4 (1996): 34–41.

Paul, Richard, and Steven Moss. "'There Was a Lot of History There': Theodis Ray." In *We Could Not Fail: The First African Americans in the Space Program*, 32–49. Austin: University of Texas Press, 2015.

Pellow, David N., and Lisa Sun-Hee Park. *The Silicon Valley of Dreams: Environmental Injustice, Immigrant Workers, and the High-Tech Global Economy*. New York: New York University Press, 2002.

Perold, Colette. "Assembling the Continental Computer: IBM's Resurgence in Cold-War Brazil." A presentation in the symposium Just Code: Power, Inequality, and the Global Political Economy of IT. CBI, University of Minnesota, October 23–24, 2020.

Perold, Colette. "IBM's World Citizens: Valentim Bouças and the Politics of IT Expansion in Authoritarian Brazil." *IEEE Annals of the History of Computing* 42, no. 3 (2020): 38–52.

Perry, Tekla S. "Foundry Father." *IEEE Spectrum* 48, no. 5 (2011): 46–50.

Peters, Benjamin. *How Not to Network a Nation: The Uneasy History of the Soviet Internet*. Cambridge, MA: MIT Press, 2016.

Peterson, William Wesley. *Computing with IBM 1620*. Taiwan: Central Book Company, 1964.

Peterson, William Wesley. *Error-Correcting Codes*. Cambridge, MA: MIT Press, 1961.

Petrick, Elizabeth. "Imagining the Personal Computer: Conceptualizations of the Homebrew Computer Club 1975–1977." *IEEE Annals of the History of Computing* 39, no. 4 (2017): 27–39.

Petrov, Victor. *Balkan Cyberia: Cold War Computing, Bulgarian Modernization, and the Information Age behind the Iron Curtain*. Cambridge, MA: MIT Press, 2023.

Philip, Kavita, Lilly Irani, and Paul Dourish. "Postcolonial Computing: A Tactical Survey." *Science, Technology, & Human Values* 37, no. 1 (2012): 3–29.

Phipps, Charles. "The Early History of ICs at Texas Instruments: A Personal View." *IEEE Annals of the History of Computing* 34, no. 1 (2012): 37–47.

Pinch, Trevor J. "Giving Birth to New Users: How the Minimoog Was Sold to Rock and Roll." In *How Users Matter: The Co-Construction of Users and Technologies*, edited by Nelly Oudshoorn and Trevor J. Pinch, 247–270. Cambridge, MA: MIT Press, 2003.

Pinch, Trevor J., and Wiebe Bijker. "The Social Construction of Facts and Artifacts: Or, How the Sociology of Science and the Sociology of Technology Might Benefit Each Other." In *The Social Construction of Technological Systems*, edited by Wiebe Bijker, Thomas P. Hughes, and Trevor J. Pinch, 17–50. Cambridge, MA: MIT Press, 1987. Originally published in *Social Studies of Science* 14, no. 3 (1984): 399–441.

Pocock, Chris. *The Black Bats: CIA Spy Flights over China from Taiwan 1951–1969*. Lancaster, PA: Schiffer, 2010.

Polanyi, Michael. *Personal Knowledge*. Chicago, IL: University of Chicago Press, 1974 [1958].

Prakash, Gyan. *Another Reason: Science and the Imagination of Modern India*. Princeton, NJ: University Press, 1999.

Prakash, Gyan. "Science 'Gone Native' in Colonial India." *Representations* 40 (Autumn 1992): 1–26.

Prieto-Ñañez, Fabian. "Assembling a Colombian-Cloned Computer: National Development and the Transnational Trade of Electronics Parts in the 1980s." *IEEE Annals of the History of Computing* 44, no. 2 (2022): 55–64.

Prieto-Ñañez, Fabian. "Postcolonial Histories of Computing." *IEEE Annals of the History of Computing* 38, no. 2 (2016): 2–4.

Rana, Shruti. "Fulfilling Technology's Promise: Enforcing the Rights of Women Caught in the Global High Tech Underclass." *Berkeley Women's Law Journal* 15 (2000): 272–311.

Rigger, Shelley. *The Tiger Leading the Dragon: How Taiwan Propelled China's Economic Rise.* Lanham, MD : Rowman & Littlefield, 2021.

Rigger, Shelley. *Why Taiwan Matters: Small Island, Global Powerhouse.* Lanham, MD: Rowman & Littlefield, 2011.

Ristanović, Dejan, and Jelica Protić. "Once Upon a Pocket: Programmable Calculators from the Late 1970s and Early 1980s and the Social Networks around Them." *IEEE Annals of the History of Computing* 34, no. 3 (2012): 55–66.

Roberts, Brian Russell. *Borderwaters: Amid the Archipelagic States of America.* Durham, NC: Duke University Press, 2021.

Rogaski, Ruth. "Deficiency and Sovereignty: Hygienic Modernity in the Occupation of Tianjin, 1900–1902." In *Hygienic Modernity: Meanings of Health and Disease in Treaty-Port China,* 165–192. Berkeley: University of California Press, 2004.

Rosner, Daniela K., Samantha Shorey, Brock Craft, and Helen Remick. "Making Core Memory: Design Inquiry into Gendered Legacies of Engineering and Craftwork." *Proceedings of the 2018 CHI Conference on Human Factors in Computing Systems (CHI'18)* (April 2018): 1–13. Association for Computer Machinery (ACM).

Rossiter, Margaret. *Women Scientists in America: Struggles and Strategies to 1940.* Baltimore, MD: Johns Hopkins University Press, 1982.

Rudan, John W. *The History of Computing at Cornell University.* Internet-First University Press, 2005. Accessed November 30, 2011. http://hdl.handle.net/1813/82.

Rushkoff, Bennett C. "Eisenhower, Dulles and the Quemoy-Matsu Crisis, 1954–1955." *Political Science Quarterly* 96, no. 3 (1981): 465–480.

Said, Edward. *Orientalism.* New York: Vintage Books, 1994 [1978].

Saxenian, AnnaLee. *The New Argonauts: Regional Advantage in a Global Economy.* Cambridge, MA: Harvard University Press, 2007.

Schäfer, Dagmar. *The Crafting of the 10,000 Things: Knowledge and Technology in Seventeenth-Century China.* Chicago, IL: University of Chicago Press, 2011.

Schaffer, Simon. "Babbage's Engines: Calculating Machines and the Factory System." *Critical Inquiry* 21, no. 1 (1994): 203–227.

Schaffer, Simon. "Instruments as Cargo in the China Trade." *History of Science* 44, no. 2 (2006): 217–246.

Schlombs, Corinna. "A Gendered Job Carousel: Employment Effects of Computer Automation." In *Gender Codes: Why Women Are Leaving Computing*, edited by Thomas J. Misa, 75–93. Hoboken, NJ: Wiley and IEEE Computer Society, 2010.

Schlombs, Corinna. *Productivity Machines: German Appropriations of American Technology from Mass Production to Computer Automation*. Cambridge, MA: MIT Press, 2019.

Schlombs, Corinna. "Toward International Computing History." *IEEE Annals of the History of Computing* 28, no. 1 (2006): 107–108.

Schwartz, Stephen I., ed. *Atomic Audit: The Costs and Consequences of U.S. Nuclear Weapons Since 1940*. Washington, DC: Brookings Institution Press, 1998.

Scott, James C. *Seeing Like a State: How Certain Schemes to Improve the Human Condition Have Failed*. New Haven, CT: Yale University Press, 1999.

Seely, Bruce. "Historical Patterns in the Scholarship of Technology Transfer." *Comparative Technology Transfer and Society* 1, no. 1 (2003): 7–48.

Seow, Victor. *Carbon Technocracy: Energy Regimes in Modern East Asia*. Chicago: University of Chicago Press, 2021.

Seth, Suman. "Putting Knowledge in Its Place: Science, Colonialism, and the Postcolonial." *Postcolonial Studies* 12, no. 4 (2009): 373–388.

Seth, Suman. "Science, Colonialism, Postcoloniality." A talk presented in the spring 2011 New Conversations Series at the Institute of Comparative Modernity, Cornell University, April 26, 2011.

Shapiro, Don. "Ronald McDonald, Meet Stan Shih." *Sales & Marketing Management* 147, no. 11 (1995): 85.

Sharma, Dinesh C. *The Outsourcer: The Story of India's IT Revolution*. Cambridge, MA: MIT Press, 2015.

Shih, Stan. *Hongqi de shiji biange* [Acer's millennium transformation]. Taipei: Tianxia wenhua, Commonwealth Publishing, 2004.

Shichor, Yitzhak. "The Importance of Being Ernst: Ernst David Bergmann and Israel's Role in Taiwan's Defense." CIRS Asia Papers no. 2. Doha, QA: Center for International and Regional Studies, 2016.

Shorey, Samantha, and Daniela Rosner. "Making Core Memory—An Experiment in Troubling Computing Histories." *Technology's Stories* 7, no. 2 (2019). https://www.technologystories.org/making-core-memory/.

Siegel, D. R. "Functional Compatibility and the Law: From the Necessity of Cleanroom Development of Computer Software to the Copyrightability of Computer Hardware." *Digest of Paper: COMPCON Spring 88 Thirty-Third IEEE Computer Society International Conference*, 361–367. San Francisco, CA, 1988.

Simon, Denis Fred. "Taiwan's Emerging Technological Trajectory: Creating New Forms of Competitive Advantage." *Taiwan: Beyond the Economic Miracle*, edited by Denis Fred Simon and Michael Ying-Mao Kau. 123-152. New York, M. E. Sharpe, 1992

Simpson, Bradley R. *Economists with Guns: Authoritarian Development and U.S. Indonesian Relations, 1960–1968*. Stanford, CA: Stanford University Press, 2008.

Small, James S. *The Analogue Alternative: The Electronic Analogue Computer in Britain and the USA, 1930–1975*. New York: Routledge, 2013.

Small, James S. "Engineering, Technology and Design: The Post-Second World War Development of Electronic Analogue Computers." *History and Technology* 11, no. 1 (1993): 33–48.

Small, James S. "General-Purpose Electronic Analog Computing: 1945–1965." *IEEE Annals of the History of Computing* 15, no. 2 (1993): 8–18.

Smith, Merritt Roe, ed. *Military Enterprise and Technological Change*. Cambridge, MA: MIT Press, 1985.

Smith, Ted, David A. Sonnenfeld, and David Naguib Pellow, eds. *Challenging the Chip: Labor Rights and Environmental Justice in the Global Electronics Industry*. Philadelphia, PA: Temple University Press, 2006.

Smura, Tomasz. "In the Shadow of Communistic Missiles—Air and Missile Defence in Taiwan." Casimir Pulaski Foundation, October 2016. https://pulaski.pl/en/in-the-shadow-of-communistic -missiles-air-and-missile-defence-in-taiwan/.

Soman, Appu K. "'Who's Daddy' in the Taiwan Strait." *Journal of American-East Asian Relations* 3, no. 4 (1994): 387–388.

Soon, Wayne. "Making Blood Banking Work." In *Global Medicine in China: A Diasporic History*, 95–124. Stanford, CA: Stanford University Press, 2020.

Sperling, Stefan. "Managing Potential Selves: Stem Cells, Immigrants, and German Identity." *Science and Public Policy* 31, no. 2 (2004): 139–149.

Stachniak, Zbigniew. "Red Clones: The Soviet Computer Hobby Movement of the 1980s." *IEEE Annals of the History of Computing* 37, no. 1 (2015): 12–23.

Su, Yu-Feng. *Kangzhanqian de qinghua daxue: Jindai zhongguo gaodeng jiaoyu yanjiu* [Tsing Hua University 1928–1937: A study of modern Chinese higher education]. Taipei: Institute of Modern History, Academic Sinica, 2000.

Sundaram, Ravi. *Pirate Modernity: Delhi's Media Urbanism*. New York: Routledge, 2010.

Suri, Jeremi. "The Cultural Contradictions of Cold War Education: The Case of West Berlin." *Cold War History* 4, no. 3 (2004): 1–20.

Szonyi, Michael. *Cold War Island: Quemoy on the Front Line*. New York: Cambridge University Press, 2008.

Tai, Paul H. "A Slice of Cold War History: The Soaring Cat." *Journal of Chinese Studies* 13, no. 1 (2006): 89–95.

Tai, Po-Fen. "Tiawan gaodeng jiaoyu kuozhang lichen zhong de jiaoyu quanli jingying fenxi" [Academic Regime Transformation: An Analysis of Power Elites and Expanding Higher Education]. *Taiwan shehuixue xuekan* [Taiwanese Journal of Sociology] 58 (December 2015): 47–93.

Takahashi, Shigeru. "The Rise and Fall of Plug-Compatible Mainframes." *IEEE Annals of the History of Computing* 27, no. 1 (2005): 4–16.

Takahashi, Sigeru. "A Brief History of the Japanese Computer Industry before 1985." *IEEE Annals of the History of Computing* 18, no. 1 (1996): 76–79.

Takahashi, Sigeru. "Early Transistor Computers in Japan." *IEEE Annals of the History of Computing* 8, no. 2 (1986): 144–154.

Takahashi, Yuzo. "A Network of Tinkerers: The Advent of the Radio and Television Receiver Industry in Japan." *Technology and Culture* 41, no. 3 (2000): 460–484.

Takahasi, Hidetosi. "Some Important Computers of Japanese Design." *IEEE Annals of the History of Computing* 2, no. 4 (1980): 330–337.

Tan, Ying Jia. *Recharging China in War and Revolution, 1882–1955*. Ithaca, NY: Cornell University Press, 2021.

Tatarchenko, Ksenia. "The Man with a Micro-Calculator: Digital Modernity and Late Soviet Computing Practices." In *Exploring the Early Digital*, edited by Thomas Haigh, 179–199. Cham: Springer Nature Switzerland AG, 2019.

Tatarchenko, Ksenia. "Our Past Is Their Future: Rip-Off, Translation, and Trust in the Soviet-American Computing Exchanges during the Cold War." A presentation at the workshop, Appreciating Innovation across Countries, organized by Lars Heide, Copenhagen Business School, November 2015, Denmark.

Taylor, Jay. *The Generalissimo: Chiang Kai-shek and the Struggle for Modern China*. Cambridge, MA: Harvard University Press, 2009.

Taylor, Jay. *The Generalissimo's Son: Chiang Ching-kuo and the Revolutions in China and Taiwan*. Cambridge, MA: Harvard University Press, 2000.

TECO Technology Foundation and Stephanie Shih. *Dongli dongyuan: Mada zhuanchu wuxian shengji* [The Empire of Motor: TECO]. Taipei: Tianxia wenhua [Commonwealth Publishing], 1996.

Tinn, Honghong. "Between 'Magnificent Machine' and 'Elusive Device': Wassily Leontief's Interindustry Input-output Analysis and its International Applicability." *Osiris* 38 (Special issue: Beyond Craft and Code: Human and Algorithmic Cultures, Past and Present) (2023): 129–146.

Tinn, Honghong. "Cold War Politics: Taiwanese Computing in the 1950s and 1960s." *IEEE Annals of the History of Computing* 32, no. 1 (2010): 92–95.

Tinn, Honghong. "From DIY Computers to Illegal Copies: The Controversy over Tinkering with Microcomputers in Taiwan, 1980–1984." *IEEE Annals of the History of Computing* 33, no. 2 (2011): 75–88.

Tinn, Honghong. "Modeling Computers and Computer Models: Manufacturing Economic-Planning Projects in Cold War Taiwan, 1959–1968." *Technology and Culture* 59 no. 4 (Supplement, 2018): S66–S99.

Tou, Julius T. "Learning Control via Associative Retrieval and Inference." In *Pattern Recognition and Machine Learning*, edited by K. S. Fu, 243–251. New York: Plenum, 1971.

Tout, Nigel. "The Story of the Race to Develop the Pocket Electronic Calculator." Vintage Calculators Web Museum, unknown publication date. Accessed April 9, 2021. http://www.vintagecalculators.com/html/the_pocket_calculator_race.html#Ref10.

Truelove, Cynthia. "Disguised Industrial Proletarians in Rural Latin America: Women's Informal Sector Factory Work and the Social Reproduction of Coffee Farm Labor in Colombia." In *Women Workers and Global Restructuring*, edited by Kathryn Ward, 48–63. New York: Cornell University Press, 1990.

Tsao, Tsen-Cha. *Gongcheng yu kexue* [Engineering and science]. Taipei: Zhing-hua Books, 1964.

Tsao, Tsen-Cha. "Jiaoda fuxiao zhi jingguo ji qi shidai jiazhi zhi zhanwang" [The process of refounding Chiao-Tung University]. In *Jiaotong daxue jiushi nian* [Ninety years of Chiao-Tung], edited by NCTU, 52–56. Hsinchu, TW: NCTU, 1986.

Tsou, Tang. "The Quemoy Imbroglio: Chiang Kai-Shek and the United States." *Western Political Quarterly* 12, no. 4 (1959): 1084–1085.

Tsu, Jing. "Chinese Scripts, Codes, and Typewriting Machines." *Science and Technology in Modern China, 1880s–1940s*, edited by Jing Tsu and Benjamin Elman, 115–151. Leiden: Brill, 2014.

Tsurumi, E. Patricia. *Japanese Colonial Education in Taiwan, 1895–1945*. Cambridge, MA: Harvard University Press, 1977.

Tucker, Nancy Bernkopf. *The China Threat: Memories, Myths, and Realities in the 1950s*. New York: Columbia University Press, 2012.

Tucker, Nancy Bernkopf. *Strait Talk: United States-Taiwan Relations and the Crisis with China*. Cambridge, MA: Harvard University Press, 2009.

Tung, Chung C. "The 'Personal Computer': A Fully Programmable Pocket Calculator." *Hewlett-Packard Journal* 25, no. 9 (May 1974): 7–14.

Turner, Fred. *From Counterculture to Cyberculture: Stewart Brand, the Whole Earth Network, and the Rise of Digital Utopianism*. Chicago, IL: University of Chicago Press, 2006.

Varol, Y. L. "Some Remarks on Computer Acquisition." *Computer Journal* 19, no. 2 (1976): 127–131.

Vinsel, Lee, and Andrew L. Russell. *The Innovation Delusion: How Our Obsession with the New Has Disrupted the Work That Matters Most*. New York: Currency, 2020.

Wadell, Lyle. "History of the Northeast Dairy Records Processing Laboratory, 1948–1985, with Additional Comments by J. D. Burke and H. Wilmot Carter." Oral and Personal Histories of Computing at Cornell. Accessed November 30, 2011. http:// www2.cit.cornell.edu/computer /history/.

Wang, An, and Eugene Linden. *Lessons: An Autobiography*. Reading, MA: Addison-Wesley, 1986.

Wang, Bailu. *Taijidian weishenme shen* [Why does TSMC triumph]. Taipei: Shibao chuban, 2021.

Wang, Chin-Shou. *Heibianfu zhi lian* [Chain of black bats]. Taipei: Linking Books, 2011.

Wang, Fan-sen. *Fu Ssu-nien: A Life in Chinese History and Politics*. New York: Cambridge University Press, 2000

Wang, Hao. *Jieke shangshi: Jiang Jieshi yu zhonghuaminguo taiwan de xingsu* [Backdoor listing: Chiang Kai-shek's construction of Republic of China on Taiwan]. Taipei: Baqi wenhua, 2020.

Wang, Jiulong. "Dianzi jisuanji wang: Dianzi jisuanji xitong de xinfazhan" [Computer network: New development of computer systems]. *Xinxin jikan* [Hsin Hsin Quarterly] 7, no. 3 (1979): 80–86.

Wang, Lijuan (Janet). *Shi Min yu shuwei shidai de gushi* [Min Sze and the stories about the digital age]. Keelung, TW: Hongjin Shuwei Keji, 2013.

Wang, Min-Chiuan, and Hsiao-Wen Zan. *Chuanqi rensheng: Zhang Junyan zhuan* [Legendary life: Chang Chun-yen]. Hsinchu, TW: National Chiao Tung University Press, 2019.

Wang, Yao-te. "Rizhi shiqi tainan gaodeng gongye xuexiao sheli zhi yanjiu" [Founding of Tainan Technical College during Japanese Colonial Era]. *Taiwan shi yanjiu* [Taiwanese Historical Research] 18, no. 2 (2011): 53–95.

Wang, Zuoyue. "Saving China through Science: The Science Society of China, Scientific Nationalism, and Civil Society in Republican China." *Osiris* 17 (2002): 291–322.

Ward, Kathryn, ed. *Women Workers and Global Restructuring*. New York: Cornell University Press, 1990.

Webster, David. "Development Advisors in a Time of Cold War and Decolonization: The United Nations Technical Assistance Administration, 1950–59." *Journal of Global History* 6, no. 2 (2011): 249–272.

Weik, Martin H. *A Survey of Domestic Electronic Digital Computing Systems*. Report No. 971. Aberdeen, MD: Ballistic Research Laboratories, Aberdeen Proving Ground, 1955.

Weik, Martin H. *A Third Survey of Domestic Electronic Digital Computing Systems*. Report No. 1115. Aberdeen, MD: Ballistic Research Laboratories, Aberdeen Proving Ground, 1961.

Weiss, Eric A. "Elogue: An Wang, 1920–1990." *IEEE Annals of the History of Computing* 15, no. 1 (1993): 60–69.

Werrett, Simon. *Thrifty Science: Making the Most of Materials in the History of Experiment*. Chicago, IL: Chicago University Press, 2019.

Westrum, Ron. *Sidewinder: Creative Missile Development at China Lake*. Annapolis, MD: Naval Institute Press, 1999.

Wilbur, John B. "The Mechanical Solution of Simultaneous Equations." *Journal of the Franklin Institute* 222, no. 6 (1936): 715–724.

Wisnioski, Matthew. *Engineers for Change: Competing Visions of Technology in 1960s America*. Cambridge, MA: MIT Press, 2012.

Woo-Cumings, Meredith Jung-En. "National Security and the Rise of the Developmental State in South Korea and Taiwan." In *Behind East Asian Growth: The Political and Social Foundations of Prosperity*, edited by Henry Rowen, 319–337. New York: Routledge, 1998.

Wu, Huey-Fang. "Tianzhujiao shengyanhui de she fuwu shiye: Yi xindian dapinglin dehua nüzi gongyu weili (1968–1988)" [The Social Services Delivered by the Catholic Divine Word Missionaries: A Case Study of Dehua Girls' Hostel at Dapinglin, Xindian, 1968–1988]. *Guoli zhengzhi daxue lishixue bao* [The Journal of History, NCCU] 44 (November 2015): 223–280.

Wu, Rong-I, and Tseng Ming-Shen. "Taiwan's Information Technology Industry." *Manufacturing Competitiveness in Asia: How Internationally Competitive National Firms and Industries Developed in East Asia*, edited by K. S. Jomo. New York: Routledge, 2003.

Wu, Wen-Hsing, Shun-Fen Chen, and Chen-Tsou Wu. "The Development of Higher Education in Taiwan." *Higher Education* 18, no. 1 (1989): 124.

Wu, Zhongxian. *Toushi Taijidian* [A thorough analysis of TSMC]. Taipei: Wunan, 2006.

Xiao, Yiling. *Jinchaiji: Qianzhen jiagongqu nüxing laogong de koushu jiyi* [Gold hairpins: Oral histories of female laborers employed at Cianjhen Export Processing Zone]. Kaohsiung: Bureau of Cultural Affairs of Kaohsiung Municipal Government and Liwen wenhua Publishing, 2014.

Xu, Guangqiu. "Americans and Chinese Nationalist Military Aviation, 1929–1949." *Journal of Asian History* 31, no. 2 (1997): 155–180.

Yamada, Akihiko. "IPSJ Third Information Processing Technology Heritage Certification." *IEEE Annals of the History of Computing* 33, no. 2 (2011): 109.

Yamada, Akihiko. "UNIVAC 120 in Japan." *IEEE Annals of the History of Computing* 35, no. 3 (2013): 73–74.

Yang, Aili. *Sun Yunxuan zhuan* [Biography of Yun-suan Sun]. Taipei: Tianxia wenhua [Commonwealth], 1996.

Yang, Chao-Chih. "Associative Memory Systems and Their Applications." PhD diss., electrical engineering, Northwestern University, 1966.

Yang, Ching-chu. *Gongchang nü'erquan* [Womanland of factory women]. Taipei: Shuiling wenchuang, 2019 [1978].

Yang, Ching-chu. *Gongchang ren* [Factory folks]. Taiwan: Shuiling wenchuang, 2019 [1975].

Yang, Ching-chu. *Waixiang nü* [Women from other towns]. Taipei: Shuiling wenchuang, 2019.

Yang, Daqing. "Colonial Korea in Japan Imperial Telecommunications Network." In *Colonial Modernity in Korea*, edited by Gi-Wook Shin and Michael Edson Robinson, 161–188. Cambridge, MA: Harvard University Asia Center, 1999.

Yang, Tsui-hua. "Hu Shi dui Taiwan kexue fazhan de tuidong: Xueshu duli mengxiang de yanxu" [Promoting Science and Scholarship in Taiwan: Hu-Shih's Dream of "Academic Independence"]. *Hanxue yanjiu* [Chinese Studies] 20, no. 2 (2002): 327–352.

Ye, Jialong. "Tongzidian keji yu shuweihua zhanchang xianqu: Zhuanfang benyuan qian zixun tongxin yanjiusuo suozhang Zheng Mingjie boshi" [Forerunner on the battlefield of telecommunications, informatics, electronics, and digitalization: An interview with Dr. Mingjia Zheng, former director of the information and communications research division, CSIST]. *Xinxin jikan* [Hsin Hsin Quarterly] 44, no. 2 (2016): 4–7.

Yeh, Pi-ling. "Taibei diguo daxue gongxue bu zhi chuangshe" [The Construction of Faculty of Engineering at Taihoku Imperial University]. *Guoshiguan guankan* [Bulletin of Academia Historica] 52 (June 2017): 73–124.

Yip, Wai. "RB-57A and RB-57D in Republic of China Air Force Service." *American Aviation Historical Society Journal* 44, no. 4 (1999): 259–274.

Yost, Jeffrey R. *Making IT Work: A History of the Computer Services Industry*. Cambridge, MA: MIT Press, 2017.

Yost, Jeffrey R. "Manufacturing Mainframes: Component Fabrication and Component Procurement at IBM and Sperry Univac, 1960–1975." *History and Technology* 25, no. 3 (2009): 219–235.

Yost, Jeffrey R. "Materiel Command and the Materiality of Commands: An Historical Examination of the U.S. Air Force, Control Data Corporation, and the Advanced Logistics System." In *History of Computing: Lessons from the Past,* edited by Arthur Tatnall, 89–100. London: Springer, 2010.

Yost, Jeffrey R. "Where Dinosaurs Roam and Programmers Play: Reflections on Infrastructure, Maintenance, and Inequality." *Interfaces: Essays and Reviews in Computing and Culture* 1 (May 2020). https://cse.umn.edu/cbi/past-interfaces.

Yost, Jeffrey R., ed. A World of Computers (Special Issue). *IEEE Annals of the History of Computing* 30, no. 4 (2008).

Young, Lewis H. "Why U.S. Companies Can Compete." *Journal of International Affairs* 28, no. 1 (1974): 81–90.

Yu, Howard H., and Willy C. Shih. "Taiwan's PC Industry, 1976–2010: The Evolution of Organizational Capabilities." *Business History Review* 88, no. 2 (Summer 2014): 329–357.

Zhang, Ruxin. *Xishuo Taiwan: Taiwan bandaoti chanye chuanqi* [Speaking of silicon in Taiwan: Legends about Taiwan's semiconductor industry]. Taipei: Pan Wenyuan Foundation and Tianxia wenhua [Commonwealth Publishing], 2006.

Zhongguo gongchengshi xiehui [Chinese Institute of Engineers]. *Yinianlai de gongcheng jianshe gaikuang* [The previous year's engineering accomplishments]. Taipei: Zhongguo gongchengshi xiehui [Chinese Institute of Engineers], 1970.

Zhou, Zhengxian. *Shi Zhenrong de diannao chuanqi* [Stan Shih and the computer legend]. Taipei: Linking Books, 1996.

INDEX

fig. denotes figures; t denotes a table

History of Computing

William Aspray and Thomas J. Misa, editors

Janet Abbate
Recoding Gender: Women's Changing Participation in Computing

John Agar
The Government Machine: A Revolutionary History of the Computer

William Aspray and Paul E. Ceruzzi
The Internet and American Business

William Aspray
John von Neumann and the Origins of Modern Computing

Charles J. Bashe, Lyle R. Johnson, John H. Palmer, and Emerson W. Pugh
IBM's Early Computers

Martin Campbell-Kelly
From Airline Reservations to Sonic the Hedgehog: A History of the Software Industry

Paul E. Ceruzzi
A History of Modern Computing

I. Bernard Cohen
Howard Aiken: Portrait of a Computer Pioneer

I. Bernard Cohen and Gregory W. Welch, editors
Makin' Numbers: Howard Aiken and the Computer

James Cortada
IBM: The Rise and Fall and Reinvention of a Global Icon

Nathan Ensmenger
The Computer Boys Take Over: Computers, Programmers, and the Politics of Technical Expertise

Thomas Haigh, Mark Priestley, and Crispin Rope
ENIAC in Action: Making and Remaking the Modern Computer

John Hendry
Innovating for Failure: Government Policy and the Early British Computer Industry

Mar Hicks
Programmed Inequality: How Britain Discarded Women Technologists and Lost Its Edge in Computing

Michael Lindgren
Glory and Failure: The Difference Engines of Johann Müller, Charles Babbage, and Georg and Edvard Scheutz

David E. Lundstrom
A Few Good Men from Univac

René Moreau
The Computer Comes of Age: The People, the Hardware, and the Software

Arthur L. Norberg
Computers and Commerce: A Study of Technology and Management at Eckert-Mauchly Computer Company, Engineering Research Associates, and Remington Rand, 1946–1957

Emerson W. Pugh
Building IBM: Shaping an Industry and Its Technology

Emerson W. Pugh
Memories That Shaped an Industry

Emerson W. Pugh, Lyle R. Johnson, and John H. Palmer
IBM's Early Computers: A Technical History

Kent C. Redmond and Thomas M. Smith
From Whirlwind to MITRE: The R&D Story of the SAGE Air Defense Computer

Alex Roland with Philip Shiman
Strategic Computing: DARPA and the Quest for Machine Intelligence, 1983–1993

Raúl Rojas and Ulf Hashagen, editors
The First Computers—History and Architectures

Corinna Schlombs
Productivity Machines: German Appropriations of American Technology from Mass Production to Computer Automation

Dinesh C. Sharma
The Outsourcer: A Comprehensive History of India's IT Revolution

Dorothy Stein
Ada: A Life and a Legacy

Christopher Tozzi
For Fun and Profit: A History of the Free and Open Source Software Revolution

John Vardalas
The Computer Revolution in Canada: Building National Technological Competence, 1945–1980

Maurice V. Wilkes
Memoirs of a Computer Pioneer

Jeffrey R. Yost
Making IT Work: A History of the Computer Services Industry

Thomas Haigh and Paul E. Ceruzzi
A New History of Modern Computing

Daniel D. Garcia-Swartz and Martin Campbell-Kelly
Cellular: An Economic and Business History of the International Mobile-Phone Industry

Victor Petrov
Balkan Cyberia: Bulgarian Modernization, Computers, and the World, 1963–1989

Jacob Ward
Visions of a Digital Nation: Market and Monopoly in British Telecommunications

Honghong Tinn
Island Tinkerers: Innovation and Transformation in the Making of Taiwan's Computing Industry